SOCIAL TEXTURES OF WESTERN CIVILIZATION: THE LOWER DEPTHS

VOLUME I

MICHAEL CHERNIAVSKY
State University of New York, Albany

ARTHUR J. SLAVIN
University of California, Los Angeles

With the assistance of John Kaufmann,
State University of New York, Albany

XEROX COLLEGE PUBLISHING

Waltham, Massachusetts / Toronto

Social Textures of Western Civilization

The Lower Depths

VOLUME I

A book of this sort is never the sole effort of those listed on the title page. Without further ado, we would like to thank the following people: our lawyer, Fred Cohen, who did his best to keep us out of trouble while this book was being prepared; and the following people for their invaluable advice and assistance: WALTER ADAMS, BARRY ALPER, CAROL ALPER, LOREN BARITZ, PAUL BREINES, WINI BREINES, PAUL S. EWEN, SAM T. EWEN, NAOMI GLAUBERMAN, SUSAN GOLDMACHER, KATHY GRAHAM, RUSSELL JACOBY, PAUL LEVINE, MARY LYNN, DIANE MACNAMARA, RISE ROTHBART, LESLIE SARTORIS

All power to the People!

To the Victims of Civilization and for the Revenge of Innocents

ACKNOWLEDGMENTS

Footnotes have been omitted except where they were
necessary for an understanding of the text.

AELFRIC OF EYNSHAM, from *English Literature from Widsith to the Death of Chaucer*, edited by Allen Rogers Benham. Reprinted by permission of Yale University Press.

ARISTOPHANES, from *The Acharnians* in *The Complete Greek Drama*, vol. 2, edited by Eugene O'Neill and Whitney J. Oates. Copyright 1938 and renewed 1966 by Random House, Inc. Reprinted by permission of the publisher.

ARISTOTLE, from *The Constitution of Athens*. Reprinted by permission of Hafner Publishing Company.

MARC BLOCH, from *Feudal Society*, translated by L. A. Manyon. Reprinted by permission of The University of Chicago Press. Copyright © 1968 by The University of Chicago. From *Land and Work in Medieval Europe*, translated by J. E. Anderson. Reprinted by permission of Routledge & Kegan Paul.

PROSPER BOISSONNADE, from *Life and Work in Medieval Europe*. Reprinted by permission of Harper & Row, Publishers, and Routledge & Kegan Paul.

NORMAN O. BROWN, from *Hermes the Thief*. Copyright 1947 by Norman O. Brown. Reprinted by permission of Random House, Inc.

JÉRÔME CARCOPINO, from *Daily Life in Ancient Rome*. Reprinted by permission of Yale University Press and Routledge & Kegan Paul. Copyright © 1940 by Yale University Press.

GEOFFREY CHAUCER, from *The Canterbury Tales*, translated by David Wright. Copyright © 1964 by David Wright. Reprinted by permission of Random House, Inc., and A. D. Peters and Company.

COLUMELLA, from *De re rustica*. Reprinted by permission of the publishers and The Loeb Classical Library from the Harrison B. Ash translation, Cambridge, Mass.: Harvard University Press, 1941.

VICTOR EHRENBERG, from *The People of Aristophanes*. Reprinted by permission of Basil Blackwell, Publisher.

ROBERT FLACELIÈRE, from *Daily Life in Greece at the Time of Pericles*. Reprinted with permission of The Macmillan Company. © 1965 by George Weidenfeld & Nicolson, Ltd. From *Love in Ancient Greece*, © 1963 by Crown Publishers, Inc. Reprinted by permission of Crown Publishers, Inc.

LUDWIG FRIEDLÄNDER, from *Roman Life and Manners Under the Early Empire*. Reprinted by permission of Routledge & Kegan Paul.

GUSTAVE GLOTZ, from *Ancient Greece at Work. An Economic History of Greece from the Homeric Period to the Roman Conquest*. Reprinted by permission of Routledge & Kegan Paul.

The Goodman of Paris, translated by E. Power. Reprinted by permission of Routledge & Kegan Paul.

CHARLES HOMER HASKINS, from *The Rise of Universities*. Copyright © 1957 by Cornell University. Used by permission of Cornell University Press.

FRIEDRICH HEER, from *The Medieval World: Europe 1100–1350*, translated by Janet Sondheimer. Copyright 1961 by George Weidenfeld & Nicolson, Ltd. English translation Copyright © 1962 by George Weidenfeld & Nicolson, Ltd.

HESIOD, from *Works and Days*, translated by Richmond Lattimore. Reprinted by permission of The University of Michigan Press.

GEORGE C. HOMANS, from *English Villagers of the Thirteenth Century*. Reprinted by permission of the publishers, Cambridge, Mass.: Harvard University Press, Copyright 1941, by the President and Fellows of Harvard College; 1969 by George Caspar Homans.

HORACE, from *The Satires and Epistles of Horace*, translated by Smith P. Bovie. Reprinted by permission of The University of Chicago Press. Copyright © 1959 by The University of Chicago.

JOHAN HUIZINGA, from *The Waning of the Middle Ages*. Reprinted by permission of St. Martin's Press, Inc., Macmillan & Co., Ltd., and Edward Arnold, Ltd.

OTTO KIEFER, from *Sexual Life in Ancient Rome*. Reprinted by permission of Routledge & Kegan Paul.

LIVY, from *History of Rome*, translated by B. O. Foster. Reprinted by permission of G. P. Putnam's Sons.

ROBERT S. LOPEZ and IRVING W. RAYMOND, eds., from *Medieval Trade in the Mediterranean World*. Reprinted by permission of Columbia University Press.

MILLARD MEISS, from *Painting in Florence and Siena After the Black Death* (copyright 1951 by Princeton University Press). Reprinted by permission of Princeton University Press.

ÉMILE MIREAUX, from *Daily Life in the Time of Homer*, translated by Iris Sells. Reprinted with permission of The Macmillan Company. Copyright Allen & Unwin, Ltd., 1959.

PETRONIUS, from *The Satyricon* in *The Complete Works of Gauis Petronius*, translated by Jack Lindsay. Reprinted by permission of Jack Lindsay and Rarity Press.

HENRI PIRENNE, from *Medieval Cities: Their Origins and the Revival of Trade* (copyright 1952 by Princeton University Press; Princeton Paperback, 1969). Reprinted by permission of Princeton University Press.

"The Plight of the French Poor," from *The Portable Renaissance Reader*, edited by James Bruce Ross and Mary Martin McLaughlin. Copyright 1953 by The Viking Press, Inc. Reprinted by permission of The Viking Press, Inc.

GERTRUDE R. B. RICHARDS, editor, from *Florentine Merchants in the Age of the Medici*. Reprinted by permission of the publishers, Cambridge, Mass.: Harvard University Press, 1932.

J. S. SCHAPIRO, from *Social Reform and the Reformation*. Reprinted by permission of Columbia University Press.

SUETONIUS, from *The Twelve Caesars*. Reprinted by permission of Collins-Knowlton-Wing, Inc. Copyright © 1957 by Robert Graves.

THUCYDIDES, from *History of the Peloponnesian War* in *Classics in Translation*, ed. Herbert Howe and Paul MacKendrick (Madison: The University of Wisconsin Press); © 1963 by the Regents of the University of Wisconsin.

PAUL ZUMTHOR, from *Daily Life in Rembrandt's Holland*. Reprinted with permission of The Macmillan Company and George Weidenfeld & Nicolson. © George Weidenfeld & Nicolson, Ltd., 1962.

Contents

PART 2
The Middle Ages and the Renaissance

CHAPTER FIVE

Land and Labor: Material Bases of a Traditional Society

Introduction

One can ring many changes upon *Social Textures of Western Civilization: The Lower Depths.* An obvious one for the historian is the Lower Depths as the history of the great inarticulate masses, those at the bottom of the social hierarchy which seems to be an inherent feature of civilization itself. Another aspect is that of the context of social texture, the study which takes us beneath the surface of political and intellectual history into the social and psychological reality of everyday life. We may likewise go back to Maxim Gorky's notions of history. The author of the novel from which we take our title viewed the past with the mind of a moralist and saw in it the nightmare of a Samoyed god sick from too much rotten pork at dinner.

Just as the moralist peels back layers of consciousness to expose the secret soul, so the editors of this book hope to reveal the hidden life of ordinary men and women. This we want to do by stripping our story of traditional elitist poses and gazing at the customs, occupations, and beliefs of the vast majority of the human population who appear in histories as "the people." The peasants and urban lower classes were in many ways makers of history. Yet they appear in it as faceless shadows of beings whose substance is hardly perceived and rarely discussed. There are good reasons for this dark space between shadow and substance. Historians' ideological preferences often led them to ignore the masses. There is also the simple fact that peasants and proletarians more often than not left no evocative memoirs and wrote no histories. Nor did they record their problems in terms of their own consciousness. Hence historians in subsequent generations know their ideas and aspirations from the distorting perspective of such notices of them as were given by the traditional elites of church and aristocracy, the men who had the learning and leisure to write what they observed in their time.

If we have little understood "the people" and written even less about them, we do know enough to doubt that for many of them our traditional periodization of Western history mattered very much. The lives of peasants in many countries altered little from medieval to Renaissance times, or from early modern to modern times. They did not participate freely in politics; nor did their thoughts appear to shape the "culture" of their societies. Modes of harvest changed slowly, if at all, in many parts of the West. Russian peasant costume in the early nineteenth century was little different from that worn by Scythian and Samartian tillers in the third millennium B.C. Indeed, it can be argued that only the advent of the railroad radically transformed the life style of peasant communities in the nineteenth century. Thus there is in records of peasant life the appearance of sameness. Yet that appearance is itself deceptive. Peasant life was tied to the cycle of the seasons. Their religious festivals mirrored the calendar. War and natural disasters, however, did trigger peasant outbursts of despair and provoke peasants to rise against the order of their day. Even savage repression did not then mask the fact that peasants shaped the world of politics. And John Ball's rude question, "When Adam delved and Eve spanned/Who was then the gentleman," left an impress on the culture of the fourteenth century in England.

Those who lived on the land were the great majority of people in every age of Western history before our own. So we are here concerned with them. But our word *civilization* itself derives from the Latin *civitas*, which means city. This fact alone is enough warrant for us to turn our attention to the urban lower classes.

Townsmen of the lower social orders wrote few memoirs or histories. Yet their lives more closely paralleled those of the urban elites we so fully discover in traditional histories. This does mean that surviving materials do tell us a good bit about what they thought and felt, what was their life style, their food, clothing, and even their politics. One thing that does emerge from such materials is the belief that the traditional periods of our history have more relevance to city masses than rural ones. Urban dress changed significantly in every period between the classical world and our own. The town mobs were a potent political force. The feudal enclosed town gave them protection and legal rights. The church needed their labor to build cathedrals and their allegiance to fill them. The rise of money economies, the development of overseas empires in the age of discoveries, changing techniques of warfare, and industrialization thrust forward the proletarians of our towns and made them significant forces for historians to ponder.

In recent years it has become almost fashionable to give emphasis to the way in which the masses of humanity have been makers of history. It is not too much to say that historians now recognize they cannot ignore the material bases of politics and culture or the masses of men who made those bases possible. Peasant rebellion and urban riots were sometimes crucial in the fate of dynasties. And folk culture kept alive in myth, tale, and legend reminds us how small the space is that separates a literary genius from his materials. The high culture whose record so much shapes our own consciousness often stands in debt to quite humble beginnings, something we perhaps recognize most easily in the literary relations linking tribal fears of incest, the tragedies of Sophocles and Freud's theories about Oedipus.

This calls to mind the ambiguity and even the irony of our title. *The Lower Depths* are meant to imply that the "heights" of Western culture are not self-suspended in space. Moreover, the *Social Textures* of our title is meant to imply a concern with all social classes and layers of society. Insofar as these shed light on the lower classes, our materials necessarily make use of data about the traditional elites and elitist history: sources about courts, aristocracies, political intrigues. This stems from our basic confidence that the texture of social life is not that of great events but of daily routines. And these routines depended in turn on the encounters of diverse social groups with each other. In such encounters men formed their own images as well as their images of others. Who will study the impact of slavery without wanting to know the mind of the slaveowner, or that of money on the merchant without knowing his supplying artisans? What student of political forms in the classical world can for long escape the problem of the family as a unit of political life? Can any reader of *The Spartan Women* (*Lysistrata*) filter out half of humanity and still make sense of war and peace? And who concerned about poverty and disease will gain much understanding without some attention to the physicians of a society and the state of the art of healing?

However much elitist historians ignore the masses, historians of the masses cannot afford to turn their backs on the elites!

Finally, an apology. We know less about ancient society than we do about societies of the last four centuries. This produces some imbalance in our use of space. It also sometimes creates a sharpness of separation in our images of peasants and proletarians in antiquity that contrasts sharply with the more blurred reality of life in a city-state or medieval town that lived by farming its hinterland. For this faulty vision there is no good remedial prescription, beyond a measure of tolerance and the zeal of students hopefully excited by our provisional labors. We count on that zeal to compensate for any appearance of our material being too academic. We would have liked to present more data from a greater variety of sources. Yet we have often not allowed the sources to speak for themselves. This is partly the result of space limitations. But another reason lies behind our choice of modern authorities who sum up the sources they study in a few pages. Just as some foods are indigestible until they are refined and cooked, so some sources are hard fare until they are assimilated. Hence the compromise we made between sources and academic work in making our book. This compromise seemed worth making in introducing students to "the great silent beast of history," the people. The intellectual necessity of doing it hardly needs justification in this age, when the beast is finding its voice in so many tongues.

MICHAEL CHERNIAVSKY
ARTHUR J. SLAVIN

Part 1
The Ancient World

Men of letters and modern historians often focus their attention on the conflicts among people over property and status. This is no new issue. Since the days of Jacob and Esau conflicts over place and possessions served as a prime mover of creative imaginations. Witness the myth of Hermes the Thief. Furthermore, conflicts often seem to lie at the base of organized life—in the institutions of family, politics, slavery, crafts, as well as in those which express sexual mores. Whether the competition be that for territory, privilege, or merely for the attentions of a handsome boy, in the classical world it was often the case that ordinary life seemed an illustration of the truth of Herakleitos' saying, "Strife is the father of all things."

One useful way of organizing widely varied materials from and about ordinary people in ancient civilizations of Greece and Rome is to watch them in their competitions. Beginning with the most fundamental institutions of property, the family, and conflict in domestic politics, and carrying forward these themes in the history of politics of Empire, we have tried to illuminate classical social life. Some very obvious facets of shared Greek and Roman historical experience provided us with categories of arrangement: the impact of commerce in an agrarian society; slavery and slaves; mobility, social and geographical; sexuality and male dominance; cities and the craftsmen classes. Finally, we thought it more instructive to place Greek and Roman material on similar themes together to enhance your grasp of the classical world as one of comparable experiences in the lives of ordinary people.

CHAPTER ONE

Property, Family, and Political Forces in the Ancient World

The Problem of Origins

In his book on the *Origin of the Family, Private Property and State* (1884), Friedrich Engels sought to extricate the origins of basic human institutions from mythology and the complacent religious ideologies of "civilized" men. Against their prejudiced use of *civilization* and *barbarism* as self-serving terms helping to stress liberal values, Engels put before his Victorian readers another view of the family and its attendant institutions. The roots of primitive peoples lay deep in the soil of communal groups or broad kindred masses. Within the kindred families were broad-based collectives centered in the needs of agriculture. Women played a dominant social role, and the descent of men was reckoned in the female line. A dramatic revolution in some early age generated out of this raw material of the inbred society the exclusive pairing of couples. Within the union, the male came to dominate groups whose central purpose was (following Aristotle's theory) the security of property and the subordination of wife, children, and servants to the head of such families. The communal kindred groupings broke down, therefore, and a plurality of groups or clans emerged, based on father-right and male supremacy. Eventually closely related groups of clans (*gens, gentes,* from the Greek *genos*) formed tribes, and the tribe became the basis of protective social institutions, including kingship and legal codes securing private property.

Engels thus considered the family to be motivated economically and consecrated by inequalities of wealth and status. Modern anthropology gives little support to this notion of the origins of civilization in the monogamous family and the institutions of private property. Engels's account, however, had the merit of recognizing how Greeks and Romans sometimes viewed their own origins. *Familia* (a Latin term) signified servile property, the thralls of a master, including domestic persons, wives, and children with a common ancestor. Aristotle in *The Constitution of Athens* presumes the same relations of persons and things as do other classical writers—Hesiod and Herodotus being good examples. Engels's work is useful, moreover, in reminding us that slavery, female subordination, and property conflicts are fundamental and related in the ancient world, whatever we think of his own myth—that the rise of civilization was the death of happy communities of free and equal beings.

From *Origin of Family, Private Property and State*

Friedrich Engels

Here (in the Old World) the domestication of animals and the breeding of herds had developed a hitherto unsuspected source of wealth and created entirely new social relationships. Until the lower stage of barbarism, fixed wealth consisted almost entirely of the house, clothing, crude ornaments and the implements for procuring and preparing food: boats, weapons and household utensils of the simplest kind. Food had to be won anew day by day. Now, with herds of horses, camels, donkeys, oxen, sheep, goats and pigs, the advancing pastoral peoples— the Aryans in the Indian land of the five rivers and the Ganges area, as well as in the then much more richly watered steppes of the Oxus and the Jaxartes, and the Semites on the Euphrates and the Tigris—acquired possessions demanding merely supervision and most elementary care in order to propagate in ever-increasing numbers and to yield the richest nutriment in milk and meat. All previous means of procuring food now sank into the background. Hunting, once a necessity, now became a luxury.

But to whom did this new wealth belong? Originally, undoubtedly, to the gens. But private property in herds must have developed at a very early stage. It is hard to say whether Father Abraham appeared to the author of the so-called First Book of Moses as the owner of his herds and flocks in his own right as head of a family community or by virtue of his status as actual hereditary chief of a gens. One thing, however, is certain, and that is that we must not regard him as a property owner in the modern sense of the term. Equally certain is it that on the threshold of authenticated history we find that everywhere the herds are already the separate property of the family chiefs, in exactly the same way as were the artistic products of barbarism, metal utensils, articles of luxury and, finally, human cattle—the slaves.

For now slavery also was invented. The slave was useless to the barbarian of the lower stage. . . . Human labour power at this stage yielded no noticeable surplus as yet over the cost of its maintenance. With the introduction of cattle breeding, of the working up of metals, of weaving and, finally, of field cultivation, this changed. Just as the once so easily obtainable wives had now acquired an exchange value and were bought, so it happened with labour power, especially after the herds had finally been converted into family possessions. The family did not increase as rapidly as the cattle. More people were required to tend them; the captives taken in war were useful for just this purpose, and, furthermore, they could be bred like the cattle itself.

Such riches, once they had passed into the private possession of families and there rapidly multiplied, struck a powerful blow at a society founded on pairing marriage and mother-right gens. Pairing marriage had introduced a new element into the family. By the side of the natural mother it had placed the authenticated natural father—who was probably better authenticated than many a "father" of the present day. According to the division of labour then prevailing in the family, the procuring of food and the implements necessary thereto, and therefore, also,

SOURCE: Friedrich Engels, *Origin of Family, Private Property and State*, in Karl Marx and Friedrich Engels, *Selected Works* (Moscow: Foreign Languages Publishing House, 1962), pp. 214–218, 259–274.

the ownership of the latter, fell to the man; he took them with him in case of separation, just as the woman retained the household goods. Thus, according to the custom of society at that time, the man was also the owner of the new sources of foodstuffs—the cattle—and later, of the new instrument of labour—the slaves. According to the custom of the same society, however, his children could not inherit from him, for the position in this respect was as follows:

According to mother right, that is, as long as descent was reckoned solely through the female line, and according to the original custom of inheritance in the gens, it was the gentile relatives that at first inherited from a deceased member of the gens. The property had to remain within the gens. At first, in view of the insignificance of the chattels in question, it may, in practice, have passed to the nearest gentile relatives—that is, to the blood relatives on the mother's side. The children of the deceased, however, belonged not to his gens, but to that of their mother. In the beginning, they inherited from their mother, along with the rest of their mother's blood relatives, and later, perhaps, had first claim upon her property; but they could not inherit from their father, because they did not belong to his gens, and his property had to remain in the latter. On the death of the herd owner, therefore, his herds passed, first of all, to his brothers and sisters and to his sisters' children or to the descendants of his mother's sisters. His own children, however, were disinherited.

Thus, as wealth increased, it, on the one hand, gave the man a more important status in the family than the woman, and, on the other hand, created a stimulus to utilise this strengthened position in order to overthrow the traditional order of inheritance in favour of his children. But this was impossible as long as descent according to mother right prevailed. This had, therefore, to be overthrown, and it was overthrown; and it was not so difficult to do this as it appears to us now. For this revolution—one of the most decisive ever experienced by mankind—need not have disturbed one single living member of a gens. All the members could remain that what they were previously. The simple decision sufficed that in future the descendants of the male members should remain in the gens, but that those of the females were to be excluded from the gens and transferred to that of their father. The reckoning of descent through the female line and the right of inheritance through the mother were hereby overthrown and male lineage and right of inheritance from the father instituted. We know nothing as to how and when this revolution was effected among the civilised peoples. It falls entirely within prehistoric times. . . .

The overthrow of mother right was the *world-historic defeat of the female sex*. The man seized the reins in the house also, the woman was degraded, enthralled, the slave of the man's lust, a mere instrument for breeding children. This lowered position of women, especially manifest among the Greeks of the Heroic and still more of the Classical Age, has become gradually embellished and dissembled and, in part, clothed in a milder form, but by no means abolished.

The first effect of the sole rule of the men that was now established is shown in the intermediate form of the family which now emerges, the patriarchal family. It s chief attribute is not polygamy—of which more anon—but "the organisation of a number of persons, bond and free, into a family, under the paternal power of the head of the family. In the Semitic form, this family chief lives in polygamy, the bondsman has a wife and children, and the purpose of the whole organisation is the care of flocks and herds over a limited area." The essential features are the incorporation of bondsmen and the paternal power; the Roman family, accordingly, constitutes the perfected type of this form of the family. The word *familia* did not originally signify the ideal of our modern Philistine, which is a compound of sentimentality and domestic discord. Among the Romans, in the beginning, it did not even refer to the married couple and their children, but to the slaves alone. *Famulus* means a household slave and *familia* signifies the totality of slaves belonging to one individual. Even in the time of Gaius the *familia, id est patrimonium* (that is, the inheritance) was bequeathed by will. The expression was invented by the Romans to describe a new social organism, the head of which had under him wife and children and a number of slaves, under Roman paternal power, with power of life and death over them all. "The term, therefore, is no older than the ironclad family system of the Latin tribes, which came in after field agriculture and after legalised servitude, as well as after the separation of the Greeks and (Aryan) Latins." To which Marx adds: "The modern family contains in embryo not only slavery (*servitus*) but serfdom also, since from the very beginning it is connected with agricultural services. It contains within itself in *miniature* all the antagonisms which later develop on a wide scale within society and its state."

Such a form of the family shows the transition of the pairing family to monogamy. In order to guarantee the fidelity of the wife, that is, the paternity of the children, the woman is placed in the man's absolute power; if he kills her, he is but exercising his right. . . .

At any rate, the patriarchal household community with common land ownership and common tillage now assumes quite another significance than hitherto. We can no longer doubt the important transitional role which it played among the civilised and many other peoples of the Old World between the mother-right family and the monogamian family. . . .

. . . The *Monogamian Family*. As already indicated, this arises out of the pairing family in the transition period from the middle to the upper stage of barbarism, its final victory being one of the signs of the beginning of civilisation. It is based on the supremacy of the man; its express aim is the begetting of children of undisputed paternity, this paternity being required in order that these children may in due time inherit their father's wealth as his natural heirs. The monogamian family differs from pairing marriage in the far greater rigidity of the marriage tie, which can now no longer be dissolved at the pleasure of either party. Now, as a rule, only the man can dissolve it and cast off his wife. The right of conjugal infidelity remains his even now, sanctioned, at least, by custom (the *Code Napoléon* expressly concedes this right to the husband as long as he does not bring his concubine into the conjugal home), and is exercised more and more with the growing development of society. Should the wife recall the ancient sexual practice and desire to revive it, she is punished more severely than ever before.

We are confronted with this new form of the family in all its severity among the Greeks. While, as Marx observes, the position of the goddesses in mythology represents an earlier period, when women still occupied a freer and more respected place, in the Heroic Age we already find women degraded owing to the predominance of the man and the competition of female slaves. One may read in the *Odyssey* how Telemachus cuts his mother short and enjoins silence upon her. In Homer the young female captives become the objects of the sensual lust of the victors; the military chiefs, one after the other, according to rank, choose the most beautiful ones for themselves. The whole of the *Iliad*, as we know, revolves around the quarrel between Achilles and Agamemnon over such a female slave. In connection with each Homeric hero of importance mention is made of a captive maiden with whom he shares tent and bed. These maidens are taken back home, to the conjugal house, as was Cassandra by Agamemnon in Aeschylus. Sons born of these slaves receive a small share of their father's estate and are regarded as freemen. Teukros was such an illegitimate son of Telamon and was permitted to adopt his father's name. The wedded wife is expected to tolerate all this, but to maintain strict chastity and conjugal fidelity herself. True, in the Heroic Age the Greek wife is more respected than in the period of civilisation; for the husband, however, she is, in reality, merely the mother of his legitimate heirs, his chief housekeeper, and the superintendent of the female slaves whom he may make, and does make, his concubines at will. It is the existence of slavery side by side with monogamy, the existence of beautiful young slaves who belong to the *man* with all they have, that from the very beginning stamped on monogamy its specific character as monogamy *only for the woman*, but not for the man. And it retains this character to this day.

As regards the Greeks of later times, we must differentiate between the Dorians and the Ionians. The former, of whom Sparta was the classical example, had in many respects more ancient marriage relationships than even Homer indicates. In Sparta we find a form of pairing marriage—modified by the state in accordance with the conceptions there prevailing—which still retains many vestiges of group marriage. Childless marriages were dissolved: King Anaxandridas (about 650 B.C.) took another wife in addition to his first, childless one, and maintained two households; King Aristones of the same period added a third to two previous wives who were barren, one of whom he, however, let go. On the other hand, several brothers could have a wife in common. A person having a preference for his friend's wife could share her with him; and it was regarded as proper to place one's wife at the disposal of a lusty "stallion," as Bismarck would say, even when this person was not a citizen. A passage in Plutarch, where a Spartan woman sends a lover who is pursuing her with his attentions to interview her husband, would indicate, according to Schömann, still greater sexual freedom. Real adultery, the infidelity of the wife behind the back of her husband, was thus unheard of. On the other hand, domestic slavery was unknown in Sparta, at least in its heyday; the Helot serfs lived segregated on the estates and thus there was less temptation for the Spartiates to have intercourse with their women. That in all these circumstances the women of Sparta enjoyed a very much more respected position than all other Greek women was quite natural. The Spartan women and the *élite* of the Athenian *hetaerae* are the only Greek women of whom the ancients speak with respect, and whose remarks they consider as being worthy of record.

Among the Ionians—of whom Athens is characteristic—things were quite different. Girls learned only spinning, weaving and sewing, at best a little reading and writing. They were practically kept in seclusion and consorted only with other women. The women's quarter was a separate and distinct part of the house, on the upper floor, or in the rear building, not easily accessible to men, particularly strangers; to this the women retired when men visitors came. The women did not go out unless accompanied by a female slave; at home they were virtually kept under guard; Aristophanes speaks of Molossian hounds kept to frighten off adulterers, while in Asiatic towns, at least, eunuchs were maintained to keep guard over the women; they were manufactured for the trade in Chios as early as Herodotus' day, and according to Wachsmuth, not merely for the barbarians. In Euripides, the wife is described as *oikurema*, a thing for housekeeping (the word is in the neuter gender), and apart from the business of bearing children, she was nothing more to the Athenian than the chief housemaid. The husband had his gymnastic exercises, his public affairs, from which the wife was excluded; in addition, he often had female slaves at his disposal and, in the hey-day of Athens, extensive prostitution, which was viewed with favour by the state, to say the least. It was precisely on the basis of this prostitution that the sole outstanding Greek women developed, who by their *esprit* and artistic taste towered as much above the general level of ancient womanhood as the Spartiate women did by virtue of their character. That one had first to become a *hetaera* in order to become a woman is the strongest indictment of the Athenian family. . . .

This was the origin of monogamy, as far as we can trace it among the most civilised and highly-developed people of antiquity. It was not in any way the fruit of individual sex love, with which it had absolutely nothing in common, for the marriages remained marriages of convenience, as before. It was the first form of the family based not on natural but on economic conditions, namely, on the victory of private property over original, naturally developed, common ownership. The rule of the man in the family, the procreation of children who could only be his, destined to be the heirs of his wealth—these alone were frankly avowed by the Greeks as the exclusive aims of monogamy. For the rest, it was a burden, a duty to the gods, to the state and to their ancestors, which just had to be fulfilled. In Athens the law made not only marriage compulsory, but also the fulfilment by the man of a minimum of the so-called conjugal duties.

Thus, monogamy does not by any means make its appearance in history as the reconciliation of man and woman, still less as the highest form of such a reconciliation. On the contrary, it appears as the subjection of one sex by the other, as the proclamation of a conflict between the sexes entirely unknown hitherto in prehistoric times. In an old unpublished manuscript, the work of Marx and myself in 1846, I find the following: "The first division of labour is that between man and woman for child breeding." And today I can add: The first class antagonism which appears in history coincides with the development of the antagonism between man and woman in monogamian marriage, and the first class oppression with that of the female sex by the male. Monogamy was a great historical advance, but at the same time it inaugurated, along with slavery and private wealth, that epoch, lasting until today, in which every advance is likewise a relative regression, in which the well-being and development of the one group are attained by the misery and repression of the other. It is the cellular form of civilised society, in which we can already study the nature of the antagonisms and contradictions which develop fully in the latter.

The old relative freedom of sexual intercourse by no means disappeared with the victory of the pairing family, or even of monogamy. "The old conjugal system, now reduced to narrower limits by the gradual disappearance of the punaluan[1] groups, still environed the advancing family, which it was to follow to the verge of civilisation. . . . It finally disappeared in the new form of hetaerism, which still follows mankind in civilisation as a dark shadow upon the family." [2] By hetaerism Morgan means that extramarital sexual intercourse between men

---

1 Punaluan: tribal family groups.
2 From Lewis Morgan, *Ancient Society*.

and unmarried women which exists *alongside of monogamy*, and, as is well known, has flourished in the most diverse forms during the whole period of civilisation and is steadily developing into open prostitution. This hetaerism is directly traceable to group marriage, to the sacrificial surrender of the women, whereby they purchased their right to chastity. The surrender for money was at first a religious act, taking place in the temple of the Goddess of Love, and the money originally flowed into the coffers of the temple. The hierodules of Anaitis in Armenia, of Aphrodite in Corinth, as well as the religious dancing girls attached to the temples in India—the so-called bayaders (the word is a corruption of the Portuguese *bailadeira*, a female dancer)—were the first prostitutes. This sacrificial surrender, originally obligatory for all women, was later practised vicariously by these priestesses alone on behalf of all other women. Hetaerism among other peoples grows out of the sexual freedom permitted to girls before marriage —hence likewise a survival of group marriage, only transmitted to us by another route. With the rise of property differentiation—that is, as far back as the upper stage of barbarism—wage labour appears sporadically alongside of slave labour; and simultaneously, as its necessary correlate, the professional prostitution of free women appears side by side with the forced surrender of the female slave. Thus, the heritage bequeathed to civilisation by group marriage is double-sided, just as everything engendered by civilisation is double-sided, double-tongued, self-contradictory and antagonistic: on the one hand, monogamy, on the other, hetaerism, including its most extreme form, prostitution. Hetaerism is as much a social institution as any other; it is a continuation of the old sexual freedom—in favour of the men. Although, in reality, it is not only tolerated but even practised with gusto, particularly by the ruling classes, it is condemned in words. In reality, however, this condemnation by no means hits the men who indulge in it, it hits only the women: they are ostracised and cast out in order to proclaim once again the absolute domination of the male over the female sex as the fundamental law of society.

A second contradiction, however, is hereby developed within monogamy itself. By the side of the husband, whose life is embellished by hetaerism, stands the neglected wife. And it is just as impossible to have one side of a contradiction without the other as it is to retain the whole of an apple in one's hand after half has been eaten. Nevertheless, the men appear to have thought differently, until their wives taught them to know better. Two permanent social figures, previously unknown, appear on the scene along with monogamy—the wife's paramour and the cuckold. The men had gained the victory over the women, but the act of crowning the victor was magnanimously undertaken by the vanquished. Adultery—proscribed, severely penalised, but irrepressible—became an unavoidable social institution alongside of monogamy and hetaerism.

THE GRECIAN GENS

Greeks as well as Pelasgians and other peoples of the same tribal origin were constituted since prehistoric times in the same organic series as the Americans:[3] gens, phratry, tribe, confederacy of tribes. The phratry might be missing, as, for example, among the Dorians; the confederacy of tribes might not be fully devel-

---

3 Americans: Iroquois Indian tribal patterns.

oped yet in every case; but the gens was everywhere the unit. At the time the Greeks entered into history, they were on the threshold of civilisation. Almost two entire great periods of development lie between the Greeks and the above-mentioned American tribes, the Greeks of the Heroic Age being by so much ahead of the Iroquois. For this reason the Grecian gens no longer bore the archaic character of the Iroquois gens; the stamp of group marriage was becoming considerably blurred. Mother right had given way to father right; thereby rising private wealth made the first breach in the gentile constitution. A second breach naturally followed the first: after the introduction of father right, the fortune of a wealthy heiress would, by virtue of her marriage, fall to her husband, that is to say, to another gens; and so the foundation of all gentile law was broken, and in such cases the girl was not only permitted, but *obliged* to marry within the gens, in order that the latter might retain the fortune.

According to Grote's *History of Greece*, the Athenian gens in particular was held together by:

1. Common religious ceremonies, and exclusive privilege of the priesthood in honour of a definite god, supposed to be the primitive ancestor of the gens, and characterised in this capacity by a special surname.
2. A common burial place. (Compare Demosthenes' *Eubulides*.)
3. Mutual rights of inheritance.
4. Reciprocal obligation to afford help, defence and support against the use of force.
5. Mutual right and obligation to marry in the gens in certain cases, especially for orphaned daughters or heiresses.
6. Possession, in some cases at least, of common property, and of an archon (magistrate) and treasurer of its own.

The phratry, binding together several gentes, was less intimate, but here too we find mutual rights and duties of similar character, especially a communion of particular religious rites and the right of prosecution in the event of a phrator being slain. Again, all the phratries of a tribe performed periodically certain common sacred ceremonies under the presidency of a magistrate called the *phylobasileus* (tribal magistrate) selected from among the nobles (*eupatrides*). . . . .

For the Grecian gens has also the following attributes:

7. Descent according to father right.
8. Prohibition of intermarrying in the gens except in the case of heiresses. This exception and its formulation as an injunction clearly proves the validity of the old rule. This follows also from the universally accepted rule that when a woman married she renounced the religious rites of her gens and acquired those of the gens of her husband, in whose phratry she was enrolled. This, and a famous passage in Dicaearchus, go to prove that marriage outside of the gens was the rule. Becker in *Charicles* directly assumes that nobody was permitted to marry in his or her own gens.
9. The right of adoption into the gens; it was practised by adoption into the family, but with public formalities, and only in exceptional cases.
10. The right to elect and depose the chiefs. We know that every gens had its archon; but nowhere is it stated that this office was hereditary in certain families. Until the end of barbarism, the probability is always against strict

heredity, which would be totally incompatible with conditions where rich and poor had absolutely equal rights in the gens. . . .

. . . The system of consanguinity corresponding to the gens in its original form—the Greeks once possessed it like other mortals—preserved the knowledge of the mutual relation of all members of the gens. They learned this for them decisively important fact by practice from early childhood. With the advent of the monogamian family this dropped into oblivion. The gentile name created a genealogy compared with which that of the monogamian family seemed insignificant. This name was now to attest to its bearers the fact of their common ancestry. But the genealogy of the gens went so far back that its members could no longer prove their mutual real kinship, except in a limited number of cases of more recent common ancestors. The name itself was the proof of a common ancestry, and conclusive proof, except in cases of adoption. . . .

Closely packed in a comparatively small territory as the Greeks were, their differences in dialect were less conspicuous than those that developed in the extensive American forests. Nevertheless, even here we find only tribes of the same main dialect united in a larger aggregate; and even little Attica had its own dialect, which later on became the prevailing language in Greek prose.

In the epics of Homer we generally find the Greek tribes already combined into small peoples, within which, however, the gentes, phratries and tribes still retained their full independence. They already lived in walled cities. The population increased with the growth of the herds, with field agriculture and the beginnings of the handicrafts. With this came increased differences in wealth, which gave rise to an aristocratic element within the old natural-grown democracy. The various small peoples engaged in constant warfare for the possession of the best land and also for the sake of loot. The enslavement of prisoners of war was already a recognised institution. . . .

Thus, in the Grecian constitution of the Heroic Age, we still find the old gentile system full of vigour; but we also see the beginning of its decay: father right and the inheritance of property by the children, which favoured the accumulation of wealth in the family and gave the latter power as against the gens; differentiation in wealth affecting in turn the social constitution by creating first rudiments of a hereditary nobility and monarchy; slavery, first limited to prisoners of war, but already paving the way to the enslavement of fellow members of the tribe and even of the gens; the degeneration of the old intertribal warfare to systematic raids, on land and sea, for the purpose of capturing cattle, slaves, and treasure as a regular means of gaining a livelihood. In short, wealth is praised and respected as the highest treasure, and the old gentile institutions are perverted in order to justify forcible robbery of wealth. Only one thing was missing: an institution that would not only safeguard the newly-acquired property of private individuals against the communistic traditions of the gentile order, would not only sanctify private property, formerly held in such light esteem, and pronounce this sanctification the highest purpose of human society, but would also stamp the gradually developing new forms of acquiring property, and consequently, of constantly accelerating increase in wealth, with the seal of general public recognition; an institution that would perpetuate, not only the newly-rising class division of society, but also the right of the possessing class to exploit the non-possessing classes and the rule of the former over the latter.

And this institution arrived. The *state* was invented.

THE RISE OF THE ATHENIAN STATE

How the state developed, some of the organs of the gentile constitution being transformed, some displaced, by the intrusion of new organs, and, finally, all superseded by real governmental authorities—while the place of the actual "people in arms" defending itself through its gentes, phratries and tribes were taken by an armed "public power" at the service of these authorities and, therefore, also available against the people—all this can nowhere be traced better, at least in it initial stage, than in ancient Athens. The forms of the changes are, in the main, described by Morgan; the economic content which gave rise to them I had largely to add myself.

In the Heroic Age, the four tribes of the Athenians were still installed in separate parts of Attica. Even the twelve phratries comprising them seem still to have had separate seats in the twelve towns of Cecrops. The constitution was that of the Heroic Age: a popular assembly, a popular council, a *basileus*.[4] As far back as written history goes we find the land already divided up and transformed into private property, which corresponds with the relatively developed state of commodity production and a commensurate commodity trade towards the end of the higher stage of barbarism. In addition to cereals, wine and oil were cultivated. Commerce on the Aegean Sea passed more and more from Phoenician into Attic hands. As a result of the purchase and sale of land and the continued division of labour between agriculture and handicrafts, trade and navigation, the members of gentes, phratries and tribes very soon intermingled. The districts of the phratry and the tribe received inhabitants who, although they were fellow countryman, did not belong to these bodies and, therefore, were strangers in their own places of residence. For in time of peace, every phratry and every tribe administered its own affairs without consulting the popular council or the *basileus* in Athens. But inhabitants of the area of the phratry or tribe not belonging to either naturally could not take part in the administration.

This so disturbed the regulated functioning of the organs of the gentile constitution that a remedy was already needed in the Heroic Age. A constitution, attributed to Theseus, was introduced. The main feature of this change was the institution of a central administration in Athens, that is to say, some of the affairs that hitherto had been conducted independently by the tribes were declared to the common affairs and transferred to a general council sitting in Athens. Thereby, the Athenians went a step further than any ever taken by any indigenous people in America: the simple federation of neighbouring tribes was now supplanted by the coalescence of all the tribes into one single people. This gave rise to a system of general Athenian popular law, which stood above the legal usages of the tribes and gentes. It bestowed on the citizens of Athens, as such, certain rights and additional legal protection even in territory that was not their own tribe's. This, however, was the first step towards undermining the gentile constitution; for it was the first step towards the subsequent admission of citizens who were alien to all the Attic tribes and were and remained entirely outside the pale of the Athenian gentile constitution. A second institution attributed to Theseus was the division of the entire people, irrespective of gentes, phratries and tribes, into three classes: *eupatrides*, or nobles; *geomoroi*, or tillers of the land; and *demiurgi*, or artisans, and the granting to the nobles of the exclusive right to public office.

---

4 Basileus: popular political meeting, like a tribal meeting, where decisions were made.

True, apart from reserving to the nobles the right to hold public office, this division remained inoperative, as it created no other legal distinctions between the classes. It is important, however, because it reveals to us the new social elements that had quietly developed. It shows that the customary holding of office in the gens by certain families had already developed into a privilege of these families that was little contested; that these families, already powerful owing to their wealth, began to unite outside of their gentes into a privileged class; and that the nascent state sanctioned this usurpation. It shows, furthermore, that the division of labour between husbandmen and artisans had become strong enough to contest the superiority, socially, of the old division into gentes and tribes. And finally, it proclaimed the irreconcilable antagonism between gentile society and the state. The first attempt to form a state consisted in breaking up the gentes by dividing the members of each into a privileged and an inferior class, and the latter again into two vocational classes, thus selling one against the other.

The ensuing political history of Athens up to the time of Solon[5] is not completely known. The office of *basileus* fell into disuse; *archons*, elected from among the nobility, became the heads of the state. The rule of the nobility steadily increased until, round about 600 B.C., it became unbearable. The principal means for stifling the liberty of the commonalty were—money and usury. The nobility lived mainly in and around Athens, where maritime commerce, with occasional piracy still as a sideline, enriched it and concentrated monetary wealth in its hands. From this point the developing money system penetrated like a corroding acid into the traditional life of the rural communities founded on natural economy. The gentile constitution is absolutely incompatible with the money system. The ruin of the Attic small-holding peasants coincided with the loosening of the old gentile bonds that protected them. Creditors' bills and mortgage bonds—for by then the Athenians had also invented the mortgage—respected neither the gens nor the phratry. But the old gentile constitution knew nothing of money, credit and monetary debt. Hence the constantly expanding money rule of the nobility gave rise to a new law, that of custom, to protect the creditor against the debtor and sanction the exploitation of the small peasant by the money owner. All the rural districts of Attica bristled with mortgage posts bearing the legend that the lot on which they stood was mortgaged to so and so for so and so much. The fields that were not so designated had for the most part been sold on account of overdue mortgages or non-payment of interest and had become the property of the noble-born usurers; the peasant was glad if he was permitted to remain as a tenant and live on *one-sixth* of the product of his labour while paying *five-sixths* to his new master as rent. More than that: if the sum obtained from the sale of the lot did not cover the debt, or if such a debt was not secured by a pledge, the debtor had to sell his children into slavery abroad in order to satisfy the creditor's claim. The sale of his children by the father—such was the first fruit of father right and monogamy! And if the blood-sucker was still unsatisfied, he could sell the debtor himself into slavery. Such was the pleasant dawn of civilisation among the Athenian people. . . .

. . . The appearance of private property in herds of cattle and articles of luxury led to exchange between individuals, to the transformation of products into *commodities*. Here lies the root of the entire revolution that followed. When

---

[5] Ca. 594 B.C.

the producers no longer directly consumed their product, but let it go out of their hands in the course of exchange, they lost control over it. They no longer knew what became of it, and the possibility arose that the product might some day be turned against the producers, used as a means of exploiting and oppressing them. Hence, no society can for any length of time remain master of its own production and continue to control the social effects of its process of production, unless it abolishes exchange between individuals.

The Athenians were soon to learn, however, how quickly after individual exchange is established and products are converted into commodities, the product manifests its rule over the producer. With the production of commodities came the tilling of the soil by individual cultivators for their own account, soon followed by individual ownership of the land. Then came money, that universal commodity for which all others could be exchanged. But when men invented money they little suspected that they were creating a new social power, the one universal power to which the whole of society must bow. It was this new power, suddenly sprung into existence without the will or knowledge of its own creators, that the Athenians felt in all the brutality of its youth.

What was to be done? The old gentile organisation had not only proved impotent against the triumphant march of money; it was also absolutely incapable of providing a place within its framework for such things as money, creditors, debtors and the forcible collection of debts. But the new social power was there, and neither pious wishes nor a longing for the return of the good old times could drive money and usury out of existence. Moreover, a number of other, minor breaches had been made in the gentile constitution. The indiscriminate mingling of the gentiles and phrators throughout the whole of Attica, and especially in the city of Athens, increased from generation to generation, in spite of the fact that an Athenian, while allowed to sell plots of land out of his gens, was still prohibited from thus selling his dwelling house. The division of labour between the different branches of production—agriculture, handicraft, numerous skills within the various crafts, trade, navigation, etc.—had developed more fully with the progress of industry and commerce. The population was now divided according to occupation into rather well-defined groups, each of which had a number of new, common interests that found no place in the gens or phratry and, therefore, necessitated the creation of new offices to attend to them. The number of slaves had increased considerably and must have far exceeded that of the free Athenians even at this early stage. The gentile constitution originally knew no slavery and was, therefore, ignorant of any means of holding this mass of bondsmen in check. And finally, commerce had attracted a great many strangers who settled in Athens because it was easier to make money there, and according to the old constitution these strangers enjoyed neither rights nor the protection of the law. In spite of traditional toleration, they remained a disturbing and foreign element among the people.

In short, the gentile constitution was coming to an end. Society was daily growing more and more out of it; it was powerless to check or allay even the most distressing evils that were arising under its very eyes. In the meantime, however, the state had quietly developed. The new groups formed by division of labour, first between town and country, then between the various branches of urban industry, had created new organs to protect their interests. Public offices of every description were instituted. And then the young state needed, above all, its own fighting forces, which among the seafaring Athenians could at first be only naval forces, to be used for occasional small wars and to protect merchant vessels. At some uncertain time before Solon, the naucraries were instituted, small territorial districts, twelve in each tribe. Every naucrary had to furnish, equip and man a war vessel and, in addition, detail two horsemen. This arrangement was a two-fold attack on the gentile constitution. First, it created a public power which was no longer simply identical with the armed people in its totality; secondly, it for the first time divided the people for public purposes, not according to kinship groups, but territorially, according to *common domicile*. We shall see what this signified.

As the gentile constitution could not come to the assistance of the exploited people, they could look only to the rising state. And the state brought help in the form of the constitution of Solon, while at the same time strengthening itself anew at the expense of the old constitution. Solon—the manner in which his reform of 594 B.C. was brought about does not concern us here—started the series of so-called political revolutions by an encroachment on property. All revolutions until now have been revolutions for the protection of one kind of property against another kind of property. They cannot protect one kind without violating another. In the Great French Revolution feudal property was sacrificed in order to save bourgeois property; in Solon's revolution, creditors' property had to suffer for the benefit of debtors' property. The debts were simply annulled. We are not acquainted with the exact details, but Solon boasts in his poems that he removed the mortgage posts from the encumbered lands and enabled all who had fled or had been sold abroad for debt to return home. This could have been done only by openly violating property rights. And indeed, the object of all so-called political revolutions, from first to last, was to protect *one* kind of property by confiscating —also called stealing—*another* kind of property. It is thus absolutely true that for 2500 years private property could be protected only by violating property rights.

But now a way had to be found to prevent such re-enslavement of the free Athenians. This was first achieved by general measures; for example, the prohibition of contracts which involved the personal hypothecation of the debtor. Furthermore, a maximum was fixed for the amount of land any one individual could own, in order to put some curb, at least, on the craving of the nobility for the peasants' land. Then followed constitutional amendments, of which the most important for us are the following:

The council was increased to four hundred members, one hundred from each tribe. Here, then, the tribe still served as a basis. But this was the only side of the old constitution that was incorporated in the new body politic. For the rest, Solon divided the citizens into four classes, according to the amount of land owned and its yield. Five hundred, three hundred and one hundred and fifty medimni of grain (1 medimnus equals appr. 41 litres) were the minimum yields for the first three classes; whoever had less land or none at all belonged to the fourth class. Only members of the first three classes could hold office; the highest offices were filled by the first class. The fourth class had only the right to speak and vote in the popular assembly. But here all officials were elected, here they had to give account of their actions, here all the laws were made, and here the fourth class was in the majority. The aristocratic privileges were partly renewed in the form of privileges of wealth, but the people retained the decisive power. The four classes also formed the basis for the reorganisation of the fighting forces. The first two classes furnished the cavalry; the third had to serve as heavy infantry; the fourth served as light infantry, without armour, or in the navy, and probably were paid.

Thus, an entirely new element was introduced into the constitution: private ownership. The rights and duties of the citizens were graduated according to the amount of land they owned; and as the propertied classes gained influence the old consanguine groups were driven into the background. The gentile constitution suffered another defeat.

The gradation of political rights according to property, however, was not an indispensable institution for the state. Important as it may have been in the constitutional history of states, nevertheless, a good many states, and the most completely developed at that, did without it. Even in Athens it played only a transient role. Since the time of Aristides, all offices were open to all the citizens.

During the next eighty years Athenian society gradually took the course along which it further developed in subsequent centuries. Usurious land operations, rampant in the pre-Solon period, were checked, as was the unlimited concentration of landed property. Commerce and the handicrafts and useful arts conducted on an ever-increasing scale with slave labour became the predominating branches of occupation. Enlightenment made progress. Instead of exploiting their own fellow-citizens in the old brutal manner, the Athenians now exploited mainly the slaves and non-Athenian clients. Movable property, wealth in money, slaves and ships, increased more and more; but instead of being simply a means for purchasing land, as in the first period with its limitations, it became an end in itself. This, on the one hand, gave rise to the successful competition of the new, wealthy industrial and commercial class with the old power of the nobility, but, on the other hand, it deprived the old gentile constitution of its last foothold. The gentes, phratries and tribes, whose members were now scattered all over Attica and lived completely intermingled, thus became entirely useless as political bodies. A large number of Athenian citizens did not belong to any gens; they were immigrants who had been adopted into citizenship, but not into any of the old bodies of *consanguinei*. Besides, there was a steadily increasing number of foreign immigrants who only enjoyed protection.

Meanwhile, the struggles of the parties proceeded. The nobility tried to regain its former privileges and for a short time recovered its supremacy, until the revolution of Cleisthenes (509 B.C.) brought about its final downfall; and with them fell the last remnants of the gentile constitution.

In his new constitution, Cleisthenes ignored the four old tribes based on the gentes and phratries. Their place was taken by an entirely new organisation based exclusively on the division of the citizens according to place of domicile, already attempted in the naucraries. Not membership of a body of *consanguinei*, but place of domicile was now the deciding factor. Not people, but territory was now divided; politically, the inhabitants became mere attachments of the territory.

The whole of Attica was divided into one hundred self-governing townships, or demes. The citizens (demots) of a deme elected their official head (demarch), a treasurer and thirty judges with jurisdiction in minor cases. They also received their own temple and a tutelary deity, or *heros*, whose priests they elected. The supreme power in the deme was the assembly of the demots. . . .

The consummation was the Athenian state, governed by a council of five hundred—elected by the ten tribes—and, in the last instance, by the popular assembly, which every Athenian citizen could attend and vote in. Archons and other officials attended to the different departments of administration and the courts. In Athens there was no official possessing supreme executive authority.

By this new constitution and by the admission of a large number of dependents [*Schutzverwandter*], partly immigrants and partly freed slaves, the organs of the gentile constitution were eliminated from public affairs. They sank to the position of private associations and religious societies. But their moral influence, the traditional conceptions and views of the old gentile period, survived for a long time and expired only gradually. This became evident in a subsequent state institution.

We have seen that an essential feature of the state is a public power distinct from the mass of the people. At that time Athens possessed only a militia and navy equipped and manned directly by the people. These afforded protection against external enemies and held the slaves in check. . . . For the citizens, this public power at first existed only in the shape of the police force, which is as old as the state, and that is why the naïve Frenchmen of the eighteenth century spoke, not of civilised, but of policed nations (*nations policées*). Thus, simultaneously with their state, the Athenians established a police force, a veritable gendarmerie of foot and mounted bowmen—*Landjäger*, as they say in South Germany and Switzerland. This gendarmerie consisted—of *slaves*. The free Athenian regarded this police duty as being so degrading that he preferred being arrested by an armed slave rather than perform such ignominious duties himself. This was still an expression of the old gentile mentality. The state could not exist without a police force, but it was still young and did not yet command sufficient moral respect to give prestige to an occupation that necessarily appeared infamous to the old gentiles.

How well this state, now completed in its main outlines, suited the new social condition of the Athenians was apparent from the rapid growth of wealth, commerce and industry. The class antagonism on which the social and political institutions rested was no longer that between the nobles and the common people, but that between slaves and freemen, dependents and citizens. . . . With the development of commerce and industry came the accumulation and concentration of wealth in a few hands; the mass of the free citizens was impoverished and had to choose between going into handicrafts and competing with slave labour, which was considered ignoble and base and, moreover, promised little success—and complete pauperisation. . . .

The rise of the state among the Athenians presents a very typical example of state building in general; because, on the one hand, it took place in a pure form, without the interference of violence, external or internal (the short period of usurpation by Pisistratus left no trace behind it); because, on the other hand, it represented the rise of a highly-developed form of state, the democratic republic, emerging directly out of gentile society; and lastly, because we are sufficiently acquainted with all the essential details.

Class and the Athenian Constitution

Nothing was more basic in the *Politics*, *Ethics*, and *Economics* of Aristotle (384–322? B.C.) than the ultimate division of society into two "classes" of men —those who rule and those who are ruled. In his historical writings about Athenian politics, Aristotle had the advantages of an "outsider." His family was Hellenic but for generations had lived at Stageira on the Macedonian frontier. Thus, while he passed many years with Plato and knew the culture of the Athenian agora (marketplace) in detail, there was in his writing about it a direct, earnest criticism devoid of artifice; hence the extraordinary interest of the modern discovery of *The Constitution of Athens*. In this work, Aristotle commented on the relationships between the growing complexity of the economic life of Athens and her political institutions. The old tribes had dominated gentilitian, agrarian, aristocratic society. Their hegemony had been undermined by the rise of a money economy, and the very basis of their power was only a memory in Solon's day. Here we see the original ground of Engels's commentaries, as the clans are strangled to death in the subtle web woven by the many classes of rich and poor craftsmen and merchants, professional soldiers and bureaucrats. The sense of civic virtue—that habit of action so basic in the *Ethics* and *Politics*—is lost in the historical reality of a complex empire. In Athens, slavery, militarism, and the uses of power finally made of the Aristotelian "political animal" something less than a man.*

From *The Constitution of Athens*

Aristotle

1. . . . With Myron acting as accuser, a court, selected from the nobility and sworn in upon the sacrifices, passed a verdict to the effect that a sacrilege had been committed. Thereupon, the bodies of the guilty were removed from the tombs, and their family was exiled forever. On account of these events, Epimenides of Crete purified the city.

2. After that,[1] there was civil strife for a long time between the nobility and the common people. For the whole political setup was oligarchical, and, in particular, the poor together with their wives and children were serfs of the rich. They were called Pelatae[2] and Hectemori ["sixth-parters"], for it was at this rent[3] that they cultivated the land of the wealthy. All the land was in the hands of a few, and if the serfs did not pay their rent, they and their children could be sold into slavery. All loans were contracted upon the person of the debtor, until the

---

SOURCE: Aristotle, *The Constitution of Athens*, trans. Kurt Von Fritz and Ernst Kapp (New York: Hafner Publishing Company, 1966), pp. 69–87, 94–95, 97–98, 114–116.

* In the *Politics*, Aristotle had declared that man was a "political animal" (*zōon politikon*); men who lived apart from the *polis* or city-state were either beasts or gods.

1 This obviously refers to the Cylonian affair itself, not to the purification of Athens by Epimenides, which is usually dated in the time of Solon.

2 Usually explained to mean *clients*, but the exact meaning is doubtful.

3 That is, one sixth of the produce.

time of Solon, who was the first to become a leader[4] of the people. The hardest and most hateful feature of the political situation as far as the many were concerned was their serfdom. But they also nursed grievances in all other respects, for they had, so to speak, no share in anything.

3. The ancient political order that existed before Draco[5] was as follows: The magistrates were selected from the noble and the wealthy. At first, they governed for life; later, for periods of ten years. The most important and the earliest offices were those of the King, the Polemarch, and the Archon. Of these, the office of the King was the earliest, for it had come down from ancestral times. Secondly, there was introduced the office of the Polemarch, which was added because some of the kings turned out unfit for war; this, by the way, is why Ion was sent for in an emergency. Last was the office of the Archon. Most authorities say that this office was introduced under Medon, but some say under Acastus. They offer as evidence the fact that the nine archons swear to execute the oaths "as under Acastus," which seems to indicate that under his rule the descendants of Codrus gave up the kingship in return for the prerogatives given to the Archon. Whichever of the two accounts is true, the chronological difference is not very great. At any rate, that the office of the Archon came last is indicated by the fact that the Archon is not, like the King and the Polemarch, in charge of any of the ancestral functions, but only of those that were added later. And this is the reason why the archonship became great only in more recent times, having been increased in importance by these added functions. . . .

4. In outline, such was the first political order. Not much later, in the archonship of Aristaechmus,[6] Draco enacted his laws. His constitutional order was the following: Full political rights had been given to those who provided themselves with full military equipment; and these elected the nine Archons and the Treasurers from persons who possessed an unencumbered property of not less than ten minae, the minor magistrates from those who owned full military equipment, and the Generals (*strategoi*) and the Commanders of the cavalry (*hipparchoi*) from those who declared an unencumbered property of not less than one hundred minae and legitimate sons over ten years old. These officers [that is, the newly elected Generals and Hipparchs] were to be held to bail by the Prytanes, as were the Generals and Hipparchs of the preceding year until the completion of their audit. The Prytanes were to accept as securities four citizens from the same class as that to which the Generals and Hipparchs belonged. The Council [7] was to consist of four hundred and one chosen by lot from those possessing full rights of citizenship. This Council and the other magistrates[8] were chosen by lot from those citizens who were more than thirty years old. The same man could not become a magistrate twice until all other citizens had had a turn. Then the whole procedure of casting lots would begin again. If anyone of the Councilmen failed to attend when there was a session of the Council or of the Assembly of the

---

4 The Greek word is προστάτης. In later times leadership of the people became almost a permanent institution, but it was never an official position.

5 Draco (ca. 691 B.C.–?) was a leading representative of the aristocratic order.

6 This Archon had not been known before the discovery of Aristotle's treatise. But the traditional date of Draco's legislation is 621 B.C.

7 Here, as elsewhere, Aristotle seems to take it for granted that there always existed an "Assembly of the People," and a "Council" different from the Council of the Areopagus.

8 That is, the "minor" magistracies, which in the preceding paragraph have been distinguished from the Archons and Treasurers, on the one hand, and from the generals and commanders of the cavalry, on the other.

People, he had to pay a fine of three drachmae if he was a Pentacosiomedimnus, two drachmae if he was a Knight, and one drachma if he was a Zeugites. The Council of the Areopagus was the guardian of the laws and also kept watch over the magistrates so as to take care that they ruled according to the laws. Anyone who had been wronged could file complaint with the Council of the Areopagus indicating the law which had been violated by the wrong done to him. But loans were secured on the person of the debtor, and the land was in the hands of a few.

5. This being the political order and the many being serfs of the few, the common people rose against the upper class. When the civil discord had become violent and the two opposing parties had been set against each other for a long time, they chose, by mutual agreement, Solon as their mediator and Archon[9] and entrusted the state to him. This happened after he had composed the elegy[10] that begins:

> I observe, and my heart is filled with grief when I look upon the oldest land
> of the Ionian world as it totters.

In this poem he fights for both parties against both parties. He tries to distinguish the merits and demerits of the one and of the other, and, after having done so, he exhorts both of them together to end their present dispute.

Solon was by birth and renown one of the most distinguished men of the country, but by wealth and occupation he belonged to the middle class. This can be inferred from many facts and is also confirmed by Solon's own testimony in the following passage of a poem in which he exhorts the wealthy not to set their aims too high:

> You who are plunged into a surfeit of many goods restrain the strong desires
> in your breast, let your proud mind be set on moderate aims.
> For we shall not submit to you, and not everything will turn out according to
> your wishes.

And, in general, he attaches the blame for the conflict to the rich; and, accordingly, he says, in the beginning of the poem, that he was always afraid of "love for money and an overbearing mind," implying that these had been the cause of the conflict.

6. As soon as Solon had been entrusted with full powers to act, he liberated the people by prohibiting loans on the person of the debtor, both for the present and for the future. He made laws and enacted a cancellation of debts both private and public, a measure which is commonly called *seisachtheia* [the shaking-off of burdens], since in this way they shook off their burdens. In regard to this measure, some people try to discredit him. For it happened that when Solon was about to enact the *seisachtheia*, he informed some of his acquaintances of his plans, and when he did so, according to the version of the adherents of the popular party, he was outmaneuvered by his friends; but, according to those who wish to slander him, he himself shared in the gain. For these people borrowed money and bought a great extent of land; and a short time afterwards, when the cancellation of debts was put through, they became very rich. It is said that this was the origin of those who later were considered to be of ancient wealth. However, the version of

---

9 The traditional date of Solon's archonship is 594 B.C.
10 At the time of the Solon, the art of reading and writing was practiced for only very limited purposes; a political pamphlet, in order to be effective, had to be conceived in poetic form; in this way it could be easily memorized and transmitted.

the friends of the people appears much more trustworthy. For it is not likely that in all other respects Solon should have been so moderate and public-spirited that, when it would have been in his power to subdue all others and to set himself up as a tyrant, he preferred to incur the hostility of both parties and valued his honor and the common good of the state higher than his personal aggrandizement, and that yet he should have defiled himself by such a petty and unworthy trick. Now, that he did have that opportunity [that is, of setting himself up as a tyrant] is proved by the desperate situation of the state at that time; he himself mentions the fact frequently in his poems, and it is universally admitted. Hence, one must regard the accusation as completely unfounded.

7. Solon set up a constitution and also made other laws. After that, the Athenians ceased to make use of the laws of Draco with the exception of those relating to murder. The laws were inscribed on the Kyrbeis[11] and placed in the portico of the King, and all swore to observe them. The nine Archons, however, regularly affirmed by an oath at the Stone that they would dedicate a golden statue if they ever should be found to have transgressed one of the laws; and they still swear in the same fashion down to the present day. He made the laws unalterable for one hundred years and set up the political order in the following way:

He divided the population, according to property qualifications, into four classes as they had been divided before—namely, Pentacosiomedimni, Knights, Zeugitae, and Thetes.[12] He distributed the higher offices, namely, those of the nine Archons, the Treasurers, the Poletae,[13] the Eleven,[14] and the Colacretae[15] so that they were to be held by men taken from the Pentacosiomedimni, the Knights, and the Zeugitae, and assigned the offices to them in proportion to their property qualifications. To those who belonged to the census of the Thetes, he gave only a share in the Assembly of the People and in the law courts. A person belonged to the census of the Pentacosiomedimni if he obtained from his own property a return of five hundred measures of dry and liquid produce, both of them reckoned together. If he had an income of three hundred measures, or, as others say, if he was able to keep horses, he was rated a Knight; and as confirmation of the latter explanation they adduce the name of the class ["Knights"] as being derived from the fact mentioned, and some ancient votive offerings. For on the Acropolis there is a statue of Diphilus with the following inscription:

Anthemion, the son of Diphilus, has dedicated this statue to the Gods, when from the status of a Thes he had been raised to the status of a Knight.

And a horse stands beside him in testimony of the fact that the status of a Knight means this [that is, the ability to keep a horse].

In spite of this, it is more probable that this class also, like that of the Pentacosiomedimni, was distinguished by measures. To the census of the Zeugitae[16]

11 Wooden tablets set up on pillars revolving around an axis.
12 These terms are explained a little below in the present chapter.
13 Officials who farmed out public revenues, sold confiscated property, and drew up all public contracts.
14 The superintendents of the State Prison.
15 A very ancient office connected with the administration of finances. The specific duties assigned to the Colacretae seem to have changed again and again in the course of time. They are still mentioned in the last decade of the fifth century. But there is no evidence of the existence of the office after the restoration of democracy in 403 B.C. oxen.
16 The word Zeugites is derived from zeugos, which means a yoke, in this case probably a team of oxen.

belonged those who had an income of two hundred measures (liquid and dry). The rest belonged to the census of the Thetes and had no share in the magistracies. Consequently, even today when the superintending officer asks a man who is about to draw the lot for an office to what census class he belongs, nobody would ever say that he is a Thes.

8. Solon established the rule that the magistrates were to be appointed by lot out of candidates previously selected by each of the four Tribes. In regard to the nine Archons, each Tribe made a preliminary choice of ten, and among these they cast the lot. Hence the custom still survives with the Tribes that each of them first selects ten by lot and then they choose, again by lot, from these men. A confirmation of the fact that the magistrates were to be selected by lot from the respective property classes mentioned is the law concerning the Treasurers, which is still in use down to the present day. This law orders that the Treasurers are to be chosen by lot from among the Pentacosiomedimni.

Such, then, was Solon's legislation in regard to the nine Archons. For, in the ancient times, the Council of the Areopagus called upon suitable persons and appointed, according to its own independent judgment, for one year to the various offices whomever it found fit for the respective tasks. There were four Tribes, as before, and four Tribe-kings. Each Tribe consisted of three *trittyes*, and there were twelve *naucrariai*[17] to each Tribe. The Naucraries were presided over by the Naucrari who were appointed to supervise the receipts and expenditures. Hence the expressions, "the Naucrari shall levy . . ." and "the Naucrari shall spend from the Naucraric fund," are frequently found in those Solonian laws which are no longer in force. Solon also established a Council of Four Hundred, one hundred from each Tribe. Yet he still made it the task of the Areopagus to watch over the laws, just as in the preceding period it had been the guardian of the political order; and this Council [that is, the Areopagus] still supervised the greater and more important part of public life and, in particular, chastised offenders, with full power to impose punishment and fines. It deposited the money exacted through fines in the Acropolis without having to indicate the reasons for the imposition of the fine. It also tried those who had conspired to deprive the people of their political rights, Solon having enacted a law of impeachment for such cases. Finally, seeing that violent political dissensions frequently arose in the city but that some citizens, out of a tendency to take things easy, were content to accept whatever the outcome of the political struggle might appear to be, Solon made a special law for persons of this kind, enacting that whoever, in a time of political strife, did not take an active part on either side should be deprived of his civic rights and have no share in the state.

9. This, then, was the order established by Solon in regard to the public offices. The three most democratic features of his constitution appear to be the following: first, and most important, the law that nobody could contract a loan secured on his person; secondly, the rule that anyone who wished to do so could claim redress on behalf of a person who had been wronged; thirdly (and, according to the prevailing opinion, this more than anything else has increased the political power of the common people), the right of appeal to a jury court. For when the people have a right to vote in the courts, they become the masters of the state. Moreover, since the laws are not written down in clear and simple terms,

---

17 Shipbuilders.

but are like the one about inheritances and heiresses, disputes over interpretation will inevitably arise, and the court has the decision in all affairs, both public and private. Some people believe that Solon deliberately made the laws obscure so that the people would be masters of the decision. But this is not likely. The reason is rather that he was not able to formulate the best principle in general terms. It is not fair to interpret his intentions on the basis of what is happening in the present;[18] it should be done on the basis of the general character of his constitution.

10. As far as his legislation is concerned, these appear to be its democratic features; but, even before his legislation, he had effected the abolition of debts and afterwards the augmentation of the measures, the weights, and the coin. For it was under his administration that the measures became larger than those of Pheidon, and the mina, which formerly had had a weight of seventy drachmae, was increased to a full hundred. The original type of coin was that of the double drachma. He also introduced trade weights corresponding to the coinage at the rate of sixty-three minae to the weight of a talent, and proportional parts of the three additional minae were apportioned to the stater and the other units of weight.

11. After Solon had established the political order described above, many people came to him and plagued him with all sorts of criticisms and questions in regard to his laws. So, since he did not wish to change them nor to become an object of invidious attacks if he stayed, he went on a journey to Egypt with the object of doing some business and of seeing the country at the same time; and, at his departure, he announced that he would not be back for ten years. For, he said, he did not consider it right for him to interpret the laws, as he would inevitably be called upon to do if he stayed, but that, in his opinion, every citizen should rather be careful to obey them to the letter. Moreover, it happened that many of the nobles had become his enemies because of the cancellation of debts, and that both parties were alienated from him because the settlement had turned out contrary to their expectations. For the common people had believed that he would bring about a complete redistribution of property, while the nobles had hoped he would restore the old order or at least make only insignificant changes. Solon, however, set himself against both parties, and while he would have been able to rule as a tyrant if he had been willing to conspire with whichever party he wished, he preferred to antagonize both factions while saving the country and giving it the laws that were best for it, under the circumstances.

12. That this was Solon's attitude is generally acknowledged, and it is also confirmed by the following passages from his own poems:

> To the common people I have given such honor and privilege as is sufficient
>   for them, granting them neither less nor more than their due.
> For those possessed of power and outstanding through wealth I had equal
>   regard, taking care that they should suffer no injury.
> Firmly I stood, holding out my strong shield over both of them, and I did not
>   allow either party to triumph over the other in violation on justice.

In another passage he makes clear in what way one should deal with the common people:

---

[18] Namely, when the unclear wording of the laws on inheritances and heiresses caused innumerable lawsuits.

The people will follow the leaders best if it is neither given too much license nor restrained too much.

For satiety breeds insolence when too great prosperity comes to men lacking right judgment.

Again, in another place, he speaks of those who wished to redistribute the land:

Those who gathered, setting their minds on plunder, nourished excessive hopes.

Everyone of them expected to win great riches, and believed that I was wheedling with smooth words but would finally come out with a revolutionary plan.

Idle were their expectations. Now they are irate against me and they look at me askance as if I were their enemy.

This they should not do. For, with the help of the Gods, I have accomplished what I promised.

Other things I did not vainly undertake.

I find no pleasure in achieving anything by the forceful methods of a tyrannical regime, nor would it please me to see the noble and the vile have an equal share of the rich soil of our fatherland.

Again, about the cancellation of debts and about those who formerly had been slaves but then were freed in consequence of the *seisachteia*, he says:

Which of the aims because of which I gathered the people did I abandon before I had accomplished it?

My best witness before the tribunal of posterity will be the great mother of the Olympian Gods, black Earth.

For I removed the markstones of bondage[19] which had been fastened upon her everywhere; and she who had then been a slave is now free.

I brought home to Athens, to their fatherland, many Athenians who, lawfully or unlawfully, had ben sold abroad, and others who, having fled their country under dire constraint of debts, no longer spoke the Attic tongue—so wide had been their wanderings.

I also restored to freedom those who here at home had been subjected to shameful servitude, and trembled before their masters.

These things I accomplished by the power which I wielded, bringing together force and justice in true harmony, and I carried out my promise.

I enacted laws for the noble and the vile alike, setting up a straight rule of justice for everybody.

Yet, if another man, of evil intent and filled with greed, had held the goad as, I did, he would not have held the people back.

For if I had been willing to do what pleased the enemies of the people at that time, or again what *their* opponents planned for *them*,[20] this city would have been deprived of many of her sons.

For this reason I had to set up a strong defense on all sides, turning around like a wolf at bay in the midst of packs of hounds.

And again, reproaching both parties for the attacks which they afterwards directed against him, he says:

---

[19] These are the stones (ὅροι) set up on the lands of the indebted farmers to indicate that the land, which could not be sold outright, was mortgaged, together with its owners, to the creditor.

[20] What .Solon means is that if, instead of taking his position between the two parties, he had sided either with the enemies of the common people or with *their* opponents, that is, with the revolutionary leaders of the common people itself, there would have been civil war and bloodshed in either case.

If I must express publicly my just rebuke, I have to say that the common people would never have seen in their dreams what they now enjoy. . . .
And those who are privileged and powerful might well praise me and call me
their friend;

for, he says, if someone else had obtained such an exalted office,

He would not have held the people back and would not have rested until, by
shaking up the state, he would have got the butter from the milk for himself.
But I set myself up as a barrier between the battleline of the opposing parties.

13. These, then, were the reasons for which he went abroad. After Solon's
departure, the state was still torn by internal dissensions. Yet, for four years, they
kept the political peace. But in the fifth year after Solon's archonship, they were
unable to appoint the Archon because of the party strife; and another four years
later they again skipped the appointment of the Archon for the same reason.
Then, still within the same period, Damasias was elected Archon, and he ruled
for two years and two months, until he was expelled from his office by force.
Then, because of their dissensions, they decided to elect ten Archons, five from
the nobility of birth, three from the farmers, and two from the craftsmen. From
this, incidentally, it is clear that the Archon had the greatest power. For the
dissensions arose always in regard to this office. In general, however, they continued in a condition of internal disorder. Some people found reason for discon-
tent in the cancellation of debts; for they had been reduced to poverty by this
measure. Others were dissatisfied with the political order because it had undergone such radical changes. Still others participated in the party strife because of
their personal rivalries with one another.

There were three parties. First, there was the party of the Shore, which was
led by Megacles, the son of Alcmeon. This party seemed to follow a middle road.
The second party was that of the Plain; their aim was oligarchy, and Lycurgus
was the leader. The third party was that of the Highlanders, which was headed
by Pisistratus, who was considered a champion of the common people. Affiliated
with this party were also those who had lost money owed them, for they had now
become poor; and those who were not of pure descent, for they were apprehensive
[concerning their rights of citizenship]. This latter observation is confirmed by
the fact that, after the overthrow of the tyrants, the Athenians revised the citizen
roll on the ground that many had assumed citizen rights without being entitled to
it. The different parties derived their names from the parts of the country in
which they had their lands.

14. Pisistratus, who was considered as the outstanding advocate of the common people and who had distinguished himself in the war against Megara, inflicted a wound on himself, and then, under the pretense that he had suffered this
injury from the hands of his political opponents, he persuaded the people to grant
him a bodyguard, on a motion presented by Aristion. With the help of these so-
called "club bearers," he rose up against the people and occupied the Acropolis,
under the Archonship of Comeas,[21] in the thirty-second year after Solon's legislation. It is said that, when Pisistratus asked for the bodyguard, Solon opposed the
demand and said that he [that is, Solon] was wiser than some and braver than
others. For, he said, he was wiser than those who were not aware that Pisistratus

___
21 561–560 B.C.

was aiming at tyranny, and braver than those who were aware of it but kept silent. But when he did not convince them by what he said, he brought his armor out and placed it in front of his door saying that he had come to the aid of his fatherland as far as was in his power (for he was already a very old man), and that he called on all others to do likewise.

This time, then, Solon had no success with his exhortations. And Pisistratus, having seized the government, administered the state in a constitutional rather than in a tyrannical fashion. But his rule had not yet taken root; and the parties of Megacles and Lycurgus made common cause and drove him into exile in the sixth year after his first accession to power in the Archonship of Hegesias. In the twelfth year after this, however, Magacles, being hard-pressed in the party struggle, opened negotiations with Pisistratus and, having reached an understanding that he [Pisistratus] would marry his daughter, brought him back by a rather primitive device. He first spread the rumor that the goddess Athena was bringing back Pisistratus. Then, having picked out a tall and beautiful woman by the name of Phya (according to Herodotus she was from the Attic deme of Paeania; according to others she was a Thracian flower-girl from the deme Collytus), he dressed her up so that she looked like the goddess and brought her into the city together with Pisistratus. And so, Pisistratus drove into the city on a chariot, the woman standing at his side, and the citizens fell down in worship and received him with awe! [22]

15. His first return to power occurred in the way described. But when, in the seventh year after his return from exile, he was exiled again—for he did not hold his rule for a long time, and slipped out of the country when he became afraid of a combination of both parties because he had refused to consummate the marriage with the daughter of Megacles—he first participated in the establishment of a colony at a place by the name of Rhaecelus, near the Gulf of Thermae. From there he went to the region around Mount Pangaeus. There he raised money with which he hired mercenaries and went again to Eretria in the eleventh year. It was then that he made an attempt for the first time to recover his rulership by force, an undertaking in which he was supported by many allies, especially the Thebans and Lygdamis of Naxos and, further, the Knights of Eretria, who held all the political power in that city.

When he had been victorious in the battle of Pallene and had captured the city of Athens and confiscated the weapons of the people, he had, at last, the tyranny firmly in his hands. He also took Naxos and established Lygdamis as ruler. The manner in which he disarmed the people was this: He held a military review in full armor at the Theseum and began to address the assembled crowd. Then, after he had spoken for a short time, when the people said they could not hear him, he told them to come forward to the gateway of the Acropolis so that his voice would carry better. While he continued to talk and talk, some men appointed for the purpose collected the arms and locked them up in the buildings adjoining the Theseum. Then they came to Pisistratus and informed him that it had been done. Upon hearing this, Pisistratus finished the rest of his speech and then told the crowd what had happened to their arms, adding that they should not be surprised or distressed, but should go home and take care of their private affairs, since in the future he would attend to all the business of the state.

---

22 Cf. Herodotus *I*, 60.

16. In this way the tyranny of Pisistratus was first established, and these were the vicissitudes which it underwent later. As said before, Pisistratus administered the state in a moderate fashion, and his rule was more like a constitutional government than like a tyranny. For he was benevolent and kind, and readily forgave those who had committed an offense; he even advanced money to the poor to further their work so that they could make a living by farming. In doing this he had a twofold purpose: first, that they might not stay in the city but live scattered all over the country; secondly, that they might be moderately well off but fully occupied with their own affairs so that they would have neither a strong desire nor the leisure to concern themselves with public affairs. Another incidental consequence was that his income was increased by the thorough cultivation of the land. For he exacted a tax of ten percent on the produce.

For the same general purpose, he also set up judges in the demes[23] and frequently made the tour of the country himself in order to inspect everything and to settle private disputes, so that the people would not have to come to the city and meanwhile neglect their farmwork. It is said that, during one of these inspection tours of Pisistratus, there occurred the famous story of the farmer from the Hymettus Mountain who cultivated the piece of land which was later called "the Taxfree Farm." Pisistratus saw a man who was working hard to dig a piece of land which was, so to speak, nothing but stones. He became curious and ordered his attendant to ask the man how much he got out of the land. The man answered: "Just so many aches and pains; and of these aches and pains Pisistratus ought to take his ten percent." He gave this answer without knowing that he spoke to Pisistratus. Pisistratus, however, liked his frankness and his industriousness and freed him from all taxes. And so, in general, Pisistratus did not impose any heavy burdens on the people as long as he ruled, but kept everything in a peaceful state both externally and internally, so that it became a common saying that the tyranny of Pisistratus had been the Golden Age. For later, when his sons succeeded in the rulership, the tyranny became much more severe. But most important of all the qualities mentioned was his popular and kindly attitude. For in every respect it was his principle to regulate everything in accordance with the laws without claiming a special privilege for himself. Once, when he was summoned to appear before the Areopagus on a charge of homicide, he even appeared in person to defend himself. But the man who brought the charge against him became afraid and stayed away. For these reasons he remained in power for a long time; and even when he was expelled, he reconquered the power easily. For the majority both of the nobles and of the common people were in his favor. The former he won over through his friendly intercourse with them, the latter through the help which he gave them in their private affairs; and he always proved fair to both of them. On the other hand, as to the Athenians themselves, it should be observed that in that period their laws concerning tyranny were also very mild. This is especially true of the law regarding specifically the establishment of a tyranny. For it reads as follows: "This is the law and the ancestral rule of the Athenians. Whoever conspires to set up a tyranny, or helps to set up a tyranny, shall lose his citizenship and so shall his whole family. . . ."

24. After this, when the Athenian state was growing in self-confidence and in the accumulation of much wealth, he [that is, Aristeides] advised the Athenians

---

[23] That is, the country districts.

to seize the leadership and to give up their residence in the countryside to come to live in the city. For they would all have their livelihood there, some by participating in military expeditions, some by doing garrison service, and still others by participating in public affairs; and in this way they would keep hold of the "leadership." They followed this advice and placed themselves in control of the empire; and from then on they got into the habit of treating their allies, with the exception of Chios, Lesbos, and Samos, as if they were their masters. These three they used as guards of the Athenian empire and, therefore, left their constitutions untouched and allowed them to rule over whatever subjects they happened to have.

They also made it possible for the masses to live comfortably, as Aristeides had proposed. For out of the income derived from the contributions made by the allies and from internal levies more than two thousand persons were maintained. For there were six thousand judges, one thousand six hundred bowmen, one thousand two hundred cavalry men, five hundred Councilmen, five hundred guards of the dockyards plus fifty guards on the Acropolis, about seven hundred state officials at home and about seven hundred abroad. In addition, when later they went to war, there were two thousand five hundred heavy-armed soldiers, twenty guard-ships,[24] and other ships carrying the guardians, that is, two thousand men chosen by lot. Finally, there were the Prytaneum,[25] the orphans, and the jail-keepers. All these persons received their livelihood from the state. . . .

. . . Pericles[26] was also the first to introduce payment for service on the law courts, a measure by which he tried to win popular favor to counteract the influence of Cimon's wealth. For Cimon, who possessed a truly regal fortune, performed the regular public services in a magnificent manner, and, in addition, supported a good many of his fellow demesmen. For anyone of the deme of Laciadae who wished to do so could go to him every day and receive a reasonable maintenance; and his whole estate was unfenced so that anyone who liked could help himself to the fruit.

Pericles' resources were quite unequal to such lavish liberality. So he followed the advice of Damonides of Oea, who was generally believed to have been the instigator of most of Pericles' measures, and was later ostracized for that reason. This man had advised Pericles to "offer the people what was their own," since he was handicapped as far as his own private means were concerned; and, in consequence of this, Pericles instituted pay for the judges. Some people blame him on this account and say that the law courts deteriorated, since after that it was always the common men rather than the better men who were eager to participate in drawing the lot for duty in the law courts. Also, after this corruption ensued; and Anytus was the first to set an example, after his command at Pylos, for he bribed the judges and was acquitted when he was prosecuted by some because he had lost Pylos.

42. The present constitutional order is as follows: the right of citizenship belongs to those whose parents have been citizens. They are enregistered on the rolls of the demes at the age of eighteen. When they come up for enrollment, their fellow demesmen decide by vote under oath the following: first, whether

---

24 That is, a number of about four thousand men, since there were normally a crew of two hundred to one trireme.
25 This means, in fact, the citizens who had done special service to the State and were honored by being entertained at public expense at the Prytaneum, that is, the city hall.
26 Ca. 451 B.C.

they appear to have reached the legal age—and if they do not appear of the right age, they return to the state of boys; secondly, whether the candidate is freeborn and of such parents as the law requires. If they [their fellow demesmen] decide that he is not free, he appeals to the law court and the demesmen choose five men from among themselves as his accusers; and if it appears that he has no right to be enrolled, the city sells him into slavery, but if he wins, the demesmen are compelled to enroll him. After this the Council examines those who have been enrolled, and if someone appears to be younger than eighteen years, the Council fines the demesmen who enrolled him. When the young men (*epheboi*) have passed this examination, their fathers assemble by tribes and, after having taken an oath, elect three of their fellow tribesmen over forty years of age whom they consider the best and the most suitable to supervise the young men. Then out of these men the people elect by vote one from every tribe as guardian (*sophronistes*), and, from the other Athenians, they elect a superintendent (*kosmetes*) for all of them. These men then call the young men together and first make the circuit of the temples. Then they proceed to the Piraeus, and one part takes garrison at Munichia, the other at Acte. The Assembly also elects two trainers for them and special instructors who teach them to fight in full armor and to use the bow, the javelin, and the catapult. They [the people] pay the guardians one drachma each for their keep and the young men four obols each. Each guardian receives the allowance for all those of his tribe, buys all the necessary provisions for their common upkeep (they have their meals together by tribes), and also takes care of all other matters. This is what they do in the first year. In the following year, when there is an Assembly of the People in the theatre, the young men give a public display of their military drill before the people. Then they receive shield and spear from the state, and from then on they patrol the country and are stationed at the guardposts. While they are in service for their two years, their uniform is a military cloak, and they are free from taxes. And, so that they will have no pretext for requesting a leave, they cannot be sued at law or bring suit against someone else. Exceptions are only cases of inheritance, of unmarried heiresses,[27] and of a man's having to take over a priesthood hereditary in the family. When the two years are over, their place is with the other citizens.

---

27 An orphaned girl could be claimed in marriage by her nearest male relative; and if her father had left an estate, this estate went to the sons born of such marriage. If the girl was poor, the nearest of kin was obliged either to marry her or to provide her with a dowry.

The Demonry of Power

In Thucydides' account of the great war between Athens and Sparta—*History of the Peloponnesian War* (431–404 B.C.)—we may read how the schoolmaster of Hellenic culture failed as a teacher. The degeneration of Aristotle's "political animal" is unfolded there in minute detail. How the golden architecture, drama, rhetoric, sculpture, poetry, science, and philosophy of Athens were transmuted into the base metal of greed and imperial devastations of weak peoples is the great theme of Thucydides. On the way to the crushing defeat at Syracuse, the Athenians extinguished the people of the tiny island of Melos. In our juxtaposition of the "Melian Dialogue" and the "Defeat at Syracuse," the student will perhaps see Thucydides' main points. In war, truth is the first casualty, along with piety and greatness of spirit. The mercantile and maritime democracy had become a pitiless giant, cruel in its conquest and control of pigmies. Yet the man of empire could not be satisfied. The "school of Hellas" transformed itself into a garrison state and was corrupted by power. So were its men. Thucydides claimed for his work that it was "a treasure for all time." Few who read his merciless accounts of how the uses of power made demonic Athens's democracy will doubt this claim.

From *History of the Peloponnesian War*

Thucydides

THE MELIAN DIALOGUE

[During the first years of the war the Peloponnesians faithfully invaded the countryside of Attica every year and destroyed the crops, but Athens itself was impregnable behind its walls and its ships. The Athenians in turn raided the coasts of the Peloponnese and its outposts regularly and gained one small but spectacular victory over the Spartans themselves at Pylos, on the southwest coast (424). Shortly afterward the Spartan Brasidas made a series of lightning conquests in the north (Macedonia and Thrace), but they fell apart at his death in 422.

After ten years of this inconclusive sparring between land power and sea power the war was a stalemate, and Nicias very sensibly made peace (421) on the basis of the status quo. But it did not last. Egged on by the irresistible young Alcibiades, Athens gradually resumed the contest. In the year 416 she determined to clean up one of the last remaining neutral spots in the southern Aegaean, the island of Melos.]

Athenian Expedition Against Melos, 416 B.C.

84. The following summer Alcibiades made a raid on Argos with a fleet of twenty ships and took prisoner three hundred of the Argives who were still under suspicion of being favorable to the Spartans; the Athenians interned them on the

---

SOURCE: Thucydides, *History of the Peloponnesian War*, in *Classics in Translation*, vol. 1, ed.

near-by islands that were their control. The Athenians also sent an expedition against the island of Melos; the force consisted of thirty of their own ships, six from Chios and two from Lesbos, twelve hundred Athenian infantry, three hundred archers, twenty mounted archers, and approximately fifteen hundred infantry from the allies and the islands. The Melians are a colony of the Spartans who had refused to become Athenian subjects like the other islanders; for some time they had remained neutral in the war, but now, under threat of Athenian invasion and pillage, they made ready for open hostilities. When the expedition arrived, with the complement mentioned above, the two generals, Cleomedes son of Lycomedes and Tisias son of Tisimachus, sent spokesmen to hold a conference before they proceeded to hostilities. The Melians did not present the envoys to the popular assembly but invited them to explain to a gathering of the officials and influential men what they had come for; and the Athenians then spoke more or less as follows:

Debate Between Athenians and Melians

85. "Since this discussion is not being held in the assembly, for fear the citizens in general might find our arguments attractive, in fact irrefutable, at first hearing and so be led astray—for we are well aware that this is why we have been brought before this select group—you who are present can make things still safer for yourselves if you like: you may judge the case item by item, without any long speeches, interrupting whenever you think one of our points is not well taken. First, then, say whether you accept this procedure."

86. The Melian representatives replied: "We cannot quarrel with the reasonableness of a leisurely exchange of views, but the warlike preparations we see already surrounding us, not merely in prospect, have a different look. It is obvious that you have come to sit in judgment on the discussion and that in all probability, if we win the argument on the score of justice and therefore do not give in, the result of our talk will be war, and if we submit it will be slavery."

87. ATH. Well, if your purpose in this meeting is to deal in vague conjectures about the future instead of planning how to save your city on the basis of present observable facts, we may as well stop now; or, if you accept our condition, we will continue.

88. MEL. It is natural and forgivable that men in a situation like ours should try many shifts, in their speech as well as their thoughts; however, the purpose of this meeting is to discuss the preservation of our city and the discussion may proceed along the lines you suggest, if you wish.

89. ATH. Well then, we will not spin out a long, wearisome round of speeches, with fine phrases about how righteously we gained our power by destroying the Persians, or how this attack was provoked by some wrong we have suffered; and in return please do not count on convincing us with arguments about your not joining us in the war because you are colonists of the Spartans, or about your never having done us any harm. Let us deal with practical possibilities, the basis of our real intentions on both sides, since you know as well as we do that in their stated arguments and conclusions about justice men are under equal constraints, but when it comes to practice the strong do as they please and the weak acquiesce.

Herbert Howe and Paul MacKendrick (Madison: The University of Wisconsin Press, 1963), pp. 247–259.

90. MEL. The way it looks to us (and we have no choice, since you have laid it down that we are to ignore justice and talk expediency), you would find it advantageous not to rule out the common good, but to leave open to all who may be in danger an appeal to reason and justice, and the hope of bettering their position by argument, even if they fall short of proof. Actually that is in your own interest, considering that if you fail disastrously others will treat you by your own example.

91. ATH. We are not worried about what may happen to our empire, even if it comes to an end. The real danger is not the threat of being beaten by another ruling nation like the Spartans—and after all we are not arguing with the Spartans here—but the possibility that subject states may revolt against their rulers and subjugate them. Anyhow, you can leave that risk for us to deal with. We will state frankly that we are here to further the interests of our empire and that what we are about to say is for the preservation of your city; that is, we want to subdue you with the least effort and leave you unharmed for your sake as well as ours.

92. MEL. But, granting that your power is profitable to you, how can slavery be so to us?

93. ATH. In that you have a chance here to submit before you suffer the worst consequences, while we stand to gain by not utterly destroying you.

94. MEL. Would it be acceptable to you if we took no part in the war and remained friendly to you but neutral towards both sides?

95. ATH. No, because your hostility is not as dangerous to us as your friendship would be; our subjects would take the friendship as a standing sign of weakness, and the hatred as a sign of our power.

96. MEL. Do your subjects consider it reasonable not to make any distinction between those who are no kin to you and those—mostly your own colonists—who have rebelled against you at one time or another and been reduced to subjection?

97. ATH. Why, so far as pleas of justice go they consider that both groups have a case, but that those who elude us owe it to their power, because we are afraid to attack them. That is why, aside from the enlargement of our empire, your subjugation will bring us security: the failure of you, an island and a rather weak one, to win against a sea power would be especially striking.

98. MEL. But don't you think there is any security in our proposal? Here again, since you have excluded arguments based on justice and have told us to consider only your advantage, it is incumbent on us to try to show you convincingly where our own interests lie, if it happens that yours lie in the same direction. Now those states that are neutral at present will inevitably join the fight against you when they see what has happened here and become convinced that some day you will attack them too. In short, what are you accomplishing by all this except to strengthen your existing enemies and force others, who had never thought of such a thing, to join them?

99. ATH. No, you see we are not much afraid of potential enemies on the mainland; they are so used to their freedom that they will procrastinate a long time before putting up a defense against us. Our chief threat is from the islanders, those who are independent like yourselves and also those who have been irritated by the constraints of our empire. They are the ones who are most likely to give way to folly and involve themselves and us in dangers that they might have foreseen.

100. MEL. Upon our word, if you are ready to take such extreme risks to

maintain your empire, and those who are already your slaves to rid themselves of it, it would be the most contemptible cowardice for us who still have our freedom not to go to any length to avoid slavery.

101. ATH. Not if you look at the matter sensibly. What faces you here is not a free and equal contest in bravery, to sustain your honor; the issue is self-preservation, that is, not to struggle against a far superior power.

102. MEL. And yet we know that the chances of war often turn out to be more evenly balanced than the difference in numbers between the two sides would indicate; so in our case immediate surrender would mean giving up hope, but if we try to do something there is hope that we may still succeed.

103. ATH. Hope is a consolation in time of danger, and those who have plenty of other resources are not ruined by it, though they may be damaged. But those who stake everything they have on hope (which is a spendthrift by nature) only recognize it for what it is after they fail: while there is still time to see through it and put themselves on guard, it retains its full strength. Now you are a weak nation and dependent on a single turn of the scale; do not let yourselves suffer the fate of so many others. In a situation where they might be saved by their own efforts, when they have their backs to the wall and visible hopes have abandoned them, they resort to the invisible kind—divination, oracles, and other things that destroy men by feeding them on hope.

104. MEL. We think as you do, make no mistake about it, that it will be hard for us to compete with your power and fortune combined, if she turns out not to be impartial. However, so far as fortune is concerned we have faith that with divine help we shall not lose out in this contest between piety and injustice, and we rely on an alliance with the Spartans to make up the deficiency in our power; they will be forced to help us because of our kinship with them and from a sense of honor, if for no other reason. So our confidence is not quite so irrational after all.

105. ATH. Well, as for enjoying the divine favor, we do not expect to come off second best either. In our demands and in our actions we are not departing in any way from the norm of what men practise towards the gods or desire for themselves. You see we believe that divine beings, and we know for certain that human beings, by a necessity of their nature, rule wherever they have the power to rule. We did not establish this principle and were not the first to use it once it was established; we found it in force when we began, expect to leave it in force after us, and meanwhile take advantage of it, knowing that you or anybody else, if you had our power, would do the same. So we have good reasons for not fearing to lose out in the competition for divine favor.

As for your expectations from the Spartans, your faith that they will come to the rescue for honor's sake, we admire your innocence but do not envy your lack of common sense. The Spartans, in their treatment of each other and their domestic institutions, are models of virtue and honor; their conduct towards others would make a very long story, but one can sum it up by saying that of all nations known to us they are the most conspicuous for identifying honor with their own pleasure and justice with their own interests. Surely such an attitude is not conducive to your present foolish hopes of salvation.

106. MEL. But that is the very reason why we place so much trust in their seeing where their interest lies; that is, we are sure they will not choose to forfeit the confidence of the Greek states that are favorable to them, and lend aid to their enemies, by betraying their own colony, Melos.

107. ATH. Apparently you aren't aware that the pursuit of one's own interest is safe enough, but just and honorable conduct is a risky business—and the Spartans are generally the last to take that risk.

108. MEL. Yes, but we think that they would consider the risks lighter in our case and be more inclined to face them for our sake than they would for others, since in case of action we are close to the Peloponnese and thanks to our kinship they can trust our way of thinking more than that of others.

109. ATH. A potential co-belligerent does not base his confidence on the good will of his would-be allies, but on the fact of a clear predominance of power. The Spartans are especially given to this way of thinking—at least they seem to distrust their own resources and attack their neighbors only at the head of a host of allies—so it is not likely that they will venture offshore to help an island while we have command of the sea.

110. MEL. There are others they might send; besides, the Cretan sea is broad and makes the dominant power's task of search and seizure more desperate than the efforts of the hunted to escape. And if their efforts should fail they can try a diversion against Attica or the rest of your allies, those whom Brasidas never got to; then you will have a hard struggle for your confederacy proper and your own territory, instead of land that does not belong to you.

111. ATH. Any of the things you mention might happen to you too, for that matter; you have had experience with them. Also you are aware that Athens has never yet abandoned a siege out of fear of anybody. But it strikes us that though you said you would discuss your own salvation here, in this whole long argument you have not said a thing that normal human beings would consider a reliable guarantee of salvation; your strongest assurances are nothing but hopes, still unrealized, and your actual resources are slim indeed to win out over those that face you. You are showing a gross lack of intelligent thinking if you do not recess this conference now, before it is too late, and come to a more sensible decision. Surely you are not going to give way to shame, the most ruinous impulse one can have when facing a really shameful and foreseeable danger. Often, that is, while men can still foresee what they are being swept into, the thing we call shame, having first overcome their resistance by the power of a seductive word, the work of a mere phrase, lures them on to fall of their own accord into irreparable disaster and incur a new shame, more shameful because brought on by folly, not fortune. If you think carefully you will avoid this error; you will consider it no disgrace to be conquered by the greatest city in Greece, when she generously offers you the chance to become autonomous tribute-paying allies, and will not make a poor decision out of mere stubornness when given a choice between war and security. In general, those who stand their ground against their equals, behave well towards their superiors, and are decent towards their inferiors, will get along best. Think seriously, then, after we withdraw, and remind yourselves again and again that the issue here is your country: that you have only one, and her success or failure hangs on a single decision.

The Melian Decision; the Seige Begins in Earnest

112. With that the Athenians withdrew from the discussion. The Melians, after conferring among themselves and finding that their sentiments had not changed from those they had expressed before, made the following reply: "To

begin with, gentlemen, our views are still the same as they were, and secondly we do not propose to sacrifice in a few hours the freedom of a city that has now existed for seven hundred years. We will try to achieve our own salvation, trusting in the good fortune from heaven that has saved us up to now, and also in human sympathy and help, particularly from the Spartans. Here is our offer: we to be on friendly terms with you but remain neutral in the war, and you to withdraw from our territory after we have arranged a truce that both parties consider acceptable."

113. This was all that the Melians said in their reply. The Athenians, as they left the conference, made this statement: "Well then, it seems to us, judging from these proposals, that unlike all other human beings you consider future possibilities plainer than present facts and view uncertainties as if they were already happening, because you want them to. You have staked everything you have, placed all your reliance, on mere hopes, good luck, and the Spartans, and your failure will be correspondingly complete."

114. The Athenian representatives then returned to the army. The two generals, seeing that the Melians were not going to submit, immediately opened hostilities and walled off the city, parcelling out the work among the various allied contingents. Then, leaving a force of Athenian and allied troops to guard the place by land and sea, they withdrew the major part of the expedition; and those who were left behind settled down to conduct the siege.

115. [Later in the summer the Melians broke through the siegeworks at one point and brought in supplies.]

116. [During the following winter] the Melians captured another section of the Athenian siege wall, where the force on guard was light. Later, however, seeing what was going on, the Athenians sent out a second expedition under the command of Philocrates son of Demeas; the Melians were now under heavy siege, treason appeared in their own ranks, and they surrendered to the Athenians with the understanding that the latter would decide their fate. The Athenians executed all the adult male prisoners they captured, enslaved the women and children, and took over the site for a colony of their own; later they sent out five hundred settlers to occupy it.

THE END OF THE WAR IN SICILY

[In spite of Nicias' prudent warnings Alcibiades lured the Athenians into a really grandiose scheme: the conquest of Sicily. The greatest Greek expeditionary force ever assembled set out for the west in 415 under Nicias, Alcibiades, and Lamachus, with golden dreams of power and fortune whirling in their heads. Here is Thucydides' description of the scene at their departure from Piraeus.]

Departure of the Expedition

Then, after the crews and complement had gone aboard and the supplies they were to take with them had all been laid in, there was a trumpet call for silence and they offered up the prayers that are customarily offered before putting to sea, not ship by ship this time but the whole fleet at once, following the herald's signal; wine was mixed in the mixing bowls throughout the entire armada, and marines and officers together poured the libations from drinking cups of gold and silver. Meanwhile the other throng on shore, made up of citizens and any others who had come down to wish them godspeed, joined in their supplications. Then, after singing the paean and completing their offerings, they stood out to sea, first issuing from the harbor in column and then racing each other as far as Aegina. They were in hot haste to reach Corcyra, where the rest of the expedition, the allied forces, were assembling.

[But the high spirits did not last. Alcibiades was summoned home on a religious charge after the fleet reached Sicily (he was too intelligent actually to go home and settled in Sparta instead), leaving Nicias to fight a war he did not believe in.

The key to Sicily was Syracuse. Athenian strategy was predicated on the hope of winning the help of the other Greek cities and the Sicels (native non-Greek population). The Sicels joined, most of the Greek cities did not, and the Athenians sat down to besiege Syracuse. Syracuse got itself a new commander-in-chief from Sparta, Gylippus, and by the end of 414 the besiegers themselves were besieged. A relief expedition came out from Athens under Eurymedon and Demosthenes (not the orator), and at the beginning of September 413 we find the combined force making a final effort to break out of containment inside the Great Harbor of Syracuse. The war is almost over. Conquest is fading; the issue now is survival. The selection begins just after Nicias' speech of encouragement to his men and Gylippus' counterspeech to the Syracusans and their allies.]

The Athenians Attempt to Escape

69. Gylippus and the Syracusan generals, after these and similar speeches of encouragement to their troops, began to man their ships as soon as they saw the Athenians doing so. But Nicias was completely unnerved by the situation; realizing what kind of danger they faced and how near at hand it was (since the fleet was just on the point of sailing), and feeling, as a man will at times of great stress, that everything they needed to do was still undone and what had been said in his speech had not been said adequately, he went around once more, speaking personally to every ship commander, addressing them ceremoniously by name, father's name, and tribe; exhorting everyone who had any reputation of his own not to betray it, and those who had illustrious ancestors not to let their family glory be tarnished; reminding them of their country, the freest in the world, and how there everybody had full license to live his own life, subject to no man's orders; and so on, repeating all the things that men will say at these crucial moments, not caring whether they may sound trite and old-fashioned, the appeals that are always brought forward in the same way on such occasions, in the name of wives and children and ancestral gods: worn phrases that are revived and shouted once more because they seem helpful in the face of terror and panic.

After these exhortations, which he considered necessary if not adequate to the situation, Nicias set out, leading his land forces down to the shore and deploying them over as wide a front as he could so as to give all possible encouragement to the men on board the ships. Meanwhile Demosthenes, Menander, and Euthydemus, the generals who had taken command of the Athenian fleet, put out from their own camp and sailed straight for the barrier at the mouth of the harbor and the opening that had been left in it, intending to force their way through.

70. But the Syracusans were too quick for them. They and their allies put out with as many ships as the Athenians, and sooner; they not only posted a detachment to guard the entrance but patrolled the entire harbor so as to be in position to attack the Athenians from all sides at once, and meanwhile their land forces were ready to come up in support wherever the enemy might put ashore. The Syracusan naval commanders were Sicanus and Agatharchus, each holding one wing, with Pythen and the Corinthians in the center.

When the rest of the Athenian forces reached the barrier they charged. Their first rush carried the ships that were stationed near it and they started trying to undo the fastenings, but at that moment the Syracusan and allied vessels bore down on them and the fighting became general, not only at the barrier but throughout the harbor. It was a hard-fought battle, the hardest of the entire campaign. For one thing, the rowers on both sides were on the alert and eager to pull the moment the word of command was given; for another, there was great professional jealousy and competitive spirit among the steersmen; and the marines were on the *qui vive*, when ship struck ship, to see to it that the deck fighting did not fall below the standard of the other services; every man was out to distinguish himself in the duty to which he had been assigned.

With so many ships engaged in such a small space (there was a record number of vessels in a very small area, almost two hundred in the two fleets combined), there were few chances to back water or break through the enemy line, and therefore few deliberate rammings, but numerous chance collisions as one ship fell afoul of another while trying either to escape or charge at a third. While another vessel was bearing down, the men on deck facing her would keep her under heavy fire with javelins, arrows, and stones; then, after they collided, the marines on each ship would move in at close quarters and attempt to board the other. The maneuvering space was so narrow that often one ship would have rammed another and been rammed by still another, and sometimes two or even more would be inextricably entangled around a single one; steersmen found they had to defend themselves on one side and carry out offensive maneuvers on the other, not one at a time but in several directions at once; and the continual crash of dozens of ships colliding resulted not only in panic among the crews but inability to hear the boatswains' commands.

For a great shouting of orders and words of encouragement was going on among the boatswains on both sides, spurred by the rivalry of the moment as well as the demands of their work: on the Athenian side the cry was to force the passage out of the harbor and strike out boldly now, if ever, for the chance to return to their country alive; on the side of the Syracusans and their allies, how fine a thing it would be to prevent the enemy's escape and win new glory, each contingent for its own country. In addition the Athenian and Syracusan high command, when they saw a ship backing water where it was not necessary, would call out the commanding officer's name and ask him a question: the Athenians wanted to know whether their men were backing off because they thought the soil of their bitterest enemies was more their own than the sea which they had mastered by so much effort; the Syracusans would ask their men whether they couldn't see for themselves that the Athenians were trying to escape at any cost: were they going to run from men who were on the run themselves?

Suspense of the Athenians on Shore

71. The two armies on shore suffered agonies of suspense and conflicting emotions as long as the fighting on the water was evenly balanced, the natives exultant and eager to add to the glory they had already won, the invaders fearful that their final state would be even worse than their present situation. Not only were the Athenians apprehensive of the outcome as never before, now that their fate depended wholly on the fleet, but their view of the battle from the shore necessarily varied with the variation in their position. That is, since the range of vision was very short and the whole army could not look at the same thing simultaneously, those who saw their compatriots winning at one point would be jubilant and fall to invoking the gods not to deny them their chance of returning home; others, with a reverse taking place before their eyes, would give way to loud wails and cries of grief and were more cast down in spirit than those in the midst of the action; still others, watching a part of the battle where the fighting was nip and tuck, and overcome by the long-drawn-out uncertainty of the struggle, would reel back and forth together in an agony of fear, bodies swaying in time with their feelings. These last were among the worst sufferers: they were always just on the verge of escaping or being cut to pieces.

So, as long as the battle was nearly even, every kind of exclamation could be heard at once in the Athenian army—wails of grief and shouts of joy, "They're winning!" "They're losing!"—all the manifold cries that would naturally be wrung from a large body of troops under stress of great danger. The men on board ship went through the same mixed emotions, until finally, after the battle had gone on for a long time, the Syracusans and their allies routed the Athenians and pursued them ashore, pressing them vigorously and cheering each other on with loud shouts of encouragement. At the same time the Athenian naval complement, those who had not been captured on the water, drove for a landing at scattered points along the shore and rushed pell-mell for camp. The feelings of the land forces ceased to be divided; they all suffered alike under the new turn of events and burst with one accord into sighs and groans; some went to lend a hand at the ships, others to help post a guard over the remainder of the wall, still others—and they were the largest number—began to think about themselves and their own salvation. The panic in the army at this point was as great as any in the whole war. They had undergone a defeat like the one they themselves had inflicted on the Spartans at Pylos, when the latter had their ships destroyed and lost the men who had crossed over to the island besides. So in this case the Athenians had no hope of getting away safely by land, unless some miracle should occur.

The Athenians Refuse to Try Another Escape by Sea

72. Thus after all the hard fighting and the loss of many ships and men the Syracusans carried the day; then, gathering up their wrecks and their dead, they sailed back to the city and erected a trophy. The Athenians on the other hand were so dazed by the magnitude of the disaster that it did not even occur to them to ask permission to retrieve the dead or the wrecked ships; their sole idea was to retreat as soon as night fell. Demosthenes, however, went to Nicias and suggested another scheme, that they man the remaining ships and force their way

out through the gap in the barrier at dawn, if possible; he pointed out that they still had more ships in serviceable condition than the enemy, about sixty as against less than fifty on the other side. Nicias agreed to the plan and they tried to man the ships; but the sailors refused to go aboard, they were so utterly dejected by their defeat and convinced that they could not win again.

The Ruse of Hermocrates

73. The Athenians, then, were unanimously agreed on the idea of retreating by land. But Hermocrates the Syracusan, suspecting their intentions and thinking it was dangerous to have so large an army fall back by the land route, settle somewhere else in Sicily, and be in a position to make war on Syracuse again in the future, went and made representations to the authorities that the Athenians should not be allowed to leave during the night. He explained his reasons, and proposed that the whole Syracusan and allied force should go out immediately to block off the roads and seize and hold the key points commanding the narrowest passes before the Athenians could reach them. He found that the authorities thought as he did and agreed that his proposals should be carried out; however, since the men were thoroughly enjoying their first rest after the long fighting and there was also a festival going on (it happened that that was the day for one of the sacrifices to Heracles), they thought it would not be easy to get them to obey: most of them, overjoyed by their victory, had taken to drinking at the festival and were likely to listen to almost anything rather than a command to fall in once more and go out on duty.

When the officials presented all these arguments and insisted the thing could not be done, Hermocrates saw that he was not winning his point and resorted to another stratagem, for he was afraid the Athenians would get away in the night and negotiate the most difficult key points without hindrance. He sent some of his personal companions to the Athenian camp with a cavalry escort, just as it was getting dark. They rode up close enough so that a man's voice could be heard and asked for certain Athenians by name, pretending they were good friends of theirs (and in fact Nicias did have informants inside the city); then they told them to warn Nicias not to move his troops during the night, because the Syracusans were guarding the roads, but to make full preparations and retire during the daytime, when he would be undisturbed. With this they rode away; the men they had spoken to passed the message on to the Athenian generals, and they held up the retreat overnight because of the warning, not suspecting that it was a trick.

The Syracusans Block the Roads

74. Then, since they had not started out immediately after all, they decided to wait over the next day as well, so as to give the men all the opportunity they could to pack up the most useful supplies; they intended to leave behind all the rest of the stores they had on hand and take along on the march only those that were needed for personal subsistence. Meanwhile the Syracusan land forces under Gylippus marched out ahead of them and barricaded the roads throughout the countryside wherever the Athenians were likely to pass; they also set guards at the stream and river crossings and posted themselves at chosen spots where they could receive the enemy and prevent their passage. The fleet sailed over also and began hauling the Athenian ships down off the beach; the Athenians had man-

aged to burn a few according to plan, but they gathered up the rest at their leisure, without interference, from the various places where they had run aground, took them in tow, and pulled them to Syracuse.

75. Not until the third day after the battle, when Nicias and Demosthenes considered that their preparations were satisfactory, did the withdrawal of the army actually begin. It was a fearful experience in more ways than one: not only were they in retreat, with all their ships lost and the prospect of real danger ahead, instead of high hopes for themselves and their country, but even in the process of departing from camp every man was confronted by things that were painful either to behold or to think about. The bodies of their dead were still unburied, and whenever a man saw one of his own comrades lying there he was overcome with grief and fear together. But the living—the wounded and sick who were being left behind—were far more pitiable objects to the living than the dead, more utterly wretched than those who had perished. They drove them frantic with the entreaties and lamentations they set up, begging to be taken along and crying out at every friend or relative they saw anywhere along the line of march, hanging on their former tent mates as they passed by, following after them as far as they could go and at last falling behind, as their physical powers failed, with a few last pathetic groans and appeals to the mercy of the gods.

With all this the whole army was reduced to tears and so distracted that they could hardly bring themselves to leave, even though the country round them was hostile and their recent disasters as well as their apprehensions of what might happen to them in the uncertain future were too overwhelming for tears. At the same time there was a general wave of dejection and self-reproach. And in fact they looked like nothing so much as the population of a city in panic flight after losing a siege—and a large city at that, for, counting all the hangers-on, there were not less than forty thousand people in the line of march. All the others among them were dragging everything usable they could carry, and the military personnel, infantry and cavalry alike, contrary to usual practice, had their own provisions tucked in with their weapons, either for lack of servants or lack of confidence in them (the servants had started deserting long since, most of them immediately after the battle). Even so, what they carried was insufficient; for there was no more grain left in the camp. But the rest of their sufferings, this new equality in misfortune, though it was somewhat lighter for being shared with so many others, seemed especially hard to bear in their present situation, when they remembered how gloriously and boastfully they had begun and saw to what a miserable end they had now come. This was in fact the greatest reverse ever suffered by a Greek army: to have come expecting to enslave others and now to depart dreading that the same thing might happen to themselves; to set out for home again uttering imprecations—how different from the prayers and paeans with which they had sailed from home; and to be travelling by land instead of by sea, relying on the infantry instead of the fleet. And yet in spite of everything, considering the magnitude of the danger that still hung over them, all this seemed endurable.

Nicias Encourages His Men

76. Nicias, seeing the army so dispirited and its morale so radically changed, went up and down the ranks trying to encourage them and cheer them up as best he could under the circumstances, raising his voice more and more in his zeal as

he moved from one group to another and feeling that it might help them some-
how if he shouted as loud as possible:

77. "Athenians and allies, even in this situation you must keep up your hopes
—others in the past have escaped from even worse dangers than this—and not
blame yourselves too much for your reverses or the undeserved misery you are
suffering now. I certainly am no stronger than any of you, in fact you can see for
yourselves how I am affected by my illness; and though I was once considered as
happy and prosperous as any man, in private life and public life too, now I am at
the mercy of the same danger as the poorest of you. And yet throughout my life I
have paid all due respect to the gods and treated men fairly, so that no one could
hold any grudge against me. Because of that I am still confident for the future, in
spite of everything, and our misfortunes do not frighten me as they might. Perhaps they may even begin to let up; the enemy is satisfied with his present success, and if by chance one of the gods was offended at our undertaking we have
amply atoned for it by now. Others before us have gone on foreign expeditions;
their actions were only human and in return their sufferings were bearable. So
in our case it is only reasonable to expect milder treatment from the gods in the
future: as we are now we have more claim on their pity than their jealousy.

"Furthermore, look at yourselves, think what good soldiers you are and how
many there are of you still in fighting order, and do not be too despondent. Count
up and you will find that you make a city all by yourselves, wherever you choose
to settle; no other city in Sicily could very well withstand an attack from you or
drive you out again once you were established anywhere. It is your business to
see to it that security and order are maintained on the march; and you can do it if
every man tells himself that whatever spot he is forced to fight in will become his
home and his fortress—if he wins. We will have to move fast on the road, both at
night and in the daytime, because we are very short of provisions, and we cannot
feel assured of being in safe country until we reach some friendly village in the
territory of the Sicels; they are still loyal to us out of fear of the Syracusans.
Messengers have been sent on ahead to them and they have been told to meet us
and bring more provisions.

"To sum it up, soldiers, make up your minds now that there is nothing for it
but to conduct yourselves like brave men. There is no strong point near by where
you can find salvation if you turn cowards; whereas if you escape the enemy now,
you in the allied forces will live to see what I know you are longing to see again,
and the Athenians will be saved to help restore the mighty power of our city,
though it has fallen so low. For a city is its men, not walls and ships without
men."

Slow Progress of the Retreat

78. So Nicias went up and down the ranks plying his men with exhortations
like these, and also reforming and tightening up the formation where he saw men
straggling or dropping out of line; and Demosthenes did the same in his command, with the same kind of speeches or others like them. The army was marching in hollow rectangle formation, with the baggage-carriers and most of the
main mass surrounded by the infantry. When they reached the crossing of the
Anapus river they found some Syracusan and allied cavalry drawn up to meet
them on the bank; they routed these units after a skirmish, gained possession of
the bridgehead, and began to move forward again. But the Syracusan cavalry

continued to ride on their flanks and the light-armed troops kept them under javelin fire.

On this first day the Athenians marched approximately five miles and bivouacked for the night near a low hill. The next day they set out early in the morning and after marching about two and a half miles came down into a level space where they pitched camp, intending to get something to eat from the houses (for it was an inhabited place) and some water to carry with them when they moved on, water being scarce for a number of miles ahead in the direction they were travelling. But while they were halted the Syracusans went on and began walling off the pass at a point on the road ahead of them; it was at a steep hill with precipitous ravines on both sides, called the Acraean ridge.

The next day the Athenians resumed the march, with the Syracusan and allied cavalry and javelin-throwers hanging on their flanks in heavy force, impeding their advance and maintaining a steady javelin fire. The Athenians kept up the running fight for a long time, then gave way and returned to the camp they had just left. This time the shortage of supplies was more acute, since they could not leave camp to forage because of the cavalry.

Blocked, the Athenians Try a New Route

79. They broke camp early and set out again, and forced their way as far as the wall that had been built from the hill across the pass. Here they found the enemy infantry drawn up in front of them behind the wall, in very deep formation because of the narrowness of the pass. The Athenians charged and tried to storm the wall, but coming under heavy fire from the hill, whose steep slope brought them within easier range of the men farther up, and not being able to force the passage, they fell back again to rest. It so happened that just then a thunderstorm and some rain came up—a common occurrence at that time of year, in late autumn, but the Athenians were more dejected than ever and felt that all this too was meant for their destruction. While they were resting, the Syracusans under Gylippus sent back a detachment to build another wall behind them, on the road along which they had just come up; but this time the Athenians sent some of their own men to head them off and managed to prevent it. They then withdrew their whole force in the general direction of the plain and bivouacked for the night.

The next day they advanced again. The Syracusans now attacked them from all sides at once and wounded a large number of men. Every time the Athenians rushed them they would fall back, only to press in again the moment they gave ground; and especially they kept falling on the rear guard, hoping that perhaps by routing them a few at a time they might throw the whole army into a panic. The Athenians maintained their defense against these tactics for a long time, then finally retreated about three-quarters of a mile into the plain; whereupon the Syracusans left them and returned to their own camp.

80. By now the army was suffering badly from lack of provisions, and there were a great many men wounded from the enemy's constant cavalry attacks. During the night, therefore, Nicias and Demosthenes decided to light as many watchfires as possible and then withdraw the army, not along the road they had planned to follow, where the Syracusans were watching for them, but in the opposite direction, towards the sea. The general direction of this new route was

not towards Catane but along the other coast of Sicily, towards Camarina and Gela and the other Greek and foreign cities on that side of the island. So they lit a great many fires and set out during the night.

Panic and Surrender of Demosthenes' Division

Then a thing happened that is common, for that matter, in all armies, and particularly in very large ones; they are all subject to sudden outbreaks of fear. So the Athenians, especially as they were travelling by night in enemy territory, and with the enemy himself not far away, fell into panic and confusion. Nicias' column, which was leading, stayed together and got a long way ahead, but Demosthenes' division—a good half and more of the whole army—began to get dispersed and out of formation. They reached the sea at dawn, however, struck into the so-called Helorine road, and continued the march, intending when they reached the Cacyparis river to cross it and follow it upstream into the interior; they were hoping that the Sicels whom they had sent for would meet them here. But when they reached the river, again they found a Syracusan patrol building a wall and palisade at the crossing. They forced their way through, crossed the river, and on the recommendation of their guides went on towards another river, the Erineus.

81. Meanwhile, when dawn came and the Syracusans and their allies realized that the Athenians were gone, the majority accused Gylippus of deliberately letting them get away. However, they set out in hot pursuit along the road which the enemy had clearly taken and caught up with them about noon. When they made contact with the rearmost troops, Demosthenes' men, who were still straggling and in ragged order from the confusion of the night before, they immediately fell upon them and began fighting; the latter, separated as they were from the rest of the army, were easily surrounded by the Syrascusan cavalry and driven together into a compact mass. Nicias' column was a full six miles or more ahead of them: he was moving faster, thinking that at this point their salvation did not lie in deliberately waiting for the enemy and courting a battle but in withdrawing as fast as they could, fighting only if and when they were forced to.

Demosthenes' circumstances were different, and so were his methods. On the whole he was much more steadily exposed to trouble by being second in the line of march and so the first target for enemy attacks; and on this occasion, when the Syracusans appeared in pursuit, he spent more time drawing up a battle line than moving forward, until finally, thanks to his delay, he was encircled and both he and the men under his command were thrown into wild confusion. Crowded together in a small space surrounded by a wall, with a road on each side and containing a number of olive trees, they were under fire from all directions. The Syracusans were very sensible in employing attacks of this kind instead of fighting at close quarters: at this stage it would have been more in the Athenians' interest than their own if they had taken serious risks against a body of desperate men, and at the same time victory was so clearly in sight that they felt a certain reluctance to throw away their lives prematurely. They were confident that they could wear down and capture the Athenians by the tactics they were using.

82. So, after keeping the Athenians and their allies under fire from all sides the rest of the day, and seeing them by now in desperate straits with wounds and

misery of all kinds, the Syracusan and allied forces under Gylippus issued a proclamation. It provided first that all islanders who would come over to them would be guaranteed their freedom; and some of the island contingents went over, but not many. Later a general agreement was reached with the rest of Demosthenes' command, specifying that they were to surrender their arms and that no one was to be put to death by execution, imprisonment, or withholding of minimum subsistence. On these terms the whole body of six thousand men surrendered. They also gave up all the money they had; it was thrown into upturned shields and filled four of them. These men were immediately sent back to Syracuse under escort; and on the same day Nicias and his column arrived at the river Erineus, crossed it, and encamped on a piece of high ground.

Nicias Is Trapped; His Surrender

83. The Syracusans overtook Nicias the next day, told him that Demosthenes and his command had surrendered, and urged him to do likewise. But Nicias was skeptical and sent back a cavalryman under truce to investigate. When the man returned and confirmed the report of the surrender, Nicias notified Gylippus and the Syracusans by herald that he was ready to offer in the name of Athens to repay all costs that Syracuse had contracted on account of the war, on condition and he and his troops be released; pending payment of the money he would post Athenian citizens as bond, one man per talent. But Gylippus and the Syracusans would not accept the terms; they closed in on the column, surrounded it on all sides, and kept it under fire throughout the day. Nicias' men also were in bad condition from lack of food and supplies in general; but still they waited till the quiet part of the night and started to move on. They had just taken up their arms when the Syracusans heard them and sounded the battle warning. Seeing that they were detected the Athenians laid down their arms again, except one group of about three hundred men who fought their way through the patrols and went off in the darkness any way they could.

84. In the morning Nicias led the army on again; and the Syracusans and their allies pressed on their flanks as they had before, showering them from all sides with missiles and javelin fire. The Athenians meanwhile pushed on towards the Assinarus river. They were suffering under the constant attack of large cavalry and other forces coming at them from all directions and thought things might be a little easier if they could cross the river; at the same time they were in a bad way from exhaustion and extreme thirst. When they reached the river bank they abandoned all formation and plunged in, every man trying to be the first across; and this if nothing else, combined with the constant pressure from the enemy, made the crossing difficult.

They were so squeezed together that they could not help falling over and treading on each other; some were killed outright in the press of spears and miscellaneous baggage, others got entangled and were swept downstream. The Syracusans stood on the far bank, which was a steep one, and fired from above on the confused mass of Athenians in the shallow water, most of whom were drinking greedily, while the Peloponnesians came straight down into the river and butchered them wholesale. The water was befouled in no time, but they kept on drinking it just as it was, muddy and full of blood; in fact most of them were ready to fight for it.

85. Finally, with corpses piled high on top of each other in the river and the army destroyed, part of it lost in the carnage at the river itself and what little had got across cut down by the cavalry, Nicias surrendered personally to Gylippus because he trusted him more than the Syracusans; he told Gylippus that he and the Spartans could do whatever they pleased with himself, but he begged them to stop slaughtering the rest of the men. Gylippus then issued an order to take prisoners henceforth; and the rest, all who had not been spirited away by the Syracusans (and that happened to many), were gathered up alive.

Treatment of Nicias and the Other Prisoners

The number of men who were thus collected as state prisoners was not very large, while those surreptitiously carried off were very numerous—later all Sicily was filled with them—because they were not covered by explicit terms of surrender like Demosthenes' men. Also, there were a considerable number killed: the slaughter at the river was enormous, as high as in any action of the war; and a good many had been killed in the constant cavalry attacks during the retreat. Still, there were also many who escaped. Some did so immediately, others ran away later after serving as slaves; and the escapees could always find refuge at Catane.

86. After the battle the Syracusan and allied forces reassembled, gathered up the booty and as many of the prisoners as they could, and returned to Syracuse. All the Athenian and allied personnel they had captured were thrown into the stone quarries, that being the safest place of confinement they could think of; but Nicias and Demosthenes were slaughtered outright, over the protests of Gylippus who thought it would be a great personal triumph, as a climax to his other achievements, if he took the generals who had been his rivals back to Sparta with him. It so happened that Demosthenes was the Athenian the Spartans hated most, on account of the Pylos-Sphacteria episode, while they were most friendly toward Nicias for the same reason; for it was he who had exerted himself, at the time when he persuaded Athens to make peace, to arrange for the release of the Spartans who had been captured on the island. Hence they had very friendly feelings towards him, and that was one of the chief reasons why he had placed himself in Gylippus' hands when he surrendered.

But some of the Syracusans, it was said, had previously been in communication with Nicias and hence were afraid that he might be put to torture because of it and make trouble for them in the midst of their success; and others, including the Corinthians particularly, were afraid he might escape by bribing certain people —for he was a rich man—and later cause them new difficulties again; so they talked their allies into executing him. Such was the cause of Nicias' death, or something very much like it; and yet of all the Greeks of my time he least deserved to come to such a miserable end, after a lifetime spent entirely in honorable, upright conduct.

87. The men in the quarries got harsh treatment during the early part of their confinement. Crowded as they were, large numbers of them, into a small pit open to the sky, they suffered at the beginning from the sun and the choking heat; then came nights that were quite the other way, with the chill of late fall, and the changes in temperature lowered their resistance to sickness still further; they had to do everything they did on one spot, and the bodies piled up alongside them,

heaped on top of one another, as men died of their wounds and the change of seasons and the like, so that the stenches were unendurable; they were constantly plagued by hunger and thirst (for eight months the Syracusans gave them only one cup of water and two cups of food apiece per day); and of all the other miseries that men trapped in such a place would naturally experience, not one was lacking. After some seventy days of this kind of existence together, all the prisoners except the Athenians and some Sicilians and Italians who had taken part in the expedition were sold as slaves. The total number of men captured, though hard to estimate accurately, was at least seven thousand.

As it turned out, this was the greatest and most decisive action of the war—and for that matter, I think, of recorded Greek history—the most brilliant for the conquerors and the most disastrous for those who were destroyed in it. They were utterly beaten at every point and suffered severely from every defeat they underwent, until finally, in utter ruin and destruction, as the old phrase goes, they lost their army, their ships, and everything they had, and few out of the many who set forth ever came home again.

So much, then, for the events that took place in Sicily.

The Old Order Changeth

We have already noticed that Roman law and the Latin language preserve evidence for Engels's theory of the origins of the family and tribal or *gentile* groupings. When the earliest Roman writers tried to grasp their own past, they insisted that the *gentes* had evolved into patriarchic, aristocratic families who dominated politics and every phase of public life. These aristocrats had developed into a broad patrician class which preserved the past in its system of clan names. As Roman society developed, however, agrarian aristocrats were constantly excluding from their ranks wealthy men of obscure origins, while at the same time holding at bay the claims on privilege being made by the numerous commercial, industrial, and professional classes. These plebeians had no gentile (clan) affiliation. They thus lacked status, whatever else they may have had, and their rise was symptomatic of the eclipse of traditional aristocratic society in the Roman Republic. Given the gentile origins of power and privilege, the aristocratic order naturally cast itself into the role of protecting traditional constitutions and their monopoly of power. Plebeians had no corporate rights, were not originally citizens, and had been for centuries excluded completely from government.

In a world increasingly molded by property and mobility, status had to be adjusted to wealth and real power, however. The patricians and plebeians were natural antagonists in the Republic. Not unnaturally, therefore, patrician historians elaborated in their histories an ideology of aristocratic virtue. To protect the purity of the clan, marriage between patricians and plebeians had once been outlawed. And to protect the monopoly of office, both law and history located in the gentes all honor, wisdom, and virtue. Perhaps the best example of this genre is the *History of Rome* by Livy (59 B.C.–A.D. 17). Born in northern Italy in the year of Caesar's first consulship, Livy of Padua enjoyed the full Roman franchise and was perhaps of the Roman gens of *Livius*. His own life was spent mostly in Rome where he had a Greek education and became a supporter of the old republican aristocracy.* But he lived through the collapse of the Republic, rued Caesar's birth, and did not hail Augustus (as did Vergil and Horace). Instead, he compensated for present realities by resurrecting a more glorious past. The *History*, which begins with the landing of Aeneas in Italy, ends with the death of Drusus in 9 B.C. Only 35 of its 142 books (*libri*) survive. But in them we have history as a lesson in aristocratic virtue and plebeian infamy. Opposed to those who would purchase peace at the price of true liberty, Livy celebrated the moral lessons of an age of aristocrats. His work as it survives is less a history than a series of pictures of great men struggling for Rome and against the degradation of her old dogmas.

---

* He was thus directly the beneficiary of the plebeian constitutional victories and the defeat of narrow-based Roman patrician rule.

From *History of Rome*

Livy

1. Marcus Genucius and Gaius Curtius succeeded these men as consuls.
It was a year of quarrels both at home and abroad. For at its commencement
Gaius Canuleius, a tribune of the plebs, proposed a bill regarding the intermarriage of patricians and plebeians which the patricians looked upon as involving
the debasement of their blood and the subversion of the principles inhering in the
gentes, or families; and a suggestion, cautiously put forward at first by the tribunes, that it should be lawful for one of the consuls to be chosen from the plebs,
was afterwards carried so far that nine tribunes proposed a bill giving the people
power to choose consuls as they might see fit, from either the plebs or the patriciate. To carry out this last proposal would be, in the estimation of the patricians,
not merely to give a share of the supreme authority to the lowest of the citizens,
but actually to take it away from the nobles and bestow it on the plebs. The
Fathers therefore rejoiced to hear that the people of Ardea had revolted because
of the unjust decision which deprived them of their land; that the men of Veii had
ravaged the Roman frontier; and that the Volsci and Aequi were murmuring at
the fortification of Verrugo; so decidedly did they prefer even an unfortunate war
to an ignominious peace. Accordingly they made the most of these threats, that
the proposals of the tribunes might be silenced amidst the din of so many wars;   B.C. 445
and ordered levies to be held and military preparations to be made with
the utmost energy, and if possible, even more strenuously than had been done
when Titus Quinctius was consul. Thereupon Gaius Canuleius curtly proclaimed
in the senate that it was in vain the consuls sought to frighten the plebs out of
their concern for the new laws; and, declaring that they should never hold the
levy, while he lived, until the plebs had voted on the measures which he and his
colleagues had brought forward, at once convened an assembly.

2. At one and the same time the consuls were inciting the senate against the
tribune, and the tribune was arousing the people against the consuls. The consuls
declared that the frenzy of the tribunes could no longer be endured; the end had
now been reached, and there was more war being stirred up at home than abroad.
This state of things was, to be sure, as much the fault of the senators as of the
plebs, and the consuls were as guilty as the tribunes. That tendency which a state
rewarded always attained the greatest growth; it was thus that good men were
produced, both in peace and in war. In Rome the greatest reward was given to
sedition, which had, therefore, ever been held in honour by all and sundry. Let
them recall the majesty of the senate when they had taken it over from their
fathers, and think what it was likely to be when they passed it on to their sons,
and how the plebs could glory in the increase of their strength and consequence.
There was no end in sight, nor would be, so long as the fomenters of insurrection
were honoured in proportion to the success of their projects. What tremendous
schemes had Gaius Canulius set on foot! He was aiming to contaminate the
gentes and throw the auspices, both public and private, into confusion, that   B.C. 445
nothing might be pure, nothing unpolluted; so that, when all distinctions had

SOURCE: Livy, *History of Rome*, trans. B. O. Foster (New York: G. P. Putnam's Sons, 1923),
pp. 83–89.

been obliterated, no man might recognise either himself or his kindred. For what else, they asked, was the object of promiscuous marriages, if not that plebeians and patricians might mingle together almost like the beasts? The son of such a marriage would be ignorant to what blood and to what worship he belonged; he would pertain half to the patricians, half to the plebs, and be at strife even with himself. It was not enough for the disturbers of the rabble to play havoc with all divine and human institutions: they must now aim at the consulship. And whereas they had at first merely suggested in conversations that one of the two consuls should be chosen from the plebeians, they were now proposing a law that the people should elect consuls at its pleasure from patriciate or plebs. Its choice would without doubt always fall upon plebeians of the most revolutionary sort, and the result would be that they should have consuls of the stripe of Canuleius and Icilius. They called on Jupiter Optimus Maximus to forbid that a power regal in its majesty should sink so low. For their parts, they would sooner die a thousand deaths than suffer so shameful a thing to be done. They felt certain that their forefathers too, had they divined that all sorts of concessions would make the commons not more tractable but more exacting, and that the granting of their first demands would lead to others, ever more unjust, would rather have faced any conflict whatsoever than have permitted such laws to be imposed upon them. Because they had yielded then, in the matter of the tribunes, they had yielded a second time; it was impossible there should be any settlement of B.C. 445 the trouble, if in one and the same state there were both plebeian tribunes and patricians; one thing or the other must go—the patriciate or the tribunate. It was better late than never to oppose their rashness and temerity. Were they to be suffered with impunity first to sow discord and stir up neighbouring wars, and then to prevent the state from arming and defending itself against the wars they had raised themselves? When they had all but invited in the enemy, should they refuse to allow the enrolment of armies to oppose that enemy; while Canuleius had the hardihood to announce in the senate that unless the Fathers permitted his laws to be received, as though he were a conqueror, he would forbid the levy? What else was this than a threat that he would betray his native City to attack and capture? How must that speech encourage, not the Roman plebs, but the Volsci, the Aequi, and the Veientes! Would they not hope that, led by Canuleius, they would be able to scale the Capitol and the Citadel? Unless the tribunes had robbed the patricians of their courage when they took away their rights and their dignity, the consuls were prepared to lead them against criminal citizens sooner than against armed enemies.

3. At the very time when these opinions were finding expression in the senate, Canuleius held forth in this fashion in behalf of his laws and in opposition to the consuls: "How greatly the patricians despised you, Quirites, how unfit they deemed you to live in the City, within the same walls as themselves, I think I have often observed before, but never more clearly than at this very moment, B.C. 445 when they are rallying so fiercely against these proposals of ours. Yet what else do we intend by them than to remind our fellow citizens that we are of them, and that, though we possess not the same wealth, still we dwell in the same City they inhabit? In the one bill we seek the right of intermarriage, which is customarily granted to neighbours and foreigners—indeed we have granted citizenship, which is more than intermarriage, even to defeated enemies—in the other we propose no innovation, but reclaim and seek to exercise a popular right,

to wit that the Roman People shall confer office upon whom it will. What reason is there, pray, why they should confound heaven and earth; why they should almost have attacked me just now in the senate; why they should declare that they will place no restraint on force, and should threaten to violate our sacrosanct authority? If the Roman People is granted a free vote, that so it may commit the consulship to what hands it likes, if even the plebeian is not cut off from the hope of gaining the highest honours, if he shall be deserving of the highest honours; will this City of ours be unable to endure? Is her dominion at an end? When we raise the question of making a plebeian consul, is it the same as if we were to say that a slave or a freedman should attain that office? Have you any conception of the contempt in which you are held? They would take from you, were it possible, a part of this daylight. That you breathe, that you speak, that you have the shape of men, fills them with resentment. Nay, they assert, if you please, that it is sinning against Heaven to elect a plebeian consul. . . .

"Shall the son of a stranger become patrician and then consul, but a Roman B.C. 44 citizen, if plebeian, be cut off from all hope of the consulship? Do we not believe it possible that a bold and strenuous man, serviceable both in peace and in war, should come from the plebs—a man like Numa, Lucius Tarquinius, or Servius Tullius? Or shall we refuse, even if such an one appear, to let him approach the helm of state? Must we rather look forward to consuls like the decemvirs, the vilest of mortals, who nevertheless were all of patrician birth, than to such as shall resemble the best of the kings, new men though they were?

4. " 'But,' you will say, 'from the time the kings were expelled no plebeian has ever been consul.' Well, what then? Must no new institution be adopted? Ought that which has not yet been done—and in a new nation many things have not yet been done—never to be put in practice, even if it be expedient? There were neither pontiffs nor augurs in the reign of Romulus; Numa Pompilius created them. There was no census in the state, no registration of centuries and classes; Servius Tullius made one. There had never been any consuls; when the kings had been banished, consuls were elected. Neither the power nor the name of dictator had ever been known; in the time of our fathers they began. Plebeian tribunes, aediles, and quaestors, there were none; men decided to have them. Within the past ten years we have elected decemvirs for drawing up the laws, and removed them from the commonwealth. Who can question that in a city founded for eternity and of incalculable growth, new powers, priesthoods, and rights of families and individuals, must be established? Was not this very provision, that patricians and plebeians might not intermarry, enacted by the decemvirs a few years since, B.C. 44 with the worst effect on the community and the gravest injustice to the plebs? Or can there be any greater or more signal insult than to hold a portion of the state unworthy of intermarriage, as though it were defiled? What else is this but to suffer exile within the same walls and banishment? They guard against having us for connections or relations, against the mingling of our blood with theirs. Why, if this pollutes that fine nobility of yours—which many of you, being of Alban or of Sabine origin, possess not by virtue of race or blood, but through co-optation into the patriciate, having been chosen either by the kings, or, after their expulsion, by decree of the people—could you not keep it pure by your own private counsels, neither taking wives from the plebs nor permitting your daughters and sisters to marry out of the patriciate? No plebeian would offer violence to a patrician maiden: that is a patrician vice. No one would have compelled anybody to

enter a compact of marriage against his will. But let me tell you that in the statutory prohibition and annulment of intermarriage between patricians and plebeians we have indeed at last an insult to the plebs. Why, pray, do you not bring in a law that there shall be no intermarrying of rich and poor? That which has always and everywhere been a matter of private policy, that a woman might marry into whatever family it had been arranged, that a man might take a wife from that house where he had engaged himself, you would subject to the restraint of a most arrogant law, that thereby you might break up our civil society and make two states out of one. Why do you not enact that a plebeian shall not B.C. 445 live near a patrician, nor go on the same road? That he shall not enter the same festive company? That he shall not stand by his side in the same Forum? For what real difference does it make if a patrician takes a plebeian wife, or a plebeian a patrician? What right, pray, is invaded? The children of course take the father's rank. There is nothing we are seeking to gain from marriage with you, except that we should be accounted men and citizens. Neither have you any reason to oppose us, unless you delight in vying with each other how you may outrage and humiliate us."

CHAPTER TWO

Country Folk and the Mobility of Money and Men

Advice to a Brother

Hesiod was the son of an immigrant seafarer of slender means who came to Boetia from Aeolian Cyme in Asia Minor. His father's experience of grinding poverty the poet transmuted into the "poverty which Zeus inflicts upon men," and in his major works (*Theogony*, *Works and Days*, and *Carmina*) this eighth-century writer speaks always to the experience of social dislocation made more poignant by being a stranger in a strange land. His brother Perses bribed the magistrates of his adopted village of Ascra and thus cheated him of his share of a meager parental fortune. The poet experienced both poverty and the extreme mental stress of not being able to provide adequately for a woman, apparently his wife. This often led him toward hostile characterizations of "the race of women," creatures who did not allow a man either dignity or a sense of partnership in making a living. Hesiod was himself not a peasant farmer but a shepherd, a man who in his poverty considers himself "a mere belly," and an "ugly shame" (*Works and Days*).

There is an autobiographical ring to the fact that the *Theogony* opens with the appearance of the Muses to the shepherd Hesiod and to the fact that the victim of the magistrates in *Works and Days* is also a sheepherder. These experiences provided the poet with the fundamental themes of his work: the defeat of justice in the world through the wiles of men; the hardship of agrarian life in a society in which the economy of money was rapidly changing social relations and legal conventions; and the debasement of religious virtue which to Hesiod seemed to accompany the development of a sopisticated economy in which avarice played too large a part. The context of truth and falsehood is basic to Hesiod's thought, the more so because he does not know whether the old myths of the gods hold "truth" in the new society. The world of Hesiod is thus permeated by dislocating experiences: poverty, the treachery of family, the deceitfulness of the gods, emigration, the vanity of women. In the *Works and Days*, from which we here print selections, the poet focuses his wrath on his cheating brother Perses in a long exhortatory speech. This *paraenesis* was a familiar device in classical poetry after Hesiod, but he formed it and had a significant role in making a tradition in which men appealed from the deceits of other men to the truth of Zeus. The poem is in fact in two main sections: the first deals with myths and apocalyptic visions, while the second is dedicated to what we may call *prudence*, the right ordering of the farmer's life, good seafaring, marriage, and other domestic themes. Yet the two parts are related; the fables show Hesiod a world in which there is divine justice, no poverty, no hunger, no sickness, old age or death, and, significantly, no women. The contrast between this paradise of the senses and the real world of the poet is evident in the following selection, beginning with the exhortation to Perses from the second part of the poem. Hesiod's preference for the "changeless order of the Gods" (*diké* or divine justice) rather than the misery of his shepherd's life and its treacheries has been called "nostalgic" and "reactionary" by some critics. But the student will perhaps see in this preference one of the central concerns of oppressed men in every age, and in Hesiod's fables hope where optimism is not warranted. His vision of justice has echoes in Isaiah and Micah and in the works of religious and revolutionary reformers from Luther to Marx.*

From *Works and Days*

Hesiod

It was never true that there was only one kind
  of strife. There have always
been two on earth. There is one
  you could like when you understand her.
The other is hateful. The two Strifes
  have separate natures.
There is one Strife who builds up evil war,
  and slaughter.
She is harsh; no man loves her, but under compulsion
and by will of the immortals men
  promote this rough Strife.
But the other one was born
  the elder daughter of black Night.
The son of Kronos,[1] who sits on high and
  dwells in the bright air,
set her in the roots of the earth and among men;
  she is far kinder.
She pushes the shiftless men to work,
  for all his laziness.
A man looks at his neighbor, who is rich:
  then he too
wants work; for the rich man presses on with
  his plowing and planting
and the ordering of his state.
  So the neighbor envies the neighbor
who presses on toward wealth. Such Strife
  is a good friend to mortals.
Then potter is potter's enemy, and
  craftsman is craftsman's

SOURCE: Hesiod, *Works and Days*, trans. Richmond Lattimore (Ann Arbor: The University of Michigan Press, 1959), pp. 19–31, 53–91, 101–109.

* The student wishing to pursue the interpretation of Hesiod and his work in its historical context may consult works of recent origin, among them: W. Jaeger, *The Theology of the Early Greek Philosophers*, and his *Paideia*, vol. 1; F. Solmsen, *Hesiod and Aeschuylus;* and F. J. Teggart, "The Argument of Hesiod's *Works and Days*," *Journal of the History of Ideas*, vol. 8 (1947); and E. Voegelin, *The World of the Polis*, chap. 5, "Hesiod."
1 Zeus.

rival; tramp is jealous of tramp,
    and singer of singer.
        So you, Perses, put all this firmly away
    in your heart,
nor let that Strife who loves mischief
    keep you from working
as you listen at the meeting place
    to see what you can make of
the quarrels. The time comes short for litigations
    and lawsuits,
too short, unless there is a year's living
    laid away inside
for you, the stuff that the earth yields,
    the pride of Demeter.[2]
When you have got a full burden of that,
    you can push your lawsuits,
scheming for other men's goods, yet you
    shall not be given another chance
to do so. No, come, let us finally settle
    our quarrel
with straight decisions, which are from Zeus,
    and are the fairest.
Now once before we divided our inheritance,
    but you seized
the greater part and made off with it,
    gratifying those barons
who eat bribes, who are willing
    to give out such a decision.
Fools all! who never learned
    how much better than the whole the half is,
nor how much good there is
    in living on mallow and asphodel.
      For the gods have hidden and keep hidden
    what could be men's livelihood.
It could have been that easily
    in one day you could work out
enough to keep you for a year,
    with no more working.
Soon you could have hung up your steering oar
    in the smoke of the fireplace,
and the work the oxen and patient mules do
    would be abolished,

. . . .

That man is all-best who himself works out
    every problem
and solves it, seeing what will be best late
    and in the end.

2 The Greek goddess of argiculture, fertility, and marriage.

That man, too, is admirable who follows one
    who speaks well.
He who cannot see the truth for himself, nor,
    hearing it from others,
store it away in his mind, that man
    is utterly useless.
As for you, remember what I keep telling you
    over and over:
work, O Perses, illustrious-born, work on,
    so that Famine
will avoid you, and august and garlanded Demeter
will be your friend, and fill your barn
    with substance of living.
Famine is the unworking man's most constant
    companion.
Gods and men alike resent that man who, without work
himself, lives the life of the stingless drones,
who without working eat away the substance
    of the honeybees'
hard work; your desire, then, should be
    to put your works in order
so that your barns may be stocked with all
    livelihood in its season.
It is from work that men grow rich and own flocks
    and herds;
by work, too, they become much better friends
    of the immortals
[and to men too, for they hate the people
    who do not labor].
Work is no disgrace; the disgrace is in not working;
and if you do work, the lazy man will soon begin
    to be envious
as you grow rich, for with riches go nobility
    and honor.
It is best to work, at whatever you have a talent
    for doing,
without turning your greedy thought toward what
    some other man
possesses, but take care of your own livelihood,
    as I advise you.
Shame, the wrong kind of shame, has the needy man
    in convoy,
shame, who does much damage to men,
    but prospers them also,
shame goes with poverty, but confidence
    goes with prosperity.

Goods are not to be grabbed; much better if God
    lets you have them.

If any man by force of hands wins him
    a great fortune,
or steals it by the cleverness of his tongue,
    as so often
happens among people when the intelligence
    is blinded
by greed, a man's shameless spirit tramples
    his sense of honor;
lightly the gods wipe out that man, and diminish
    the household
of such a one, and his wealth stays with him
    for only a short time.
It is the same when one does evil to guest
    or suppliant,
or goes up into the bed of his brother, to lie
    in secret
love with his brother's wife, doing acts
    that are against nature;
or who unfeelingly abuses fatherless children,
or speaks roughly with intemperate words
    to his failing
father who stands upon the hateful doorstep
    of old age;
with all these Zeus in person is angry,
    and in the end
he makes them pay a bitter price
    for their unrighteous dealings.
Keep your frivolous spirit clear of all such actions.
As far as you have the power, do sacrifice
    to the immortals,
innocently and cleanly; burn them the shining
    thighbones;
at other times, propitiate them with libations
    and burnings,
when you go to bed, and when the holy light
    goes up the sky;
so They may have a complacent feeling and thought
    about you;
so you may buy someone else's land, not have someone
    buy yours.

Invite your friend to dinner; have nothing to do
    with your enemy.
Invite that man particularly who lives close to you.
If anything, which ought not to happen, happens
    in your neighborhood,
neighbors come as they are to help; relatives
    dress first.
A bad neighbor's as great a pain as a good one's
    a blessing.

One lucky enough to draw a good neighbor
    draws a great prize.
Not even an ox would be lost, if it were not
    for the bad neighbor.
Take good measure from your neighbor,
    then pay him back fairly
with the same measure, or better yet,
    if you can manage it;
so, when you need him some other time,
    you will find him steadfast.
No greedy profits; greedy profit is a kind
    of madness.
Be a friend to your friend, and come to him
    who comes to you.
Give to him who gives; do not give to him
    who does not give.
We give to the generous man; none gives to him
    who is stingy.
Give is a good girl, but Grab is a bad one;
    what she gives is death.
For when a man gives willingly, though he gives
    a great thing,
yet he has joy of his gift and satisfaction
    in his heart,
while he who gives way to shameless greed and takes
    from another,
even though the thing he takes is small,
    yet it stiffens his heart.
For even if you add only a little to a little, yet if
you do it often enough, this little may yet
    become big.
When one adds to what he has,
    he fends off staring hunger.
What is stored away in a man's house
    brings him no trouble.
Better for it to be at home, since what is abroad
    does damage.
It is fine to draw on what is on hand, and painful
    to have need
and not have anything there; I warn you
    to be careful in this.
When the bottle has just been opened, and when
    it's giving out, drink deep;
be sparing when it's half full; but it's useless
    to spare the fag end.

Let the hire that has been promised to a friend
    be made good.
When you deal with your brother, be pleasant,
    but get a witness; for too much

trustfulness, and too much suspicion,
    have proved men's undoing.
        Do not let any sweet-talking woman beguile
    your good sense
with the fascinations of her shape. It's your barn
    she's after.
Anyone who will trust a woman is trusting flatterers.
        One single-born son would be right to support
    his father's
house, for that is the way substance piles up
    in the household;
if you have more than one, you had better live
    to an old age;
yet Zeus can easily provide abundance
    for a greater number,
and the more there are, the more work is done,
    and increase increases.
        If the desire within your heart is for greater
    abundance,
do as I tell you: keep on working with work
    and more work.

At the time when the Pleiades, the daughters
    of Atlas, are rising,
begin your harvest, and plow again when they
    are setting.
The Pleiades are hidden for forty nights and forty
days, and then, as the turn of the year reaches
    that point
they show again, at the time you first sharpen
    your iron.
This is the usage, whether you live in the plains,
    or whether
close by the sea, or again in the corners
    of the mountains
far away from the sea and its tossing water,
    you have your rich land;
wherever you live: strip down to sow, and strip
    for plowing,
and strip for reaping, if you wish to bring in
    the yields of Demeter
all in their season, and so that each crop
    in its time will increase
for you; so that in aftertime you may not be in need
and go begging to other people's houses,
    and get nothing;
as you have come now to me; but I will give to you
    no longer;
no further measure: Perses, you fool, work for it,

with those works which the gods have arranged
    men shall do,
lest some day you, with your wife and children,
    in anguish of spirit,
have to look to your neighbors for substance,
    and they not heed you.
Twice you may get help, and three times even,
    but if you plague them
further, you will get nothing more,
    and your pleading will fall flat.
Your style with words will do you no good; rather,
    I urge you
to work out some way to pay your debts, and escape
    from hunger.

First of all, get yourself an ox for plowing,
    and a woman—
for work, not to marry—one who can plow
    with the oxen,
and get all necessary gear in your house
    in good order,
lest you have to ask someone else, and he deny you,
    and you go
short, and the season pass you by, and your work
    be undone.
Do not put off until tomorrow and the day after.
A man does not fill his barn by shirking his labors
or putting them off; it is keeping at it that gets
    the work done.
The putter-off of work is the man who wrestles with disaster.
    At the time when the force of the cruel sun
    diminishes,
and the sultriness and the heat, when powerful Zeus
    brings on
the rains of autumn, and the feel of a man's body
    changes
and he goes much lighter, for at this time
    the star Seirios
goes only a little over the heads
    of hard-fated mankind
in the daytime, and takes a greater part
    of the evening;
at this season, timber that you cut with your ax
    is less open
to worms, now when it sheds its leaves to the ground,
    and stops sprouting.
Now, remembering your tasks in their season,
    is your time to cut wood.
Cut a three-foot length for a mortar and a pestle

of three cubits,
and a seven-foot length for an axle; that would be
    quite enough for you,
except if you made it eight feet you could cut a maul
    from the end.
For a wagon of ten palms cut a quarter-felly
    of three spans.
Cut many curved pieces; and look on the mountain
    and in the meadows,
for a good piece of holm oak to make your plow-beam,
    and bring it
home when you find it; this is the strongest
    for plowing oxen,
once you have taken it to the carpenter,
    Athene's apprentice,
and he fixes it in the share and bolts it to the pole
    with dowels.
You can work the plows in your house,
    and you should have a pair of them,
one in a single piece, one composite;
    this is the better way,
for if you break one of them, you can put the oxen
    to the other.
Poles of laurel or elm are least likely
    to be worm-eaten.
The share should be oak, the beam holm oak.
Get yourself two oxen,
males, nine years old, for their strength
    will be undiminished
and they in full maturity, at their best to work with,
for such a pair will not fight as they drive
    the furrow, and shatter
the plow, thus leaving all the work done
    gone for nothing.
And have a forty-year-old man, still young enough,
    to follow
the plow (give him a full four-piece loaf to eat,
    eight ounces);
such a man will keep his mind on his work,
    and drive a straight
furrow, not always looking about for company,
    but keep
his thoughts on business. A younger man
    will be no improvement
for scattering the seeds and not piling them
    on top of each other.
A younger man keeps looking for excitement
    with other young people.

At the time when you hear the cry of the crane
   going over, that annual
voice from high in the clouds, you should take notice
   and make plans.
She brings the signal for the beginning of planting,
   the winter
season of rains, but she bites the heart
   of the man without oxen.
At this time keep your horn-curved oxen indoors,
   and feed them.
It is easy to make a speech: "Please give me two oxen
   and a wagon."
But it's also easy to answer: "I have plenty of work
   for my oxen."
And a man, rich in his dreams, sees his wagon
   as built already,
the idiot, forgetting that the wagon has
   a hundred timbers,
and it takes some work to have these laid up at home,
   beforehand.
      At the first moment when the plowing season
   appears for mankind,
set hard to work, your servants, yourself,
   everybody together
plowing through wet weather and dry
   in the plowing season;
rise early and drive the work along, so your fields
   will be full.
Plow fallow in spring. Fallow land turned in summer
   will not disappoint you.
Fallow land should be sown while the soil
   is still light and dry.
Fallow land is kind to children, and keeps off
   the hexes.
      Make your prayers to Zeus of the ground
   and holy Demeter
that the sacred yield of Demeter may grow complete,
   and be heavy.
Do this when you begin your first planting, when,
   gripping the handle
in one hand, you come down hard with the goad
   on the backs of your oxen
as they lean into the pin of the straps.
   Have a small boy helping you
by following and making hard work for the birds
   with a mattock
covering the seed over. It is best to do things
   systematically,

since we are only human, and disorder
    is our worst enemy.
Do as I tell you, and the ears will
    sweep the ground in their ripeness,
if the Olympian himself grants that all
    shall end well;
and you can knock the spider-webs from your bins,
    and, as I hope,
be happy as you draw on all that substance
    that's stored up.
You will have plenty to make it till the next
    gray spring; you need not
gaze longingly at others. It is the other man
    who will need you.
        But if you have waited for the winter solstice
    to plow the divine earth,
you will have to squat down to reap, gathering it
    in thin handfuls,
down in the dust, cross-binding for the looks of it,
    not very happy;
you will bring it home in a basket,
    and there will be few to admire you.
Yet still, the mind of Zeus of the aegis changes
    with changing
occasions, and it is a hard thing for mortal men
    to figure.
Even if you plant late, here is one thing
    that might save you:
at that time when the cuckoo first makes his song
    in the oak leaves,
and across immeasurable earth makes glad
    the hearts of mortals,
if at that time Zeus should rain three days
    without stopping,
and it neither falls short of, nor goes over,
    the height of an ox hoof;
then the late planter might come out even
    with the early planter.
Be careful and watch everything well. Let not
    the gray spring
go by unnoticed in her time, nor the rain
    in its season.
        Walk right on past the blacksmith's shop
    with its crowds and its gossip
for warmth, in the winter season, when the cold
    keeps a man from working.
A lively man can do much about the house
    in this season.
Winter can be a harsh time of helplessness;

let it not catch you
in need, as you try to warm a thick foot
    with a thin hand.
The unworking man, who stays on empty anticipation,
needing substance, arranges in his mind
    many bad thoughts,
and that is not a good kind of hopefulness
    which is company
for a man who sits, and gossips, and has not enough
    to live on.
While it is still midsummer, give your people
    their orders.
It will not always be summer. The barns
    had better be building.

Beware of the month Lenaion, bad days,
    that would take the skin off
an ox; beware of it, and the frosts, which,
    as Boreas,
the north wind, blows over the land, cruelly develop;
he gets his breath and rises on the open water
    by horse-breeding
Thrace, and blows, and the earth
    and the forest groan, as many
oaks with sweeping foliage, many solid fir trees
along the slopes of the mountains his force bends
    against the prospering
earth, and all the innumerable forest
    is loud with him.
The beasts shiver and put their tails
    between their legs, even
those with thick furry coats to cover their hides,
    the cold winds
blow through the furs of even these, for all
    their thickness.
The wind goes through the hide of an ox,
    it will not stop him;
it goes through a goatskin, that is fine-haired;
    but not even Boreas'
force can blow through a sheepskin to any degree,
    for the thick fleece
holds him out. It does bend the old man
    like a wheel's timber.
It does not blow through the soft skin
    of a young maiden
who keeps her place inside the house
    by her loving mother
and is not yet initiated in the mysteries of Aphrodite
the golden, who, washing her smooth skin carefully,

and anointing it
with oil, then goes to bed, closeted
    in an inside chamber
on a winter's day
                that time when old No-Bones the polyp
gnaws his own foot in his fireless house,
    that gloomy habitat,
for the sun does not now point him out any range
    to make for
but is making his turns in the countryside
and population of dusky
men, and is dull to shed his light
    upon Hellenic peoples.
Then all the sleepers in the forest,
    whether horned or hornless,
teeth miserably chattering, flee away
    through the mountainous
woods, and in the minds of all
    there is one wish only,
the thought of finding shelter, getting behind
    dense coverts
and the hollow of the rock; then like
    the three-footed individual
with the broken back, and head over, and eyes
    on the ground beneath
so doubled, trying to escape the white snow,
    all go wandering.
        Then you had better cover your skin well,
    as I instruct you.
Put on both a soft outer cloak, and a fringed tunic,
and have an abundant woof woven across a light warp;
put this on you, so that your hairs will stay quiet
    in their places
and not bristle and stand up shivering
    all over your body.
Upon your feet tie shoes made of the hide
    of a slaughtered
ox; have them fit well, and line them with felt
    on the inside.
Take skins of firstling kids, when the cold season
    is upon you,
and stitch them together with the sinew of an ox,
    for a cape to put over
your back, and keep the rain off,
    and on your head you should wear
a hat made out of felt, to keep your ears free
    of the water.
Daybreaks are cold at the time when Boreas
    comes down upon you,

and at dawn there comes down from the starry sky,
    and spreads all over
the land, a mist, helping growth
    for fortunate men's cultivations.
This, drawn up from rivers that flow forever,
    and mounting
to a high level over earth on the turn
    of the windstorm,
comes down in the form of rain toward evening
    sometimes, but sometimes
blows as wind, when Thracian Boreas is chasing
    the thick clouds.
Beat this weather. Finish your work
    and get on homeward
before the darkening cloud from the sky
    can gather about you,
and soak your clothing through to the skin,
    leaving you wet through.
Better keep out of its way; of all months
    this is the hardest,
full of stormy weather, hard on flocks,
    hard on people.
At that time the oxen should have half rations,
    but a man
more than usual, for the nights add up and are longer.
Keep all these warnings I give you, as the year
    is completed
and the days become equal with the nights again,
    when once more
the earth, mother of us all, bears yield
    in all variety.
        Now, when Zeus has brought to completion
    sixty more winter
days, after the sun has turned in his course,
    the star
Arcturus, leaving behind the sacred stream
    of the ocean,
first begins to rise and shine at the edges
    of evening.
After him, the treble-crying swallow,
    Pandion's daughter,
comes into the sight of men when spring's just
    at the beginning.
Be there before her. Prune your vines.
    That way it is better.
        But when House-on-Back, the snail,
    crawls from the ground up
the plants, escaping the Pleiades, it's no longer
    time for vine-digging;

time rather to put an edge to your sickles,
   and rout out your helpers.
Keep away from sitting in the shade or lying in bed
   till the sun's up
in the time of the harvest, when the sunshine
   scorches your skin dry.
This is the season to push your work and bring home
   your harvest;
get up with the first light so you'll have enough
   to live on.
Dawn takes away from work a third part
   of the work's measure.
Dawn sets a man well along on his journey,
   in his work also,
dawn, who when she shows, has numerous people going
their ways; dawn who puts the yoke upon many oxen.
      But when the artichoke is in flower,
   and the clamorous cricket
sitting in his tree lets go his vociferous singing,
   that issues
from the beating of his wings, in the exhausting
   season of summer;
then is when the goats are at their fattest,
   when the wine tastes best,
women are most lascivious, but the men's strength
   fails them
most, for the star Seirios shrivels them, knees
   and heads alike,
and the skin is all dried out in the heat; then,
   at that season,
one might have the shadow under the rock,
   and the wine of Biblis,
a curd cake, and all the milk that the goats
   can give you,
the meat of a heifer, bred in the woods,
   who has never borne a calf,
and of baby kids also. Then, too, one can sit
   in the shadow
and drink the bright-shining wine, his heart
   satiated with eating
and face turned in the direction where Zephyros
   blows briskly,
make three libations of water from a spring
   that keeps running forever
and has no mud in it; and pour wine
   for the fourth libation.
      Rouse up your slaves to winnow the sacred yield
   of Demeter
at the time when powerful Orion first shows himself; do it

in a place where there is a good strong wind,
    on a floor that's rounded.
Measure it by storing it neatly away in the bins.
    Then after
you have laid away a good store of livelihood
    in your house,
put your hired man out of doors, and look
    for a serving-maid
with no children, as one with young
    to look after's a nuisance;
and look after your dog with the sharp teeth,
    do not spare feeding him,
so the Man Who Sleeps in the Daytime won't be
    getting at your goods.
Bring in hay and fodder so that your mules
    and your oxen
will have enough to eat and go on with. Then,
    when that is done,
let your helpers refresh their knees,
    and unyoke your oxen.
Then, when Orion and Seirios are come to the middle
of the sky, and the rosy-fingered Dawn
    confronts Arcturus,
then, Perses, cut off all your grapes, and bring
    them home with you.
Show your grapes to the sun for ten days
    and for ten nights,
cover them with shade for five, and on the sixth day
    press out
the gifts of bountiful Dionysos into jars.
    Then after
the Pleiades and the Hyades and the strength of Orion
have set, then remember again to begin
    your seasonal plowing,
and the full year will go underground,
    completing the cycle.

    . . . .

You are of age to marry a wife and bring her
    home with you
when you are about thirty, not being many years
    short of
that mark, nor going much over. That age
    is ripe for your marriage.
Let your wife be full grown four years,
    and marry in the fifth.
Better marry a maiden, so you can teach her.
    good manners,
and in particular marry one who lives close by you.

Look her well over first. Don't marry what will
    make your neighbors
laugh at you, for while there's nothing better
    a man can win him
than a good wife, there's nothing more dismal
    than a bad one.
She eats him out. And even though her husband
    be a strong man,
she burns him dry without fire, and gives him
    to a green old age.

Always observe a due regard for the blessed
    immortals.
Do not put some friend on equal terms
    with your brother;
but if you do, never be the first to do him an injury.
Do not tell lies for the sake of talking.
    If your friend begins it
by speaking some disagreeable word,
    or doing some injury,
remember, and pay him back twice over. Then,
    if he would bring you
back into his friendship, and propose
    to give reparation,
take him back. A mean man's one
    who is constantly changing
friend for friend. Do not let appearance
    confound perception.
        Do not be called every man's friend.
        Do not be called friendless,
nor companion of bad people, nor one who quarrels
    with good ones.
        Never be so hard as to mock a man for hateful,
        heart-eating
poverty. That's a gift given
    by the blessed immortals.
The best reserve of resource that men can have
    is a sparing
tongue, and they are best liked when that
    goes moderately;
if you say a bad thing, you may soon hear a worse
    thing said about you.
        Never be disagreeable at a feast
        where many guests
come together; there good feeling's greatest,
    expense is slightest.
        Never, from dawn forward, pour a shining libation
of wine to Zeus or the other immortals,
    without washing your hands first.

When you do, they do not hear your prayers;
   they spit them back at you.
      Never stand upright and make water
   facing the sun,
but only, remember, when he has set,
   or before his rising.
Nor do it when you are on the road,
   nor yet turning out from
the road, nor showing yourself. For nights
   belong to the Blessed Ones.
A devout man, one who has learned the right way
   to do things,
will huddle down, or go to the wall
of a courtyard enclosure.
      Do not, when in your house, ever show yourself
   near the hearthside
when you are physically unclean,
   but keep away from it.
Do not, when you have come back
   from an ill-omened burial,
beget children, but when you come from a feast
   of the immortals.
      Never wade through the pretty ripples
   of perpetually flowing
rivers, until you have looked at their lovely waters,
   and prayed to them,
and washed your hands in the pale enchanting water.
   For if one
wades a river unwashed of hands
   and unwashed of wickedness
the gods are outraged at him, and give him pains
   for the future.
      Never, at a happy festival of the gods, cut off
the dry from the green on the five-branch plant
   with shining iron.
      Never put the wine-ladle on top
   of the mixing bowl
when people are drinking. This brings
   accursed bad luck with it.
      Never, when you are building a house,
   leave rough edges on it,
for fear a raucous crow may perch there,
   and croak at you.
Never take up, without an offering, a piece of pottery
and eat or wash from it. There is a forfeit
   on these also.
      Never let a twelve-year-old boy sit on anything
not to be moved; better not; it makes a man
   lose his virility;

nor a twelve-month-boy either, for this will work
    in the same way.
      A man should never wash his body in water a woman
has used, for there is a dismal forfeit
    that comes in time also,
for this act.
          Nor, if you chance on sacred offerings
    burning,
must you make fun of the rites. The god, naturally,
    resents this.
      Never make water into the outlets
    of rivers meeting
the sea, nor in their springs, but altogether
    avoid this;
nor plunge in them to cool off; it means no good
    if you do this.
      Do as I tell you. And keep away from
    the gossip of people.
For gossip is an evil thing by nature,
    she's a light weight to lift up,
oh very easy, but heavy to carry, and hard
    to put down again.
Gossip never disappears entirely once many people
have talked her big. In fact, she really is
    some sort of goddess.

Homer's Cosmopolitan Wanderers

In the centuries before Hesiod, Greek society was already of a mixed agrarian-commercial character. Herodotus reported that Homer was Hesiod's contemporary. Texts in which it is hard to distinguish myth from fact record also that Homer was born in Cyme, the birthplace of Hesiod's sailor father. What is less disputable is that the social world of Homer's ordinary people is, like Hesiod's, not merely one of peasants and shepherds. There was already a full array of mercenary soldiers, pirates, and sailors living from commerce, craftsmen skilled in dozens of specialties, and a rich underworld of beggars, thieves, expatriate wanderers, and heroic adventurers. Their activities were central in Homer's epics, and from the poet's work Émile Mireaux has built a history of daily life in that shadowy time when the simple patterns of agrarian society were disrupted by the movements of commercial men. Mireaux is able to illustrate from Homer the growth of early commercial institutions out of the often illicit work of piratical salesmen. Homer's people traded extensively throughout the Mediterranean, possessed coins, knew the Phoenicians, Egyptians, and other African folk, and in other ways betrayed the poet's feat of picturing a society far more archaic than his own. The world of commerce pictured on the painted wine vases which were the chief item of Greek trade was already cosmopolitan.

From *Daily Life in the Time of Homer*

Émile Mireaux

WANDERERS AND EXPATRIATES

Homeric poetry is the poetry of adventure in far countries. The first of these miraculous masterpieces, which were to fix the destiny of Greek epic as a whole, were probably in the first instance, towards the end of the eighth century, addressed to a public of sailors and pioneers who were preparing to set sail for the conquest of barbarian coast-lands. Two privileged ports of call had been chosen as the scenes of their fabulous and varied stories, both on the margin of the Greek world properly so-called: the entry to the Hellespont which gave access to the north-eastern seas, and on the other hand the two ports of Corcyra, in the isle of the Phaeacians, which was on the threshold of western navigation. The choice was significant and is no doubt revealing.

All or nearly all the heroes of these poems themselves figure as wanderers and even expatriates. At the point where the *Iliad* opens, it has been nearly ten years since the heroes, at the head of their followers, have left hearth and home; and after the final victory, Odysseus wanders over the seas for another ten years and returns to Ithaca only to set out on another adventure, at the bidding of Tiresias. Menelaus, whose ships had been thrown onto the coast of Crete, embarks on a new expedition, this time to Egypt and Phoenicia, and only returns to Sparta

Source: Émile Mireaux, *Daily Life in the Time of Homer*, trans. Iris Sells (New York: The Macmillan Company, 1959), pp. 241–259.

seven years later. How unsettled and mobile were all these heroes! Agamemnon, Menelaus, Aegisthus, Diomed, Ajax, Achilles and Alcinous are all grandchildren of exiles. Patroclus and Phoenix had been compelled to flee their country and take refuge with Peleus, as Theoclymenus does with Telemachus in the *Odyssey*. The next generation, moreover, had to leave their native land, in their turn, under the violent pressure of the "return of the Heracleidae." It was they who went to colonize the islands and the coasts of Asia Minor, and plant new dynasties descended from Nestor and Agamemnon.

In this way the heroic world of the epics appears in our eyes as something mobile, effervescent and tumultuous. We must certainly make an allowance for dramatic imagination; nevertheless, the reading of the poems warns us that we should be taking a great risk if we supposed that here is an incorrect and distorted picture of the actual life of the times when the poems were written, and if we ignored the fact that Homeric society, no doubt like all human societies, comprised a mobile element, a contingent that was wandering and itinerant, often by profession, but also from a taste for change, a desire for gain, a love of risk, and sometimes from necessity. On the margin of the established routine of the community, there was a picturesque and unexpected fringe, which perhaps represented only the inevitable reaction of liberty, chance, and individual initiative to the quieter virtues.

The Seafarers

We must naturally place seafarers in the front rank of those whom we see leading an adventurous life.

In Homeric society the sailors do not represent a social category or a clearly-defined profession, and this is a most original feature. They have in this respect nothing in common with the *demiourgoi* who occupy as such a well-marked social position and prefer to exercise certain activities in the service of the public. On the other hand a "king," a landowner, or again the son of a "king" or landowner, may apply himself to seafaring even as his principal occupation, as Menelaus does during his seven years of enterprise in Egypt and Phoenicia. He none the less remains, socially speaking, a sailor by chance, a temporary and in some sort an accidental seaman. Or perhaps it would be truer to say that, if no one was specifically a sailor in Greece, everyone was more or less a sailor, at least in the seaboard towns, and that is nearly everywhere.

There was therefore no naval "demiourgia," [1] or class of specialized navigators. This was so true that the cities were obliged to organize, for the service of the state, a kind of maritime conscription. We discover from the Homeric texts that such in fact existed in Phaeacia and at Ithaca; but we know most about it from the very ancient Athenian institution of the naucrary, which there has already been occasion to cite. In each of the forty-eight naucraries in Athens, among which the citizens were distributed, one great family was required to equip and maintain a ship which the head of the family as *naucraros* would eventually command; the other members of the naucrary had to furnish the rigging and the crew. But this was an official organization designed to operate only in the event of mobilization or when it was a question of providing overseas transport.

---

[1] Class of craftsmen.

It is not, save by exception, in the ranks of the officially enrolled that we should look for the promoters of maritime enterprise or adventure, or for those who took part in them; and among these men all were not of the same rank nor did all possess the same capacity.

In a passage of the *Works and Days* which Hesiod devotes to navigation, he introduces us to the humblest of such men, those who aspired only to a modest profit at the cost of a limited risk. When Hesiod's father was living at Cymé, in Eolis, he had gone in for seafaring, the poet tells us, in the delusive hope of making a fortune; but in the end, and to avoid penury, he had crossed the Aegean on "a black ship" and taken refuge at Ascra in the heart of Boeotian territory. Hesiod confesses that he himself has no leaning for sea life. Perhaps his brother Perses had inherited the paternal vocation; however that may be, the poet is lavish of counsel, a counsel of prudence. He first recommends him to choose a broad-beamed vessel to carry his merchandise, and not a light one; which shows that Perses did not himself own a ship. For a landsman it would not have been normal to do so.

These details throw light, indirectly, on some of the forms of maritime trading. The latter must have been carried on, in part, by means of boats which were rented from the shipowners on the coast; we are here concerned only with the question of coastal trading or short crossings. Indeed in Hesiod's view, the season for navigation lasted only for fifty days after the summer solstice, when the winds were regular and the sea without peril. There was another good season at the beginning of spring, but it was very short and rather uncertain. Such periods, in any case, did not allow of long expeditions, and those who engaged in trading under these conditions had not ceased to be landsmen, and, as Hesiod recommends, they would hasten to pull the ship up on the beach before the first equinoctial rains set in. They would then surround it with heavy stones to prevent its being overturned by the winter storms. They would open the bung-hole in the hold so that the vessel should not fill with rain-water, which would rot the boards; they would carry the rigging and sails up to the house and hang the rudder in the smoke above the hearth.

But there were on the high seas men of a different character and of another stamp.

In the story he invents for the benefit of Eumaeus, Odysseus poses as a Cretan bastard of good family. On the death of his father, he had been allotted the small portion due to him by the legitimate sons, but he had then married a rich heiress. He was great in war, he says,

> But labour in the fields I never loved,
> Nor household thrift, that nurse of goodly children:
> But ever to my taste were ships of oars,
> And war, and polished spears and darts—grim things
> Whereat most others shudder. Well, no doubt
> I loved the things the gods put in my heart.

Nine times, at the head of a fleet and a company of gallants, he had sought fortune in foreign lands; by which we understand that he had taken part in nine expeditions of rapine and piracy.

Maritime ventures of a commercial kind were not yet materially distinguishable from piratical incursions. Opportunity created the robber. Piracy was moreover a very honourable occupation and one readily boasted of it, as does Odysseus who does not hesitate to pose as a former pirate. When Polyphemus finds Odysseus and his companions in his cavern, he asks them quite simply: "Are you not pirates?" The question was natural. Pirates went in for trade; they had to, in order to convert the value of their takings. Merchant seamen, on the other hand, dabbled in piracy when occasion offered. They were armed, if only to defend themselves against pirates; and being equipped for violence, they were naturally ready to take advantage of every good opportunity for pillage, like all men of pluck. In the course of a fairly long cruise, was it not necessary to pillage here and there in order to obtain the necessary provisions as cheaply as possible?

In any event, we learn from Odysseus' narrative in Book XIV of the *Odyssey*, how a maritime venture was organized, what sort of men the crew were, and how they were recruited. In the circumstances in question the promoter is the illegitimate son of a great noble who has only received a paltry share of the estate and who, after marrying a rich young woman, has tried to make his fortune outside the ordinary or traditional walks of life. Our man has decided to try cruising along the coast of Egypt. He begins by securing the assistance of some "godlike comrades," that is, men of noble birth but in a social and moral position similar to his own. They are to be the officers under him, and this enables them to fit out nine vessels. It was easy to collect the crews. The narrative tells us that the sailors were recruited from among the "warriors," that is from the class of peasant-soldiers which must also have counted in its ranks a number of younger sons ready for adventure. The men "come in quickly," we are told. For six days they are entertained and they carouse with the commander; on the seventh they embark and sail for Egypt, where they arrive after a crossing of five days. The enterprise, in point of fact, goes very much amiss owing to the indiscipline of the men and their excessive thirst for pillage. There is no question of dividing the booty, which was usually done by allotting equal portions to the men after the leader of the expedition had first received "what liked him best."

Commercial ventures to distant lands were planned in much the same way. The ships had to sail as a fleet and on a semi-military basis, if they wished to avoid unpleasant surprises at sea, and especially when moored at anchor, even in Greek waters and, *a fortiori*, on barbarian coasts.

The composition of a commercial convoy was a complex affair. The Homeric texts and the contemporary paintings or drawings which have survived acquaint us with three kinds of ship. There were two sorts of "hollow" and "swift" boats, with keels. The lighter one carried a crew of twenty-two men: the captain, the pilot who managed the rudder, and twenty rowers. The more powerful was manned by fifty-two men, of whom fifty were rowers. This was the *pentecontoros*, a warship as much as a trader, and often armed with a ram. Finally, there was the *phortis*, broad in the beam and flat-bottomed: this was the trader par excellence. All three were navigated by oar and sail. The sail was a large sheet of square or rectangular canvas which was fastened to a horizontal yard and attached to the deck-planks. Handling it was difficult and awkward. It could scarcely be raised except in the open where it was easy to catch the wind. The oars were invariably used to bring the ship in and out of port.

A commercial convoy almost necessarily comprised these three kinds of ship. The "swift" ships served as scouts and defended the others in case of need; but their capacity for lading was poor. A passage-way or half-deck ran from end to end of them, and there was practically no space for cargo except under this half-deck or under the rowers' benches. It was therefore only the presence of some real trading ships that allowed of profitable transport. Unfortunately these were slow and hard to manoeuvre. If the weather were ever so slightly uncertain, one had to land and wait for the sky to clear completely. So the convoy would trail from anchorage to anchorage, the daily stages in calm weather not exceeding fifty sea miles. One never sailed by night, because navigation near the coasts was always dangerous in the dark, even for a lone ship. For a convoy it was impossible.

No doubt Telemachus sails by night on his journey from Ithaca to Pylos and back; but it is a question of escaping the attention of the suitors; and, besides, he is under the special protection of Athena, both ways. Eurylochus roughly reminds Odysseus of the sound rule when the latter refuses to stop at nightfall off the Island of the Sun. "The nights beget fierce winds, the bane of ships," he says. One should stop towards sunset, go ashore and prepare supper near the vessel; and only put to sea next morning.

The best landing place was a deep cove, with a good beach on which the ships could be pulled up. Failing that, one might take refuge in a bay hemmed in with cliffs, provided it were deep and well protected, sheltered from the waves and from wind-squalls. The ships would then be moored side by side and roped to jutting rocks: this is how Odysseus' fleet proceeded when they stopped at the "fair haven" of the Laestrygonians.

The first precaution, immediately on landing, was to send scouts to the nearest look-out place so as to raise the alarm in case of danger. Hence the favourite landing places were on small islands, deserted or thinly inhabited, just off the mainland: these offered the maximum of security. On such an islet Odysseus hauled up his boats, facing the land of the Cyclops. This was the site of the islet of Ortygia at the entrance to the bay of Syracuse: it was a very ancient resting place for the seamen of Chalcis and Corinth.

When one had to cruise for several weeks along barbarian shores, such improvised halts in hostile or unknown territory were inadequate for a convoy. It needed organized stopping places where the men could obtain some rest, take in stores and carry out repairs. From the last third of the eighth century onwards, this need was to be supplied by the colonies that had been regularly founded all along the coastlands of southern Italy and Sicily, and on the shores of the Black Sea. These colonies were like organized swarms of bees; they had been sent out by the mother-city to hold a strategic point or provide a safety-valve for the turbulence of a surplus population. Built in accordance with ancient rites and with the offering of "splendid hecatombs" to the gods, they were established by a founder who belonged to one of the holy families of the home city, and who brought with him some of the soil of the motherland and fire taken from the civic hearth. At the beginning of Book VI of the *Odyssey* we learn that the city of the Phaeacians had been founded in this way by Nausithous.

These official colonies, however, were not planted at hazard. We know that in many places they had been preceded by the trading posts of adventurers, like that of the Eretrians on Corcyra or of the Chalcidians at Syracuse, prior to the foundation of the Corinthian colonies. These settlements were not real "cities"; because, as Fustel de Coulanges says, "a band of adventurers could never found a city." They were mere fortified camps, places of refuge for sea convoys, and held by a little group of sedentary traders. We can imagine what they were like from the picture in the *Iliad* of the "camp of the Achaeans" which had been pitched on the shore near the entrance to the Hellespont. Now in the bay of the Scamander, at this important stage in the sea route to the north-east, an encampment of this kind must have preceded the foundation of the Greek Ilium, which took place about 700 B.C. This early settlement perhaps served as a model for the author of the older poem on *The Wrath of Achilles*. Protected by a ditch and a "wall" which was only a strong palisade, and built by a stream which supplied drinking water, it covered a landing place for the ships that were drawn up on the beach and included a group of huts and a place for public meetings, where also cases were judged by the tribunal, and where there were altars to the gods. The author of the *Iliad*, however, takes great care to inform us in Book XII that his camp had been built "in despite of the immortal gods" and without the sacrifice of those "splendid hecatombs" which accompanied the foundation of every true city.

We must suppose the existence, at least at the outset, of a whole series of similar settlements, half counting-houses, half refuges, along the two great sea routes that led to the sources of supply for the trade in tin, that rare metal indispensable for the making of bronze. These were the Caucasus in one direction and Etruria in the other. In Book I of the *Odyssey* it was the latter route that Athena, disguised as Mentor, Prince of the Taphians, pretended to be following when she, or rather perhaps he, said he was carrying a cargo of iron to be exchanged for bronze at Temesa, on the shores of the Tyrrhenian sea.

How long might these distant cruises last? The convoys that left the various ports on the Aegean Sea assembled some at Corcyra, others at the entrance to the Hellespont where they awaited the definite return of good weather, which was at the summer solstice. From those places they needed twenty or thirty days if the former were to reach Cumae in Campania, the oldest of the Greek colonies, or the coasts of Etruria; and if the latter were to reach the further shores of the Black Sea and the land of Colchis. If there were no accidents or serious delays, they might in theory be back home just before the autumnal equinox, which put an end to navigation. In fact, the business of discharging cargo, bartering and reloading must have taken a fairly long time. The return voyage was, as a general rule, only made in the following year.

Maritime ventures to Egypt, which became frequent after the expulsion of the Assyrians and the establishment, in the second third of the seventh century, of the Twenty-Sixth Dynasty, followed a somewhat different pattern. Only bold sailors ventured to contemplate such a voyage. From Crete to Egypt one had to reckon with a crossing of at least five days and five nights, a terrifying venture for such poor navigators as were the Greeks of Homer's time, men who would not readily lose sight of land. The crossing was only possible in summer, thanks to the Etesian winds which then blew steadily from the north. No one dreamed of crossing in the opposite direction in winter, when the winds were blowing from the south. The return voyage was effected in spring by following the shores of Phoenicia, of Cyprus and of southern Asia Minor. Egypt was not a country where one was satisfied with simply landing in order to take on cargo and set off again. One stayed there for months, even years; one traded; one made a fortune. The Greek merchants lived in concessions, of which the most important was to be Naucratis. Menelaus stayed in Egypt for six years.

Commerce and Coinage

The only passage where Homer depicts merchant seamen does not concern Greeks but Phoenicians. In Book XV of the *Odyssey*, Eumaeus relates how he has been kidnapped, as a small child, with the complicity of a slave from Sidon, by the crew of one of their ships which had come to trade at the isle of Syros. He naturally speaks of them without indulgence:

> Thither Phoenicians, famous seamen, came,
> Rapacious rascals, bringing countless trinkets
> In their black ship. . . .

These merchants hauled their ship up on the beach, camped nearby and went from house to house until they had disposed of all their junk, glass trinkets, jewels, textiles, embroidered veils and so on. Homer describes them in the *megaron* of a "fine house" offering to the mistress surrounded by her maids a necklace of gold and amber; the jewel is weighed and examined; they bargain. While this is going on, the servant, who is in league with the Phoenicians, picks up three gold cups, takes little Eumaeus by the hand, and escapes to the ship.

Prior to this the merchants, as they sold their stock, had been making up a new cargo by purchasing local products which they would go and sell elsewhere. These operations, systematically recorded by an accountant on the ship, had been going on for a year. When the hold is full again and good weather has returned, they put out to sea, in secret, because they have taken advantage of the last few days to enrich themselves by theft.

This realistic description no doubt reveals to us what the general pattern of maritime trading was like.

What now, we may ask, did Greek commerce deal in, during the age of Homer?

It handled, as regards exports, the products of Greek industry, which was now rapidly expanding: pottery, weapons, woollen goods and linen, but also and perhaps especially wine. This was not what we call wine, which would have occupied an exorbitant amount of room in the diminutive boats of the eighth century, but a thick black juicy liquid which could only be drunk after being copiously diluted with water and which, being transported in well-stoppered *amphorae* (wine-jars) was not excessively cumbersome. The sailors of Chios had at an early date become specialists in the wine trade.

A first priority in the import trade was placed on the metals that were lacking in the Aegean basin, especially tin. Egypt supplied salt, soda, alum, alabaster, certain medicinal specialties celebrated in the *Odyssey*, corn also, and what was perhaps essential in the history of civilization—papyrus. The northern seas provided corn, slaves and timber, which as Greece became progressively deforested was more and more indispensable.

The transport of timber by sea set an even more difficult problem than that of wine. The ships of the eighth and seventh centuries were scarcely capable of taking any considerable cargoes of wood, which was cumbersome and hard to stow away. Fortunately the sixth book of the *Odyssey* informs us that Greek navigators possessed a highly advanced technique for the construction of rafts, which were formed of a score of well-dried tree-trunks, grouped in an oval like the bottom of a flat boat and well ballasted to resist wind and wave. Herein perhaps lay the secret of the oldest sea commerce in timber.

There is no doubt whatever that barter was still the principal instrument in the technique of exchange. The operation involved no major difficulty. By reason of the comparatively small number of categories of merchandise, it cannot have been hard to establish a ready-reckoner of their relative values. Immense progress, however, in the form of an invention that was greatly to facilitate the expansion of overseas trade, had been made at the very beginning of the Homeric era. This was the appearance of coinage.

Historians have hotly debated, and will continue to argue, about how money was invented. Who were the people or the ruler who first had the notion of circulating little stamped ingots of gold, silver or electron (a mixture of gold and silver) coins of a specified weight and a value guaranteed by the imprint of the place of origin? If we are to believe Herodotus, it was the Lydians, and therefore Gyges their first king, who first struck a coinage in gold and silver. This testimony has lost much of its value since the discovery of a deposit of coins in the foundations of the first temple in Ephesus, which certainly antedated the accession of Gyges. It seems in fact that money was invented fairly early in the eighth century by the Greek cities in Asia Minor and that the use of it spread into continental Greece quite quickly, thanks to Pheidon of Argos.

Homer was certainly acquainted with money. His affected archaism, natural in a poet who was depicting the heroes of very ancient times, merely prevented his mentioning as a standard of value, apart from the ox, anything but the golden talent. But this clearly figures as a unit of value of a fixed but inconsiderable weight. In Book XXIII of the *Iliad*, the second prize in the foot race is an ox, the third a half-talent of gold; in the chariot race the third prize is a cauldron of four measures, the fourth, two gold talents. We may recall here that in the treasure which was discovered in Aegina and which dates from about the beginning of the eighth century, the excavators discovered five gold rings without ornamentation, four of which weighed about 132.7 grains, that is, half the weight of the Babylonian shekel which had been adopted by Phocaea as a unit of value, and which was also, in the future, to be the weight of the Athenian gold stater. It was perhaps the prototype of the Homeric talent.

In this field, as elsewhere, there was decidedly nothing primitive about Homeric civilization.

The Strange Life of Beggars

Among the host of wanderers in the Homeric world, a place apart should be reserved for the beggars. Not that the profession—for it was a real profession that men adopted either from necessity or a sense of vocation—was in principle a vagabond's occupation. But strangers who had been or seemed to have been particularly unlucky adopted it readily enough: as witness Odysseus. It was, as we shall see, a genuine function which one sometimes had to gain, and also defend, with all one's might. If one lost it, one risked having to go, once again perhaps, into exile.

However surprising it may appear at first glance, we must accept the evidence. Homer, and after him Hesiod, count beggars among the *demiourgoi*, that is, the servants of the community or "public workers." There is no ambiguity in the texts. When Eumaeus enumerates the *demiourgoi* who were sometimes sought out among foreigners, he mentions prophets, healers, carpenters and bards; but, he adds, "none would call a beggar in." So the latter is included among the genuine *demiourgoi*. And we find the same inclusion in Hesiod, who inserts the beggar in a list of men who were unquestionably such:

> So potter with potter contendeth: the hewer of wood with the hewer of wood:
> the beggar is jealous of the beggar, the minstrel jealous of the minstrel.

What then was this strange "service" that an official beggar performed?

At the beginning of Book XVIII of the *Odyssey*, Homer describes the life, appearance and manner of the "official" or "public" beggar who collects alms all over the city of Ithaca. He is an insatiable eater and drinker; tall, young and of fine appearance. The young fellows call him Irus because, like Iris, he carries their messages. He inspires no pity. Although well fed, he wears the uniform of his profession: clad in rags, with only a stick and a wallet, for the dress and the attitude of a beggar are a matter of ritual. The beggar sits in the doorway, with his back to the doorpost against which he "rubs his shoulders." Here he waits until someone brings him his portion of bread and meat or, if a banquet is proceeding, until he is invited to go from table to table, collecting. His response is to beg Zeus to fulfil the desires of the master of the house.

In respect of a stranger newly arrived, who presents himself as a beggar, as Odysseus does, and who may be one of those gods who go from city to city sounding the hearts of men, feelings of humanity and a consciousness of the duties of hospitality may well play a part; they would obviously not be felt for a professional beggar like this Irus, who is healthy and prosperous. The part he plays and the reception reserved for him must have been due to other causes and admit of another meaning, connected no doubt with his paradoxical function as a *demiourgos*.

A probable clue is afforded by the word *apolymanter* which was applied to beggars as a natural epithet, in an expression usually translated by "scourge of banquets" or "kill-joy." This term is derived straight from a Greek verb which belongs to the Homeric vocabulary and admits of only one precise meaning: "to purify by eliminating pollutions." It can in our opinion only mean "purifier." On the occasion of meals or banquets, the beggar plays the part of one who takes pollutions upon himself in order to remove them.

On the plane of daily life, this was the function of the scapegoat. We have elsewhere studied its manifestations in public life, where it had a solemn and collective character. That the part should be played by a ragged beggar was perfectly natural. It was not unusual for the scapegoat, when he figured in official ceremonies, to be represented by a man ridiculously clad in old toggery. The attitude of the public toward the scapegoat was dual: on the one hand he was loaded with gifts and at the same time with blows and insults. Those were two ways of transferring pollutions in a rite of purification. Now beggars were received at banquets in exactly the same manner; for the insults and blows, and notably the hurling of a stool, were traditional, at least in the same degree as the gifts of food. When Eumaeus is bringing Odysseus disguised as a beggar to his own manor house, Melantheus the goatherd predicts as much; and the prediction was a sound one, and it was fulfilled, because while one section of the suitors treated Odysseus well, the others insulted him; Antinous and Eurymachus hurling stools at him while Ctesippus aimed a leg of beef at his head.

By his presence and behaviour as a daily purifier, the beggar is in fact a source and guarantee of prosperity. He begins, moreover, by praying Zeus to grant prosperity to those who receive him; and he is himself the living image of prosperity. This is why the beggar is, and professionally must be, a heavy eater and great drinker, a yawning gulf for food and wine. It explains the allusions to the demands of his stomach of which Odysseus, disguised as a beggar but knowing the trade, is so lavish in his addresses; allusions which he repeats, and which have surprised and shocked many commentators who have wanted to exclude them from the text. Such an exclusion would remove from the work a good part of its meaning.

The beggar is, in fine, the ragged bringer of good luck; an office he has not even today entirely ceased from filling in popular belief.

As a symbol and pledge of prosperity, he must naturally be a worthy representative of it. He must be the best man, at least in that branch of mendicancy which has fallen to his lot. This means that, when a competitor arrives, he must defend his position. This misadventure is exactly what befalls Irus, the official beggar, when he finds Odysseus seated at the very door which is Irus' domain. The question must first be settled by a regular bout as fisticuffs. Irus is vanquished and dragged by one leg out of the palace. But this is not the end of his misfortunes. Fallen as he has from the dignity of public beggar, he is to be expelled from Ithaca and landed on the mainland, in the domains of king Echetus who, it appears, gives a very rough reception to newcomers of this kind. So he is now, at the very least, condemned to the wandering life which his competitor Odysseus is supposed to have been leading prior to his arrival in Ithaca; and these wanderings will end only when he wins by main force a new position as a mendicant *demiourgos*.

Exiles and Mercenaries

The exiles, men of all classes who had lost their homes, were to go on increasing as social and political conflict became more exasperated. The man who had been banished owing to civil strife was to become a familiar type during the classical age.

We see him appear with the advent of the tyrannies in the second third of the seventh century. Cypselus, the first tyrant of Corinth, banished a great number of his adversaries and confiscated their property. The Bacchiads, an aristocratic family who had ruled the city for nearly a century, took refuge, some in Thebes, some in Sparta, others in Corcyra, in Sicily, and even in Etruria where, according to a legend, Demaratus, the father of Tarquin the Old, found a dwelling place. A few years later Theagenes in his turn drove the noble families from Megara.

But in any event, before the era of political exiles, large contingents of men had, during the Homeric age; fled their homeland after being conquered by some enemy. Thus the men of Asiné in Argolis who had been driven out by the Argives, wandered about for years before finding a refuge in a new Asiné which was conceded to them by their Spartan friends after the conquest of Messenia. In the reverse direction, the Messenians, who had just been heroically resisting the

Spartans, made their way to Argos, to Sicyon, and the friendly cities of Arcadia.

To the number of these expatriates must be added all those who, in consequence of an act of vengeance, a crime, or a sometimes accidental homicide, had been forced to leave hearth and home and seek, often for good, the hospitality of a land that might be far distant. For when the guilty fugitive was only a bastard or even a younger son, there were not many families disposed to assume the burden of a fine in order to relieve a member who, as a general rule, was himself only an additional burden on the family patrimony.

However varied their origin, all these refugees lived for long periods in the shadow of the great aristocratic families; and here, under the traditional shelter of hospitality, they generally became integrated into the feudal system, thanks to the bond of "companionship." There was nothing of the romantic rebel about these outcasts and outlaws. A new development was, however, taking shape; new openings for such men were appearing, outside the patriarchal system and even beyond those maritime ventures which might attract a certain number of them.

On the edges of the Greek world and yet in close relations with it, new forces were arising and becoming centres of attraction, powers born of conflict and maintaining themselves by conflict. Such, in Asia Minor, was the monarchy of the Mermnadae which had overthrown the old dynasty of the Heracleidae at Sardis, and founded the great Lydian power, such too, in Egypt, the Twenty-Sixth Dynasty which recovered national independence from Assyrian rule. Gyges the Lydian and Psammetichus the Egyptian were moreover allies, and both were attracted by Greek civilization. Both also had recourse to the military technique of Greece which was now asserting its superiority under the influence of the armourers of Chalcis, Corinth and Boeotia, and of the Spartan tacticians.

The era of the Greek mercenary, which was to reach its zenith three hundred years later with the expedition of the Ten Thousand, had begun. At Sardis Gyges surrounded himself with a Greek bodyguard. In Egypt, Psammetichus sent to the Aegean to recruit an army of Greek and Carian mercenaries; to these men he allotted lands, and he posted them on the desert frontiers. In both cases the soldiers were hoplites, or heavy-armed infantry. Clad in the panoply of helmet, breastplate, buckler and greaves, this armoured infantry with its discipline and its serried ranks was a formidable adversary for the light-armed Asiatics who were generally provided only with helmet and buckler.

Even in Greece itself the new tyrants were before long to recruit personal bodyguards. Some of the cities also would be organizing troops of mercenaries to send as punitive expeditions into barbarian territory, with a view to covering the foundation of colonies. It was in one of these that the poet Archilochus enlisted, after he had lost his money. It was a question of punishing a Thracian tribe. Archilochus himself records this adventure in the verses where he tells how he had thrown away his shield when beating a hasty retreat.

Thus it was that, to ensure means of subsistence, the individual no longer needed to be inside the traditional social framework. Adventure ceased to be a matter of collective effort: it was individual, and at the same time the field of action was growing markedly wider. At the beginning of the sixth century the brother of the poet Alcaeus was to take service in the armies of Nebuchadnezzar.

The Greek world was moving towards individualism, democracy and cosmopolitanism.

Hermes the Thief

Norman O. Brown has for many years championed the use of psychoanalytic techniques in understanding the nature of civilization itself. A classicist by vocation, his books on the power of Eros (*Life Against Death* and *Love's Body*) are well-known studies of unconscious forces and processes and the repressive impact of "civilization" on them. In his earlier book *Hermes the Thief* (1947), Professor Brown tried to show how the myth of Hermes may be made to yield data about the consciousness of early Greek people. Specifically, since the thief Hermes is nowhere made the object of disapproval in the *Homeric Hymn* about his thefts, Brown argues the god had been "appropriated" by the classes of craftsmen and merchants then flourishing in Athenian society. Hermes found a home among the men of the agora—where men lived by a sort of legal stealing. The sociopsychological "type" represented by the god is in fact that of Reynard the Fox in feudal society: "the little Prometheus," the man of wit, inventive power, daring, enterprise, and the desire to turn a profit at the expense of his more tradition-bound countrymen. Hermes was "the philosopher of the acquisitive life," as Brown remarks, and thus a guide to the conflict of values between merchants and farmers, traditionalists and innovators, those who live by custom and those who dare to break it. The Athenians made Hermes over in their image. Elsewhere in Greek society, Hermes remained the messenger of the gods, king of the dead, or a fertility god; but within the city gates he was *agoraios*, the demon of its marketplace.

The Homeric Hymn to Hermes

Norman O. Brown

The *Homeric Hymn to Hermes* is the canonical document for all subsequent descriptions and discussions of Hermes the Thief. Before we analyze the component elements in its synthesis of the mythology of the god, we must first survey the plot of the *Hymn* as a whole.

A brief preface (lines 1–19) informs us of the subject of the *Hymn*—Hermes, the son of Zeus and Maia, whom Zeus used to visit for as long as he could while his lawful wife Hera was sleeping. The fruit of this clandestine union was an unusual child who was shifty, cunning, and thievish, and highly precocious: on the very day of his birth he stole the cattle of Apollo. The rest of the *Hymn*, apart from five valedictory lines at the end, is a narrative of the events of this exciting day.

As he crossed the threshold of the cave on Mount Cyllene in Arcadia where his mother lived, he found a tortoise. Realizing at once the use to which he could put this find, he fashioned it into a lyre, thus becoming the inventor of the tortoise-shell lyre. After accompanying himself on his new instrument in a song about the love of Zeus and Maia—by which he was begotten—he left the lyre in his cradle and, feeling hungry, proceeded on his way after the cattle of Apollo (lines 20–

SOURCE: Norman O. Brown, "The Homeric Hymn to Hermes," in *Hermes the Thief* (New York: Random House, 1947), pp. 66–89.

62). These he found in the region of Mount Olympus, with the cattle of the rest of the gods. It was nighttime by now. Hermes drove away fifty cows of Apollo's herd, taking many precautions to throw the pursuit off the scent: he drove the cattle backward so that their footprints would point to the meadow from which he had stolen them, and he made himself a pair of sandals so constructed as to cover up his own footprints. On his way back to Arcadia he met only one person, an old man working in his vineyard at Onchestus in Boeotia; Hermes advised him that if he knew what was good for him he would keep his mouth shut about what he had seen (lines 63–93).

Upon reaching the ford across the Alpheus (in Elis) he foddered the cattle and put them away in a cave. Then he collected some wood and lit a fire with firesticks, thus becoming the inventor of this method of creating fire. Next he dragged two of the cows out of the cave, threw them on the ground, and made a sacrifice, dividing them into twelve portions. After throwing away his sandals, he smoothed the sand and returned to his mother's home on Mount Cyllene without being observed. He entered the house through the keyhole, like a wisp of cloud, and nestled down in his cradle, tucking the tortoise-shell lyre under his arm, like a baby with his toy (lines 94–153). But he had not fooled his mother. She asked him what he had been up to, and she took a pessimistic view of his chances of getting away with his first venture on a career of thievery. "Alas," she sighed, "when your father begot you, he begot a deal of trouble for mortal men and for the immortal gods." Hermes' reply was definitely in character: "Why do you try to scare me as if I were nothing but a silly child? I shall follow the career that offers the best opportunities, for I must look after my own interests and yours. It is intolerable that we alone of the immortals should have to live in this dreary cave, receiving neither offerings nor prayers. Would it not be better to spend our days in ease and affluence like the rest of the gods? I am going to get the same status in cult as Apollo. If my father does not give it to me, I will become the prince of thieves. If Apollo hunts me down, I will go and plunder his shrine at Delphi; there is plenty of gold there—just you see" (lines 154–181).

Meanwhile Apollo was in pursuit of the thief. Aided by information from the old man of Onchestus and by the flight of a bird—Apollo was a master at interpreting such omens—he identified the culprit and arrived at Maia's home. When Hermes saw him, he curled up in his cradle and pretended to be asleep. Apollo searched the place for his cattle. Failing to find them, he brusquely ordered Hermes to tell where they were. "Why, son of Leto," Hermes asked, "what means this rough language? I never even saw your cattle. Do I look like a cattle-raider? I am only two days old, and all I am interested in is sleep and warm baths and my mother's milk." "You certainly have won the title of prince of thieves," replied Apollo, as he picked Hermes up. But Hermes also knew about omens; as he was being lifted up, he let out an omen, "an unfortunate servant of the belly, an impudent messenger," and sneezed for good luck. Apollo dropped him at once. After further mutual recriminations, the matter was referred to Zeus for judgment (lines 182–324). "And what is this fine prize you have carried off?" Zeus asks Apollo as he sees him carrying a new-born baby under his arm. "It is not fair to accuse me of carrying things off," Apollo replied; "he is the thief, and a most cunning one too." Then he told Zeus about Hermes' devices for covering up his traces, and how he had pretended ignorance about the stolen cattle. At this point Hermes spoke in his own defence. "Father," he said, "you know I cannot tell a lie.

He came to our house looking for some cattle and began threatening me—and he is grown-up, whereas I was born only yesterday. I swear by the gates of heaven that I never drove the cattle to our house, and that I never stepped across our threshold. I will get even with this fellow for so violently arresting me; you must defend the cause of the weak and helpless." Zeus laughed heartily when he heard his dishonest son's ingenious denials; but his judgment was that Hermes should show Apollo where the cattle were (lines 325–396).

So Hermes took Apollo to the ford across the Alpheus and drove the cattle out of the cave where he had hidden them. Outside the cave were two cowhides which Hermes had laid out on a rock after the sacrifice. Apollo was amazed that a new-born baby should have been able to skin two cows. "You don't need to grow up," he said, as he began to twine a rope of withies to lead away the cattle.[1] But Hermes did not want him to lead away the cattle; so, to Apollo's amazement, he used his magic powers to make the withies twine over the cattle and take root in the ground. He then produced the lyre and began playing on it, singing of the origin of the gods and of the offices assigned to each. Apollo was overcome by the sweetness of the music. "What you have there is worth fifty cattle," he said to Hermes; "I know about music; I accompany the Muses when they dance to the sound of flutes; but never have I heard music such as this, music full of invitations to gaiety and love and sleep. Tell me the secret of your instrument; I will see to it, I swear, that you get a position of wealth and honor among the gods." Hermes replied with characteristic shrewdness, "I am not selfish; it would be a pleasure to teach you the secret of my instrument, just as Zeus taught you the art of prophecy. It is indeed a marvellous instrument in the hands of a true artist. In return you must be generous and share your patronage over cattle with me." And so a bargain was struck: Hermes received the neatherd's staff from Apollo, and Apollo received the lyre from Hermes. The two brothers drove the cattle back to the meadow at the foot of Mount Olympus, lessening the tedium of the journey with music on the lyre. To the delight of Zeus, they were friends ever after. As a neatherd, Hermes invented another instrument, the rustic pipe (lines 397–512).

Then Apollo said to Hermes, "I am afraid you may steal my lyre and bow, for Zeus has put you in charge of establishing the art of exchange on earth. I won't feel secure until you take a solemn oath." So Hermes swore he would not steal Apollo's property, or go near his house. In return Apollo swore he would consider no friend dearer than Hermes; he also promised to give him a magic wand empowered to execute all the good decrees pronounced by Apollo in his capacity as the oracular interpreter of the will of Zeus. "But as for this matter of prophecy which you are always referring to, Zeus has ordained that this province must belong to me alone; it is a difficult and responsible position. There is, however, a type of divination which three old witches taught me in my childhood when I was tending cattle on the slopes of Mount Parnassus. Zeus does not think much of it, but you are welcome to it. In addition I put you in charge of the whole animal kingdom, wild and domestic, and you alone shall be messenger to Hades." These favors, the poet goes on to say, show how much Apollo loves Hermes; their friendship was blest by Zeus (lines 513–575). The last few lines of the *Hymn* give a final judgment of the god: Hermes associates with all sorts and conditions

---

[1] I interpret the object which Apollo wants to bind, mentioned in the lacuna after line 409, as the cattle (see Allen and Halliday, *The Homeric Hymns*, 330–332), not as the lyre (see Radermacher, *op. cit.*, 145–147).

of men; he does little good; he spends his whole time playing tricks on mankind (lines 576–580).

The subject of the *Hymn* is Hermes the Thief—in the words of the invocation, "a plunderer, a cattle-raider, a night-watching and door-waylaying thief"—who stole Apollo's cattle on the very day he was born. He is also the Trickster, showing cunning in the execution of his theft, and guile in his verbal exchanges with Apollo and Zeus. His tricks are sometimes magical, as when he transforms himself into a wisp of cloud to pass through the keyhole, or when he makes Apollo's rope of withies take root in the ground. But in the plot of the *Hymn* Hermes the Trickster-Magician fades into the background, and Hermes the Thief occupies the center of the stage.

Hermes is a thief because he appropriates the property of Apollo; the notion of theft in the *Hymn* is firmly based on the recognition of individual property rights. His theft is, moreover, represented as a crime: as such, Apollo refers it to the judgment of Zeus, and Zeus adjudicates in his favor. Property rights are no longer derived from the autonomous family but are protected by a judicial process which enforces the general will of society as a whole. In terms of this code of justice, Hermes the Thief is a criminal.

Criminal though he is, Hermes has the devotion and admiration of the author of the *Hymn*. The repetitious emphasis on Hermes' thievishness in the invocation has the air of a defiant challenge—*Honi soit qui mal y pense*. Nowhere is moral disapproval expressed. It is indeed recognized that thieving does harm to those who are its victims—in the words of Maia, it is a "nuisance"; but of the idea that crime does not pay—in Hesiod's words, that "wrongful gains are baneful gains"—or of the doctrine of Theognis that it is better to be poor but honest there is no trace in the *Hymn*.[2] On the contrary, crime pays Hermes rich dividends. In his *apologia pro vita sua* to his mother, Hermes dismisses her scruples as childish, and justifies thieving in terms of the moral philosophy of egoism—"I will take up whatever business is most profitable." His arguments are left unanswered. Particularly revealing is the poet's handling of the scene which presents an obvious opportunity for vindicating the moral law: the judgment of Zeus— Zeus who, according to Hesiod, has thrice ten thousand detectives at work tracking down crime. Zeus's first reaction is to laugh heartily over his "evil-minded" son's sophistic oath. As was inevitable, he orders Hermes to give up the cattle, and Hermes hastens to obey. This attitude of obedience he maintains for the space of thirteen lines of the *Hymn*. Then he is up to his old tricks again, preventing Apollo from leading away the cattle. For the rest, Zeus is only mentioned as being delighted that Hermes and Apollo finally came to terms.

How are we to explain this tolerant and admiring attitude toward theft? Since the authorship of the *Hymn* is unknown, the problem reduces itself to a definition in general terms of the type of milieu within Greek culture to which the author and his audience can plausibly be assigned, on the ground that in such a milieu the glorification of Hermes the Thief would be both appropriate and acceptable. This task is less simple than it is sometimes taken to be. For example, the standard authorities all regard the *Hymn* as the expression of the uncivilized mores of primitive pastoral life in backward parts of Greece, such as Arcadia, where cattle-

---

2 Strictly speaking, this is true only of lines 1–512; later in this chapter it will be shown that lines 513–580 are the work of a different author, with a different attitude. See *Hymn*, 160; Hesiod, *Works and Days*, 352; Theognis, 145–146.

raiding remained the honorable exploit it was in the Homeric age. This interpretation rests primarily on the assumption that, except for the element of magic, the exploits of the hero are a faithful transcription of the mores of the audience for which the *Hymn* was written. It is indisputable that myths must originally have had some such simple and direct relation to the behavior of the myth-makers, and no one will dispute the primitive origin of the myth-motifs of the trickster and the cattle-raider. But it is also true that myths may be transplanted into an environment different from the one in which they originated and that they can survive, by subtle adaptation, all manner of changes in a culture in which they have once taken root. This truth is ignored by those who regard the *Hymn* itself as primitive. They tell us that "the idea of a trickster-god is one which appeals to the primitive mind," and forget that the same idea also appeals to minds that are far from primitive; a case in point is the medieval epic of Reynard the Fox. Similarly, while it is true that "the extraordinary feats of a tiny and apparently helpless person is a familiar subject of savage humor," it is also a popular subject in the folklore of the American Negro; witness the Brer Rabbit stories, which, whatever their origin, became the vehicle for comment on the relations between slave and master.

Actually, we know that the myth of the *Hymn* did survive the changes which elevated Greek culture above the primitive level, and survived not merely as a tradition, but as a living inspiration for new imaginative creations. From the archaic period (the sixth century B.C.)—in which the *Hymn* itself is generally placed—we have two vase-paintings, one depicting Apollo demanding the cattle from the baby in the cradle, the other depicting Hermes tucked up in the cradle with the cattle in the background. The first of these is from one of the Caeretan Hydriae, a group of vases famous for the sophisticated sense of humor they embody, which are ascribed to the most cultured areas of the Greek world. The second is credited to the Attic master Brygos. In the same period the equally sophisticated poet Alcaeus wrote a hymn to Hermes, in which he told of Hermes' theft of the cattle, capped by an attempt to steal Apollo's bow. In the invocation of this hymn Alcaeus says, "Hail, thou who rulest over Cyllene: for the spirit moves me to sing of thee"; the spirit moved him, he was attracted by the subject. The primitive origin of the myth does not prove that the *Hymn* itself is the product of a primitive environment. There is no reason why we cannot attribute to the author of the *Hymn* the same kind of interest in Hermes the Thief as was shown by Alcaeus, Brygos, and the painter of the Caeretan Hydria, all of whom belong to the artistic *avant-garde* of the urban and commercial culture that was maturing in the most advanced areas of the Greek world.[3]

The intention of the author is revealed not in the substance of the myth—traditions which he is not free to change at will—but in his portrait of Hermes as a socio-psychological type. The realism of this portrait is universally acclaimed; it is based on observation, and hence reveals the sort of environment in which the *Hymn* was written and the human type whose patron and ideal was Hermes the Thief. For although Hermes is represented as a new-born babe, he is no more a study in infant psychology than Reynard is a study in animal psychology. Just as certain qualities attributed to the fox in medieval folklore made Reynard a good

---

[3] The Caeretan Hydria is Louvre E702, *Corpus Vasorum Antiquorum*, France 14, Louvre 9, III.F.a, Plates 8 and 10. The Caeretan Hydriae are regarded as having been produced in either Ionia or southern Italy about the middle of the sixth century.

vehicle for portraying the psychology of the middle classes under feudalism, so the *Hymn* projects into the mythical concept of the divine thief an idealized image of the Greek lower classes, the craftsmen and merchants. Hermes is, as one critic has said, "the little Prometheus." The references to Hermes as an inventor are frequent, vivid, and elaborate. In all of them the individual and original genius of the inventor is emphasized; this is the typical conceit of the Attic craftsmen, as displayed in the proud signatures of the potters. The praise of the lyre— "This is marvellous music that I hear now for the first time"—compares with the exultant vase-inscription, "Euphronius never equalled this." In the description of the invention of the sandals Hermes' skill at improvisation is emphasized; improvisation is the talent in which Themistocles, the genius of the industrial and mercantile party, excelled all, according to Thucydides. From the observation of craftsmen at work are derived such vivid touches of psychological portraiture as Hermes' joyous laughter, his "eureka" when he gets the idea of the tortoise-shell lyre. In three different passages the *Hymn* mentions the sparkle in Hermes' bright eyes, the first time when Hermes is making the lyre: it is the gleam in the eyes of a craftsman enjoying his work. The craftsman also suggested the idea of the bustle that constantly surrounds the activities of Hermes, "who, as soon as he had issued from his mother's womb, did not long remain lying in the sacred cradle, but up he jumped and went hunting for the cattle of Apollo"; the lyre was constructed "no sooner said than done"; while he is playing it "new plans occupy his mind"; in his herculean labors to prepare two of the cattle for sacrifice, "work piled on work."

And it is not only Hermes' technical ability and his delight in technique that are modeled on the craftsman type, but also his moral philosophy, and even his manners. Hermes expounds his creed in his speech to his mother: he tells her that her scruples about his activities are childish; that he intends to put his own interests first, and follow the career with the most profit in it; that a life of affluence and luxury would be better than living in a dreary cave; that he is determined to get equality with Apollo—by illegal means if he cannot get it by legal means (that is, by gift of Zeus): he will go so far as to break into Apollo's Delphic treasury. What is this if not the businessman's creed, the philosophy of the acquisitive way of life which the Greek philosophers of the fifth century discuss; to use a phrase coined in the archaic age, what is it if not the doctrine that "money is the man?" This philosophy inspires not only Hermes' theft, but also his inventions. In his first speech, addressed to the tortoise, the idea of the profit that can be got from the tortoise is repeated three times; Hermes is particularly pleased over being the pioneer in the business—"I will be the first to get profit from you," he says. He makes a mocking allusion to the traditional and rustic use of the tortoise as a charm—"While you live you will be a good charm, if you die you will become a pretty singer"—and then he proceeds to kill her. Most pointed of all is the delightful parody of the line in Hesiod already quoted as an epitome of Hesiod's rejection of the new commercial culture; Hermes applies it to the tortoise, in the same way as the spider might apply it to the fly, "You come along with me; it is better to stay at home since the outside world is noxious." Such sophistication in the art of parody is a significant indication of the type of audience for which the *Hymn* was composed; even more significant is the selection of the passage to be parodied—a maxim expressing Hesiod's rejection of the new commercial culture.

And then there are Hermes' manners and morals in the more personal sense. They are on the vulgar side. For his first song on the lyre he selects a subject which a critic delicately refers to as unhomeric. Shelley translates:

> He sung how Jove and May of the bright sandal
>   Dallied in love not quite legitimate;
> And his own birth, still scoffing at the scandal,
>   And naming his own name, did celebrate.

He is litigious, skillful at making the worse appear the better reason. He lies brazenly to Apollo. He tries a mixture of trickery, bluffing, flattery, and cajoling to persuade Apollo to let him keep the cattle, and it succeeds. These are the essential traits of the impudent and smooth-talking self-seeker that haunted the Athenian agora, portrayed by Aristophanes in the Sausage-Seller of the *Knights*, the Unjust Reason of the *Clouds*, and the litigious type satirized in the *Wasps*. And as for Hermes' shameless omen, the "unfortunate servant of the belly," where do the commentators turn for analogies except to Aristophanes? Although the type is best portrayed by Aristophanes, it is also found in the archaic period. Already Hesiod has his brother Perses typed as a man who hangs around the agora and prefers to make money dishonestly, particularly by legal chicanery. But the best example is the sixth-century Ionian poet Hipponax, of whom a critic says, "The moralistic anthologists found little of value in his poems. He hardly ever rises above the level of his own personal squabbles and needs, and stories from the lowest type of everyday experience. In general his topics are personal abuse, threats, complaints, direct begging—for warm clothes and shoes, food, money. This gentleman, the last word in realism, individualism, and vulgarity, found even the simple iamb of Archilochus too exalted for his purposes." Hipponax, significantly enough, found Hermes the most congenial god; he is in fact the only personality in Greek literature of whom it may be said that he walked with Hermes all the days of his life. Hermes in the *Hymn* is an idealized Hipponax.

The portrait of Hermes as the ideal of the new commercial culture is projected into the traditional concepts of Hermes the Trickster and Thief. Sometimes the *Hymn* relates Hermes the Trickster to Hermes the Craftsman by preserving the original notion of trickery as magic skill: the sandals are described as the work of a mighty demon; the lyre, which is generally personified, is represented as a miraculous creation; Apollo's rope of withies magically takes root in the ground "due to the will of Hermes the stealthy-minded." In general, however, Hermes' trickery symbolizes the self-interested cunning that is characteristic of Aristophanes' agora type. His speech to Maia, his lying denial to Apollo, his speech before the judgment seat of Zeus, and his speech to Apollo in the negotiations leading up to the exchange are all described as "cunning." In his speech to Maia he is not trying to trick her; what is "cunning" is the acquisitive philosophy expressed in the speech. In his denial to Apollo and his speech to Zeus he is cunning in the use of courtroom sophistry. In the negotiations with Apollo he shows shrewdness in bargaining.

Not only as trickster, but also as thief Hermes symbolizes the new commercial culture. In his speech to Maia, which, as one commentator has said, contains the gist of the whole *Hymn*, Hermes deduces his justification of a career of theft from the ethical principles of acquisitive individualism—the duty of self-help and the doctrine that money is the man. An even more obvious clue to the meaning of the *Hymn* is contained in the reason advanced by Apollo for demanding that Hermes swear an oath not to steal his property—"Son of Maia, messenger full of shifty guile, I am afraid that you may steal from me both my lyre and my curved bow; for you have received from Zeus the office of establishing the practice of commerce among mankind." Apollo explicitly identifies commerce with theft.[4]

This equation of commerce with theft has been compared to the attacks on the profit motive in some modern economic theories. Whether or not the comparison is justified, the point of view expressed in the *Hymn* is virtually axiomatic in Greek moral philosophy. Everyone is familiar with the aristocratic prejudice against retail trade and manual labor, rationalized by Plato into the ethical doctrine that all professions in which the end is profit are vulgar and incompatible with the pursuit of virtue. The prejudice is ultimately derived from the conflict between the traditional patriarchal morality, sustained by the aristocracy, and the new economy of acquisitive individualism—the conflict of Metis and Themis in Hesiod. One of the results of this attitude was to identify trade with cheating, and the pursuit of profit with theft. As we saw in the preceding chapter, Hesiod regards acquisitive individualism as "theft" and "robbery." Solon uses the same terminology in his indictment of those who pursue wealth without regard to the common weal: "The very citizens, in their folly, are willing to contribute to the destruction of our great city, yielding to the temptation of riches. They do not have the sense to set limits to their superabundance. They grow rich through yielding to the temptation of unjust practices, and sparing neither sacred nor public property, they go stealing and robbing wherever they can." In Aristophanes' *Plutus*, when Poverty argues that "Good manners dwell with me, while insolence goes with Wealth," Chremyles answers, satirizing the enlightened ethics of mercantile Athens, "It is perfectly good manners here to steal and bore walls." Theft by wall-boring, of which Apollo accuses Hermes in the *Hymn*, is used metaphorically by Demosthenes to mean sharp practice in commerce; Plato argues that wall-boring is only a bolder expression of the same love of wealth which animates the merchant and the craftsman. Socrates in the *Gorgias* describes the "life of desire" (a more abstract expression for the acquisitive way of life) as the "life of a robber."

Hesiod, Solon, and Plato all use "theft" and "robbery" as interchangeable metaphors in their denunciations of acquisitive individualism, thus ignoring the distinction between forcible and fraudulent appropriation. The *Hymn* also ignores the distinction, by attributing a cattle-raid to Hermes the Thief, and by describing him as a "robber" and "plunderer." The mythical symbol for acquisitive individualism is thus composed of exactly the same ingredients as the verbal symbol. This coalescence of the notions of theft and robbery is due not to an obliteration of the distinction between force and fraud, but to the fact that rob-

---

[4] *Hymn*, 514–517. The point is entirely missed by the modern commentators: Allen and Halliday *op. cit.*, 342; Radermacher, *op. cit.*, 162–163; Eitrem, *op. cit.*, 278; Robert, "Zum homerischen Hermeshymnos," *Hermes*, 41 (1906): 413–414. They all take "practice of commerce" or "acts of exchange" to be an euphemism for stealing itself, thus giving the sense "since Zeus placed you in charge of stealing, I am afraid lest you steal my bow and lyre." Note that the *Hymn* makes clear that the stealing of which Hermes is the patron is an urban as well as a rural phenomenon. In line 15 Hermes is called a "door-waylaying thief"—a reference to the practice described in Aristophanes' *Birds*, 496–497: "I just stuck my head outside the wall and a bandit clubbed me in the back." In line 283 Apollo prophesies that Hermes will often "bore into rich men's houses"; theft by wall-boring is frequently alluded to by Aristophanes and the Attic Orators.

bery has ceased to be the honorable exploit it was in the Homeric age, and has come to be regarded, along with theft, as a crime. Hesiod calls robbery "wrongful," and specifically condemns cattle-raiding: "Not an ox would be lost, if there were no wicked neighbors." By the time of Hesiod the kings had ceased to be the leaders of marauding bands; they now formed a landed aristocracy which had a vested interest in the suppression of all attacks on property, including both robbery and theft. This change of heart is reflected in Hesiod's *Shield of Heracles*, in which Heracles, one of the great cattle-raiders of Greek mythology, is represented as a reformed character now applying his prowess to the task of ridding the earth of such nuisances as Cycnus, who "lay in wait for the hecatombs on the way to Pytho and robbed them by force." At the same time that Heracles renounces cattle-raiding, Hermes takes it up. Combined with theft, Hermes' robbery completes the mythical symbol of the pursuit of wealth without regard to the dictates of justice and Themis.[5]

In a society which shares Benjamin Franklin's opinion that commerce is generally cheating, the merchant is a thief whatever he does; it is only natural for him to react by justifying and idealizing theft. In the Middle Ages the Church's doctrine on usury confronted the merchant with the same dilemma: as one wit said, if you practise usury you end up in hell, and if you don't you end up in destitution. The medieval merchant accepted his own equation with the thief: he carried a thief's thumb as a talisman to help him in his business, shared his patron Saint Nicholas with the thief, and made Reynard the Fox his hero and ideal. In Greece the philosophic defenders of individualism, lacking the doctrine that

> Thus God and Nature formed the general frame,
> And bade self-love and social be the same,

proclaimed the war of every man against every man, and attempted to justify what Socrates called the life of a robber. The average tradesman found his self-justification in Hermes. Thus fortified, the impudent Sausage-Seller in Aristophanes' *Knights* not only admits that he steals, but wants to perjure himself by Hermes of the Agora to prove that he steals.[6]

The connection with commerce and craftsmanship persists throughout the various stages of the mythology of Hermes; and the *Hymn*, despite superficial appearances, is no exception. At the same time the *Hymn* grafts new themes ónto the parent stem of the myth—themes derived from experience, which give the

---

5 . . . Hermes first appears as cattle-raider in a poem of the Hesiodic school (Hesiod, Frg. 153, ed. Rzach), which may be as late as the sixth century B.C.; cf. Schmid-Stählin, *op. cit.*, Part I Vol. I, pp. 268–269. This poem first launched the myth which is told in the *Hymn*.

6 Aristophanes, *Knights*, 296–298. Diodorus (V.75) says Hermes invented "measures and weights and commercial profits and how to appropriate other people's property by stealth." Plato (*Cratylus*, 407E) says Hermes symbolizes "theft and verbal deceit and the ethics of the agora." See also T. Zielinski, *The Religion of Ancient Greece* (Oxford, 1926), 53: "Among the ancients, as in our own time, trade was of two sorts: wholesale import and export trade (*emporike*) and local retail trade (*kapelike*); the first enjoyed much respect, the second very little. The fact that Hermes extended his protection even over the second, with its inherent knavery, could not help lowering the significance of the god himself." Zielinski shares the Greek attitude toward trade; hence he urges us (*ibid.*, 115) to "forget as completely as may be" Hermes the god of thieves, and concentrate on Hermes the god of international law, etc., calling these other aspects alone expressions of "genuine religion." Quite apart from the fact that these aspects were all interconnected, . . . this hypostatization of selected phenomena as "genuine religion" is entirely subjective and sacrifices the facts to a sentimental urge to assimilate Greek to modern religious values. A Hermes without the Thief is a chimerical abstraction. In contrast with Pope's dictum quoted in the text, the orthodox Greek attitude is stated by Euripides (*Phoenissae*, 395): "We are slaves of profit contrary to the laws of nature."

myth new life by renewing its contact with the ever-changing reality it symbolizes. In addition to its novel portrait of Hermes' personality, the *Hymn* contains two new themes which radically alter the meaning of the myth—the theme of strife between Hermes and Apollo, and the representation of Hermes as a newborn baby.

The responsibility for the strife between Hermes and Apollo falls on Hermes: he is clearly the aggressor. His ambitious aggressiveness is the mainspring of the whole plot of the *Hymn*. As he explains to his mother, ambition was his motive for stealing the cattle. His determination to hold what he has makes him prevent Apollo from leading away the cattle, the episode which leads to the revelation of the lyre and the subsequent exchange. It is Hermes' aggressiveness that makes Apollo feel insecure even after the exchange and leads him to extract a further oath from Hermes and make further concessions to him. This trait in Hermes' character is in sharp contrast with Homer's picture of Hermes the loyal subordinate of Zeus.

The goal of Hermes' ambition is equality with Apollo. It is the cattle of Apollo that Hermes chooses to steal, though all the gods have herds and all the cattle of the gods are grazing together when Hermes separates fifty of them belonging to Apollo. Hermes and Apollo are contrasting figures in Greek mythology; the poet exploits this contrast, particularly when he brings the two gods together for the first time. Apollo is a majestic figure as he approaches Hermes' home, "his ample shoulders curtained in a purple cloud." When this majestic figure, "far-darting Apollo in person," appears, Hermes makes himself as small as possible. But Apollo cannot be deceived: "The son of Leto knew, and did not fail to know, the nymph and her son." This is the formula which Hesiod uses when Zeus unmasks Prometheus: "Zeus, whose mind is full of immortal wisdom, knew and did not fail to know the trick." With the power as well as the knowledge of Zeus, Apollo threatens to hurl the infant to the depths of Hell—the same threat is used in the *Iliad* when Zeus delivers a speech to the rebellious gods. This is the familiar contrast between Power and Helplessness, as in the Brer Rabbit stories; there is in it the same invitation to the reader's sympathies which Hermes addresses to Zeus—"Uphold the cause of the young and helpless."

Hermes' ambition is to secure the "status" and "privileges" that will place him on a par with Apollo, the aristocrat of Olympus. The result is not merely strife, in the sense in which Hesiod uses the term to designate competitive individualism; it is "civil war within the community of kindred," to use a phrase of Solon's. The theme of strife between Hermes and Apollo translates into mythical language the insurgence of the Greek lower classes and their demands for equality with the aristocracy. The *Hymn* thus reflects the social crisis of the archaic age—the crisis depicted by Solon when he says that the unrestrained pursuit of wealth has brought Athens to the verge of "civil war within the community of kindred," and by Theognis when he says that no city remains long at peace "when this becomes the aim of evil men, individual profits at the expense of the common weal; thence come civil wars and the shedding of kindred blood and tyrannies." It is the crisis that Solon attempted to solve by a redistribution of "status" and "privilege": "To the common people I have given a sufficient amount of privilege, not taking away from their status, nor adding to it superfluously." Theognis laments that a similar solution was applied in his home city of Megara: "Our city is still a city, but the folk are not the same. Those who before knew nothing of judgments or laws, but rubbed their ribs bare with the goat-skins they used for clothing and stayed outside the city like wild deer, now they are the nobility, and those who were noble before, now they are nobodies. Who can endure this sight?" If Hermes is "the little Prometheus," then the *Hymn* brings us to a period not far distant from the release of that Prometheus whom Hesiod left bound in adamantine chains.

The drama of the contemporary social scene is also infused into the representation of Hermes as a baby. By this device the poet accentuates the contrast between Power and Helplessness, between the established authority of the aristocracy and the native intelligence of the rising lower classes; as in the Brer Rabbit stories, our sympathies are enlisted on the side of the underdog. Furthermore, the baby Hermes "makes good" on the very day of his birth. To emphasize Hermes' meteoric rise in status the poet exploits the widespread theme of the marvelous child who proves his divinity by precocious prodigies. It is the symbol of the birth of a new world in which, as a result of the redistribution of status described by Solon and Theognis, the lower classes come into their own. Hermes is the Pantagruel of the Greek Renaissance.[7]

Did Hermes get that equality with Apollo which was his ambition? This question is answered in the exchange scene. Hermes gave Apollo the lyre, and Apollo gave Hermes the neatherd's staff. Most critics feel that Hermes got the worst of the bargain. Hermes, they say, forfeits his marvelous invention, and Apollo passes on to Hermes the menial task of tending cattle. Strangely enough, however, neither the poet nor Hermes seems to regard the exchange as a setback for Hermes. The initiative throughout the exchange scene is in his hands. When Apollo is about to lead the cattle away, Hermes "freezes" them. It is Hermes who produces the lyre, and shows it off to Apollo like a merchant in a bazaar. Apollo is swept off his feet; in the first flush of his enthusiasm he says the lyre is worth fifty oxen, the number Hermes had stolen; he apparently thinks that to exchange the lyre for oxen would be not a bad bargain for Hermes but a real tribute to the value of his invention. His actual offer, however, is considerably toned down—he only makes vague promises of "wealth and honor." In his reply Hermes proposes the terms on which the bargain was actually struck, and the speech is described as a sample of Hermes' shrewdness in bargaining. After flowery compliments to Apollo and praise of the lyre, he ignores the vague offer of wealth and honor and asks specifically for what he got, charge over the cattle. Hermes knows what he wants, and gets what he wants.

If the lyre is worth fifty oxen, Hermes, in losing the lyre and gaining the oxen, comes out even in cash value—no great achievement for the genius of theft and trickery—and Apollo comes out ahead in social prestige. Was this the goal of Hermes' ambition? The truth is that the terms of the bargain have been misunderstood. Hermes does not lose the lyre, Apollo does not lose the cattle; they

---

[7] On the theme of the marvelous child, see Allen and Halliday, *op. cit.*, 269; Radermacher, *op. cit.*, 64, 197. Radermacher interpreted the form of the myth in the *Hymn* as reflecting a situation in which the cult of Hermes was actually a new arrival intruding into an area previously monopolized by the cult of Apollo. This hypothesis is altogether improbable for any part of the Greek world even in the seventh century B.C.: those who stratify the Greek Pantheon into different chronological strata put Hermes in the oldest, and Apollo in the latest, stratum. Hence Wilamowitz (*Glaube der Hellenen*, I, 328) reversed the interpretation of the *Hymn*, maintaining that Apollo is intruding into an area previously monopolized by Hermes. In this, as in many other cases in Greek mythology, the effort to explain the myth in terms of cult-diffusion yields contradictory results and should be abandoned. Both interpretations neglect the possibility of a conflict between two already established cults (see below, pp. 89–101). A rise in the status of one cult at the expense of the other is sufficient to explain the form of the myth in the *Hymn*.

agree to share both lyre and cattle. Each initiates the other into his own art. The poet does not define the nature of the exchange explicitly because his audience knew that both Hermes and Apollo were in fact patrons of both the musical and the pastoral worlds; indirectly, however, his narrative points to the correct interpretation. Thus in his final speech Hermes lectures Apollo on the use of the lyre, as a teacher would a pupil; while willing to share the patent, he takes pains to point out that he remains the inventor—"I am not stingy about your learning *my* art. . . . Just as Zeus has initiated you into his oracular secrets, so I will initiate you into *my* new art. . . . Enjoy the lyre, receiving it from *my* hands; only let the glory be *mine*." Apollo for his part subsequently refers to the lyre as a "token" of friendship between him and Hermes; he is referring to the custom of sealing an agreement by breaking a token and giving each party half of it. As for the cattle, what Hermes actually says is this: "I will give you the lyre; in return let *us* herd the cattle. Then the cows will mingle with the bulls and breed sufficiently." Hermes did not steal Apollo's whole herd, but only fifty cows, leaving the bull; he envisages a union of the fifty cows with the rest of Apollo's herd, which at no point is involved in the exchange. Finally the poet says: "They *both* turned the cattle to the meadow, and went back to Olympus, *both* amusing themselves with the lyre." Hermes has every right to be content with the exchange; he has achieved exact equality with Apollo.

What does the poet mean by attributing to Hermes equality with Apollo? Hermes and Apollo are symbols of rival forces in the social and political conflict of the archaic age; the myth credits the lower classes with having achieved the equality they fought for. But Hermes and Apollo are not symbols invented by the poet; he is writing about two recognized Greek cults. His mythical description of the relations between Hermes and Apollo is not only an interpretation of the social scene but also an interpretation of the relations between the two cults.

The People of Aristophanes

It is by now obvious that conflicts between country and city values and habits of life provided generations of Greek writers with thematic material. Victor Ehrenberg, the distinguished Czech-German émigré classicist (now emeritus professor of classics, London University), has written extensively on Greek life and literature. One of the themes in his study of *The People of Aristophanes* (1962) is that of the dependence of the town on its agrarian hinterland. This produced marked social tensions between the wholly rural peoples and city folk who were still more rustic than their pride admitted. Using Aristophanes' comedies as source material, Ehrenberg outlines the modes of agricultural production. He also comments on working-class country life and the social values of farmers. Aristophanes claimed for his plays novelty—"Always fresh ideas sparkle, always novel jests delight"—and criticized Euripides' plagiarism.* But this novelty was built on comic genius playing over stock figures and stock situations. It is this facet of the Aristophanic art which enabled Ehrenberg to mine from the plays gems of Greek social life. Similarly, some future historian of everyday life in twentieth-century America may well use Jules Feiffer's strip-cartoons to tell us what people ate for breakfast, why they paid psychiatrists, or what tensions in society demagogues exploited.

From *The People of Aristophanes*
Victor Ehrenberg

THE FARMERS

On innumerable occasions we find the comic poets describing peasants and farmers; praising, or slightly ridiculing, their life and work, and emphasizing their importance to people and State. In consequence, the reader receives a general impression, the truth of which must be discussed. No doubt, Aristophanes was, so to speak, in love with those modest and industrious small farmers and vine-dressers who formed a large part of the Attic population. The question is whether that liking was more than personal, more than a view based chiefly on a private, primarily ethical, bias. It is true that many of the phrases in which the hard-working peasants are contrasted with the idlers, sycophants and snobs of the town are the expression of such a personal opinion; but to recognize this fact is not enough. Who were the peasants who play such a large part in comedy? To find an answer in comedy, an answer which is neither tendentious nor distorted, it will not suffice to regard figures such as Dikaiopolis or Trygaios as typical representatives of their fellow peasants as a whole. The importance of these "heroes" of the comedies is, at any rate, exceptional. Whether they are typical in other respects, is a question which can only be answered after considering the arguments which will form the subject of this chapter.

SOURCE: Victor Ehrenberg, *The People of Aristophanes* (Oxford: Basil Blackwell, 1962), pp. 73–89.

* *Clouds*, 547. In Aristophanes, *fragments*, 581 K., he says Euripides licked "Sophocles' honeyed lip."

We begin with the economic basis of farming, the cultivation of the soil. "The earth bears everything and takes it back." The first fact which emerges is that the cultivation of vine, olive and fig tree predominated, and that corn-growing was much less important. Ever since the days of Solon, Attic agriculture had been undergoing a process of transformation which had led to this result, which is confirmed by the evidence of comedy. It is interesting to hear a farmer say almost the opposite of what we in our northern climate should expect a peasant to say: "I had sold my grapes, and, with my mouth full of coppers, I went off to buy flour in the market." A fragment which runs: "One man gathers grapes, the other picks olives," is probably meant to describe the two chief kinds of crop. Similarly grapes and figs are mentioned together. The vines were either supported by stakes or grew between, and climbed up, the olive trees or fig trees. Thus the owner of even a small estate was able to cultivate all the chief fruits of the country within a small space. We realize that they all grew together when we hear, for instance, that the slave who has stolen some grapes is led to the olives in order to be flogged. The stump of an olive tree could be an obstacle to the growing of vines. Old and young vines, young fig-tree shoots and olives grew next to each other. For work among the olives, figs and vines there were special words which can be paraphrased, but not translated. Wine, figs (either fresh or dried) and olives represent, together with myrtles and fragrant violets, the established natural life of the country. The cultivation of these three fruits, above all of the vine, needed great and intensive care, and the character of the Attic peasant, who himself worked and cultivated his soil, was strongly influenced by this fact. The goddess of Peace is called "giver of grapes," and she has another name which is also applied to the peasant himself: "vine-loving."

The three fruits, of course, were not the only food; but besides bread and fish (and a little meat), they provided the staple diet. The main point, shown by the words already quoted of the peasant in the market, is that the average farmer grew no corn, or, at any rate, less than he needed for himself. It is estimated that Attica produced herself about a quarter of the grain she consumed. Much corn came from Euboia, and after the occupation of Dekeleia this amounted to more than what Attica supplied. There was, of course, some corn-growing, chiefly of barley, in the fertile plains, and elsewhere either in small fields or in the space between the rows of fruit trees or vines. Trygaios, who is a vine-dresser, prays to the gods that they may give to the Greeks (that is to say, not only to the Athenians, but perhaps also to some corn-growing districts like Boeotia or Thessaly) wealth such as barley, wine, figs and children. The farmer had not much to do after sowing time, when the ground was too wet for working in the vineyard. "I know," says a peasant, "how to tend goats, how to dig, to plough and to plant." Here we have the whole scope of farming, and "ploughing" means corn-growing. Occasionally we hear of someone carrying sheaves, or of a boy being bound with a sheaf-band. Phrynichos knows the song which people sang when winnowing the grain. The tilling of the soil, which was necessary for corn-growing, was the hardest part of all the hard work of husbandry, especially in the Greek climate, where the farmer performed his work almost naked; the soil of Attica was, to a large extent, poor, stony and often still uncultivated. Deforestation was far advanced; yet, charcoal-burning, as the *Acharnians* shows, was still being practised and important. Swelling land which could be graphically described as "the buttocks of the field" was rare, in spite of the famous phrase of "rich Athens," or the beautiful patriotic outburst of Aristophanes: "O beloved city of Kekrops, native-born Attica, hail, thou rich soil, udder of the good land!" When Wealth comes, says someone to Poverty, we shall have no further need "of thy ploughmen or yoke-makers, thy sickle-makers or blacksmiths, of mowing or fencing the fields." Here again, the activities named refer to the tilling of the soil, and it was only in the dreamland of fairy-tale that ample crops would grow without hard labour.

When ploughing and tilling the peasant used oxen or cows. The name and cult of one of the oldest Athenian families, the *Bouzygai*, symbolized this, especially in the plain of Athens. Various breeds of cattle were known in Attica. "The ox in the stable" was a proverbial phrase for some useless person. In the early morning the poor peasant drives out his oxen to sow his fields. The farmer who looks forward to peace and work, remembers above all his yoke of oxen. This "yoke of two oxen" was a fixed and much-used expression, and represented the usual modest number of cattle the farmer owned. Euelpides was the owner of a "puny pair," a two-oxen man. A peasant from the mountainous district of Phyle, where ploughing and tilling were especially hard, has lost his two oxen and with them the support of his farm. We never hear of larger numbers of cattle, although they must have existed; the property of a rich man could be described as fields, sheep and goats and cattle.

The peasant, who did some corn-growing, needed his two oxen. Milk and cheese, however, were usually taken from goats and sheep, not from cows, and the farmer from the Mycenaean mountains—where, as on those of Attica, few cattle could be kept—is expressly called "milk-drinking." It can be said as a general rule that cattle were of little importance in the holding of the average farmer. "Ox-loosing time" as an hour of the day was certainly a Homeric reminiscence rather than a practical expression of time used in Attica, and in general one expects to find that in a country where the cultivation of fruit trees and vines predominated, the usual domestic animals were donkeys, mules and goats. It is still the same in many parts of Greece today. The comedians, too, mention the donkeys, the animal most frequent in Greek proverbs, and the sheep and goats of Attica. Attic sheep and wool were highly valued. Pigs, too, were common. It was known to be most lucrative to kill a pig: "its meat is delicious, and nothing in a pig is lost except the bristles, the mud and the squeal." The mule is mentioned occasionally. More frequently we hear about horses; they were mainly used for aristocratic sport, and we may assume that the horses of the State cavalry came from native stock. Horse-breeding was practised, for instance, in the plain of Marathon, but, as a rule, horses came from abroad. Horses and, even more, cattle needed green food, and that was scarce in Attica. It is clear why stock-farming could never be undertaken except on a small scale.

A great deal of rural work consisted in market-gardening by the country people and especially by the yeomen burgesses who lived in the town or nearby and brought their produce to market. Most of the gardens were situated outside the walls; later Epicurus' garden was said to be the first inside the town, but this statement is an exaggeration. Gardening on a somewhat larger scale is revealed by the comic writers' knowledge of a great number of different vegetables. An abundance of flowers and fruit grew well in the mild climate and were available for a longer season than elsewhere; "no one any longer knows what season of the year it is," though this was largely due to the extension of trade. The saying: "You are not yet on the celery and rue" was used to mean: "You are scarcely at the edge of a thing," for these plants formed the borders in gardens. Further allusions to the products of the vegetable garden are made, when a stye on the eyelid is compared with a pumpkin, or when Odysseus is said to have bought in Samos a seed-cucumber. Bees and poultry were included in this kind of farming. People were naturally proud of hens who laid well, and seem to have believed that it was possible to force them to lay "wind-eggs." Geese and pigeons apparently were usually imported from Boeotia. The frequent combination of agriculture and gardening is shown in a lively fashion by the proverb "A pig among roses." It had the same meaning as our "A bull in a china-shop"; the Athenian expression, however, is not taken as it were from a story, but from the possibilities of everyday life.

The picture of husbandry and gardening gains in vividness when seen in its dependence on nature. The farmer loved all the seasons, each of which had its beauty, its pleasures and its advantages; but they could also do much harm. "The fruits are spoiled by hoar-frost, and I give my sweat to the winds": the picture is harsh; it may be an allusion to some mountain district of Attica. The winds certainly made the peasant's work very hard, especially the growing of corn; this is perhaps confirmed by the saying "to till the winds" as a proverbial phrase for useless toil. The blessings of rain and wind as well as the damage they cause are frequently mentioned. In general, the farmers' labours were heavy. "No idle man, however much he may talk of the gods, can gather a livelihood without toil," says the hard-working peasant in Euripides. The soil on the hills was poor, although the Attic mountains were not so inaccessible as, for example, Mount Taygetos, and it was thought better to live in a remote place among the rocks than to have unkind neighbours in the plain. We hear too of the fear of wild animals, such as the wolf carrying off sheep and goats. The dependence of human beings on nature is expressed in a general sense, when the birds, who are specially important to the peasant, as, for instance, in heralding the seasons, promise reward or punishment to mankind. Either they will kill all the animals that harm the plants, such as locusts, ants and wasps, or the sparrows will devour the seeds of grain, and the ravens will pluck out the eyes of the ploughing oxen and the grazing sheep. As a rule, the Attic farmer did not try very hard to force more than its natural yield from the soil. He had his busy times, but a passage already quoted proves that there were intervals. Farming was largely a seasonal occupation in Attica, and in winter after the olive harvest, as at midsummer, there was very little work. It might happen that the value of an estate was doubled by intensive labour, but that was exceptional, and it would be wrong to regard as a usual feature the gradual intensification of the methods of farming.

Whether he grew more corn or gave the greater part of his time to market-gardening, whether he cultivated chiefly vines and olives which certainly paid best, or earned his living mainly by charcoal-burning—the Attic farmer, as far as we know him and his rather primitive methods and modest aims, was as much a *petit bourgeois* as was the tradesman or craftsman. We shall also see that the vocations overlapped. The ideal of the small holder as of every *petit bourgeois* was for the most part the peaceful and care-free enjoyment of simple pleasures, of food and drink and love. Aristophanes again and again gives expression to such hopes. This is, on the whole, a realistic picture, although in comedy such longings may sometimes be idealized, or at least have a touch of unreality. In wartime many peasants reached the depth of poverty and misery like the man who asks Dikaiopolis for a small share in his peace, while in general it is true that modest and scanty conditions, and modest and simple men, were characteristic features of rural life.

On the other hand, the fact that almost all these small people kept at least one or two slaves bears witness to a certain *bourgeois* standard of life. There is very little evidence indeed for slave labour in the fields, except on comparatively large estates; and they were not frequent. Still, one slave girl at least, chiefly employed in the house, can in general be regarded as a minimum. There were also free workmen who worked for a wage as seasonal workers, for example as olive-gatherers, although we do not know how many of the olive-orchards were so large that their owners could not do without paid workers during the harvest. The "diggers, donkey-drivers and mower-women" who are mentioned together may have been day-labourers of this type, and a poor cottage will be the home of such "a delver or herdsman." We know almost nothing about the extent to which the individual farmer enjoyed social and economic security, but during the fifth century his position was not too bad. When someone spoke of "measuring the land," the peasant did not think of big estates to be divided into allotments, but rather of cleruchies, or colonies of Athenian citizens abroad, which, from the point of social economy, were attempts not so much to improve the conditions of Attic peasants as to provide with land the surplus of Athenian citizens, and thus to relieve the overpopulated town. The Attic farmers were not much affected by such measures, at least those of them who remained at home. There was, in fact, a process of gradual deterioration, largely by the war, and the clear evidence of the last plays of Aristophanes is the best illustration of the change from earlier conditions, and shows the direction of the development. In the *Ekklesiazousai* and the *Ploutos* we find the situation of the peasants very much deteriorated. Then, and not until then, they were on their way to becoming something that could almost be called "proletarians."

As late as the *Ekklesiazousai*, in the 'nineties of the fourth century, it is said that naval expenditure is good for the poor, but not "for the rich and the farmers." Only the poor townsman is called a poor man, just as he was a generation earlier. The peasant, however, is not described as wealthy, but as naturally averse from naval service. So we cannot draw a definite conclusion from that statement. In the same comedy the dramatic contrast between rich and poor forms the very basis of the communist programme. One man has extensive landed property, the other not enough for his grave. One has many slaves, the other not even one; the communist ideal, in this case, does not abolish slavery, as slaves are needed for the common agricultural work. In those years the number of comparatively large estates must have slowly increased. Their owners were the "gentlemen-farmers" for whom Xenophon wrote his *Oikonomikos*. In the *Ploutos* the social tension has further increased, and we find an almost revolutionary atmosphere. The peasants of this play are a poor, hard-working class, the community of those without bread, or those who like to chew the roots of thyme growing by the wayside, a habit which indicates a very simple and frugal life, or even severe poverty. To them the miracle of the coming of Wealth meant the true and only release from their misery.

Among the social and economic contrasts in the life of the farmers, that between town and country was the most obvious. No doubt the evidence of comedy on this question is rather one-sided. To arrive at the true facts, we must first ask if townsfolk and countryfolk were as rigidly separated as many quotations seem to prove.

It need hardly be specially emphasized that town and country depended on each other, and that both of them, "town- and country-folk," suffered alike in war. A living proof of the connection of both sections of the populace were, for instance, the artisans who were glad when, after the war, they could sell their wares "to the country." Kleon is said to have sold bad leather to the peasants. On the other hand, the farmers naturally tried to sell their vegetables, poultry and other products in the town market. Especially on festival days a great number of peasants used to come to town, and some used to stay with friends.

However, the relation between town and country involves more than the plain fact of mutual economic dependence. The barriers between them were not nearly so high as some scholars seem to believe. It was altogether exceptional when during the rule of the Thirty some people were expelled from the town, but apparently lived in the country without difficulty; this could occur only at that time, when the populace was profoundly divided. At other times town and country were very closely connected. One who as a boy had been a poor shepherd might later become a demagogue, and one who was educated in town as the son of wealthy parents might cultivate his estate when he had grown up. The chorus of the *Clouds* tries to get the judges on its side; as clouds can do nothing except send rain, it is natural that they should address those to whom the rain brought benefit or damage, the farmers. Their fallow land, their crops and vineyards and olives, even the home-made tiles are threatened if the vote of the judges is given against the poet. This speech would be entirely meaningless if the people addressed did not feel themselves concerned. We are entitled to assume that a large part of the audience had a hand in farming.

Even more distinctly than in the theatre we recognize in the assembly the rural part of the town population, or to put it more accurately, the inseparability of the two spheres. Clearly the men who often arrived in the assembly at the last moment, dusty and smelling of garlic, were peasants. The men "from the fields," though a minority, voted against the motion to give the rule to women. The assembly was composed of both townsmen and countrymen, but the former were in the majority, and when they were absent, the assembly more or less ceased to function; in later years they only turned up regularly after the fees had been raised to three obols. Nevertheless, when the voters who had elected bad officials are cursed, the words of the curse are: "neither shall their cattle produce offspring, nor their soil bear crops." This, again, would be nonsense if the majority of the people in the *ekklesia* were only townsfolk, as is generally believed. They were peasants as well.

Other sources tell us, for instance, that the wealthy citizen used to own not only a house in town, but also houses in the country, which were either let or administered as an estate by a paid steward. It was possible even, though probably not before a rather late date, for an *oikos* to be distributed over the territories of more than one State. Less wealthy people also lived in town, but cultivated their holdings and kept there various kinds of property, especially agricultural tools. Strepsiades asks pardon for his kicking at the door, because he lives "far off in the country"; in fact, he lived in town as a yeoman burgess, but his original home was in the country. Others lived so near—Trygaios, for example, whose vineyard was at Athmonia, about six miles out of town—that they might easily turn up at the pnyx or in the theatre or even unexpectedly in their own house; but they might also be so late that they would find no supper ready any longer. Euelpides, invited for the so-called *dekate*, the name-day feast of a boy, "first drank in town," then slept and went back to Halimus; on his way home (about four miles) he was robbed of his coat. It is difficult to say whether he belonged to town or country, and it seems almost impossible to draw a sharp line of demarcation between the two sections of the people.

The rural outlook also predominated in the town. The chorus-leader of the *Wasps*, a townsman if ever there was one, thinks only of the effect on the fruit, when rainy weather comes. The average town household at least owned poultry, and other birds, which might be kept in special cages, also a donkey, goats and other animals. The baker Nausikydes possessed pigs and even cattle. People who could not bear the crowing of the cock were clearly townsfolk. It is significant, too, that we occasionally come across comparisons or metaphors from country life, such as a modern townsman would hardly understand, and certainly not use.

In the first years of the great war the farmers who were crowded together between the Long Walls naturally longed to get back to their fields. "Now let us leave the town for the country; it is high time we were taking our ease there after a bath in the bronze tub." Many, however, had settled in town. There were intermarriages like that of Strepsiades. The question arises whether we have here the signs of an actual migration from country to town, which Aristotle believed to have occurred as early as the 'sixties of the fifth century. It may be taken as significant that the owners or tenants of some estates or farms changed more than once within a comparatively short time; but it is impossible to give definite figures for those who left the country for good. Sometimes comic characters give utterance to bitter remarks about people who wanted easier work or a fine style of living. He may well become refined as he lives in town." "The countryman standing at the perfume shop" was a proverbial saying. By and by, it seems, people left the country and went to town. There were enough honest professions besides toilsome "digging." In the comic ideal of women's communism the slaves were to cultivate the fields. It is advisable, however, to take into account the tendencies and sympathies of the comic poets, and not to overestimate such remarks. There was no real migration from country to town, no *Landflucht*, simply because the majority of the townspeople still had their farms or at least their gardens which were economically important. The process of man's alienation from the soil was still in its beginnings, and the evidence from comedy in general does not disprove the view that in the last part of the fifth century not more than a quarter of the Athenian citizens was without landed property. Beyond doubt, Athens and Attica were a State and a country with a large class of small farmers.

In spite of all these facts, there was an undeniable cleavage between townsfolk and countryfolk. The peasants would incline to pride themselves on rarely going to town and knowing little of the evil things going on there. A peasant, wearing perhaps his warm cap, the rural *kyné*, attracted attention in town. Someone who seems to have come to town, along with a very typical peasant, was asked: "Are you going to bring to town this 'rest-harrow,'" this weed from the countryside? There actually was an opposition between town and country, caused chiefly by differences of social position and intellectual level. Here we must be specially careful not to take as valid evidence what is tendentious or satiric in comedy. Aristophanes frequently gives idealized pictures of a delightful bucolic life, and, on the other hand, dwells on and exaggerates the wickedness of town life. But even in such one-sided pictures, when he jokes, for instance, about the differences of language and behaviour, the poet has to keep close enough to real facts in order to be understood and to evoke the right kind of laughter. Euripides was abused because of his "agrarian" mother. It was indeed a clever touch to make Strepsiades stress the difference of smell between himself and his wife; he had the peasant's inferiority complex. The word *asteios*, which indicated the townsman, became an expression for a "fine man," and "urban" meant something like "refined" or *comme il faut*. At the same time the word *agroikos* developed from meaning a peasant to meaning a bucolic and uneducated man, even a "barbarian," or a man "making rude jokes, and telling idle tales in a stupid fashion, relevant to nothing." The god who took no notice of the fact that in his temple someone loudly broke wind was called a true peasant. Demos himself, when finally dealing with the demagogues, will be on their trail as a "fierce *agroikos*"; he will be a peasant again then, but he will also be rough and rude. Under the influence of the sophists, town language, above all among the younger generation, became both refined and affected. When speaking in public, the peasant had to face the arrogance of the townspeople, and the point is specially stressed when he did not speak like a rustic. If a man found it difficult to proceed in his speech, he sometimes used the proverbial phrase of an ox standing on his tongue—a bold allusion to a countryman's inhibitions. It was possible to distinguish three sorts of Attic pronunciation: "the midway speech of the Polis, the town-speech which had a flavour of effeminacy, and the rugged speech with a flavour of the country." It is understandable that as a rule the farmer did not like to speak in town before the public, and that the educated townsman did not care for country people, nor they for the man "who had tramped the town and had the knack of words. If you search a bit, you will find in the country the anti-heliast's seedling," but those who always go to the courts and make speeches before the juries no longer care for rural life and least of all for its hard labour. There is no trace in comedy of the attitude well known from many ancient writers, when the peasant is contrasted with every other kind of manual worker, when he alone is not a despised *banausos*. This is a very remarkable fact in view of the general partiality of the comedians in favour of the peasants.

The passages mentioned prove that the deeper reason for the opposition between town and country, which developed in spite of their close connection, was based on actual differences in social and economic conditions. The "most pleasant country life," which is so often described in its modest happiness and care-free peace, was, at the same time, hard and dirty and poor. The idealization of a peaceful and sensual life is not the romantic glorification of bucolic existence as with Theokritos; that was unknown to the earlier Greeks. Even the comedians of the fifth century, though they praised to the utmost the peasant's life, did not deny its hardships and difficulties. You must be content with porridge and olives. If you get into debt, the demarchos as a bailiff "bites you from the mattress." The farmer, on the whole, still adhered to the old *oikos*-economy, and he hated all trade where he was always cheated. Here older and modern forms of economy met, and they could not easily work together. The peasant also felt himself somewhat harshly treated by the State, worse, at any rate, than the townsman, for instance, when he was called up for active service; townspeople always found a trick to get out of it. "What in town seems golden, becomes lead again in the country," runs a saying. Often the peasants had the feeling that they fought or suffered for a cause unknown to them; "there is a lot we don't know." The poor husbandman, even if by chance he was not ignorant, could not be concerned with public affairs, because he had to work so hard. Undoubtedly the farmers had to suffer more than anybody else during the Spartan invasions. When they were forced to settle inside the walls, it was as if each of them "had left his own Polis." The country people longed to leave the safety of the town and return to their homes, though they had lost a great deal of their property, while the townsfolk "lived without fear." On the other hand, the people left behind in the country suffered even more, and after the occupation of Dekeleia, Eupolis could say that "those inside the Long Walls" had a much better breakfast than the demes in the country. Even in Sparta, it is said, it was the peasants and not the "big people" who suffered in war, though we may wonder to whom the poet here refers—certainly not to the helots who cultivated the fields of the Spartiates, perhaps to the perioeci, and even other Peloponnesians, who had suffered from Athenian raids.

These varied references show that the ordinary Attic farmer had very little money, though he needed it for buying seeds, manure and even food, and, after the invasions, for restoring his farm; but that (at least before 404) he was not wholly impoverished, and had just enough to live on. The modesty or even meanness of rural life was not caused by the accumulation of land in the hands of a few. There were some wealthy men whose estates were cultivated by slaves or tenants; but the large estates, in fact never very large, were not of decisive economic importance. The characteristic feature of Attic agriculture was a far-going partition of the soil rather than the reverse. The small peasant, though not oppressed by big landowners, was oppressed by poverty and the growing difficulty of living on the yield of his piece of land. The population, on the whole, was growing, and so were the people's economic demands. In an ever-increasing degree the economic life of Attica was shifting to the town where political and social life had always been concentrated. The soil was too scanty and too poor, and there was no important intensification in farming methods; so that among the farmers poverty increased steadily, and the social and intellectual level sank.

Trimalchio Described

Petronius "Arbiter" is the name by which we know the wholly obscure author of the surviving fragments of a long work known as *Satires* or *Satyricon*. On every question of his biography uncertainty reigns. There is reason to think he was a native of Marseilles from the references in Tacitus which assign the *Satirae* to one C. Petronius, "elegentiae arbiter" (the last word on good living). But he certainly lived in Rome. Tacitus knew him well enough to paint a vivid picture of him in *Annals*, 18–19. Yet most later Latin writers ignore him, perhaps because his great, ribald book did not fit their notions of what should be available for the purposes of education.* Be that as it may, Petronius hides himself behind the masks of his characters, men who are strangers in southern Italy where most of the actions take place, well educated in literature, and all bearing Greek-sounding names.

In the following selections from the episode the "Banquet of Trimalchio," we meet Petronius's greatest creation. The man himself is vulgar, ostentatious, altogether illiterate, but somewhat good natured. Yet, he is also arbitrary, fond of humiliating clients and dependents, likely to take spleen for humor, and given to flights of bad verse of monstrous debauchery. Petronius wrote in Nero's time, when there flourished Trimalchio's models, the freed slaves (freedmen). Men of this class rose in the imperial service to positions of vast wealth and power or prospered in the commercial matters they handled for senatorial families to whom they became attached. Having gained their freedom, they often made millions in the corrupt, inflated economy of the Empire. They were in fact the epitome of what every age means by *nouveaux riches* (*novi homines*): given to that conspicuous consumption which was at once a creation of unrivaled coarseness and also a parody or satire of the lives of decadent patricians. Opulence was confused with status in a world where virtue was not its own reward and could look for none from the emperor. Imitative of court life in everything, Trimalchio had clients, slaves, poets, harlots, and gladiators in motley array, each bent on riches and competition for power. In his own life, truly elegant and mocking of luxury, Petronius knew intimately the language of the vulgar rich, their slang, and habits. It was a triumph of literature to record their style, embedded in the medium of his own pure Latin prose. Petronius created the form of the prose romance and also a withering portrait of what upward mobility could mean in the first century A.D.

---

* On Petronius see E. Paratore, *Il Satyricon di Petronio* (Florence 1933). There is no good English life, apart from the sketch in W. B. Sedgwick's translation of the book (Oxford 1925), where parody of Ulysses's adventures in the Encolpius episodes is fully discussed with other questions of style and authorship.

From *The Satyricon*

Petronius

"What's this?" he said. "Don't you know whose house you're going to visit to-day? Trimalchio's. He's stinking with money. Why, the man's got a clock in his dining-room with a live trumpeter attachment, uniform and all, as part of the works, so he can discover at any second how fast life's dripping."

We jettisoned all our troubles, put on our clothes as dandyishly as we could, and told Gito, who didn't seem to object to perpetually being a temporary slave, to follow us to the Baths. Meanwhile, as we were dressed up, we idled, or rather fooled, about, wandering from group to group. All of a sudden we struck an old fellow with a bald pate and a rosy shirt who was playing ball with a pack of long-tressed boys. But our attention was drawn not so much by the boys (though they well repaid a good stare) as by the slippered old gentleman who was exercising himself by tossing green balls around and who wouldn't deign to pick one up once it touched the ground. He simply turned round and took another from a slave who stood at his elbow with a bagful of them ready for the players. Another novelty that amused us was that he had two eunuchs posted at different spots in the group. One held a silver pisspot, the other was totting up the balls—not those that were caught and sent on from hand to hand, but those that fell to the ground. While we were marvelling at this egregious vanity Menelaus rushed up.

"That's the man you're to dine with," he told us, "and this here is in fact the push-off of the evening."

He had hardly got this out when Trimalchio snapped his fingers. This was a signal for the eunuch to hold the chamberpot in front of his master, who continued the game, nonchalantly pissing. When this elegant micturition was completed to the last dribble Trimalchio called for water to wash his hands; and after immersing the ends of his finger-nails he towelled his hands dry on the hair of one of the boys' heads.

It is not worth while to linger over details. Suffice to say that we entered the Baths, stayed briefly in the Sweating-room, and passed on into the Cold-house. Trimalchio was already flowing from head to foot with unguents, and being rubbed down, not by linen clouts, but by napkins of the finest textured wool. Nearby three practitioners of some massage-quackery were guggling down Falernian, pushing each other about in their rivalry for the most wine, and spilling more than they drank. Noticing this, Trimalchio said they were drinking to his health with plentiful libations. Then he was wrapped up in a scarlet dressing-gown and hoisted on to a litter. Four footmen, gorgeously liveried, strutted in the van, as well as a sedan-chair in which his Delight was borne, a decayed boy, bleary and more damaged by Time's ugliness than his handler, Trimalchio. As this array started off a musician with minute pipes came near to his master's head and whispered music furtively all the way as if he blew some mysterious information into his ear.

In a proper state of amaze by now, we took our places and followed, and by

SOURCE: Petronius, *The Satyricon*, in *The Complete Works of Gaius Petronius*, trans. Jack Lindsay (New York: Rarity Press, 1932), pp. 16–61.

Agamemnon's side reached the doorway of the house. There on one of the pillars was a notice announcing:

ANY SLAVE
GOING OUT OF DOORS
WITHOUT LEAVE OF ABSENCE
WILL GET 100 LASHES.

Right in the entrance lounged a porter in a green uniform with a cherry-coloured belt, shelling peas into a basin of silver. Above the door swung a golden cage, and in it a speckled magpie which chattered a welcome at us.

While I was gaping at all these things my knees almost gave way under me as though tapped from behind by terror. On the left, close by the porter's lodge, was a monstrous dog chained up . . . in paint on the wall, with BEWARE THE DOG printed in capital letters over it. My companions, needless to say, didn't let me off their derision, but I pulled myself together and went on inspecting the rest of the frescoes. One depicted a Slave-market, price-tickets and everything complete; another Trimalchio himself with a mop of hair on his head and Mercury's wand in his hand, being ushered into Rome by Minerva; the next related how he had learned book-keeping and been installed as bailiff. To the series the conscientious artist had appended an elaborate verbal analysis. Towards the end of the gallery Mercury was shown elevating Trimalchio by the leverage of a hand under the chin into an imposing judgment-seat, about which hovered the Goddess Fortune with her Horn that bubbles out eternal plenty, and the Trio of Fates spinning with thread of gold. I saw also in this hall a flock of runners being put through their paces by a trainer; and in one corner a great cupboard-arrangement where in a miniature shrine were silver images of the House-gods and a marble Venus-statue and a gold casket of no puny size in which, I was told, was preserved the first Snippets of the Beard of Trimalchio.

I began to ask the butler questions about the middle pictures.

"Iliad and Odyssey," he reeled off in answer, "and Gladiatorial Show Laenas gave."

But that was all the time we were given for contemplation. Already we had neared the dining-room, by the door of which sat the steward making up his accounts. But, strangest of all, stuck on the doorposts of the room were the Rods and Axes twined together at one end in a design meant to look like the brazen prow of a ship. This was the inscription:

A PRESENT
TO
GAIUS POMPEIUS TRIMALCHIO
PRIEST OF AUGUSTUS
FROM HIS STEWARD CINNAMUS

Under this dedication a twi-flamed lamp was strung from the ceiling, and there were two tablets, one to each doorpost. Unless my memory errs, one had lettered on it:

ON 30TH & 31ST DEC.
OUR GAIUS IS DINING OUT

The other calendared the phases of the moon and the seven planets, and lucky and unlucky days were differentiated by coloured buttons.

We had extracted all the humour we could from these items of interest and were about to step into the room, when a slave, placed there for this particular office, bawled out, "Right foot first, please." We stood in a tremble of indecision, afraid that one of us would lift the wrong foot and transgress the ordinance. And when we all at last gingerly moved our right feet forward as if we were being drilled, a slave, stript for the whip, grovelled before us, begging us to intercede for him. Only a little fault, that was all, he said, that had ruined him. The steward had had his clothes stolen at the Baths, and they were not worth more than a shilling or two. So we withdrew our right feet, and put in a word for him to the steward who was counting out gold in the hall. Let him off the flogging, we asked.

With majestic deliberateness the steward raised his head to look at us. "It's not that the clothes are gone," he said, "I don't mind that. It's the negligence of the shiftless trash. It was my dinner-suit he let go, and I value it—a birthday gift from a client, you know, a hanger-on of mine—genuine Tyrian it was, and not been sent to the laundry more than once. Well, what does it matter? Sirs, the man is yours. I give him to you."

The patronizing air of magnanimity with which he conferred this benefit left us speechless. We drifted back to the dining-room, and there came upon the very slave we had rescued. He surprised us with a fluster of grateful kisses and exclamations. "You'll soon find," he said, "who it was you were so good to. The master's private wine is the waiter's present."

At last we lay down in our places. Water cooled with snow was sluiced over our hands by Alexandrian slave-boys, while others applied themselves to our feet, trimming the toe-nails with the ease of accomplished artistry. But even this difficult pedicure did not obsess their minds; as they worked they sang. I wondered if the whole household were choristers also, and out of curiosity asked for a drink. A boy at once burst into song in a piercing treble and brought it to me. Indeed, any request you made was carried out to the same melodious setting. It was much more like being among a ballet of warbling vaudevillists than at dinner in a reputable person's place.

However, in they brought a relish, a very appetizing dish too. By now Trimalchio was the sole absentee. Following the newest fashion, the first seat was kept for his special occupance. Among the dinner-plate stood a small wrought ass of Corinthian bronze, panniered with olives—the black one side, the white the other. Over the ass were fixed two salvers, around the edges of which were engraved Trimalchio's name and the weight of the silver. And there were two contraptions of iron soldered on, formed like bridges, which offered dormice sauced with poppy-seeds and honey. Sausages steamed on a silver gridiron, underneath which lay black Damascus-plums and pomegranate-seeds from Carthage.

We were eating away when Trimalchio in person was borne aloft into the room, welcomed with a noise of music and tucked into a lot of dainty little cushions. Taken off their guard, some laughed. The globe of his shaven poll was balanced loosely on the scarlet mass of his cloak, his neck was hidden in shawls, and he had dressed up in a linen robe with a broad stripe and a fringe of tassels dingle-dangling. A gilt ring bulged from his left little finger, but on the last joint of the finger next to it he wore a smaller ring which I thought at first flash to be solid gold, but which was really spangled with star-tips of steel. To make it quite clear that he owned more jewellery still, his right arm was bared and clasped with a circlet of gold and an ivory bangle jointed with a glitter of gold-metal.

First of all he picked his teeth thoroughly with a silver tooth-pick. Then he spoke:

"My dear friends, had I consulted my own inclinations I would not have come so early to dinner, but as I feared my absence might cast a blight over the party, I put my personal wishes aside and came along. Still, I don't expect you'll object to me finishing my game."

Behind him entered a slave with an inlaid board of terebinth and a set of crystal men; and I noted in chief one typical lavish detail. He had substituted gold and silver money for the black and white counters. While he was still swearing away heatedly to himself and we were finishing off the relishes, a tray with a basket on it was fetched in. Inside the basket reposed a hen carved in wood with her wings cuddled round her as though she were on the cluck of dropping an egg.

Straightaway two slaves came up, and amid the squawks of the protesting orchestra began to feel through the straw and pick out pea-hens' eggs, which they handed round among the company. Trimalchio looked up over his game at this by-play.

"My friends," he said, "I gave orders for that hen to sit on a clutch of pea-hens' eggs. I assure you I'm half of a mind that she's gone and all but hatched them. The best thing to do is to break the shells bravely and see if your stomachs can stand what you find."

We took up our spoons (incidentally a clear half-pound each) and struck the eggs—only to discover that they were imitations kneaded from a rich paste. For myself, I almost threw my egg away, for it seemed to have a chick's foetus in it. But I heard a guest, who knew the house's routine of surprises, saying, "There's sure to be something pretty good in here." And so I continued peeling the layers until I came across a tiny but fleshy fig-pecker buried in the midst of a well-peppered yoke.

Here Trimalchio concluded his game, called for everything he had missed, and announced loudly that there was a second cup of mead ready for anyone with an empty glass. Then, abruptly, at a signalling bang from the band, the choric waiters bustled all the course away. But in the scramble a side-dish happened to be bumped to the ground, and a slave lifted it up. Trimalchio perceived the action and commanded that the boy's ears be tweaked and the dish dropped back. A groom came in with a broom and swept up the silver dish as well as the scraps of food spilled from it.

Next two long-haired Ethiopians appeared with small skins similar to those employed for strewing sand down the arena, and drenched our hands with wine. No one was offered anything like vulgar water. We commended the rare refinement of our host's taste.

"Fair play's a jewel," he replied. "So think Mars and I. And that's why each of you had a table to himself. It's less stuffy if the slaves can get about without sweating against each other."

Glass wine-bottles, elaborately sealed with gypsum, came in at the tail of this discourse. There was a label about their necks:

FALERNIAN
CONSUL OPIMIUS
100 YEARS OLD

As we perused these inscriptions Trimalchio slapped his hands together and ejaculated, "Alas, seeing is believing, wine outspans us poor mannikins. So gargle it down. Wine is Life, it is. And this is guaranteed Opimian I'm giving you. I didn't produce anything so good yesterday though the people I had here were much better class than any of you."

We drank, and were felicitating him on his munificent hospitality when a slave whisked out a skeleton made of silver, so craftily articulated in its joints and its backbone that it could be twisted any way. He tumbled this puppet about on the table, making it show off all the contorted postures its hinges permitted. Then he recited pat:

Man's wretched All is nothing but a Cypher.
This is how we'll each appear some day—
Murderer Death's a very certain knifer . . .
So let us all live highly while we may.

When we had got over our applause the second course was served. In luxury it didn't come up to expectations, but its originality had us all staring. . . .

. . . The centre-bit was a square of turf cut entire with its grass, on which rested a honeycomb. An Egyptian boy gave out bread from a little silver oven; and even this fellow squeezed out some excremential ditty in a grating undertone.

Rather downhearted, we began on this common enough fare; but Trimalchio exhorted, "Set to, this is where the eating really begins." As he spoke the music lashed out, four slaves tripped forwards and lifted up the top part of the charger. Then we discovered that underneath there was a second layer made up of stuffed fowls, sows' udders, and a hare garnished with fins so as to represent Pegasus. At the corners of the dish also could be seen little Marsyas-images spouting out from their bladders a stream of piquant sauce over the fish that swam about as if in an estuary's eddies. The servants applauded, and we politely joined in and inclined smilingly towards these appealing dainties.

Trimalchio beamed with hearty enjoyment at the ingenuity of his cook. "Carve-O," he cried; and the carver took up his tools. He worked away in time with the slicing music, rhythmically dissecting the meat and making one think of a war-charioteer jogging into battle to the jig of a barrel-organ.

During all this Trimalchio kept on murmuring to himself, "Carve-O . . . carve-O . . ." I guessed that some jest was intended by this iteration, and ventured to ask the man on my left what it signified. This fellow, who knew all these witticisms by heart, answered, "You see that man carving the victuals, eh? Well, he's been given the name Carvo. So when Trimalchio says Carvo, he utters name and order in one breath."

When I had eaten all I could I turned again to my new acquaintance to get as much information as possible from him. I managed to start some small talk in the hope of some disclosure, and asked him who was the woman gadding about the hall.

"That's Trimalchio's wife," he told me. "Fortunata's her name, and she counts out her fortune by the shovelful."

"And has she always been like that?"

"No offence meant, but she was one of those you wouldn't touch with a pitch-fork if she was windward of you. Now—the gods above know why or what for—she's a heavenly creature and Trimalchio's Can't-do-without. It's this way; if she said it was bedtime at noon he'd lie down and snore on the word. He doesn't know what he's got, he's so crawling with money; but she keeps a shrewd watch on everything, even where you'd least expect it. She's a thrifty housewife, hard as nails, with her eyes screwed on the pennies, though she can talk the leg off an iron pot and has him tied to her apron strings. If she likes you, she likes you; and if she doesn't, she hates you. As for Trimalchio, he's got more estates than a crow could fly over, and simply sweats gold. Why, the silver in his potter's cupboard would make you or me rich. Slaves! don't mention them—I suppose there's not one in ten has even seen him. Just to show you: I give you my oath he could buy up anyone here lock, stock, and barrel and not wink an eyelid.

"And don't think he spends a ha'penny in a shop either. He grows everything he wants himself. Wool, citrons, pepper, hen's milk, if you want it. Ask for any-thing, there you are! Just to show you: a while ago he found that his flocks were going off in the quality of their fleeces. What do you think he did? Bought some pedigree rams from Tarentum and turned them in on to the ewes with a kick in the rear. He wanted some real Attic honey. So over the bees hum from Athens at his orders and the cross-breeding results in a better swarm all round. And just think of this. Three days past he sent to India for some mushroom seed. He wouldn't have a mule in his stables that wasn't foaled of a wild ass. And just take these cushions. The wadding in them is dyed right through the best purple or scarlet. They're all amateurs next to him in extravagance.

"Not that you should make the mistake that his brother freedmen are a stingy mob next to him. They're all lousy with gold. You see the fellow lowest on the lowest couch. To-day he's worth eight hundred thousand or nothing. He was down and out at the start, he humped timber about on his shoulders. How did he manage it? they ask. There's a story, I won't vouch for it, but it goes that he pulled the cap off a bogey and found the hidden treasure. Well, I don't grudge anyone God's help, God knows. He's still bruised with the manumitting cuff, but he's not one to bear malice against himself. So the other day he placarded the door of his hovel with this:

> GAIUS POMPEIUS DIOGENES
> WILL LET THIS ATTIC
> FROM JULY 1ST
> AS HE HAS TAKEN A HOUSE

But what's your opinion of the fellow lounging on the freedman's seat? He was well off, but I don't blame him though his fortune once ran into the millions and he soon played ducks and drakes with it, and now the hair on his head's mort-gaged. And not his fault either, not a bit. He's just a damned nice fellow; but those scoundrelly freedmen pulled the shirt off his back. Take my tip on this: when you've put your last faggot under the pot and when your affairs get shaky on their legs, good-bye friends. An undertaker he was, but a king in the dining-room. Whole carcasses of boars, finicky pastry, game—he had expert bakers and special confectioners. More wine leaked under the table in his house than most men could tap in their cellars. He was like someone out of a book, such an exaggeration. When all was up with him and he feared the duns had got wind of his going on the rocks, he put out an auction notice:

GAIUS JULIUS PROCULUS
SALE OF SURPLUS GOODS.

But Trimalchio here interrupted our pleasant chat. For the course had now been cleared away, and the guests, feeling very happy, were occupied in pouring out wine and general conversation. Leaning on his elbow, he expatiated:

"Now I expect the wine to go to your heads in a steam of bons mots. It's good stuff. Fishes must swim, wine must talk. Do you think I was satisfied with the food you saw partitioned in the charger?

And is Ulysses no more known than this?

Fie now, sirs, we must show we can work our minds as well as our jaws. My reverend patron—peace to his bones—saw to it that I was as good as any man. No one can tell me anything I don't know. Take the amount of learning employed in that dish there for example. That's the Sky, see! and there are twelve gods in it, and each of those emblems stands for one of them. First let's take the Ram. Anyone born under that influence has many cattle, stacks of wool, an iron will, a brazen face, and plenty of bastards. Under it are born schoolmasters, lechers, and suchlike who ram matters into us."

We applauded these astronomical niceties. Whereupon he continued, "Then heaven's dome arches into a Bull. Under him are born irascible galled persons, cowherds, and men who always find their breakfast under a gooseberry bush.

"Under the Twins are brought forth two-wheelers, yoked oxen, testicles, and two-faced women.

"Under Cancer I was born, which is the reason why I stand on so many feet and have estates on land and in the sea. For the crab takes to either with an impartial scuttle. On this account I placed no morsel of food there lest I beshit my horoscope.

"Under the Lion appear cannibals and lordly strutters.

"Under the Virgin come forth females, runaway slaves, and fellows with their legs tied together.

"Libra weighs the fates of butchers, perfumers, and men that think twice.

"The Scorpion brings in poisoners and thugs.

"The Archer ushers in squinters and those who stare the cabbage out of countenance while palming the bacon fat.

"The Goat produces the woebegone streaks of misery whose foreheads get horny with too much thinking.

"Aquarius is over innkeepers and people with water in the top storey.

"The Fishes look after caterers, speechifiers, and such fishy folk.

"So round like a mill wheel rolls the world, always starting some trouble off, with the result that there's continuously someone dying or being born. But the turf with the honeycomb perched on it which you saw in the middle, that too has a moral which I'll tell you. That's Mother Earth which is in the middle of the universe like a monstrous round egg, and with all kinds of good things inside it like honeycomb."

"What a man!" was our unanimous cry of admiration; and with raised hands we swore that he made pygmies of Hipparchus and Aratus. Then the servants approached and over the couches laid out coverlets painted with nets and all the circumstance of the hunt. We were still undecided what this heralded, when outside the dining-hall a bellow sounded, and whisht! some Spartan hounds bounded in out of the noise and began skurrying round and round the table. A charger came in after them, a bulky wild boar dumped upon it, the cap of liberty cocked on its head, with two plaited palm-leaf baskets, one packed with Syrian, the other with Theban dates, hanging from its tusks. Round the boar lay sucking-pigs, snout to teat, moulded out of pastry, to show that it was flesh of a female.

This brood were parting gifts for the guests. The boar was not to be divided by Carvo, the demolisher of the poultry; this dissection was reserved for a big bearded fellow with strips of cloth wound round his legs and a hunting cape of spangly damask. He unsheathed a woodman's knife and buried it with a deft stroke in the boar's side. At the blow a covey of thrushes flew out, to be snared after a few bewildered flutters round the room by fowlers who stood prepared with limed twigs.

Trimalchio ordered each man to receive one and went on, "Now we'll see what pretty kind of acorns our forest porker has been grubbing."

At once boys went to the baskets swinging from the tusks, and gave out some of both the fresh and dry dates to each guest. Meanwhile I had withdrawn into a quiet corner of speculation: why was the boar cockaded with the sign of Liberty? But all the suggestions that drifted through my mind were so idiotic, and as the question persisted in worrying me, I decided to ask my worldly wise neighbour. "O your humble servant," he said, "can interpret that, for there's no catch anywhere. Just the obvious answer. This boar was the prizewinner yesterday, and the guests gave him his dismissal. So he returns to-day a Freed Boar."

I felt a blockhead, and asked no more questions. They would be thinking I had never dined among people of any quality.

A comely lad came twining through our conversation, garlanded with ivy and vine leaves, posing one moment as Bacchus the ecstatic one, then Bacchus Liberator the drunken, then Bacchus rapt among acclaiming Evohes. He moved along, posing and reciting some of his master's compositions with a grasshopper's accent. Trimalchio turned towards the sound and said, "Dionysus, I liberate thee."

The boy lifted the symbolic cap from the boar's head and put it on his own. Trimalchio concluded, "Now you can't say that Bacchus Liberator isn't my father."

We applauded the verbal play, and the boy made the rounds of our kisses. We thoroughly kissed him.

After this dish Trimalchio rose and retired to the privy. His despotic discourse removed, we could talk as we liked, and so started drawing out our neighbours. Dama was the first. He called for full-sized bumpers.

"Day's but a wink of sunlight," he said. "Night pounces on you before you can yawn twice. Consequently I'm all for going to bed straight after dinner. And what freezing weather we've been having, a hot bath hardly peels the chill off me. But give me good liquor before wool any day for keeping cold away; I've lost count of the drinks I've taken, lots and lots anyway, I'm well sozzled, you know, wine gone to head, eh?"

Seleucus followed on. "Do you know what," he said, "I don't wash every day. And I'll tell you why. The bathman's got the walloping hand of a fuller, water's got teeth and bites you, and your strength's scrubbed out of you. But you just give me a tankard or two of mead, and I say to the cold Up-you! And anyhow I couldn't have had a wash to-day, I had to go to a funeral. He was a good fellow— Chrysanthus. A good fellow, and now he's gone. Only a week ago he called out to me in the street. I can hear his voice now. Well, well, how we bladders get up on our hind legs. We're not as good as flies, that's my honest opinion. There's a lot to be said for flies, but what are we if we aren't nothing but a mob of bubbles? And how would he have got on if he hadn't gone on the fasting racket? For five days not a drop of water or a crumb of food passed his lips. And yet he's dead, he's with the majority now. It's my belief the doctors killed him—no, not the doctors, just bad luck. You only call a doctor in to have someone telling you you're getting better. Still, he had a really magnificent funeral. Carried out on a bier with a very expensive pall. The mourning was excellent too—his will freed slaves by the dozens—even if his wife did seem to have to squeeze the tears out. But I dare say he led her a rotten life, it's not for us to judge. Though if you ask me, women one and all are a pack of vampires, it's simply no use to be nice to any of them. You might as well chuck your kindness in the gutter. But worn-out love pinches like an old shoe. . . ."

But Gánymede interrupted this discourse of Phileros, "Here you go on talking about things with no relation whatever to heaven or earth, and not giving a single thought to the way the price of food's rising. Good God, I couldn't find a mouthful of bread to-day. And the damned drought goes on and on and on. The food scarcity's been getting worse and worse for a year now. I say: lynch the commissioners. They're hand in glove with the bakers. Scratch my back and I scratch yours. And so the wretched public pays, while the jaws of the battening capitalists sound a champing carnival of gluttony. . . .

"What will happen if neither god nor man takes pity on our miserable town? May my children turn against me in my old age, if I don't see the hand of providence in all this. The gods are only swear-words now, not a fast is taken seriously, and Jupiter isn't as good as a scarecrow. All people do is to shut eyes and reckon up incomes. In the old days matrons, dressed up in their best, used to climb the hill barefoot, with their hair let down and their hearts clean, and pray to Jupiter for rain. And down it came on the tick, bucketfuls of it; it was then or never, and by the time the women reached their doorsteps they were as soaked as drowned rats. But as things are, we're too clever for all that, and consequently the gods have gout in their feet. And so the fields lie arid—"

"Mercy on us," said Echion, the rag-and-bag merchant, "don't get so worked up about it. Things come and things go, as the farmer said when he lost his piebald pig. Where to-day lacks, to-morrow will have too much—it's always the way. And so we amble on through life. I hold there's nothing wrong with the country. It'd be a fine place if it wasn't for the people in it. And it's not the only one with a dust-heap of troubles of its own. There's no call to be squealing, the skies are as near to you as me or anyone else. If I were to plump you down somewhere away from here, you'd go round telling everybody that roast pork walked in our streets. And don't forget we're soon to have a three days' really tiptop spectacle. It won't be just the ordinary troupe of performing gladiators; most of them will be freedmen. Our friend Titus thinks in a big way, ideas simply bubble out of him once he starts. It'll be something extraordinary, rest assured. I know, because we're bosom friends, and I can tell you he's not doing things half-heartedly. He'll give you the clang of the finest blades. You see. No dodging off; genuine slaughter in the middle of the arena where every one can watch it comfortably. And he can afford it. When his father passed over he left him an odd thirty million. So four hundred thousand won't make much of a hole in his pocket, and he'll establish a reputation out of it that'll outlast him. He's begun collecting a few choice items already. Clowns, a woman who'll fight from a chariot, and Glyco's steward who was caught doing his best to please Glyco's wife. You'll see squabbles among the audience when they take sides, cuckolds versus lovebirds. But Glyco, a fellow of the most measly means, is throwing his steward to the wild animals. Sheer waste, only to give himself away. It's not as if it was the slave's fault. Didn't he have to obey when his mistress said: come here? It's the slut of a wife who deserves to be gored by the bull. But if a man's scared to touch his donkey, he lays bravely into the saddle. And how was Glyco such an imbecile as to think any sprig of Hermogenes' would come to a good end? I'd put nothing beyond that man, he'd pare the nails of a hawk in full flight. And snakes don't beget harmless hanks of rope. Glyco, the fool Glyco, has befouled his own bed and he'll carry the stink about on him—death's the only soap that'll wash it off. Still, no man's a saint, and it's his own look-out. . . .

Conversations of this sort were rubbing sides together everywhere when Trimalchio returned. He mopped his brow, and laved his hands in unguents. After a stertorous pause, he spoke:

"My excuses, gentlemen. For several days now my bowels have stopped work; none of the doctors can diagnose the case. However, a concoction of pomegranate-rind and resin stewed in vinegar has relieved me somewhat. I hope my stomach will now learn to behave itself better; once I never had to say a harsh word to it. Besides, such gurglings rumble inside me you'd think I had a bull there. So if any of you want to retire too, don't let him be shy about it. There's none of us made of cast iron, and I hold there's no torture on earth like sitting tight on a griping desire to get to the stool. It's the one thing Jupiter himself daren't stand in the way of. What are you giggling at, Fortunata? Aren't you the one that's always waking me by getting up in the middle of the night? I wouldn't say a word if someone even took his couch for a privy. Doctors assure us it's most dangerous to hold it in. But if the matter gets really serious, outside you'll find every little thing your heart could want—water and towels and close-stools. Believe me, the fumes from constipation go straight to the brain and put a rot through the whole body. I've known lots of people who've died that way, and all because they refused to face the state they were in."

We thanked him for the generous care he had for our well-being, and hurriedly poured wine down our throats to keep the laughter from spluttering out. We had not yet realized that this was merely a lull in the festivities and that we were, as they say, with the stiffest climb before our appetites. The tables were cleared to the clatter of the music, and three white pigs, muzzled and hung with capering bells, were pulled into the room. "This one's two years old," the keeper informed us, "this one's three years, and the other's a six years' growth of bacon." I imagined we were to be given a tumbler's display, and the pigs would perform some antics as you see them at the street corners. But Trimalchio put an end to all suppositions by asking:

"Which of these would you like to come back as the next course in a minute or two? Any bumpkin cook can serve you a fowl or a hash *à la* Pentheus or such like gallimaufries; but a cook of mine thinks sending up a calf boiled whole in a cauldron all in the day's work."

Whereupon he bade them fetch the cook at once, and without waiting for any vote to be taken gave orders for the oldest pig to be stuck. Then, lifting up his voice, he turned to the cook.

"What subdivision of the household do you belong to?"

The cook replied that he belonged to the Fortieth.

"Did I buy you or were you whelped on the estate?"

"Neither, sir, I was left to you by the will of Pansa, sir," said the cook.

"Well, then," said Trimalchio, "take heed that you dress this meat specially nicely, or I'll have you degraded to the messenger-boy corps."

The man, duly admonished on his master's power, bore the pig off to the kitchen. This stern task done, Trimalchio beamed round upon us. . . .

Before he had finished speaking the cook's back was bared and he stood there shivering between the two floggers. Everyone commenced to put in a word for the defaulter. A mistake anybody might make, they said, let him off this time and if he does it again there's not one of us will lift a finger to save him. For myself, I felt none of this tenderness towards him. Indeed, I could not restrain my indignation at the scene, and leaning across I whispered to Agamemnon, "This must be a perfectly hopeless fellow; hang him, how could the veriest idiot forget to clean a pig out? I think a whipping would be his due if he'd sent in a sprat ungutted."

Not so Trimalchio. A large grin crackled up his face. "Well then," he said tolerantly, "as you've such a bad memory, out with the beast's bowels here where we can keep an eye on you."

The cook scrambled back into his shirt, snatched up a knife, and cut at the pig's belly with a shaky hand. Immediately the slashes yawned open with the pressure from behind and out wriggled a mass of sausages and black puddings.

At this the slaves went off into clockwork howls of applause and shouted, "That was a good one, Gaius." The cook too was rewarded with a bumper and a silver crown; and it was on a Corinthian dish that the goblet was presented to him. Noticing that Agamemnon was looking uninterestedly at this piece of ware, Trimalchio announced, "I'm the only man in the world with genuine Corinthian."

I expected that with his usual rash vanity he would declare he had his cups imported direct from the Corinthian workshops. But that was to underestimate the man.

"Perhaps," he proceeded, "you're saying to yourselves: how is it Trimalchio has the only genuine Corinthian? . . ."

As he was rambling on a boy let a tankard fall. Trimalchio glared round in his direction. "It seems you can't keep your head, go and have it removed then, quick!" he remarked.

The boy's lip fell, he quavered out an appeal for his life.

"What's wrong now?" demanded Trimalchio. "I only want you to take full precautions for not repeating your clumsiness. It's not my fault you're scatter-brained." However, we all came in as an expostulating chorus and prevailed on him to pardon the culprit. As soon as he was declared pardoned, the lad ran gleefully round the table.

Then Trimalchio shouted, "Pour water out, but pour wine in!" We took up the jest as intended; Agamemnon most conspicuously, for he was a past master in getting himself reinvited to dinner. But the fumes of flattery had mounted to Trimalchio's head, and he kept drinking excitedly, till at last in a thick and tipsy voice he exclaimed, "But nobody's yet asked dear old Fortunata to give us a dance. Take my word for it, she's a proper twinkletoes when she gets going."

At which he waggled his arms archly about over his head and gave us an exhibition dance *à la* Syrtus the actor while all the slaves chanted in unison:

O you Medea, you pretty Medea.

He had almost waggled himself out into the middle of the room when Fortunata went up and whispered in his ear—I suppose that he was demeaning his dignity. You could never be sure how he would take such things. At one moment he was afraid of her, the next he would let himself go without thinking twice about her.

Anyhow, he was completely diverted from these gambols by the steward of his estate entering and glibly droning out as though from the minutes of a County Council:

"*July 26th* Cumae, Trimalchio's estate, 30 boys and 40 girls born, 500,000 pecks of wheat loaded from the threshing-side to the barns, 500 oxen broken in.

"*Ditto* Slave Mithridates crucified for saying: Gaius be damned.

"*Ditto* 10,000,000 sesterces non-invested profits put back into the reserve capital.

"*Ditto* Fire occurred in gardens at Pompeii, starting from house of the bailiff Nasta—"

"Hey there," interrupted Trimalchio, "when did I buy any gardens at Pompeii?"

"Last year," replied the clerk, "consequently they are not yet entered up in the ledgers."

Trimalchio's face flamed with anger. "I'll not pass the receipts for any property," he roared, "bought in my name unless I'm notified of it within six months."

We then had a list of by-laws compiled for the Civil Service officers; the wills of some foresters in which they disinherited Trimalchio in a fawning codicil; then a roll-call of bailiffs; then the divorce case of a freedwoman, a night-watchman's wife, who had been surprised underneath a bathman; the case of a janitor exiled to Baiae; a prosecution against a steward; and the decision in a dispute among some valets.

But at last the acrobats bounded in. A slapstick jokester stood lamely by, propping up a ladder and making a small boy hop from rung to rung and skip about on the top in time to some popular airs, and then jump through blazing hoops and lift up a large wine jug with his teeth. Everyone was bored but Trimalchio, who shook his head and said, "Ah, it's a thankless job, poor fellows. It's my opinion there's only two things on the whole earth that's really worth going across the road to see, and one's acrobats and the other's a tune on the trumpets. All other shows are twaddle. I know, because I once bought an entire Greek Comedy Company and I'd never let them do anything but the good old Roman vaudeville. Look here, I said to the conductor, I want something with a tune in it—"

He was just warming to the theme when the acrobat's boy landed on top of him; the ladder had slipped. The slaves wailed with horror, and so did all the rest of us—not on behalf of the wretched lad, whose broken neck we would have cheered heartily in itself, but because we had no wish to terminate the dinner prematurely in mourning for the death of someone we didn't even know.

Trimalchio groaned with the groans of a dying man, gingerly nursing his arm as if it were damaged for life. Doctors rushed up from all sides with Fortunata jostling through them, her hair pulled down over her face, a cup in her hand, crying that she was the most miserably afflicted unfortunate woman in the world. As for the boy who had caused the mischief by slithering on to him, he was crawling round among our feet by this, puling for mercy. But I suspected darkly that his prayers were only another ruse to betray us into a bad joke. The memory of the cook who had neglected to gut the pig was still rankling. So I first made a detailed inspection of the dining-room, to see if I could detect any surprise apparatus about to emerge out of the walls. My suspicions were heightened when I saw that blows were being administered to a slave who had irreverently wrapped the bruise on his master's arm with white wool instead of purple. And I was not altogether out in my surmises. The decree finally came that so far from being punished, the boy should be liberated, so that no one should have the chance to say the great man had been wounded by a slave.

The necessary applause was given, and we expressed various banalities on the mutability of human affairs.

"Ah," Trimalchio was struck with an idea, "such an episode must have its epigram." He called at once for his tablets and with scarcely a knitting of the brow dashed off these hobbling verses:

> You always get the thing you don't expect
> Whether the weather's tempestuous or fine—
> Fortune's the lord of all, to save or wreck,
> So quickly pass me the Falernian wine.

. . . It was too much altogether for Ascyltos. He gave up trying to applaud decorously, lifted up his hands in helpless derision and laughed till the tears wet his cheeks. One of Trimalchio's fellow freedmen, the diner whose place was next above mine, took umbrage at his behaviour.

"What are you guffawing at, muttonpate?" he burst in. "Isn't our host's entertainment up to your standard? I suppose you're so much better off and used to a sweller type of evening, eh?

"As I hope the Spirits of this House may love me, I'd have put a muzzle on his bleating jaws by this if I were sitting next him. A fine condescending blot on humanity he is, to be mocking at his betters—he's only a good-for-nothing bit of raggletaggle night-scum. If I could get him in a corner he wouldn't know his own face. By heavens, it takes a lot to work me up as a rule, but far-gone meat breeds worms and if you carry things too far you must abide the result. There he is laughing. What at? What's he got to laugh about? Was he born under a lucky star?

"A Roman Knight, are you? Well, I'll have you know I'm the son of a King. How did I become a slave then? Because I went into service of my own accord, I preferred the chance of becoming a Roman citizen to being blue-blooded with taxes to pay. And now there's my life spread out like an open book, and there's no one, I trust, who can point to a speck on it. I'm a man among men, I walk with my head uncovered like the free man I am, I owe no man a penny. I've never had a writ served on me, you've never heard a man come up to me out o' doors and say, 'What about that money you owe me?' I've invested in a little stretch of land and got a few valuables as ready capital, I've twenty bellies and a dog to feed. I bought the old woman's freedom so that no one's the right to lift a hand on her but myself. My own freedom cost me a thousand in silver. I was elected a Warden of the worship of Augustus with remission of fees, and it's my earnest hope that when I'm dead I won't have a single thing to look back on with a blush. But you're so busy hustling along you've no time to look behind you. You see the louse on your neighbour, not the bug on yourself. There's no one here thinks we're funny but you. There's your master, glance at him, wiser and older than you, and he's satisfied with us. You're still an unweaned titterer, you can't squawk out *mu mu ma* yet, you lump of sloppity clay, like a strip of wash-leather all soggy, softer if anything, that's all. Are you richer than I am? Then have two breakfasts a day, and two dinners. I'd rather have my good name than a sackful of gold. To put it bluntly, who has ever had to ask me for anything twice? I was slave for forty years, yet no one knew for sure whether I was a slave or free. When I came here I was a little boy with long curly hair; they hadn't started putting up the Town Hall then. I did my level best to be a good servant to my master, and a fine upstanding gentleman he was, his little finger was worth more than the whole of you. And there were others in the house who tried to catch me tripping, and weren't above giving a push for that purpose. But I won through, thanks be to the Spirit of the House! That's something to boast about, but any fool can be born free. Well now, what are you gaping at, you, like a goat stuck in the middle of a field full of clover!"

It was here that Gito, who was standing near my feet, exploded into laughter, all the more raucous because he had been holding it in for some time. Ascyltos' abuser caught the sound and turned fiercely on him. "Ho," he said, "you're laughing too, you little mop-haired turnip? This is a Saturnalia. Is it December yet that the slaves are so merry? When did you pay the five per cent. to the Treasury for being freed? He doesn't know what to do, the gallows' fledgling, the little crows' snack. He'll take the pair of you, you and your master who can't make you behave yourself. May all food turn to dust in my mouth, but I'd make you sit up if I wasn't so polite in company. We're getting on finely, all except those ticklish asses who can't keep you in your place. Like master, like man: it's a sound adage. Why, it's all I can do to hold myself in, and I'm a quiet and easy-tempered man by nature; but start me going properly and I'll knock my own mother down if she gets in the way. No matter; I'll meet you somewhere outside, you insect, you sprout of a muddy puddle. I shan't begin to live again till I've rubbed your master's nose in a nettle-bed, and I shan't spare you either, believe me, though you call down more gods than are in heaven. Those sweet corkscrew curls shan't save you, no, nor that threepenny worth of dog's meat, your master. O ho, somebody with big boots on is going to tread on you. Either I don't know my own name or you're soon to cackle on the other side of your face though there's a goldy down on your cheeks like a godkin's. I'll draw Athene's anger down whack on to you and the man who made a pretty boy out of you. . . ."

On Roman Agriculture

De re rustica, written in the first century A.D., is the work of a Spanish "Roman" of Cadiz (Gades), Lucius Junius Moderatus Columella. It is the most complete classical treatise on agriculture. The authority of Columella on his subject was great. He possessed estates in either Spain or Sardinia and traveled extensively in other imperial grain-producing areas, although he was himself an absentee landlord who spent most of his life at Rome. He was thus directly familiar with the agrarian basis of Roman society, patrician absentee ownership of vast estates (*latifundia*), and the various systems of labor then prevailing.

The fundamental image of agrarian society in the provinces of the Empire presented in *De re rustica* is not benign. The military conscription of an expanding Empire forced peasants off their holdings. Those who could not sell land let it go to voracious patricians and freedmen anxious to build estates legally or by force, fraud, and opportunism. Farmers who remained on the land could not compete with large holders. Thus began the cycle by which many impoverished farmers went bankrupt and fled the countryside to Rome to swell the proletarian population living by bread and circuses. Those who remained behind as tenant farmers were semi-dependent cultivators living among the oppressed slave labor force which constituted the bulk of agrarian labor. The lives of slaves and tenants were unfree in various ways, dictated by the managerial elements employed by the estate owners. While Columella advocated the benefits of residence on one's estates, he was not moved to take his own advice. Indeed, even his famous "generosity" shown toward tenants and slaves appears in his own writings to be motivated by the profits of rational management. His treatise is more distinguished by its verse form than by humanity. We can read in it the origins of what Marc Bloch called "the rise of dependent cultivation" in the Roman Empire.

From *De re rustica*

Columella

7. After all these arrangements have been acquired or contrived, especial care is demanded of the master not only in other matters, but most of all in the matter of the persons in his service; and these are either tenant-farmers or slaves, whether unfettered or in chains. He should be civil in dealing with his tenants, should show himself affable, and should be more exacting in the matter of work than of payments, as this gives less offence yet is, generally speaking, more profitable. For when land is carefully tilled it usually brings a profit, and never a loss, except when it is assailed by unusually severe weather or by robbers; and for that reason the tenant does not venture to ask for reduction of his rent. But the master should not be insistent on his rights in every particular to which he has bound his tenant, such as the exact day for payment, or the matter of demanding firewood and other trifling services in addition, attention to which causes country-

SOURCE: Columella, *De re rustica*, vol. 1, trans. Harrison B. Ash (Cambridge, Mass.: Harvard University Press, The Loeb Classical Library, 1941), pp. 79–101.

folk more trouble than expense; in fact, we should not lay claim to all that the law allows, for the ancients regarded the extreme of the law as the extreme of oppression. On the other hand, we must not neglect our claims altogether; for, as Alfius the usurer is reported to have said, and with entire truth, "Good debts become bad ones if they are not called." Furthermore, I myself remember having heard Publius Volusius, an old man who had been consul and was very wealthy, declare that estate most fortunate which had as tenants natives of the place, and held them, by reason of long association, even from the cradle, as if born on their own father's property. So I am decidedly of the opinion that repeated letting of a place is a bad thing, but that a worse thing is the farmer who lives in town and prefers to till the land through his slaves rather than by his own hand. Saserna used to say that from a man of this sort the return was usually a lawsuit instead of revenue, and that for this reason we should take pains to keep with us tenants who are country-bred and at the same time diligent farmers, when we are not at liberty to till the land ourselves or when it is not feasible to cultivate it with our own servants; though this does not happen except in districts which are desolated by the severity of the climate and the barrenness of the soil. But when the climate is moderately healthful and the soil moderately good, a man's personal supervision never fails to yield a larger return from his land than does that of a tenant—never than that of even an overseer, unless the greatest carelessness or greed on the part of the slave stands in the way. There is no doubt that both these offences are either committed or fostered through the fault of the master, inasmuch as he has the authority to prevent such a person from being placed in charge of his affairs, or to see to it that he is removed if so placed. On far distant estates, however, which it is not easy for the owner to visit, it is better for every kind of land to be under free farmers than under slave overseers, but this is particularly true of grain land. To such land a tenant farmer can do no great harm, as he can to plantations of vines and trees, while slaves do it tremendous damage: they let out oxen for hire, and keep them and other animals poorly fed; they do not plough the ground carefully, and they charge up the sowing of far more seed than they have actually sown; what they have committed to the earth they do not so foster that it will make the proper growth; and when they have brought it to the threshing-floor, every day during the threshing they lessen the amount either by trickery or by carelessness. For they themselves steal it and do not guard against the thieving of others, and even when it is stored away they do not enter it honestly in their accounts. The result is that both manager and hands are offenders, and that the land pretty often gets a bad name. Therefore my opinion is that an estate of this sort should be leased if, as I have said, it cannot have the presence of the owner.

8. The next point is with regard to slaves—over what duty it is proper to place each and to what sort of tasks to assign them. So my advice at the start is not to appoint an overseer from that sort of slaves who are physically attractive, and certainly not from that class which has busied itself with the voluptuous occupations of the city. This lazy and sleepy-headed class of servants, accustomed to idling, to the Campus, the Circus, and the theatres, to gambling, to cookshops, to bawdy-houses, never ceases to dream of these follies; and when they carry them over into their farming, the master suffers not so much loss in the slave himself as in his whole estate. A man should be chosen who has been hardened by farm work from his infancy, one who has been tested by experience. If, however, such

a person is not available, let one be put in charge out of the number of those who have slaved patiently at hard labour; and he should already have passed beyond the time of young manhood but not yet have arrived at that of old age, that youth may not lessen his authority to command, seeing that older men think it beneath them to take orders from a mere stripling, and that old age may not break down under the heaviest labour. He should be, then, of middle age and of strong physique, skilled in farm operations or at least very painstaking, so that he may learn the more readily; for it is not in keeping with this business of ours for one man to give orders and another to give instructions, nor can a man properly exact work when he is being tutored by an underling as to what is to be done and in what way. Even an illiterate person, if only he have a retentive mind, can manage affairs well enough. Cornelius Celsus says that an overseer of this sort brings money to his master oftener than he does his book, because, not knowing his letters, he is either less able to falsify accounts or is afraid to do so through a second party because that would make another aware of the deception.

But be the overseer what he may, he should be given a woman companion to keep him within bounds and yet in certain matters to be a help to him; and this same overseer should be warned not to become intimate with a member of the household, and much less with an outsider, yet at times he may consider it fitting, as a mark of distinction, to invite to his table on a holiday one whom he has found to be constantly busy and vigorous in the performance of his tasks. He shall offer no sacrifice except by direction of the master. Soothsayers and witches, two sets of people who incite ignorant minds through false superstition to spending and then to shameful practices, he must not admit to the place. He must have no acquaintance with the city or with the weekly market, except to make purchases and sales in connection with his duties. For, as Cato says, an overseer should not be a gadabout; and he should not go out of bounds except to learn something new about farming, and that only if the place is so near that he can come back. He must allow no foot-paths or new crosscuts to be made in the farm; and he shall entertain no guest except a close friend or kinsman of his master.

As he must be restrained from these practices, so must he be urged to take care of the equipment and the iron tools, and to keep in repair and stored away twice as many as the number of salves requires, so that there will be no need of borrowing from a neighbour; for the loss in slave labour exceeds the cost of articles of this sort. In the care and clothing of the slave household he should have an eye to usefulness rather than appearance, taking care to keep them fortified against wind, cold, and rain, all of which are warded off with long-sleeved leather tunics, garments of patchwork, or hooded cloaks. If this be done, no weather is so unbearable but that some work may be done in the open. He should be not only skilled in the tasks of husbandry, but should also be endowed, as far as the servile disposition allows, with such qualities of feeling that he may exercise authority without laxness and without cruelty, and always humour some of the better hands, at the same time being forbearing even with those of lesser worth, so that they may rather fear his sternness than detest his cruelty. This he can accomplish if he will choose rather to guard his subordinates from wrongdoing than to bring upon himself, through his own negligence, the necessity of punishing offenders. There is, moreover, no better way of keeping watch over even the most worthless of men than the strict enforcement of labour, the requirement that the proper tasks be performed and that the overseer be present at all times; for in that case the foremen in charge of the several operations are zealous in carrying out their duties, and the others, after their fatiguing toil, will turn their attention to rest and sleep rather than to dissipation.

Would that those well-known precepts, old but excellent in morality, which have now passed out of use, might be held to to-day: That an overseer shall not employ the services of a fellow-slave except on the master's business; that he shall partake of no food except in sight of the household, nor of other food than is provided for the rest; for in so doing he will see to it that the bread is carefully made and that other things are wholesomely prepared. He shall permit no one to pass beyond the boundaries unless sent by himself, and he shall send no one except there is great and pressing need. He shall carry on no business on his own account, nor invest his master's funds in livestock and other goods for purchase and sale; for such trafficking will divert the attention of the overseer and will never allow him to balance his accounts with his master, but, when an accounting is demanded, he has goods to show instead of cash. But, generally speaking, this above all else is to be required of him—that he shall not think that he knows what he does not know, and that he shall always be eager to learn what he is ignorant of; for not only is it very helpful to do a thing skilfully, but even more so is it hurtful to have done it incorrectly. For there is one and only one controlling principle in agriculture, namely, to do once and for all the thing which the method of cultivation requires; since when ignorance or carelessness has to be rectified, the matter at stake has already suffered impairment and never recovers thereafter to such an extent as to regain what it has lost and to restore the profit of time that has passed.

In the case of the other slaves, the following are, in general, the precepts to be observed, and I do not regret having held to them myself: to talk rather familiarly with the country slaves, provided only that they have not conducted themselves unbecomingly, more frequently than I would with the town slaves; and when I perceived that their unending toil was lightened by such friendliness on the part of the master, I would even jest with them at times and allow them also to jest more freely. Nowadays I make it a practice to call them into consultation on any new work, as if they were more experienced, and to discover by this means what sort of ability is possessed by each of them and how intelligent he is. Furthermore, I observe that they are more willing to set about a piece of work on which they think that their opinions have been asked and their advice followed. Again, it is the established custom of all men of caution to inspect the inmates of the workhouse, to find out whether they are carefully chained, whether the places of confinement are quite safe and properly guarded, whether the overseer has put anyone in fetters or removed his shackles without the master's knowledge. For the overseer should be most observant of both points—not to release the shackles from anyone whom the head of the house has subjected to that kind of punishment, except by his leave, and not to free one whom he himself has chained on his own initiative until the master knows the circumstances; and the investigation of the householder should be the more painstaking in the interest of slaves of this sort, that they may not be treated unjustly in the matter of clothing or other allowances, inasmuch as, being liable to a greater number of people, such as overseers, taskmasters, and jailers, they are the more liable to unjust punishment, and again, when smarting under cruelty and greed, they are more to be feared. Accordingly, a careful master inquires not only of them, but also of those who are not in bonds, as being more worthy of belief, whether they are receiving what is due to them under his instructions; he also tests the quality of their food and drink by tasting it himself, and examines their clothing, their mittens, and their foot-covering. In addition he should give them frequent opportunities for making complaint against those persons who treat them cruelly or dishonestly. In fact, I now and then avenge those who have just cause for grievance, as well as punish those who incite the slaves to revolt, or who slander their taskmasters; and, on the other hand, I reward those who conduct themselves with energy and diligence. To women, too, who are unusually prolific, and who ought to be rewarded for the bearing of a certain number of offspring, I have granted exemption from work and sometimes even freedom after they had reared many children. For to a mother of three sons exemption from work was granted; to a mother of more her freedom as well.

Such justice and consideration on the part of the master contributes greatly to the increase of his estate. But he should also bear in mind, first to pay his respects to the household gods as soon as he returns from town; then at once, if time permits, if not, on the next day, to inspect his lands and revisit every part of them and judge whether his absence has resulted in any relaxation of discipline and watchfulness, whether any vine, and tree, or any produce is missing; at the same time, too, he should make a new count of stock, slaves, farm-equipment, and furniture. If he has made it a practice to do all this for many years, he will maintain a well-ordered discipline when old age comes; and whatever his age, he will never be so wasted with years as to be despised by his slaves.

9. Something should be said, too, as to what tasks we think each kind of body or mind should be assigned. As keepers of the flocks it is proper to place in charge men who are diligent and very thrifty. These two qualities are more important for this task than stature and strength of body, since this is a responsibility requiring unremitting watchfulness and skill. In the case of the ploughman, intelligence, though necessary, is still not sufficient unless bigness of voice and in bearing makes him formidable to the cattle. Yet he should temper his strength with gentleness, since he should be more terrifying than cruel, so that the oxen may obey his commands and at the same time last longer because they are not worn out with the hardship of the work combined with the torment of the lash. But what the duties of shepherds and herdsmen are, I shall treat again in their proper places; for the present it is sufficient to have called to mind that strength and height are of no importance in the one, but of the greatest importance in the other. For, as I have said, we shall make all the taller ones ploughmen, both for the reason I have just given and because in the work of the farm there is no task less tiring to a tall man; for in ploughing he stands almost erect and rests his weight on the plough-handle. The common labourer may be of any height at all, if only he is capable of enduring hard work. Vineyards require not so much tall men as those who are broad-shouldered and brawny, for this type is better suited to digging and pruning and other forms of viticulture. In this department husbandry is less exacting in the matter of thrift than in the others, for the reason that the vine-dresser should do his work in company with others and under supervision, and because the unruly are for the most part possessed of quicker understanding, which is what the nature of this work requires. For it demands of the helper that he be not merely strong but also quick-witted; and on this account

vineyards are commonly tended by slaves in fetters. Still there is nothing that an honest man of equal quickness will not do better than a rogue.

I have inserted this that no one may think me obsessed of such a notion as to wish to till my land with criminals rather than with honest men. But this too I believe: that the duties of the slaves should not be confused to the point where all take a hand in every task. For this is by no means to the advantage of the husbandman, either because no one regards any particular task as his own or because, when he does make an effort, he is performing a service that is not his own but common to all, and therefore shirks his work to a great extent; and yet the fault cannot be fastened upon any one man because many have a hand in it. For this reason ploughmen must be distinguished from vine-dressers, and vine-dressers from ploughmen, and both of these from men of all work. Furthermore, squads should be formed, not to exceed ten men each, which the ancients called *decuriae* and approved of highly, because that limited number was most conveniently guarded while at work, and the size was not disconcerting to the person in charge as he led the way. Therefore, if the field is of considerable extent, such squads should be distributed over sections of it and the work should be so apportioned that men will not be by ones or twos, because they are not easily watched when scattered; and yet they should number not more than ten, lest, on the other hand, when the band is too large, each individual may think that the work does not concern him. This arrangement not only stimulates rivalry, but also it discloses the slothful; for, when a task is enlivened by competition, punishment inflicted on the laggards appears just and free from censure.

But surely, in pointing out to the farmer-to-be those matters for which especial provision must be made—healthfulness, roads, neighbourhood, water, situation of the homestead, size of the farm, classes of tenants and slaves, and assignment of duties and tasks—we have now come properly, through these steps, to the actual tilling of the soil; of this we shall presently treat at greater length in the book that follows.

CHAPTER THREE

The Ethics of Total Subordination: Sex and Slavery

On Slaves and Metics

In the classical societies of Europe slavery was everywhere in A.D. 1. Its survival in Christian society was a problem for the church fathers. Romans had less trouble with it: Varro said free labor was best in unhealthy posts, because the freeman's death cost his employer nothing. Nor were slaves foreign to the Greek experience. Indeed, despite recent debates about its importance in Greek economic life, it is clear that slaves formed the broad base of the Hellenic and Hellenistic social pyramids. Their economic functions were various. In both public and private life, the worst jobs fell to them: galley rowing, the mines, and the heaviest labor in some industries. They also performed jobs deemed undignified for freemen—sanitation and police work! Absolute control was exercised over their lives, since they were in law chattels and in theory mere living tools. Thus, even if the Greek world possessed nothing like the vast masses of slave gangs working *latifundia*, slavery was so common as to make ownership an attribute of social status. Whether the institution was profitable or not is another matter. Slaves were a form of capital with only a modest yield: they required upkeep even when ill and not productive. History records their flights and rebellions, as if to mark the alienation of capital from masters, in contrast to Marx's celebrated sentence about the alienation of workers from the means of production.

Alongside slaves worked the numerous resident aliens or metics. Barred from political life, they shared only the burdens of citizenship. But, unlike the slaves, metics (those who "live with" citizens) had social rights and a status clearly protected in law, including the privilege of marriage with citizens. They were discriminated against in various ways, however, especially in legal procedures. Yet they could perform certain public offices and were allowed a wide latitude in the crafts and commerce. They were thus outsiders whose place in Athenian society was in many ways comparable to that of Jews in medieval, Christian Europe.* The word itself had a pejorative meaning and reflected some of the racial feeling carried by the word *barbarian*.

Metics and slaves were the special concern of the social historian Robert Flacelière in the following selection from *Daily Life in Greece at the Time of Pericles*.

---

* See page 321.

From *Daily Life in Greece at the Time of Pericles*

Robert Flacelière

Metics

The ability of the Athenians to devote so large a proportion of their time to the city's political affairs was due to the fact that many of them were freed from all domestic activities by the other two classes in their society: resident aliens, and slaves. The city-state of antiquity, as I said at the beginning of this chapter, was totalitarian by nature: totalitarian in its curtailment of individual liberty among citizens, and even more totalitarian in all its dealings with foreigners, whom it regarded, *a priori*, as enemies. Any foreigner living in a Greek city was more often than not a prisoner-of-war, a slave. At Sparta there took place a periodical expulsion of foreigners, known as a *xenelasia*. Athens was more liberal in her attitude, and allowed numerous non-Athenian Greeks to reside within her boundaries—even to enjoy quite appreciable rights there. These domiciled foreigners were known as "metics" (those who "live with"): it is hardly surprising that this word, like that other Greek term "barbarian," used to designate all non-Greeks, has kept its pejorative sense down the centuries. The intense national pride felt by each individual city-state is more than enough to explain it. Most Athenian metics were Greek by birth, though Phoenicians, Phrygians, Egyptians, and even Arabs were sometimes found amongst them.

Metics formed a very considerable proportion of the total Athenian population during the fifth century: there were about 20,000 of them, or something like half the total number of citizens. They were liable to nearly all an ordinary citizen's financial obligations, including the bulk of the *leitourgiai* (public services); for instance, the *chorégia* associated with the Lenaean Festival. Those chosen by the archon were required to maintain and rehearse a dramatic chorus for this festival, at their own expense. On the other hand, they were exempt from the *trierarchia*, since this involved the command of a warship. Metics also paid a special tax—a very light one, it must be admitted—called the *metoikion:* twelve drachmas per annum for men, and six for women, or the equivalent of six and three days' work respectively. Metics were debarred from ephebic training, but they were allowed to use the public *gymnasia* (from which slaves were excluded); they served in the Athenian army as hoplites or light infantry, and above all in the fleet, as rowers; their main military function was to help in the territorial defence of Attica. Marriages between citizens and metics were, doubtless, sanctioned by law; but from 451 B.C. onwards even the child of an Athenian father and a resident alien mother did not qualify for citizenship, much less one born of an Athenian woman who had married a metic. They could acquire household chattels and own slaves, but were not permitted to buy houses or property, unless granted this right of acquisition (*enktesis*) as a very exceptional favour. In such cases it was generally accompanied by another privilege, *isotelia*, which, from the financial viewpoint, put them on exactly the same footing as any ordinary citizen.

Legally they could be put to the torture, though this provision was seldom

SOURCE: Robert Flacelière, *Daily Life in Greece at the Time of Pericles*, trans. P. Green (New York: The Macmillan Company; London: George Weidenfeld & Nicolson, 1965), pp. 55–83.

enforced. If they appeared in court they were represented by a citizen who acted as their *prostatés*, or patron. A man who murdered a metic could be punished by exile, but not executed, as he was liable to be for the murder of a citizen: it follows that the law did not regard a citizen's and a metic's life as of exactly equal value. Their possessions, however, were safeguarded by the *polemarchos*, whose word was final in all lawsuits where metics were involved. Resident aliens were entirely at liberty to celebrate the cults of their own country, and for this purpose could form religious associations known as *thiasoi*. Such foreign deities as the Thracian goddess Bendis or the Great Mother of Phrygia won numerous converts amongst the Athenians themselves. A special place was found for metics in the celebration of certain official festivals, such as the Hephaesteia or the Panathenaic Games, where they appeared side by side with the allies and cleruchies (that is, groups of citizens resident in an Athenian colony). We may conclude, then, that metics were treated fairly liberally in Athens—at least as compared with the attitude taken towards them by Sparta and many other Greek cities— and we can more readily understand the claim that Pericles makes in the pages of Thucydides: "Our city is open to all men: we have no law that requires the expulsion of the stranger in our midst, or debars him from such education or entertainment as he may find among us."

The metic population was distributed through the demes, and thus incorporated, for administrative purposes, with the main citizen body; but, needless to say, metics possessed no political rights. They were allowed to perform certain strictly subordinate public duties, such as those of town crier, public medical officer, tax farmer, or public works contractor; but most of them were in business or industry, or else practised what we nowadays call the liberal professions. They were often to be found in industrial crafts, particularly weaving, tanning, pottery-making and metal-working. In the last-named field, indeed, they appear to have enjoyed a virtual monopoly. . . .

Metics also took first place in trade, both retail and wholesale. They dealt extensively in textiles, grain, and vegetable products themselves: they also acted as middlemen and import-export agents. It was a metic, nearly always, who organized the cargoes which reached Athens: timber from Macedonia, gold-leaf from the Orient, wheat and salted fish from the Black Sea. The largest salt-curing business in Athens was that owned by Chaerephilus and his sons: Chaerephilus himself was to win Athenian citizenship, and make an ex-voto offering at Delphi afterwards with this inscription: "In fulfilment of a vow, Chaere-philus, son of Pheidias, an Athenian, has set up this offering to Pythian Apollo." But we know, from other evidence, that he was not Athenian by birth. The biggest bankers (*trapezitai*) in Athens were either metics or ex-slaves, who acquired metic status on enfranchisement.

Many metics, having thus by their business activities earned themselves a comfortable sufficiency (or even, at times, considerable wealth) were able to give their sons a first-class education, so that afterwards they shone in one of the professions, as artist, doctor, or public speaker. For instance, the Syracusan arms manufacturer Cephalus, already mentioned above, had his son Lysias educated with the best families in Athens, and in due course the boy became a well-known orator. But there were also many gifted men, men who had already acquired fame in their own country, who came to Athens, drawn by the city's incomparable brilliance and the certain fulfilment it offered to their talents. Often they settled there for good. Among such we may note three great painters: Polygnotus of Thasos, Zeuxis of Heraclea, and Parrhasius of Ephesus. Hippocrates of Cos, the Father of Medicine, enjoyed a great success in Athens; and the Father of History, Herodotus of Halicarnassus, gave public readings (or, as we would say, lectures) there, spending long years as a resident before he finally took ship to Southern Italy, as one of the pan-Hellenic colonists of Thurii—a project approved by Pericles and under Athenian organization.

To judge from Pericles' law of 451, aimed at restricting any extension of the citizenship, he might well have had xenophobic tendencies; but in fact he surrounded himself with metics, including his teacher, Anaxagoras of Clazomenae, and his mistress, Aspasia of Miletus. The architect of the Piraeus, Hippodamus (also a Milesian), and Phaeinus the astronomer, Meton's teacher, were likewise metics. As for the "Sophists," itinerant scholars or instructors in the art of rhetoric who stopped off at Athens and sometimes stayed for good, we learn about them from their enemy, Plato. There was Protagoras, from Abdera in Thrace; Gorgias, from Sicilian Leontini; Prodicus, from the island of Ceos; Hippias, from Elis. The list of what were considered the ten greatest Athenian orators included three metics: Isaeus of Chalcis, Deinarchus of Corinth, and Lysias, the son of the Syracusan Cephalus. Thanks to his wealth, Lysias was also to play some part in the restoration of the democracy in 403, and came very close at this point to winning his citizenship, just as Chaerephilus did. He surely deserved the honour, considering the fact that his speeches are regarded as the best example of pure Attic in existence.

There can be no doubt whatsoever that the metics contributed largely to Athens' economic strength no less than to her intellectual and artistic prestige. Plato, however, remained suspicious of them, and wanted to see their activities severely curtailed: one can only suppose that this attitude was inspired by his admiration for the constitution of Sparta—the xenophobic city *par excellence*. However, another, more realistically minded Athenian (also a great enthusiast for Spartan institutions) advised his fellow-citizens, in their own best interests, to extend even greater facilities to metics. This was Xenophon. He had lived as an expatriate for so long (even fighting against his own country on one occasion) that he strikes us rather as a precursor of the cosmopolitanism which characterizes the Alexandrian period. Despite their unswerving loyalty to the city of their adoption, the presence in Athens of so many intensely active resident aliens must have prepared people's minds for this universalizing trend. But such an attitude, as Plato knew very well, was in direct opposition to the Greek city-state's basic totalitarianism. The way in which the Athenians actually treated their metics was an honourable compromise between their traditional political principles and their would-be liberal character.

Slaves

A metic who failed to pay the *metoikion*, or who made false claims to citizenship, was reduced to slavery—a fact which reminds us of the Greek city-states' first, and most instinctive, attitude to foreigners. But metics could not be all-sufficient in themselves; and we have already seen that many of them were, in fact, entrepreneurs, employers with a labour force working under them. This labour force was largely composed of slaves.

The great philosophers of the fourth century accepted slavery as a fact and made no kind of protest against the fundamental injustice which it represents. Plato, in *The Laws*, merely advised against the enslavement of Greeks, and exhorted masters to treat their slaves decently; and, as we shall see, when he wrote thus he was basing his argument on the humane attitude adopted by the great majority of his fellow-countrymen to their own servants. But Aristotle, in Book One of his *Politics*, chapters I and II, takes a much tougher line. He refers to those who claim that "law alone lays down the distinction between the free man and the slave, and that nature plays no part in it; they assert, further, that this distinction is an unjust one, since it is violence—violence in battle particularly—which has brought it about"; but he, Aristotle, is very far from sharing such a view, which was already quite well known by his day: "The human race," he declares, "contains certain individuals as inferior to the rest as the body is to the mind or brute beasts to men; these are the men who will respond best to the use of physical force. Such persons are destined by nature itself for enslavement, since there is nothing that suits them better than to obey." He even goes so far as to state: "Warfare is, in some sense, a legitimate method of acquiring slaves, since it allows for one's need to hunt, not wild beasts only, but also men, who are born to obey yet refuse to submit to the yoke."

Slave-hunts were a living reality in Laconia, where they went by the name of *krypteia*. But Aristotle knew very well that the only justification for slavery was the necessity, since the entire economy of life in a Greek city-state rested upon slave labour:

> If every tool could, at the word of command, go to work by itself—if the looms wove untended, and the cithara played of its own accord—then employers would dispense with workers, and masters dismiss their slaves.

Homer had already pictured the abode of the Gods as equipped with marvellous "robots" fashioned by Hephaestus, tripods which "on wheels of gold move by themselves to the palace where the gods are assembled, and then return to their own place again"; and especially two handmaidens that he had wrought in gold, who attended him and supported his lame leg as he walked. The bellows of his forge, too, worked spontaneously at Hephaestus' command. We may say, then, that the Greeks had forseen the phenomenon which today we call "automation," just as Icarus' wings were a foreshadowing of human flight; but the only energy they had at their disposal was provided by the arms of their slaves.

The main source of slave labour, as in Homeric times, was still warfare. The vanquished warrior whose life had been spared was enslaved by the man who had defeated him, and unless his relatives were able to pay an agreed sum in ransom, he remained in that man's service for good. When a town was captured, those of the inhabitants who were not killed found themselves reduced to slavery: such was the fate that befell Hecuba, Andromache and Cassandra. Piracy also ensured a good supply of slaves: Eumaeus, in the *Odyssey*, recounts how a band of Phoenician pirates, who doubled the roles of merchants and kidnappers, stole him away from his father's palace. By the fifth century Athenian thalassocracy had almost eliminated the pirates, but wars went on more or less non-stop. Thucydides, for instance, after reporting the tragic discussion between the Athenians and the inhabitants of the little island of Melos (the so-called "Melian Dialogue"), whose only crime was their desire to remain neutral, gives a brief account of the capture of Melos by an Athenian squadron, and concludes: "The Athenians killed every Melian of an age to bear arms, and reduced the women and children to slavery." The Melians, we should remember, were Greeks.

In time of peace the supply of slaves was still kept up from various sources. Among the barbarians and even in Greece (except in Attica after Solon's time) the head of a family had the right to sell his children. Dealers in "human cattle" found their best areas for supplies were Thrace, Caria, and Phrygia: slaves with these countries of origin are particularly numerous. Even at Athens, a father who, whether because of poverty or mere selfishness, did not wish to rear a child had the right to "expose" it at birth, i.e., to throw it out on a rubbish-heap. The infant either died, or, if rescued (like Oedipus and several other heroes) became a slave. An unemployed member of the proletariat, who was dying of hunger, could likewise sell himself as a slave to some master in return for bed and board. We even hear of a certain doctor, Menecrates of Syracuse, who only agreed to treat desperate cases on one condition: they were to sign a contract agreeing to become his slaves if they recovered. An undischarged debtor—again, except in Athens after Solon's reforms—was sold into slavery, and the price he fetched paid to his creditor.

The philosopher Plato had a decidedly unpleasant adventure in 388 B.C. He had gone to Sicily as a guest of Dionysius of Syracuse, and, having fallen out of favour with the tyrant, was shanghaied aboard a Lacedaemonian boat, the captain of which intended to sell him as a slave in Aegina. Luckily, he was bought up by a Cyrenaean, who sent him back home to his friends and his life of philosophy. Even in Attica, either out in the country or, sometimes, in the city itself, children and adolescents were "snatched" by slave kidnappers (andrapodistai): we know this through the existence of a law in Athens providing for just such a contingency.

The main slave marts were those of Delos, Chios, Samos, Byzantium and Cyprus. There were two in Attica: one at Sumium, designed to keep the nearby mines of Laurium supplied with an adequate labour force, and the other in Athens itself, a monthly auction held in the Agora at the time of the new moon. The price of a slave varied considerably according to the period, and was also, obviously, conditioned by individual qualities and skills. The ransom price of prisoners-of-war during the fifth century was about two minas, i.e., two hundred drachmas; by the end of the fourth century this price had risen to five minas. Unskilled labourers went for about two minas; women, as a rule, cost slightly more, and a skilled craftsman could fetch anything between three and six minas.

A distinction must be drawn between house-born and purchased slaves. It should not be supposed that masters encouraged unions between slaves (which did not count as true marriages or gamoi) for the purpose of acquiring more human livestock on the cheap, since this would involve them in supporting the children for long years before they showed any productive return on the investment. It was rather as a means of securing the loyalty of good slaves that they were, sometimes, permitted to start a family. In Xenophon's *Oeconomica* we find Ischomachus saying:

> I shall show my young wife the women's apartments, which are separated from those of the men by a locked door: thus no one can smuggle anything out improperly, and the slaves are unable to have children without our express permission. As a general rule we may say that if good slaves are allowed to have a

family, their loyalty increases; whereas when bad ones do so, they merely have greater opportunity for trouble-making.

In rural districts, except for the Laurium mining area, the slave population was relatively small, since small landowners (*autourgoi*) were not wealthy enough to maintain many dependents. But those in fairly comfortable circumstances, such as Ischomachus, had several slaves, under the direction of an overseer who was a slave himself. Smallholders could also hire out slaves for seasonal work, since many rich citizens and metics invested their capital in the purchase and maintenance of a slave-labour pool, which they then leased out as occasion demanded, for an agreed period.

It was industry, beyond any doubt, which absorbed the greatest number of slaves: the Laurium silver-mines, at the southern tip of the Attica peninsula, had a permanent labour force of over 10,000, which may at certain periods have risen to almost twice that figure. As Aymard writes:

> Their lack of technical knowledge meant that the work was done with rudimentary tools, in narrow galleries lit only by the flames of smoky oil-lamps. Above ground, the smelting of the ore, which was strongly tainted with sulphur, produced poisonous fumes that destroyed vegetation and gave the surrounding countryside a grimly desolate appearance. It was here that the slaves were lodged, in squalid camp-sites, and without any families—this last provision being imposed to save the extra cost of feeding useless mouths.

But such a concentration of labour seems to have been unique in Attica; as regards other industries, we hear of no industrialist who employed more than a hundred and twenty slaves—this being the number working in the arms factory of Lysias' father, the metic Cephalus.

Athenian commerce, in its healthily flourishing condition, also needed a large labour force, particularly at the Piraeus, for the loading and unloading of cargoes: these dockers, or the greater number of them at least, were undoubtedly slaves. From certain speeches by Demosthenes we learn of two slaves advanced to positions of trust in the banking world, who subsequently did extremely well for themselves. One Pasion, a bank employee, so distinguished himself by his diligence and sound head for business that his master enfranchised him; and later, having been able—thanks to the funds at his disposal—to do the State some service, Pasion actually acquired Athenian citizenship, an extremely rare honour for a man of servile extraction. When he died, in 370, he left a widow and two sons, one aged twenty-four, the other ten. As he had no great confidence in the older boy's financial abilities (he was a good deal better at spending money than making it) he laid down in his will that his trusty employee Phormio—an enfranchised slave like himself—should take over the direction of the bank, and that of a shield factory which he ran as a sideline. He further decided that Phormio was to marry his widow and act as the tutor of his younger son. As might be expected, this will was contested, on legal grounds, by Pasion's elder son. Here we have a classic instance of how talented slaves could, on occasion, rise to most enviable positions in the world.

All domestic work in the city was, there can be little doubt, ordinarily performed by slaves. An extremely wealthy citizen, such as the statesman Nicias, owned more than a thousand house-slaves, many of which he rented out, since obviously he could not find employment for all of them himself. A comfortably off

Athenian such as Plato would have an average of fifty. The ordinary middle-class citizen seems to have run to a dozen: a porter, a cook, the *paidagogos* who escorted his children to school and generally supervised them, and, lastly, the women who swept the house, fetched in water, ground corn in the mill, and spun and wove under their mistress' direction. Many poor Athenians, however, had no slaves at all. Such was the position of Lysias' "Cripple," a barber or cobbler in all likelihood (we do not know his exact occupation), who declared: "I have a trade, but it brings me in very little; I already find it a strain to keep going on my own, and I can't save enough to purchase a slave to do the work for me."

Finally the State itself owned slaves, as did the temple sanctuaries, in the form of *hierodouloi*. Of the public slaves known to us we mention, first, the clerks, beadles, and other underlings who served the Ecclĕsia, the Boulé, the law-courts, or the offices of the various civil authorities: indispensable cogs in the administrative machine, and the nearest we get, in Athens, to any sort of permanent civil service. There were also such state employees as the public executioner (e.g., the minion of the Eleven who prepared and administered the hemlock to Socrates), the streetsweepers, the skilled workers in the Mint, who struck Athens' drachmas, and, lastly, the police force, those Scythian archers we have already seen in action. This corps was created in 476, and must not be confused with Athens' normal auxiliary troops, who were also armed with the bow. The Scythians were purchased directly by the State, and were responsible for policing the streets, the Agora, the Assembly, and the law-courts. There were about one thousand of them, and they had their barracks on the Areopagus, whence they had a convenient bird's-eye view of the Agora and the city generally.

Slaves, whether public or private, had in theory no legal rights. They were regarded, in law, as *objects*, chattels, which could be sold, hired out, or pawned: in a juridical sense they did not exist as persons, and could not even give evidence in court. Yet a slave-owner, when involved in a trial, would frequently offer to have his slaves put to the torture, so that their testimony, given under such conditions, should add weight to his own statements. Slave unions had no legal status, and were authorized by the master of the house, to whom any resultant children belonged. A runaway slave was severely punished, and branded with a red-hot iron.

It is quite clear that the slaves at Sparta, known as "helots," led a most miserable existence: in this connection the institution of the *Krypteia* and the story of the intoxicated helot are highly significant. The helots also took advantage of the great earthquake in 464 B.C. to stage a rebellion. In Athens, however, custom—and consequently the laws—gradually became less inhumane. The slave who escaped from a cruel master could seek refuge in a sanctuary, either that of Theseus or the Erinyes; he was then protected by the right of asylum, and his master had no option but to offer him for sale. Further, the law protected a slave, just as it did a free man, against outrage and violence (*diké hybréôs*): it even guaranteed the slave a *synégoros*, that is, a kind of advisory lawyer, to help in any contested points to do with his enfranchisement. The savings a slave was allowed to build up belonged in theory to his owner, but the latter most often let him enjoy the profits of his thrift undisturbed.

Young slaves born in the house did not, as a rule, receive any education. They could not frequent the *gymnasia*, which were reserved for free citizens and their children. But purchased slaves, when brought to Athens, were put through a ceremony which made them one of the family: they were made to sit by the hearth, and the mistress of the house then scattered figs, nuts, and sweets over their head. At the same time they were given a name. This ritual was not dissimilar to those of marriage and adoption; no doubt it signified that the new arrival would henceforth be a stranger no longer, but a member of the family, sharing the same religion. So it came about that slaves participated in family prayers, and attended religious festivals; and this was also why slaves had to be buried in the family plot, a practice we have already observed in the Ceramicus cemetery. Slaves of Greek origin could also be initiated into the Eleusinian Mysteries. The enfranchised slave was not freed of all obligations towards his former master's family: the ties of religion remained.

It seems, then, that the position of domestic slaves in Athens was, on the whole, quite tolerable. As for the slaves in public service, they led a life scarcely to be distinguished from that of any other *petit fonctionnaire*. They lived where they pleased (except for the Scythian archers), they received a salary, and were allowed to contract non-legal marriages. Similarly, many slaves in commerce and industry could live where they pleased and build up businesses on their own: though all the profits, legally, belonged to their master, the latter frequently found it highly advantageous to grant them an "interest" in the venture, and let them keep a percentage of the profits. It was in this way that Pasion, and his successor Phormio, contrived to build up so solid a financial edifice. . . .

But there were at least two categories of slave whose condition remained grim in the extreme: those employed to grind corn in the mills, and the Laurium miners, whom we have already discussed above. Very often a dishonest or recalcitrant slave would be sent by his master, as a punishment, for a spell in the mills or the mines. During the Peloponnesian War, Spartan raids into Attica allowed slaves from Laurium to desert *en masse* and scatter through the countryside, where they proceeded to terrorize the inhabitants.

Numerical Proportions of the Classes

Fifth-century Athens had a population of some 40,000 citizens and 20,000 metics. If we add the women and children to both categories, we arrive at a possible "free" population of about 200,000. Now, though we have no means of estimating their number, even approximately, the slaves were at least as numerous, and may indeed have totalled 300,000, or even more.

So we see that out of a total population in Attica of some half a million souls, only two-fifths were free citizens. The men who possessed full political rights and took part in the government of the city constituted no more than a fractional minority. In any discussion of Greek democracy, this fact should never be forgotten. We should remember, too, that the Greeks of the classical period had inherited from an earlier age considerable contempt for "servile labour"—that is, for any work which made a man depend on another man for his food and income. Trade and commerce were particularly looked down on, and it was for this reason that the Athenians so readily abandoned such occupations to metics. We know that in Lacedaemonia any kind of business activity was forbidden to Spartiates enjoying full civic rights, who lived off their inalienable estates, which were worked for them by helots. At Athens, it is true, there had been since Solon's time a law which made it illegal for citizens to have no occupation—the *diké argias*. Plutarch tells the following anecdote on the subject:

A certain Spartan, finding himself in Athens on a day when the courts were in session, heard that a citizen had been condemned on a charge of 'non-activity': this citizen had returned home in a greatly distressed condition, accompanied by his friends, who were all commiserating with him and sharing his unhappy burden. The Spartan thereupon asked those who were with him to point out this person—"condemned merely for living as befits a free man." So convinced were the Lacedaemonians that it befitted none but slaves to follow a trade or work for their living.

The Lacedaemonians were by no means the only people to hold such an opinion, which was shared, even in Athens, by a great number of people—in defiance of Solon's laws.

The fact of the matter is that manual labour was regarded by the Greeks as a strictly banausic occupation, beneath the dignity of any free citizen. Plato and Aristotle considered the fabrication (*poiesis*) of any object, and even the creation of a work of art, as a strictly second-class activity; the wise man should follow no pursuits apart from *praxis* and *theôria*, that is, the business of politics and leadership on the one hand, and, on the other, the study of philosophy. In the myth of the *Phaedrus*, Plato classes ways of life, according to their value, in nine groups: the labourer and the artisan are to be found in the seventh division, just above the demagogue and the tyrant, who are, in the philosopher's opinion, humanity's worst scourge, and the most contemptible of men.

On the Status of Women

Whatever the origins of male dominion in classical civilization or the truth of Engels's account of it, women were clearly subordinate creatures. In every sphere of organized life it was clear that men prescribed for women a place based on assumptions of female inferiority to men. Women had no political life or economic initiative. Where they had occupations beyond the hearth, necessity ruled and dictated the fact, not any sense of opportunity. Even in aristocratic life, women were more usually objects of thought rather than the subjects of sentences conveying independence or initiative.

This was especially clear in Xenophon's *Oeconomicus* (*Economics*). Its author was an Athenian aristocrat famous as a soldier, statesman, and associate of Socrates. His account of economics details the management of farm and home in an age of declining Athenian power and culture.* Its focus is on the woman as manager. But even where Xenophon is pleasant and amusing, he leaves no doubt about the ideal wife. She should have no intellectual aspirations. Hard work and exercise are better than cosmetics in preserving good looks. Devotion to housework was the perfection of a woman's nature. Hence, even the highly charged account of perfect "partnerships" between men and women makes explicit the woman's subjection to her mate. Not for her Xenophon's artful writings on horsemanship and hunting; nor the moral tales of Spartan kings and cavalry heroes; nor even the delights of talk with Socrates in *The Symposium*, where she might have heard of heavenly love and its earthly counterfeits. Her world *was* her home.

From *Oeconomicus*

Xenophon

XENOPHON'S PICTURE OF AN IDEAL HOUSEHOLD
The story is supposed to be narrated by Socrates.

It chanced one day that I saw my friend Ischomachus seated in the portico of Zeus Eleutherios [in the Athenian agora], and as he seemed to be at leisure I went up to him, and sitting down by his side, accosted him: "How is this? As a rule, when I see you, you are doing something, or at any rate not sitting idle in the market place."

"Nor would you see me now so sitting, Socrates," said he, "except that I had promised to meet some strangers, friends of mine, here."

"And when you are not so employed," said I, "where, in heaven's name, do you spend your time, and how do you employ yourself? I am truly very anxious to know from your own lips by what conduct you have earned for yourself the title 'beautiful and good'?"

SOURCE: Xenophon, *Oeconomicus*, in *Readings in Ancient History*, trans. H. G. Dakyns (Boston: Allyn and Bacon, 1912), pp. 265–271.

* Xenophon's dates are 430?–354 B.C.

[Ischomachus laughed at the compliment, and said that when he was called on for any public service] "nobody thinks of asking for the 'beautiful and good' gentleman, it is plain 'Ischomachus the son of so-and-so' on whom the process is served. But I certainly . . . do not spend my days indoors, if for no other reason than because my wife can manage all our domestic affairs without my aid."

"Ah!" said I, "and that is just what I dearly want to learn about. Did you educate your wife yourself, to be all that a wife should be, or [when you married her] was she already proficient?"

"Well skilled?" he replied—"Why, what skill was she likely to bring with her? Not yet fifteen when she married me, and during her whole previous life most carefully trained to see and hear as little as possible, and to ask the fewest questions. Shouldn't anybody be satisfied, if at marriage her whole experience consisted in knowing how to take wool and make a dress and see that her mother's handwomen had their daily spinning tasks assigned? For (he added) as regards control of appetite and self-indulgence, she had the soundest education, and that I take it is the chief thing in the bringing up of man or woman."

"Then all else, (said I) you taught your wife yourself, Ischomachus, until you had made her capable of attending carefully to her proper duties?"

"That I did not do (he replied) until I had offered sacrifice, and prayed that I might teach, and she might learn all that could conduce to the happiness of us twain."

SOCRATES: And did your wife join you in the sacrifice and prayer to that effect?
ISCH: Most certainly, and with many a vow registered to heaven to become all that she ought to be.

(*Socrates now asks Ischomachus to tell how he educated his wife.*)

"Why, Socrates (he answered), when after a time she had been accustomed to my hand, that is, tamed sufficiently to play her part in a discussion, I put her this question, 'Did you ever stop to consider, dear wife, what led me to choose you, and your parents to intrust you to me? It was surely not because either of us would have any trouble in finding another consort. No! it was with deliberate intent, I for myself, and your parents for you, to discover the best partners of house and children we could find. . . . If at some future time God grant us children, we will take counsel together how best to bring them up, for that, too, will be a common interest, and a common blessing if haply they live to fight our battles and we find in them hereafter support and succor for ourselves. But at present here is our house, which belongs to both alike. It is common property, for all that I own goes by my will to the common fund, and in the same way was deposited your dowry. We need not stop to calculate in figures which it is of us who has contributed the most: rather let us lay to heart the fact that whichever of us proves the better partner, he or she at once contributes what is most worth having.' "

(*Ischomachus's wife now asks more pariticularly what her duties are to be; and her husband answers:*)

"You will need to stay indoors, and dispatch to their toils such of your servants whose work lies outside the house. Those whose duties are indoors you will manage. It will be your task to receive the stuffs brought in, to apportion part for daily use, and to make provision for the rest, to guard and garner it so that the outgoings destined for a year may not be expended in a month. It will be your duty when the wools are brought in, to see that clothing is made for those who have need. You must also see that the dried corn is made fit and serviceable for food. Then, too, there is something else not altogether pleasing. If any of the household fall sick, it will be your care to see and tend them to the recovery of their health."

WIFE: Nay—*that* will be my pleasantest task, if only careful nursing can touch the springs of gratitude and leave them friendlier than before. . . . But mine would be a ridiculous guardianship and distribution of things indoors without your provident care to see that the importations from without are duly made.

ISCH: Just so, and mine would be a pretty piece of business, if there were no one to guard what I brought in. Do you not see how pitiful is the case of those unfortunates who pour water into their sieves forever, as the story goes, and labor but in vain?

WIFE: Pitiful enough, poor souls, if that is what they do.

ISCH: But there are other cares, you know, and occupations, which are yours by right, and these you will find agreeable. This, for instance—to take some girl who knows nothing of carding wool, and to make her skillful in the art, doubling her usefulness; or to receive another quite ignorant of housekeeping or of service, and to render her skillful, loyal, serviceable, till she is worth her weight in gold; or again, when occasion serves, you have it in your power to requite by kindness the well-behaved whose presence is a blessing to your house; or maybe to chasten the bad character, should such appear. *But the greatest joy of all will be to prove yourself my better;* to make me your faithful follower; knowing no dread lest as the years advance you should decline in honor in the household, but rather trusting that though your hair turn gray, yet in proportion as you come to be a better helpmate to myself and to the children, a better guardian of our home, so will your honor increase throughout the household as mistress, wife, and mother, daily more dearly prized.

(*The wife carried out Ischomachus's instructions marvelously well: later they undertook to go through the house together, with a view to putting it in the best of order.*)

ISCH [*continuing*]: We proceeded to set apart the ornaments and holiday attire of the wife, and the husband's clothing both for festivals and war: the bedding both for the women's and for the men's apartments; next the shoes and sandals for them both. There was one division devoted to arms and armor, another to instruments used for carding wool, another to implements for making bread; another for cooking utensils, one for what we use in the bath, another for the things that go with the kneading trough, another for the service of the table. . . . We selected and set aside the supplies required for the month, and under a separate head we stored away what we computed would be needed for the year.

(*Ischomachus adds that at another time he told his wife not to use cosmetics, nor to think that she made her face more handsome with white enamel or rouge, and to leave off high-heeled shoes: she promised to comply, but asked her husband if he could advise her how she might become not a false show, but really fair to look upon? To which he replied:*)

Do not be forever seated like a slave, but, with Heaven's help, to assume the attitude of a true mistress standing before the loom, and where your knowledge gives you the superiority, there give the aid of your instruction, and where your knowledge fails, as bravely try to learn. I counsel you to oversee the baking woman as she makes the bread; to stand beside the housekeeper as she measures out her stores: to go on tours of inspection to see if all things are in order as they should be. For, as it seems to me, this will be at once walking exercise and supervision. And as an excellent gymnastic I urge you to knead the dough, and roll the paste; to shake the coverlets and make the beds; and if you train yourself in exercise of this sort you will enjoy your food, grow vigorous in health, and your complexion will in very truth be lovelier. The very look and aspect of the wife, the mistress, seen in rivalry with that of her attendants, being as she is at once more fair and more becomingly adorned, has an attractive charm, and not the less because her acts are acts of grace, not services enforced. Whereas your ordinary fine lady, seated in solemn state, would seem to court comparison with painted counterfeits of womanhood.

[*Ischomachus concludes by saying to Socrates*], And I would have you to know that still to-day my wife is living in a style as that which I taught her, and now recount to you.

How an Athenian Gentleman Passed His Morning

Ischomachus, who has narrated to Socrates how he trained up his wife to be a model helpmate, tells how he spends his own time in the morning and leads the life of a very prosperous and successful Athenian gentleman.

"Why, then, Socrates, my habit is to rise from bed betimes, when I may still expect to find at home this, that, or the other friend whom I may wish to see. Then, if anything has to be done in town, I set off to transact the business and make that my walk; or if there is no business to transact in town, my serving boy leads on my horse to the farm; I follow, and so make the country road my walk, which suits my purpose quite as well or better, Socrates, perhaps, than pacing up and down the colonnade [in the city]. Then when I have reached the farm, where mayhap some of my men are planting trees, or breaking fallow, sowing, or getting in the crops, I inspect their various labors with an eye to every detail, and whenever I can improve upon the present system, I introduce reform.

"After this, usually I my mount my horse and take a canter. I put him through his paces, suiting these, so far as possible, to those inevitable in war —in other words, I avoid neither steep slope, nor sheer incline, neither trench nor runnel, only giving my uttermost heed the while so as not to lame my horse while exercising him. When that is over, the boy gives the horse a roll, and leads him homeward, taking at the same time from the country to town whatever we may chance to need. Meanwhile I am off for home, partly walking, partly running, and having reached home I take a bath and give myself a rub—and then I breakfast—a repast that leaves me neither hungry nor overfed, and will suffice me through the day."

On Homosexuality

The relegation of women to the role of objects in upper-class domestic economy had a distorting effect on the world beyond the hearth. The most basic bipolarity of life, that of man's sexuality, was cloistered, mewed up at home, and so barred from all other scenes of human contact. Cohesive experiences of intellectual or political life were shared only by men in Greek society, and this fact found its expression in high culture. The practice of homosexual love among wealthy Athenians and intellectuals was the subject of a literature of praise.

As Robert Flacelière makes clear in his *Love in Ancient Greece*, the institution of pederasty derived from the natural love boys had for each other. Its meaning was never simply that of physical homosexuality, however, and what he calls "pedagogic pederasty"—the affection of master and pupil or students among themselves—has innocent echoes. Think of Henry V's loving appeal to his "band of brothers" on Saint Crispan's Day: "We few, we happy, few. . . ." Or recall for a moment the deeply charged comradeship of other brothers-in-arms and those which show in the affectionate language of academic discipleship (which may yet survive our time) and the vocabulary of patient-analyst relations in some modern psychotherapy.

The experiences of naked boys exercising in the schools (*gymnasia*) had erotic components for the boys themselves and between them and their teachers or other admiring adults. Given the class basis of gymnasium education in Greek society, it is not likely that homosexual practices were widespread among common people. But among the wealthy the world of both mental and physical exercise reinforced the exclusion of women from public life. Thus between men regular flirtations, confessions of love, heartbreaks, and jealousies were conventional and accepted. While it would be too much to say that pederasty and education were related as cause to effect, the cultivation of the minds and bodies of young men was erotically charged and open to homosexual practices. When Sophocles thought of his own tastes, what he regretted was that as an old man he took a young boy—not the homosexual love itself.

From *Love in Ancient Greece*

Robert Flacelière

HOMOSEXUALITY

Thyrsis and Corydon mingled the flocks that they kept.
The first had ewes, the second she-goats heavy with milk.
Both herdsmen were twin flowers of their Arcadian youth.
                    VERGIL, *Bucolics*, VII

The word pederasty is derived from the Greek *paiderasteia*, meaning literally the love of boys. In English pederasty has come to signify almost exclusively the practice of sexual inversion. But in Greek literature *paiderasteia* is used to refer

SOURCE: Robert Flacelière, *Love in Ancient Greece*, trans. James Clough (New York: Crown Publishers, 1963), pp. 33–58.

both to pure, disinterested affection and to physical homosexual relations. In the present chapter we shall employ the word in its Greek sense. It will therefore at times imply, as in the case, for instance, of "pedagogic pederasty," perfectly honourable affection between an older master or tutor and a younger disciple or pupil. It is impossible to do otherwise without deviating from the ancient texts by a change which might distort their meaning. The reader would be well advised to bear in mind that we shall occasionally give "pederasty" this extended connotation. For otherwise he may not fully understand what we have to say on so delicate a subject.

In the first place it appears extremely likely that homosexuality of any kind was confined to the prosperous and aristocratic levels of ancient society. The masses of peasants and artisans were probably scarcely affected by habits of this kind, which seem to have been associated with a sort of snobbery. The available texts deal mainly with the leisured nobility of Athens. But they may give the impression that pederasty was practised by the entire nation. The subject, however, of the comedy by Arostophanes entitled *Lysistrata* suggests that homosexuality was hardly rampant among the people at large. It would be an error of perspective to think so, and the mistake may as well be pointed out here and now. We shall return to it at the end of the present work.

There was nothing particularly "Greek" about homosexual feeling. The nation in antiquity was by no means alone in providing illustrations of inversion, which has been practised at almost all times and in almost all countries. In our own day the productions of Verlaine, Proust and Gide, to mention only French writers, as well as a number of others, are sufficient evidence of the fact. In the pre-Christian era the case of Sodom is well known. Nor were the Persians, the Etruscans, the Celts or the Romans ignorant of homosexuality. But its existence among these people was kept more or less secret on account of the discredit which attached to it. But in Greece, though pederasty was forbidden by law in most of the cities, it had become so fashionable that no one troubled to conceal it. On the contrary, such tendencies were respected and even approved. Plato himself recommended their cultivation as a necessary preliminary to the successive stages of a philosophic understanding of Being.

Many Greeks, moreover, did not feel in the least ashamed of admitting that homosexuality was held in more honour among them than anywhere else in the world. They even affirmed that other nations which practised it were their pupils in this field. Herodotus, for instance (I, 135), alleges that the Persians learned the habit from the Greeks. Xenophon in the *Cyropaedia* (II, 2, 28) makes Cyrus ask:

"Is it your intention to teach Greek customs to that young man lying next you, since he is so handsome?"

In any case the Greeks themselves considered that pederasty was of relatively recent origin among them. In the *Erotes* attributed to Lucian there is a dialogue in which the defender of homosexuality admits that it is not a very ancient custom.

"At former epochs," he says, "male love-affairs were unknown. In those days it was thought indispensable to couple with women in order to preserve the human race from extinction. . . . Only with the advent of divine philosophy did homosexuality develop. We should be careful not to condemn an invention merely because it came late. . . . Let us agree that the old customs arose from necessity, but that subsequent novelties due to the ingenuity of man ought to be more highly regarded."

In Plutarch's *Erotikos*, on the other hand (751 F), the champion of heterosexual love exclaims:

"Homosexuality resembles a son born late, of parents past their maturity, or a bastard child of darkness seeking to supplant his elder brother, legitimate love. For it was only yesterday or at best the day before yesterday that the pederast came slinking into our gymnasia, to view the games in which youths then first began to strip for exercise. Quite quietly at first he started touching and embracing the boys. But gradually, in those arenas, he grew wings—" Eros being always represented winged—"and then there was no holding him. Nowadays he regularly insults conjugal love and drags it through the mud!"

This passage by Plutarch notes an important fact. There can be no doubt that the development of homosexuality was connected with the rise of gymnasia and arenas in which boys practised the five exercises of the *pentathlon*, which comprised wrestling, the foot-race, leaping, throwing the discus and hurling the javelin. Others were boxing and the *pancration*, a mixture of fist-fighting and wrestling. The competitors were always naked and watched by admiring spectators. In the same work (749 F) Plutarch tells us that Pisias, in love with a certain youth, Bacchon, whom a rich widow wanted to marry, "imitated ill-conditioned lovers of the ordinary sort in trying to prevent his friend from marrying. The man's only object was to prolong the pleasure he took in watching the boy strip in the arena, while he still retained his virgin beauty." Most gymnasia contained not only a statue of Hermes but also one of Eros. We have already quoted Athenaeus to the effect that there was an image of Eros at the Academy, which was the gymnasium in which Plato met his disciples.

The Greeks were at all times, as we noted in the case of Homer's Helen, most sensitive to physical beauty, whether masculine or feminine. This susceptibility was felt even in the most ascetic of friendships, when the lover desired nothing more from his beloved than the pleasures of the eye. It should be borne in mind that women were almost entirely excluded from Greek social life, which resembled a man's club. This was especially so at Athens, for at Sparta girls and women had more freedom of movement. In the following chapter, dealing with marriage customs and the lives of women confined in the *gynecea*, we shall dwell more explicitly on the extraordinary fact, so amazing at first sight, that many of the ancient Greeks lavished all their sexually rooted affections upon boys. For they considered members of the other sex inferior beings, lacking all education and refinement, good for nothing but to ensure a posterity.

It was in the arenas above all that "special friendships" were formed. Moreover, if Plato's dialogues are to be credited in this connection, on certain festive occasions the establishments in question were the scenes of positive Courts of Love, in which youths and mature men discussed "Cupid's country," as it came to be called in later times. In the *Lysis* Socrates meets a certain Hippothales, whom he immediately recognises as a lover. For Socrates, as he himself declares, is an expert in such matters. The philosopher asks the other whom he loves. Hippothales blushes and refuses to answer. He is too modest and timid to say. But one of his friends reveals the secret.

"It's all very well, Hippothales," he says, "to blush and stammer over the name. But Socrates will need no more than a few seconds of talk with you before you deafen him with incessant repetitions of it." Turning to the philosopher, he adds: "As for us, Socrates, we are perfectly sick of hearing the name of Lysis. Our ears are quite stunned with it. The moment Hippothales has taken a few drinks, he starts singing it out at such a rate that when we wake up next morning we think we can still hear it. . . . Worst of all, he sometimes takes it into his head to recite his verse and prose compositions to us. He thunders out his love-songs in such terrifying tones that we can't get away from them. Yet now he blushes when you ask him with whom he is in love!" (204 c–d.)

The group formed by Socrates, Hippothales and their companions then visits the arena where Lysis practises. It is a feast-day.

"When we came in," said Socrates, "I saw that the boys had just finished sacrificing and were playing at knuckle-bones in their festival costumes. Most of them were in the courtyard. But some were in a corner of the dressing-room. They were taking knuckle-bones from a basket and playing at odd and even with them. Others stood round in a circle watching. Among these last Lysis attracted general attention. He was standing in a group of boys and young men with a crown on his head which he had worn for the religious ceremony. His bearing not only justified his reputation for beauty but showed that he was also 'fair and good,' that is to say, of a noble nature." (206 e, 207 a.)

Lysis, with charming timidity, hesitates to join the group with Socrates. For the youth had caught sight of Hippothales, his lover, with them. But at last he decides, supported by a companion, to approach. Conversation begins. It is entirely on the subject of *philia*, a name which in Greek was applied to all affectionate sentiments, whether in the family or elsewhere. The English word friendship is a very imperfect translation of it. The most important point at issue was, who best deserved the glorious name of *philos*, the lover or his beloved. The question was also raised whether flattery of a beloved person won him over or on the contrary aroused his contempt.

All the same, at Athens a whole body of laws existed for the purpose of restraining the spread of pederasty. This legislation probably dated back to the time of Solon. It aimed among other things at keeping male lovers out of the schools and exercising arenas so far as possible. (See Aeschines, *Against Timarchus*, 9–11.) But laws can do very little to suppress widely disseminated and inveterate habits.

The age of a beloved boy seems always to have been between twelve and twenty. A Greek epigram declares:

"Desirable is the bloom of a boy of twelve. But that of thirteen is much more delightful. Even sweeter is the flower of love that blossoms at fourteen years. Its charm increases still more at fifteen. Sixteen is the divine age. A boy of seventeen I would not dare to woo. Zeus alone would have the right to do so."

As a rule the first signs of down on the chin of the beloved deprived him of his lover. But there were exceptions to this convention. Plutarch writes in his *Erotikos* (770 B–C): "It is generally believed that a single hair would be enough to cut the amorous connection in two like an egg and that lovers of boys resemble nomads who set up their tents in green and flowery meadows in the spring but desert it as if it were hostile territory once the season is over. Yet we may recall a famous phrase of Euripides, spoken amid kisses and caresses to the handsome Agathon, who had already grown a beard: 'Beauty is still fair even in its autumn.' "

As a rule the lover in these associations was a mature man less than forty years of age. But Aeschines was forty-five when he made his speech *Against Timarchus* (No. 49), observing (136–7):

"I myself no more disdain the pleasures of love today than I did formerly. I freely confess it. . . . To be fond of good-looking and well-behaved young people is a natural tendency of all sensitive and liberal minds."

If we are to believe Athenaeus (XIII, 605), the poet was sixty-five when he experienced the following misadventure. "One day he left Athens with a handsome boy, intending to take his pleasure with the lad. The boy laid his own shabby cloak on the grass and they covered themselves with the poet's own fine warm woollen cloak. After the consummation of the affair the youth seized the cloak of Sophocles and made off with it, leaving his own in its place."

This incident, it appears, elicited an exchange of pungent epigrams between Sophocles and Euripides. We need only add that the former's habits by no means prevented him from being considered throughout classical antiquity as a man of exemplary piety, "beloved of the gods." The domains of religion and sexual morality were then regarded as completely separate. It was certainly not in the name of religion that Pericles, himself a "free-thinker," addressed the above-mentioned reproach to Sophocles. The reference was simply to the dignified conduct expected of a high State functionary. In any case the mighty gods of Olympus themselves, from Zeus and Apollo downwards, were represented in classical times as ardent pederasts.

Homosexual love, like heterosexual, affected the whole mind of the lover if the sentiment was strong. He paid assiduous court to the object of his choice and sometimes ruined himself by making all sorts of gifts to the lad. Such behaviour was known as the "chase" of love. Plato, in the *Symposium* (183 a), mentions "everything that lovers do for the boys they cherish, the prayers and entreaties with which they support their suit, the oaths they swear, the nights they spend on the doorstep of the beloved one and the slavery they endure for his sake, which no real slave would put up with." Plato also refers in the *Phaedo* (73 d) to the profound agitation which lovers feel at the mere sight "of a lyre, a garment or any other object in constant use by their loved ones, any of which are enough to call up the image of their darling, even if he is absent."

In Xenophon's *Symposium* Critobulus describes his love for the young Clinias in forcible terms:

"I would rather look at him than anything else in the world. I would cheerfully agree to be blind to all other objects. I resent night and sleep because they remove him from my sight. I am grateful to the sun and daylight because they allow me to behold him. . . . Though I am fond of luxury I would rather give all my possessions to Clinias than receive their equivalent from someone else. I would prefer slavery under Clinias to my freedom as a citizen, and work for his sake to rest for my own. . . . I would follow him even through fire. His image is so deeply graven in my heart that if I were a painter or sculptor I could produce an absolutely faithful portrait or bust of him in his absence."

The lover followed his beloved about everywhere, sometimes spent the night in front of his house, serenaded him, composed verses and songs in his honour, carved his name on walls, doors and trees, hung up garlands of leaves or flowers, like religious offerings, in his porch, and sent him all kinds of presents, such as fruit, a bag of knuckle-bones, a cock, a hare or a dog, as well as painted vases on which the artist had been instructed to engrave the boy's name, followed by the adjective *kalos*, "fair." Many such vases have survived. On the Acropolis at Athens a small stone cut to the shape of a wedge was discovered. It bore an inscription of the fifth century B.C. which read: "Lysitheos declares that he loves Mikion more than all the other boys in the city, for he is courageous." (Coll. Inscr. Graec., Ed. 3, 1266.)

This statement by a pederast is a simple avowal of affection based upon a laudable motive. There is nothing coarse about it, as there was in so many which were cut at an earlier date on the rocks near the gymnasium in the island of Thera, now Santorin, in the Aegean Sea. (*See Sammlung Griech. Dial. Inskr.* 4787–4797.)

In the game of *kottabos* played at banquets after the pouring of libations the lover threw the dregs of his wine in the direction of a copper basin, calling aloud upon the name of his beloved. If the drops fell into the receptacle the omen was favourable and the player would expect that his love would be returned.

When two or more lovers courted the same youth they felt the same kind of mutual jealousy as would be experienced by the lovers of a girl. Plutarch in his life of Themistocles (3) reports the rumour that the hostility between that distinguished statesman and the equally illustrious Aristides originated in the love of both these politicians for the same lad, the handsome Stesilaos of Ceos. According to the *Phaedo* of Plato (232 c and 239 d) the jealousy of the successful lover would cause him to deprive his beloved of all society other than his own and render him "destitute of friends."

The prosecutor in the speech composed by Lysias for delivery against Simon begins as follows: "You may well consider love for a young lad to be unreasonable in a man of my age. But I beg you not to conceive a bad opinion of me. For we are all slaves of passion, as you know. . . . Both Simon and myself fell in love with the youth Theodotus of Plataea. But while I hoped to advance my suit by treating him with consideration, Simon believed he could force the boy to submit to him by violent and cruel means. . . . Simon also claims that he gave Theodotus three hundred drachmae after drawing up a contract with him and that I alienated the lad's affections by my plots."

The whole speech is full of Lysias's picturesque descriptons of the squabbling and rioting, both indoors and out, which broke out in consequence of the rivalry of the two lovers of Theodotus.

A beloved youth might also be jealous of his lover. When Alcibiades arrives at Agathon's dinner-party he notes with vexation that Socrates is sharing a couch with the host, whose reputation is decidedly that of a "pretty boy," and tells the philosopher: "Naturally, you moved heaven and earth to get a place next to the best-looking person here!" Socrates exclaims irritably: "See to my defence, Agathon. For my love of that fellow is constantly getting me into trouble. Ever since I fell in love with him I'm not allowed even to glance at a single good-looking youth, much less talk to one. He gets jealous at once, envies me and permits himself unbelievable impudence to and abuse of me. I'm afraid that one of these days he'll positively go for me!"

Quarrels and brawls, and also tender reconciliations, were common between such lovers and their loved ones. For "even in disputes and disagreements a great deal of pleasure and delight may be found." (Xenophon, *Hiero*, 1, 35.)

Many crimes of passion were committed in consequence of homosexual love-affairs. A few could be regarded as acts of heroism. For Greek tyrants were often murdered by young men whom they had seduced and who afterwards came to hate them. Archelaus, King of Macedonia, for example, was assassinated by Crateas, who had been his favourite. Alexander of Pherae was killed by Pytholaus. Plutarch reports (*Erotikos*, 768 F) that "when Periander, tyrant of Ambracia, jestingly asked his 'boy friend' if he were pregnant, the enraged lad drew a dagger and slew him."

But the most celebrated of these stories is that of the Athenian tyrannicides Harmodius and Aristogiton, who enjoyed the reputation of champions of freedom, like that of Brutus in Rome. In 514 B.C. they killed Hipparchus, son of the tyrant Peisistratus and brother of Hippias, who had succeeded his father. The Sixth Book of Thucydides (54–59) tells the tale in detail.

"The daring feat executed by Aristogiton and Harmodius was the result of a chance development in a love-affair. . . . Harmodius was a lad of the most striking beauty. Aristogiton, a citizen of the middle class, had fallen in love and lived with him. Then Hipparchus, the son of Peisistratus, made advances to Harmodius, who repulsed him and reported his behaviour to Aristogiton. The latter, incensed, and fearing that the all-powerful Hipparchus would try to achieve his purpose by violence, immediately resolved to take advantage of his own good standing by conspiring to overthrow the Government. . . ."

It is a fact that homosexuality made great strides both at Athens and elsewhere in Greece during the fifth century.

Solon, archon and reformer of the Athenian Constitution in 594 B.C., wrote verses which were quoted by the devotees of both homosexual and heterosexual love. The former called attention to the lines:

> Boys in the flower of their youth are loved;
> the smoothness of their thighs and soft lips is adored.

The pederasts argued that Solon only forbade slaves to practise homosexuality and gymnastics because by so doing he intended to encourage free men to engage in them. On the other hand, the advocates of heterosexual love quoted a couplet from another elegy by Solon, composed at a later date, in which he appeared to have renounced the love of boys for that of women, wine and song:

> The works of Cypris, Bacchus and the Muses
> are the founts of joy which I prefer today.

. . . It is quite possible that many such friendships between men and youths were perfectly chaste. We saw just now that Alcibiades showed himself intemperately jealous of Socrates. But later on, in a somewhat improper passage of Plato's *Symposium* (217 a–219 e), Alcibiades himself relates an anecdote which may be called "The Temptation of Socrates," where the common procedure in such affairs is reversed. The young man in this case has already been corrupted by a number of love-affairs. It is he who makes advances—and very shocking ones—to the mature man whose knowledge and wisdom he greatly admires. "I had supposed," says Alcibiades, "when I heard him talk of the 'flower' of my beauty, that I was in for a windfall, a wonderful piece of luck. I thought that if I allowed him to be familiar with me I should learn from him everything he knew."

Accordingly, he arranged to be alone with the philosopher. But the latter was far from taking advantage of the situation to speak of love. Alcibiades then invited him to take part in some physical exercises, wrestling with him for instance. This proposal recalls the important part played in pederasty by gymnastics and the arena. Socrates agreed to the suggestion but still maintained his reserve. Thereupon the younger man took even greater risks.

"I asked him to dinner, behaving exactly like a lover who means to come to the point with his beloved. It was some time before he would accept my invitation. But at last he consented to come. Nevertheless, on his first visit he rose to go as soon as he had eaten. As I felt rather ashamed of myself I let him depart there and then. But the next time he came I exerted myself to entertain him with conversation far into the night. Then, when he said he wished to leave, I objected that it was very late and persuaded him to remain."

After still more talk in the dark Alcibiades felt the moment propitious for a final test.

"It was as though I had been shooting at him half the night and I believed I had hit him. So I got up and, without giving him time to object, I covered him with my cloak—for it was winter-time—and flung myself down beside him under the same old cloak that he had on this evening. I threw both my arms round his godlike person and spent the whole night in that attitude. . . . The most cunning of my efforts merely increased his triumph. For he disdained the 'flower' of my beauty, scoffed at it and insulted it. . . . I call the entire company of heaven to witness that throughout that night I passed beside Socrates nothing more out of the way occurred than if I had slept with my father or an elder brother."

One may wonder, however, how many replicas of Socrates there were in Athens or the rest of Greece.

The poet Aristophanes provides the clearest evidence of the character of his period. But it is necessary to take his exaggerations with a pinch of salt. For his exuberant comic spirit was more concerned to produce laughable caricatures than a precise picture of contemporary habits. He is given a strange speech, a sort of anthropological fantasy, to which we shall return later, in Plato's *Symposium*. The passage seems to be a defence of all kinds of love. Nevertheless, it appears evident enough from his comedies that Aristophanes was a determined opponent of homosexuality. In the very first of his plays, *The Guests*, now lost, we know that he contrasted a decently behaved young man with an invert, treating the latter to a flood of sarcasm. From the *Acharnians* to the *Frogs* he continually attacks both active and passive pederasts with the harshest abuse. They were called respectively *paedicones* and *pathici* at a later date by the Romans. Cleisthenes, Agathon—whom we have already met—and Cleonymus, together with several others, were thus pilloried by Aristophanes. He compares the perversion of his contemporaries, to their disadvantage, with the far purer morals, in his view, of the previous generation, that of the gallant "veterans of Marathon."

"In those days the gymnastics master compelled the boys, when they sat down, to stretch out their legs, so as to prevent any shocking exposure of themselves to the spectators. Again, when they rose, they had to smooth down the sand, so as not to leave any imprint of their masculinity to meet the eyes of lovers. Not one of the lads ever rubbed himself with oil below the navel, with the object of producing a soft kind of down, like that of a quince, upon his organs. Nor did any approach his lover with affected intonations and tempting glances." (*Clouds*, 973–980).

Manners may have been better, more refined, at that period. But were the morals of the age so superior as all that? The example of Harmodius and Aristogiton, dating back to the sixth century, and a good deal of other testimony, entitle one to doubt it.

In any case, when Aristophanes in the *Birds* (137–142) presents an Athenian of his own day dreaming of an ideal city, he makes him talk as follows:

"I should like to live in a town where the father of some pretty boy might come up to me and say angrily: 'Well, this is a nice state of affairs, you damned swaggerer! You meet my son just as he comes out of the gym, all fresh from his bath and you don't kiss him, you don't say a word to him, you don't hug him and you don't feel his testicles! And yet you're supposed to be a friend of ours!"

At Athens the morals of boys were protected not only by their fathers but also by legislation. The laws in question were summarised by Aeschines in his speech *Against Timarchus* (13–20).

"If a father, brother, uncle or tutor sells a boy to a man of licentious character, the child will not be prosecuted. But the two parties to the transaction will be charged, the one with delivering the boy and the other with having paid for receiving him. When the child grows up, he will not be obliged to provide his aged father with either food or shelter, since his parent had sold him into prostitution. The boy will be required only to bury his father in accordance with the usual formalities.

"The Law of Outrage stipulates that the assailant of a child, whether the victim be free or a slave, shall be charged and the due penalty applied or else he shall pay a fine determined by the court.

"Any Athenian citizen who prostitutes himself shall be excluded from exercising any public function or even expressing his opinion in the Assembly or the Council. He will be charged with *hetairesis*"—lack of principle—"an offence punishable with the utmost rigour of the law."

We learn from this passage that masculine prostitution, like feminine, existed at Athens. One might, indeed, already have suspected it from certain anecdotes related in the present work, especially that concerning Sophocles.

It was considered shameful, in cases of true and honourable affection, to allow oneself to be tempted by the wealth or even the political influence of a lover. (See Plato, *Symposium*, 184 a.) The favours of corrupt and venal persons, on the other hand, were bought for cash. But such youthful male prostitutes incurred general contempt. Plutarch (*Erotikos*, 768 E) states: "Boys who voluntarily agree to act as accessories to debauchery of this kind are regarded as the most degraded of beings."

Some were "peripatetic," soliciting their patrons in the street or at cross-roads. They tried their hardest to attract attention by their dress and cosmetics, the object being to look as handsome and at the same time as effeminate as possible. They were so conspicuous that, in the words of a current proverb, it was "easier to hide five elephants in one's armpit than a single invert."

Such youths would follow a client to his house or submit to his caresses in the open air, at night, in deserted outskirts of the town, on Mount Lycabettus or even in the centre of the city itself, near the Pnyx.

Others, especially young and good-looking slaves, were kept by their masters in brothels and visited there by patrons. These houses paid an official tax annually confirmed by the Council.

It has been suggested in the foregoing pages that pederasty seems to have begun by being a "communion of warriors." On this view it is intelligible that the ancients saw the practice as encouraging valour and endurance, even in times of peace, as when, for example, it was desired to rid the State of a tyrant. For it is reported that the tyrant Polycrates of Samos had all the exercising arenas in the island burnt to the ground, believing that such "hotbeds of comradeship were so many citadels erected against him and constituted a menace of his power." (Athenaeus, XIII, 602 D.)

Affection between males had its origin in the gymnasium, where youths were trained in such techniques as hurling the javelin, which were really in the nature of preparation for active service. They were continued in military camps and finally practised on the actual field of battle.

"At Thebes, when a lad associated with a lover reached the age of enrolment, his protector presented him with a complete fighting outfit. Pammenes, who understood the character of masculine love, drew up his men in accordance with an entirely new principle. He set pairs of lovers side by side in the ranks. For he knew that love is the only unconquerable general. The tribe or the family may be deserted by their members. But once Eros has entered into the souls of a pair of male lovers no enemy ever succeeds in separating them. They display their ardour for danger and risk their lives even when there is no need for it. The Thessalian Thero, for instance, once laid his hand against a wall, drew his sword and cut off his thumb, challenging his rival in love to do the same. Another such lover, having fallen face downwards in battle, begged his enemy to wait a moment before stabbing him, lest the youth whom he loved should see him wounded in the back." (Plutarch, *Erotikos*, 761 B–C.)

The famous "sacred band" of Thebes was composed of pairs of lovers. Such was even the reason, according to Plutarch in his life of Pelopidas (18), why this unit was so called, "for Plato said that lovers were friends inspired by a god." The reference is to the *Phaedo*, 255 b. This body of troops is stated to have remained invincible until the battle of Chaeronea, where it was defeated by Philip of Macedon. The latter, when he inspected the field after the engagement, halted at the spot where the three hundred members of the "band" had fallen. Every one of them had his wounds, inflicted by pikes, in front. They had all retained their weapons and lay close together. Philip was lost in admiration. When he was told that they were all homosexuals, he wept for them, exclaiming, "Accursed be those who imagine that such heroes could ever do or suffer a deed of shame!"

Plato makes Phaedo say in the *Symposium* (179 a–b): "A handful of lovers and loved ones, fighting shoulder to shoulder, could rout a whole army. For a lover to be seen by his beloved forsaking the ranks or throwing away his weapons would unquestionably be less bearable than to do so in the presence of a crowd. He would a thousand times rather die than be so humiliated. As for abandoning his loved one on the field or refusing to rescue him when in peril, the worst of cowards would be inspired by the god of love on such occasions to prove himself the equal of any man naturally brave. For the gallantry which, as Homer says, a deity 'breathes into the hearts of heroes' is truly a gift of Love to lovers, one which they owe to him alone."

It is beyond dispute, therefore, shocking as the fact may appear, that "homosexuality contributed to the formation of the moral ideal which underlies the whole practice of Greek education. The desire in the older lover to assert himself in the presence of the younger, to dazzle him, and the reciprocal desire of the latter to appear worthy of his senior's affection, necessarily reinforced in both persons that love of glory which has always appealed to the competitive spirit of all mankind. Love-affairs accordingly provided the finest opportunities for noble rivalry. From another point of view the ideal of comradeship in battle reflects the entire system of ethics implied in chivalry, which is founded on the sentiment of honour." (H.-I. Marrou, *Histoire de l'Éducation dans l'Antiquité*, pp. 58–9.)

But the apprenticeship to courage and the love of honour and glory, important as they were to the Greeks, comprised only a part of Greek education. For lovers claimed that they participated actively in all the moral and intellectual development of their loved ones.

In the age of Pericles young Athenians attended schools in which they learned to read, write and calculate, as well as musical academies and gymnasia, where physical exercises were taught. Such was the elementary or, as we should say, "primary" instruction provided. More advanced tuition and moral training in the full sense of the phrase had to be sought elsewhere.

In the second half of the fourth century the "sophists" were the first to supply Greece with secondary education. They were sages who travelled about the country followed by the disciples whom they had attracted. They taught both the higher branches of knowledge and character-building. But their lessons were expensive and occasionally disappointing.

Family life could contribute little in this field. The "pedagogue" who took charge of the children of rich families was a slave. He could teach his young master scarcely more than a few maxims acquired by experience or derived from folklore. The mother would be herself distinctly ignorant and in any case ceased to supervise her sons after the age of seven. The father could indeed, in most cases, attend to his sons' intellectual and moral discipline, if he felt genuinely impelled to do so and above all if he had the time. But he was always engrossed in public life and his own profession.

At the beginning of the *Laches* of Plato two worthy fathers of families are introduced. Lysimachus is the son of the great Aristides and his friend Melesias is the son of Thucydides, not the historian, but the political rival of Pericles. Both fathers want to find tutors for their sons, so as to give the boys a better chance than they had themselves. For, as Lysimachus remarks with touching humility, "Neither of us is distinguished, since our illustrious fathers didn't take much trouble over our education. We're a bit ashamed of that in front of our sons. We blame our fathers for it. They let us run wild when we were young, as they were so busy with other people's affairs themselves."

This serious gap in the Athenian curriculum was normally—if one can use such an adverb in referring to abnormal personages—filled by pederasts. For pederasty functioned as a branch of pedagogy, as both Xenophon and Plato testify, the older man being naturally at pains to develop the character of his pupil, his "beloved," and pass on everything he knew to the boy, just as Heracles was said to have behaved in regard to Hylas, thus raising both his intellectual and moral standards.

The tradition was certainly an aristocratic one, since it allowed pederasts to transmit from generation to generation "an exclusively male ideal, involving contempt for women."

The *Memorabilia* of Xenophon are unquestionably more realistic than the dialogues of Plato in their details of the daily life of Socrates. It is clear that he was selflessly devoted to his friends. He willingly acted, in his own words, as "agent" and even "mediator" in their affairs. He organised among them a sort of mutual aid society. Xenophon gives a number of instances of the services thus provided, adding:

"Whenever his friends were in any difficulty, he did all he possibly could to show them how to help one another." (*Mem.* 11, 7, i.)

In short, he interpreted literally the Pythagorean principle already cited, "All is common among friends." This idea played a leading part in Greek moral philosophy.

It was by no means confined, of course, to material considerations. Thoughts and feelings were also subject to it. For this reason Socrates claimed to have an expert knowledge of love. But he did not, like the sophists, represent himself as a master in it. For he used to say that there was only one thing he knew, and that was that he knew nothing. On the contrary, he professed to act only out of passionate friendship, entirely dedicated to the happiness of those he loved.

All the schools of ancient philosophy adhered to this noble principle. They were essentially societies of friends. Epicurus even made friendship a fundamental institution of his Garden.

This outlook no doubt resulted from a profound intuition. Teachers succeed only in so far as they love. Socrates affirmed that he loved, though in a purely spiritual sense, those whose minds he wished to develop. When he went in search of disciples, he said he was "hunting down good-looking young fellows." It was a widespread notion in ancient Greece that a tutor needed the stimulus of beauty in his fertilising operations. Plato regarded education as "a spiritual bringing to birth of beauty." (*Symposium*, 206 b.)

. . . It is important to stress at this point the disturbing factor introduced by the idea of beauty into Greek educational theory. For it was considered that a mysterious affinity existed between physical and moral excellence. The Stoics asserted:

"The wise should love the young. For the latter's beauty proves that they are well equipped for virtue."

It may well be objected, however, that since not all young Athenians were handsome, the education of the less attractive must have suffered.

The Greek philosophers all defined homosexuality with pedagogy in view. Plutarch (*Erotikos*, 750 D) declares:

"Love for a young and talented mind leads to virtue by the road of friendship."

In the same author's life of Romulus Polemon says (30, 6):

"Love is provided by the gods to ensure the profit and well-being of the young."

The Greeks mistrusted knowledge derived entirely from books. They suspected written notes of any kind even more. In their opinion the best teaching was always oral, a communication between minds by word of mouth. In this respect too Socrates supplied the outstanding example. He wrote nothing. He only talked, gaining his effects to some extent, no doubt, by his famous irony, but still more by the inspiration of his extraordinary personality and the utter devotion it elicited from his youthful disciples, including even Alcibiades, whose faults, however, he did not hesitate to censure. Any lover worth the name, therefore, was for the Greeks one who initiated others into the life of the mind and acted as their guide and model. At Sparta, moreover, he was held morally responsible for the behaviour of his loved one.

Socrates furnishes the supreme, almost inimitable, instance of a pedagogue actuated by love of the purest kind. After his time Plato became the lover— though apparently not in the "Platonic" sense—of Alexis and Dion. Those who succeeded his nephew Speusippus as heads of the Academy were often selected, so to speak, in accordance with the principle of "pederastic adoption." Diogenes Laertius states that Xenocrates was the lover of Polemon, who in his turn fell in love with Crates. The Stoics followed the example of their founder Zeno in being for the most part ardently addicted to homosexuality.

We may conclude, with H.-I. Marrou, that:

"The relation between master and disciple always had some resemblance in ancient times to a love-affair. In principle the instruction imparted was not so much technical as represented by the sum of all the pains taken by a tenderly solicitous elder to promote the development of a pupil fervently desirous of responding to such treatment by showing himself worthy of it." (*Histoire de l'Éducation dans l'Antiquité*, p, 62.)

Women in Roman Society

In early Rome, women attained a dignity of station well witnessed in literature. But there was a singular lack of historical evidence of legal regulations to protect the concept of free relationships between man and woman. All unions might be dissolved without public intervention. Furthermore, a woman came under her husband's power in marriage without public process. From the time of the Twelve Tables of the Law this clearly meant male dominance. But it also made for ambiguities. The widow of Gaius Gracchus recovered her dowry from his heirs. Augustus secured laws which clearly stated the principle that a bride was only on loan to the husband from her paterfamilias. Women were free to end marriages at their own discretion early in the Empire. This gain in the status of women applied to those of the patrician class and ran parallel to the decline of patrician rule in Rome. Changes in the status of dowry rights, inheritance, and divorce thus reflect a transformation in which the position of woman in society changed markedly.

Early Republican law had presumed a high degree of subjection to the husband, coupled with extensive rights of fathers in their daughters' person and property. Examples are easy to cite: a woman had no legal redress against an adulterous husband, while she might be put to death by the husband without penalty if detected in adultery herself. By the end of the Republic, however, patrician women had in practice obtained independence, but they proved themselves poor guardians of their freedom. Many became infamous by the political marriages they made. Catullus and Ovid reflect the recoil of not unworldly men from the exaggerated, immoral actions of patrician women.

A little freedom proved a dangerous thing. Patrician bachelors favored substitutes for marriage in a variety of relationships with enfranchised slaves over whom they still had the authority of patrons. The distinction between marriage and cohabitation, already fine, was thus weakened. This seriously endangered family stability, and the prospect was that the pleasure of private parties would take precedence over the interests of the community, at least among the aristocracy. The evidence about other classes does not show this aberration. Hence the enduring paradox of sexual life in the early Empire: the freest women became the least secure; while among plebeians and proletarians, the sacredness of marriage continued to subject women to the fancy of their husbands. Otto Kiefer's *The Sexual Life of Ancient Rome* (1964) traces these developments and seems to justify the opinion of Simone de Beauvoir:

> The Roman woman of the decline was a typical product of false emancipation, having only an empty liberty in a world in which man remained. . . . the sole master; she was free—but for nothing.

From *Sexual Life in Ancient Rome*

Otto Kiefer

WOMEN IN ROMAN LIFE

Marriage

If we wish to understand the nature of regular marriage in Rome (*iustum matrimonium*) we must first differentiate between marriages in which the woman comes "into the hand" (*in manum*) of her husband and those in which she does not. What is the meaning of this singular phrase? It is this.The woman stood while she was a girl under the parental authority of her father, as all children did. Her father had *patria potestas* over her. If she is married to a man "into whose hand" she goes, that means that she leaves the authority of her father and enters the authority, the *manus*, of her husband. If she is married *sine in manum conuentione*, without entering the authority of her husband, she remains under the authority of her father or of his legal representative—in practice her husband is given no rights over her property. In later ages, as Roman women gradually emancipated themselves, it was to their advantage to be independent of their husbands with regard to property rights; accordingly, they made a point of avoiding marriages where they entered the *manus* of their husbands.

Marital authority, *manus*, was acquired only through the three forms of marriage recognized by a civil court—*confarreatio, coemptio*, and *usus*. We must now examine these forms in detail, in so far as they appear to bear on our subject; the more intricate details—some of which are the subject of much dispute—properly belong to a history of Roman law.

The oldest and most ceremonious form of marriage, corresponding to our church wedding, is *confarreatio*. The name is derived from a sort of meal-cake, *farreum libum*, which was used in the ceremony. Dionysius speaks of *confarreatio* as follows (ii, 25): "The Romans of ancient times used to call a wedding which was confirmed by ceremonies sacred and profane a *confarreatio*, summing its nature up in one word, derived from the common use of *far* or spelt, which we call *zea*. . . . Just as we in Greece consider barley to be the oldest grain, and use it to begin sacrifices under the name *oulai*, so the Romans believe that spelt is the most valuable and ancient of all grain, and use it at the beginning of all burnt offerings. The custom survives still; there has been no change to some more costly initial sacrifice. From the sharing of *far* was named the ceremony whereby wives share with their husbands the earliest and most holy food, and agree to share their fortune in life, too; it brought them into a close bond of indissoluble relationship, and nothing could break a marriage of that kind. This law directed wives to live so as to please their husbands only, as they had nowhere else to appeal, and husbands to govern their wives as things which were necessary to them and inalienable. . . ." In later ages this form of marriage was still obligatory for the parents of certain priests, but it was felt to be burdensome (Tac., *Ann.*, iv, 16). Certainly it was the oldest and most aristocratic; it was originally the customary form for patricians, and it long survived beside the other simpler and less ritual types.

SOURCE: Otto Kiefer, *Sexual Life in Ancient Rome* (London: Routledge & Kegan Paul, 1934), pp. 7–55.

The relation of the other forms of marriage to the old *confarreatio* is still a subject of dispute among scholars. It is now generally assumed that the second form, *coemptio*, was originally introduced for the marriages of the common people among themselves, since the plebeians could not employ the aristocratic *confarreatio*. The distinguished legal authority, Karlowa, in his book on the history of Roman law, asserts that *coemptio* dates back to Servian times, and was invented as a legal form of marriage for plebeians. At first a marriage by *coemptio* did not cause the wife (if she were a plebeian) to enter the family (*gens*) of her husband. This aroused the animosity of the commons, resulting in the law of the tribune Canuleius, which made the effects of *coemptio* similar to those of *confarreatio*. But *confarreatio* survived as the prerogative of the patrician class.

The third form of marriage was that by custom, or *usus*. It was laid down in the legislation of the Twelve Tables that cohabitation lasting for a year without interruption should be considered as a regular marriage. The peculiarity of this type of marriage lies rather in the exception than in the rule. For the effect of an interruption of cohabitation for three nights running (the *trinoctium*) was that *manus* did not exist; that is to say, the marriage was regular enough, but the wife did not leave the authority of her father and enter that of her husband. This was established by the legislation of the Twelve Tables (Caius, *Inst.*, i, 111). This form of marriage by custom was intended in the opinion of Karlowa, to regularize unions between foreigners and Romans where they were intended to be permanent. It was only later that it was used to emancipate the wife from her subjection to the husband. As Karlowa says, the widespread popularity of this form, whereby a wife could remain free from her husband's authority by an annual *trinoctium*, dates to a "time when, after the conquest of Italy, Rome began to look about for foreign conquests, to abandon her religious outlook, and break down the old morality." We shall give more detailed consideration later in this book to what may be called woman's struggle for emancipation in Rome; we shall therefore discuss Karlowa's opinion no further here. No one knows whether this type of marriage "without *manus*" was first introduced by legal enactment or simply by naturalization through time. It is certain that the poet Ennius knew it in the time of the first Punic war.

The three forms we have discussed differ in this point. In *confarreatio* the High Priest was present and marriage and *manus* came into being together. In *coemptio* the husband acquired *manus* by a special legal ceremony which was not in itself necessary for the celebration of the wedding. In *usus* a year's cohabitation was equivalent to marriage, but there was no *manus* unless that year had been unbroken by the separation called *trinoctium*.

The legal ceremony of *coemptio* was a mock purchase: the husband bought his wife for a trifling sum like a peppercorn rent. The prefix *co* emphasizes the fact that the husband acquires his rights over the wife as a kinswoman coordinate with himself (Karlowa). By it the wife gives herself into her husband's power—she is not a passive figure in the ceremony, but takes an active part in it.

Marriage by *coemptio* was the commonest form in later times. We know that *confarreatio* was archaic, and fell into disuse because it was so troublesome to perform. The jurist Caius tells us that marriage by *usus* had become obsolete in his day, partly by legislation and partly by custom (*Inst.*, i, 111). . . .

Divorce, Adultery, Celibacy, Concubinage

A marriage concluded by *confarreatio* could not, in early Rome, be dissolved. But in early Rome *confarreatio* was the only recognized form of marriage. It follows that divorce was unknown at that period. Dionysius says (ii, 25): "Authorities are agreed that no marriage was dissolved at Rome for the space of five hundred and twenty years. But in the 137th Olympiad, in the consulship of M. Pomponius and C. Papirius, one Spurius Carvilius (a man of some distinction) is said to have separated from his wife, being the first who did so. He was obliged by the censors to swear that he could not live with his wife for the purpose of having children, since she was barren—but for the divorce (although it was necessary) he was always hated by the commons." Dionysius tells us also that if the wife committed adultery or drank wine, she was punished with death by a family council attended by the husband. But according to Plutarch (*Romulus*, 22): "Romulus made several laws—one a severe one which forbids a wife to leave her husband, but allows a husband to divorce his wife for poisoning her children, for counterfeiting his keys, and for adultery." It is certainly true that (since Rome was in those early days a state ruled by men for men) wives could not divorce their husbands, but husbands could divorce their wives, chiefly for adultery.

In the legislation of the Twelve Tables, dissolution of marriage occurs in the form of a repudiation of the wife by the husband; and according to Valerius Maximus (ii, 9, 2) such a dissolution occurred in 306 B.C. The following were the misdeeds which gave a husband the right to divorce his wife: adultery, drinking wine, and a *peruerse taetreque factum* (perverse and disgusting conduct) which cannot be described more particularly. Much was left to the husband's discretion; but, as the above-mentioned passage in Valerius Maximus shows, he was obliged to summon a council of his family or friends before repudiating his wife. Here is Gellius's comment on the first divorce (iv, 3): "Tradition says that for about five hundred years after the foundation of Rome there was no form of legal process concerning marriage nor any stipulation contemplating divorce, either in Rome or in Latium—there was no call for such a thing, since no marriages were dissolved in that period. Servius Sulpicius writes in his book on Dowries that the first legal injunctions dealing with marriage had to be made when Spurius Carvilius (also called Ruga), who was a distinguished man, divorced his wife because she was physically unable to bear children." This account shows, then, that the first dissolution of marriage in Rome was occasioned by the wife's barrenness. According to Becker-Marquardt, it was not the first; but it was the first which did not involve the disgrace and condemnation of the wife. In this case the wife would keep her dowry, while the husband would retain it after the divorce if she had been guilty of misconduct. (The juristic formula for repudiation without misconduct was *tuas res tibi habeto;* "keep your property for yourself.")

All these accounts seem to agree in making it certain that in early Rome marriages were very seldom dissolved. But can we infer from that to a high degree of morality in married life? That is another question. We must not forget that the husband could commit no action which the law recognized as a breach of his marriage-tie; he had an entirely free hand. And the liberty of wives was so restricted that they seldom had the opportunity to commit misconduct—especially since they were faced with terrifying punishments if they were convicted. Her punishment was not only to be driven with disgrace and infamy from the home in which she had lived, as well as that, she could be put to death by the family council in co-operation with the husband.

In those early times there were no statutory penalties for adultery—probably because the husband took the matter into his own hands, or called on the family council to inflict the punishment. For example, Valerius Maximus (vi, 1, 13) mentions several cases in which a detected adulterer was flogged, or castrated, or handed over as *familiæ stuprandus*—which last penalty meant that the servants and retainers of the injured husband inflicted sexual dishonour on the adulterer. Accordingly, a husband who committed adultery with a married woman was liable to severe punishment; but not if he made love to slaves or prostitutes, although we should consider that also to be adultery. For instance, Valerius Maximus (vi, 7, I) tells the story of the elder Scipio Africanus: "Tertia Æmilia, his wife . . . was so kind and patient that when she knew one of her maids pleased him she pretended to notice nothing, in order that she should not cast guilt on Africanus, the conqueror of the world." And in Plautus (*Men.*, 787 ff.) a father meets his daughter's complaints thus:

How often I've told you to honour your husband, not to watch what he does, where he goes, what he thinks of?

When she complains further of his fickleness, he replies:

And wisely!
Since you watch him so closely, I'll help his lovemaking.

And later he adds:

He keeps you well jewelled and dressed, and he gives you your food and your maids. Better come to your senses!

Cato, in concise and prosaic language, describes the contrasting situations of an adulterous wife and an adulterous husband (*ap.* Gell., x, 23): "If you take your wife in adultery you may freely kill her without a trial. But if you commit adultery, or if another commits adultery with you, she has no right to raise a finger against you." Yet if a husband committed adultery with a slave, a determined wife knew what to do. That is shown by passages like Plautus, *Men.*, 559 ff., *Asin.*, v, 2, and Juvenal, ii, 57; Juvenal speaks of the "uncouth concubine" who is "sitting in the stocks" and working at the bidding of the wife.

We have seen, then, that in early Rome there was no statutory punishment for adultery, whether committed by the wife or by the husband. That is corroborated by Cato's remark (*ap.* Quint., v, II, 39) that every adulteress was a poisoner. Since there was no law aimed directly at adultery, the crime was attacked in this curious indirect way. The first legal penalties for adultery occur in Augustus's moral reforms, of which we shall speak later. The penalties were banishment and loss of certain property rights; corporal punishment could be inflicted on members of the lower classes. In later times, as the tendency was, these penalties were increased. Constantius laid down that adultery should be punished by burning alive or drowning in a sack, and Justinian compelled adulterous women to be shut up in convents. These further developments can be described in Mommsen's words as "pious savagery."

In the later Republic, divorce became easier and more general as the status of women improved. An important point was that a marriage without *manus* could be announced simply as an agreement between two parties. This, of course, led to many frivolous results. Valerius Maximus (vi, 3, 12) speaks of a marriage which was dissolved because the wife had visited the games without her husband's knowledge. And Cicero in one of his letters relates that a wife obtained a quick divorce before her husband came home from the provinces, simply because she had made the acquaintance of another man whom she wanted to marry. There is no reason to be surprised when we hear that Sulla was married five times, Pompeius five times, and Ovid three times. We cannot say, then, that easy divorce came in only with the Empire. Still, marriage and divorce were then regarded with ever increasing frivolity. Seneca writes (*De ben.*, iii, 16, 2): "Surely no woman will blush to be divorced now that some distinguished and noble ladies count the years not by the consuls but by their own marriages, and divorce in order to be married, marry in order to be divorced! . . ."

But it would be very unfair to blame only the women for this so-called decline in marriage. We know that even in early times the men had no great enthusiasm for the responsibilities of fatherhood. If this were not so, we could not understand why a man who obstinately refused to marry should be punished by the censors with the infliction of certain pecuniary disadvantages. Cicero says (*De leg.*, iii): "The censors are to prevent celibacy." According to Valerius Maximus (ii, 9, I) there was a censorial decree against it as early as 403 B.C. Livy (*epit.*, lix) and Gellius (i, 6) tell us that in 131 B.C. the censor Metellus made a famous speech on this matter; it contains some very significant words, which throw a lurid light on the Roman conception of marriage: "If we could live without wives we should not have all this trouble. Since nature has brought it about that we can neither live with them in peace nor without them at all, we must ensure eternal benefit rather than temporary pleasure." The most interesting thing about this remark is that the speaker was happily married, and had four sons, two daughters, and eleven grandchildren; he spoke from experience. From Gellius (i, 6, 6) we know the official point of view: "The state cannot be safe unless marriages are frequent."

After the war with Hannibal, the poorer classes increased in numbers. Writers now spoke frankly about the flight from marriage. Plutarch writes (*De amore prolis*, 497e): "The poor do not rear their children, because they are afraid that if they are badly fed and educated they will grow up to be slavish and boorish and to lack all the graces." Besides that, there was the consideration stated thus by Propertius (ii, 7, 13):

> How can I furnish boys for family triumphs?
> My blood will never breed a soldier son.

Seneca adds another discouragement (*fr.*, xiii, 58): "The most fatuous thing in the world is to marry to have children so that our name is not lost, or so that we have support in our old age, or certain inheritors." In the end even the state lost its strongest motive for encouraging marriage; it ceased to need a constant supply of young soldiers for its interminable wars. In the long peace of the first centuries of our era Rome demanded no such quantities of spear-fodder to maintain its position or extend its power. At that time it was much easier to live like the man whom Pliny describes (*ep.*, iii, 14)—he was an ex-prætor who lived in his villa with some concubines. (He was, of course, unmarried.) And, finally, if a man had leanings to philosophy, a family was nothing but a burden to him. Cicero said so (*ap*. Sen. *fr*. xiii, 61): "Cicero was asked by Hirtius if he would marry Hirtius's sister now that he was separated from Terentia; he replied that he would never marry again, for he could not cope with philosophy and a wife at the same time." And Cicero says in his *Paradoxa:* "Is he free who is subject to a woman? who is ruled and regulated by her, who is told to do or not to do whatever she wishes?"

We see, then, that as the individual was gradually freed from the bonds of traditional morality and the demands of the community, his reasons for not marrying increased. This repeats itself throughout history.

Naturally, the state sometimes endeavoured to check this development through legislation; its existence was at stake. Augustus was the first to make the endeavour. His moral ordinances were bold and radical, but had little effect, for state legislation has little effect in such matters. Mommsen describes them in remarkable language: they were, he says, "one of the most impressive and long-lasting innovations in penal law which is known to history." They were known as the *Juliæ rogationes*, and included the *lex sumptuaria*, the *lex Julia de adulteriis et de pudicitia, the lex Julia de maritandis ordinibus*, and the *lex Papia Poppæa*— passed between 18 B.C. and A.D. 9. We may sum up their purpose in the words of Becker-Marquardt: "to impose property disqualifications for celibacy on men between twenty and sixty and on women between twenty and fifty and for childlessness on men over twenty-five and women over twenty; to confer various rights and privileges as encouragements on parents of three or more children; to bring about suitable marriages between people of senatorial families; and to regulate divorce by certain rules and ordinances."

Augustus rigidly enforced these laws. What effect did they produce? Let us hear the evidence of a few of his contemporaries. Suetonius (*Aug.*, 34), writing of the law to encourage marriage within the various social classes, says: "He had emended this law far more severely than the others, but so many protested that he could not carry it through, unless when he abolished or relaxed some of the penalties, granted three years' general immunity and increased the rewards. But even so, once at a public show the knights shouted for its total abolition: Augustus called the children of Germanicus to him and took them into his own embrace and gave them to their father's arms, signifying by gesture and look that the grumblers should not be reluctant to follow the example of the young Germanicus." We read in Cassius Dio (54, 16): "Loud complaints were raised in the senate about the disorderly conduct of the women and young men; to that conduct they attributed the prevailing reluctance to marry, and they tried to induce Augustus to correct it by personal abuse, hinting at his many love affairs. His first reply was that the essential points had already been settled, and that it was impossible to regulate everything in that way. Then when they still held him to it he said: "You yourselves ought to give orders and directions to your wives as you wish. I certainly do." When he said this they pressed him all the more, and demanded to know what the directions were which he said he gave Livia. So he was compelled to make a few remarks on women's dress and finery and public appearances and modest conduct—not caring that his actions did not agree with his words." In another passage Cassius Dio tells us that the emperor made a very full and detailed speech in defence of his legislation. Although the oration as given by Dio may not be authentic in every word, it contains the fundamental ideals and purposes of the Julian legislation: we shall therefore quote a few extracts from it (Cassius Dio, 56, I ff.): "During the triumphal games the knights insisted vehemently that he should repeal the law about celibacy and childlessness. Augustus therefore assembled in different parts of the forum those knights who were unmarried and those who were married, including those who had children. When he saw that the married men were much fewer than the others, he was grieved and addressed them something after this fashion:

"Rome was at first a mere handful of men; but when we bethought us of marriage and had children we came to excel the whole world not only in strength but also in numbers. We must remember this, and console our mortality by handing on our stock like a torch to a never-ending line of successors—so that we may help one another to change our mortality (the one side of our nature which makes us less happy than the gods) to eternal life. It was for this end above all that our Creator, the first and greatest of the gods, divided the human race into two sexes, male and female, and instilled into both of them love and the desire for sexual intercourse, and made their association fruitful—in order that new generations might make even mortal life immortal. . . . Surely there is no greater blessing than a good wife who orders your house, watches over your possessions, rears your children, adds happiness to your days of health and tends you when you are ill, shares your good fortune and consoles you in trouble, controls the wild passion of your youth and softens the harshness of old age. . . . These are some of the private advantages enjoyed by those who marry and have children. As for the state—for whose cause we ought to do many things against our inclination—without doubt it is honourable and necessary (if cities and peoples are to exist and if you are to rule the others and the whole world is to obey you) that a large population should in peace time till the soil and sail the seas and practise arts and apply themselves to handicrafts, and in war defend their possessions with all the more zeal because of their families and raise up others to replace the slain. . . .

"He now addressed the unmarried men as follows:

"What am I to call you? Men? You have not yet proved yourselves to be men. Citizens? As far as you are concerned the city is perishing. Romans? You are doing your best to destroy the very name. . . . A city is men and women, not buildings and colonnades and empty market-places. Consider the righteous anger which would seize the great Romulus our founder if he compared the time and circumstances when he was born with your refusal to beget children even in lawful marriage. . . . Those old Romans begot them even on foreign women, but you disdain to make Roman women the mothers of your children. . . . You are not such recluses that you live without women: not one of you eats or sleeps alone. All you wish is to have liberty for sensuality and excess. . . ."

That was the anti-Malthusian ideal which lay behind Augustus's legislation. But it found no enthusiastic supporters; all classes of society had long been endeavouring to increase their personal freedom. The measures were bound to miscarry—especially since everyone knew that the princeps himself had till then done little to live up to an austere moral code. The results were, in sum, the introduction of a hitherto unheard-of police espionage into the most intimate details of private life—and a number of marriages contracted from purely mercenary motives. Seneca says: "What shall I say of the men of whom many married and took the name of husband in order to make fun of the laws against the unmarried state?" And according to *Dig.*, xlviii, 5, 8, husbands must often have benefited by the adultery of their wives, and actually been their procurers. Tacitus writes (*Ann.*, iii, 25): "More and more citizens came into danger, since every household was shaken by the inferences made by spies; people now suffered by the laws as they had once suffered by their own crimes." In addition, a law was passed which we shall discuss elsewhere—that no woman whose grandfather, father, or husband was a knight should sell herself for money. So small was the real effect of Augustus's legislation.

One of the principal facts which kept it from real achievement was the circumstance that the law applied only to freeborn citizens. It did not, therefore, cover slaves and the various classes of women who did sell themselves. This allowed the men to find sexual satisfaction outside marriage quite as freely as before. Also the freedom of prostitutes must have had great attractions for the so-called respectable women, who now fell under legal restrictions, so that many of them assumed the prostitute's dress in order to live undisturbed by the law. (Cf. *Dig.*, xlvii, 10, 15, 15.)

We may conclude our discussion of Augustus's legislation by noticing that it gave the first legal recognition to concubinage, that is, to cohabitation without the status of husband and wife. The code had had, as one of its chief aims, the encouragement of suitable marriages among the senatorial families. It was necessary that the law should take into account marital relationships which were not "suitable"—a senator, for example, might desire to marry a freedwoman or a former prostitute, or have lived with her as husband and wife. All such cases were legally recognized as concubinage. A man could live in concubinage with the woman of his choice instead of taking her to wife; but he was compelled to give notice of this to the authorities. This type of cohabitation was externally in no way different from marriage, and its practical results were purely legal; the children were not legitimate, and could not make the claims of legitimate children on their father. Therefore, men of high rank often lived with women in concubinage after the death of their first wife in order not to prejudice the claims of the children they had had by her. For instance, the emperors Vespasian, Antoninus Pius, and Marcus Aurelius lived in this way. The principle of monogamy was not affected by concubinage, since (Paulus, ii, 29, I) it was impossible to have a wife and a concubine at once. Accordingly, the title of concubine was not derogatory, and it appears on tombstones.

The Emancipation of Roman Women

As we have often said, the early Roman republic was, as far as antiquity allows us to see, a masculine state, a state run by men for men. We may refer here to the important propositions established by Dr. M. Vaerting, in his book *The Character of Women in a Masculine State and the Character of Men in a Feminine State* (Karlsruhe, 1921). When he says that the "standards of social conduct in a masculine state are inverted in a feminine state," his remark can be applied without reservation to early Rome. The ruling sex—men—had all property rights; at marriage the wife brought a dowry to her husband; the man had the "tendency to assign to the subordinate sex, woman, the house and the home as her own province." But Dr. Vaerting establishes many other peculiar marks of a masculine state, in connection with married life; and they may all be applied to early Rome—especially the regulations on feminine chastity, the "double standard of morality."

Now Vaerting lays down that if one sex frees itself in a state dominated by the other sex, "simultaneously with the loss of power by the ruling sex, the peculiar functions and natures of the sexes also change." That is to say, the man has hitherto appeared only as a stern lord and master, as a rough soldier, and as a powerful and energetic statesman. He now becomes softer and more human—although softness and humanity would once have been regarded as unmanly. The woman has hitherto been nothing but a chaste and discreet housewife and mother. She now shows herself as an independent personality; she disregards the ties which once bound her, vindicates her own right to happiness and pursues it with all her might. When she does, her actions are regarded as degeneracy by those who know only the masculine state and its ideology.

That is exactly the change which occurred in the history of Rome. And it prompts us to ask why the old republic, which was dominated by men, should have evolved into the state which we see fully developed under the emperors. . . .

. . . In ancient Rome the case was different. The old republican institution of the family gradually altered its nature; but in our belief the cause of that alteration was purely economic. We shall expound our reasons.

It can hardly be an accident that all ancient writers mark the end of the second Punic war as the turning-point of morality and social tradition—and so as the beginning of the emancipation of Roman women. That was the period when Rome ceased to be a state of yeomen-farmers. A well-known passage of Appian describes the beginning of that ominous change (*Bell. Civ.*, i, 7): "As the Romans gradually conquered Italy, they took portions of the conquered territories and built cities on them, or sent their own citizens to colonize previously existing cities. These settlements served to garrison the conquered countries. They took the cultivated land which they had won, and immediately divided it up among their settlers, either gratis or for purchase-money, or for rent. But they did not take time to draw lots for the great areas of land which were uncultivated because of the war; they made public proclamation that anyone who wished might cultivate it, the rent being a yearly percentage of the crop—10 percent of the seed crops and 20 percent of the fruit. Those who engaged in stock-rearing were obliged to pay a proportionate rent for large and small stock. This they did in order to help the spread of the Italian population; they saw that the Italians were a hard-working race, and wished to have them as friends and allies.

"But the result was the opposite of their intentions. The rich seized most of the unoccupied land. Circumstances made them confident that no one would deprive them of it, and so they acquired the land surrounding their own, and all the small farms owned by poor men, partly by purchase and partly by force, until they were farming wide plains instead of estates. They used slaves to till the land and raise the stock, so that they should not be mobilized to serve in war, not being of free birth; also, the possession of slaves brought great profit to their owners, since slaves, being immune from war-service, multiplied with impunity.

"Consequently, the ruling class accumulated all the wealth for themselves, and the slave-population filled the country, while the real Italian population decreased terribly, worn out by poverty, taxation, and military service. And when there was a respite from these things they found themselves unemployed, because the land was owned by rich men who used slaves instead of freemen on their farms."

Whatever the origin of this passage may be, it shows the necessary result of the military expansion of Rome. The real representatives and furtherers of that policy—the old Roman families—were gradually eliminated, and replaced by slaves; and such small farmers as survived the numerous wars sank to the position of an unemployed proletariat in the cities.

And the great conquests in the West and the East had other results, which are described by other authors. Farmers found it unprofitable to grow grain in Italy since the Roman market was flooded with masses of imported grain which forced down the price (Liv., xxx, 26). And the victorious armies had brought home (especially from the East) enormous wealth and luxuries. Livy writes as follows (xxxix, 6): "The beginnings of foreign luxury were brought to Rome by the army from Asia in 186 B.C. They were the first to import couches of bronze, costly draperies, tapestries, and other woven things, as well as one-legged tables and sideboards, which were then considered costly articles of furniture. At that time banquets were first graced by the addition of girls to play the lyre and the harp, and other entertainments; and more care and expense were devoted to the banquets themselves. The price of cooks rose—although they had formerly been the cheapest and least regarded of slaves—and what had been a menial office was now regarded as an art." Polybius corroborates this (xxxi, 25, as quoted by Athenaeus, 6, 274 f.): "Cato gave vent in public to his displeasure that many people were introducing foreign luxury into Rome: they bought a keg of salt fish from the Black Sea for three hundred drachmæ, and paid more for handsome slaves than for estates." Again, we read in Velleius Paterculus (ii, I) of a slightly later period: "The elder Scipio prepared the way for Rome's power, the younger for Rome's luxury. When the fear of Carthage was removed and Rome's rival cleared from her path, the passage from virtue to vice was not a gradual process but a headlong rush. The old moral code was abandoned and a new one supplanted it. Rome gave herself up to sleep instead of watching, to pleasure instead of the use of weapons, to leisure instead of business. It was at that time that Scipio Nasica built the colonnades on the Capitol, Metellus, those others we have described, and Cneius Octavius, the handsomest of all in the Circus; and private luxury followed hard on public ostentation."

If we examine all these accounts without prejudice we must come to this conclusion: what happened was the economic conversion of a small and limited state of simple farmers into a powerful oligarchy of prosperous but uneducated landowners, merchants, financiers, with a class of proletarians. It is easy to understand that in the course of this economic change there should have been civil disturbances and the usual battles between the classes; for the new wealth and luxury blinded men's moral perceptions, and opened up unimagined possibilities for those who could seize and retain power. The civil wars of Marius and Sulla, of Pompey and Cæsar, were bound to follow. Although the two Gracchi made one more vain endeavour to set the old Roman farmer-state on its feet, the contest in Sulla's time was only a wrangle for power and the wealth of Rome. Velleius writes (ii, 22): "A new horror appeared in later times. Greed was another reason for cruelty, and a man's guilt was measured by his possessions—anyone who was rich was a criminal, and paid the price of his own life and safety; nothing was dishonourable if it was profitable."

The old organization of the family, with all its restrictions on individual freedom through the predominating *patria potestas*, was bound to break up—although it had guaranteed a certain limited standard of manners and morals. No one can wonder at its dissolution; think of the parallel instances of the boom in Germany after the Franco-Prussian war, or even of the period after the Great War. When an entire economic epoch is breaking up, it is impossible for women not to change their nature and outlook; especially since new wealth and new opportunities have a more powerful effect on the spirit of women than on men. The average woman at that time in Rome saw new and unprecedented possibilities of satisfying her innate vanity, ambition, and sensuality. But women of deeper natures welcomed the opportunity to acquire a new and better education, to develop their talents for dancing, music, singing, and poetry. There are some examples of this in ancient literature. Sallust has left us a brilliant picture of an emancipated woman of this kind (*Cat.*, 25). He says: "Among the women who supported Catiline was Sempronia, who had often shown herself as brave as any man. She was blessed by fortune in her rank, her beauty, her marriage, and her children. She was learned in Greek and Latin literature; she played the harp and danced with more grace than an honest woman should; she had many other accomplishments which pertain to a luxurious life. But she valued her honour and her chastity least of all. It was difficult to decide whether she cared less for money or for reputation. Her lust was so overpowering that she courted men more often than she was courted by them. In the past she had often committed perjury, misappropriated property entrusted to her, and been accessory to murder; she had sunk to terrible depths by her extravagance and poverty. Still, her talents were considerable. She could write verses, make jokes, and talk modestly, tenderly, or daringly; she was full of high spirits and very charming." There is a certain partiality in Sallust's account of the lady; but we can see that Sempronia must have been an unusually cultured woman, far above the level of the average Roman matron. She was such a figure as we read of in the German romantic period. In fact, she had become conscious of her rights as a woman, and cared nothing for the prejudices of her honest but dull sisters. Naturally, such a woman sometimes acquires the reputation of immorality, extravagance, debauchery; it happens to-day. To judge her properly we must remember that she came from a distinguished family, and was the wife of the consul D. Junius Brutus, and mother of D. Junius Brutus Albinus, one of Cæsar's murderers.

It is certainly wrong to hold that education and culture were responsible for making the serious matron of ancient times into a voluptuous and frivolous hetaira. That is proved, for example, by a charming passage in Pliny. He is praising his wife for her intellectual alertness (*Ep.*, iv, 19): "Her mind is keen and her tastes simple. She loves me, which proves her chastity. Besides, she likes literature, to which she was led by her affection for me. She keeps my books, reads them, and even learns passages from them off by heart. She is painfully anxious when I am to conduct a case, and delighted when I have completed it. She appoints people to tell her what applause and shouts I have received, and what the verdict was. If I am reading my work in public she sits near by, behind a curtain, and drinks in the praise of my audience with expectant ears. She also sings my verses, and even sets them to music, taught not by a musician but by love, the best master."

But the accusations of immorality against Roman women were of old standing.

It is not by chance that one of the first complaints dates almost exactly to the period when the emancipation began. The elder Pliny (*N.H.*, xvii, 25 [38]) tells us that the consul L. Piso Frugi lamented that chastity had disappeared in Rome. That was about the middle of the second century B.C. And the oldest Roman satirist, Lucilius (who lived in the same period), is said to "have blamed the excesses and vices of the rich" (*Schol.* Pers., 3, I). Similar complaints continued for centuries. We could fill books with them; a few characteristic examples will be enough.

Sallust (*Cat.*, 13) observes that after Sulla's time "men gave themselves to unnatural vice, and women publicly sold their honour." There is a famous jeremiad in Horace's sixth Roman ode (iii, 6):

> Ages fertile in crime defiled
> first pure marriage, the home, the breed:
>   thence a deluge of sheer disaster
>   burst on the land and people.
> Each ripe maiden has learnt to love
> soft Greek dances, and knows the arts
>   taught by shame, and is early practiced
>   body and soul in lewd loves.
> then seeks younger adulterers,
> while her husband's at wine; she gives
>   any man the forbidden favours
>   hastily in the dark room,
> nay, she rises obediently
> (not unknown to her husband) when
>   pedlars call, or a Spanish sailor
>   purchases her dishonour.

. . . We could, of course, see the whole development as nothing but the progressive sexual liberation of women; but the new freedom was not expressed in sexual life alone. It was principally an *economic* freedom which women then achieved.

Earlier in this book we have explained that in the old republic women were economically dependent on men. Originally, marriages ended in *manus*, which, as we saw, meant complete domination by the husband. As the old type of marriage, with its predominance of the husband, gradually changed into free marriage, women came to achieve economic independence. In a free marriage the wife kept all her property, except that her husband took her dowry. If her father died she was *sui iuris*—for she had hitherto been under his authority, but now she either remained sole mistress of her property or else took a guardian to help her in its administration. The guardian frequently entered an even closer relationship to her, and, as we know from various cases, occasionally became her lover. In time women must have come to own a very considerable amount of property. If this were not so the attempt would never have been made to decrease it by the *lex Voconia*, which in the year 169 B.C. forbade women to receive legacies. Gellius (xvii, 6) tells us that Cato recommended the adoption of the law in these words: "First of all a wife brings you a large dowry. Then she receives a great sum of money which she does not make over to her husband, but gives to him as a loan. And finally, when she becomes angry, she orders her debt-collector to follow her husband about and dun him." The law is still a subject of controversy among scholars. Certainly it cannot have had much effect, for the laws of inheritance became constantly more and more favourable to women, until finally, under Justinian, the two sexes were given almost equivalent rights. Woman had at last come of age, legally and economically. But these last stages in her development lead into the period when Christianity was supreme, and so fall without the scope of our book.

Besides the sexual and economic freedom achieved by women in early Rome, there was also a political emancipation. It is of less far-reaching importance than the emancipation in sexual and economic life; still, it is interesting enough to merit a short discussion here, for the picture of Roman womanhood would be incomplete without it.

Women in Rome had absolutely no political rights. We read in Gellius (v, 19) that "women are debarred from taking part in the citizen-assembly." But, in contrast to this, the Roman matron enjoyed a much higher degree of personal independence than the Greek wife. She took part, as we have seen, in meals with the men; she lived in the front part of the house, and she could appear in public, as Cornelius Nepos says in his preface. According to Livy (v, 25) the women freely sacrificed their gold and jewels for the state at the time of the Gallic invasion, and were consequently granted the right to drive to religious festivals and games in four-wheeled carriages and to travel about on ordinary festivals and working days in two-wheeled vehicles. Besides, there were certain religious services which were attended by women alone—we shall have more to say of them later. We may remind our readers of the conduct of the women when Coriolanus was attacking Rome. As women freed themselves gradually from the restrictions of the old patriarchal family they allied themselves with one another to further their common interests. We have no exact evidence on this stage of their evolution, but about the time of Tiberius authors speak of a previously existing *ordo matronarum*, a class, almost a society, of married women (Val. Max., v, 2, I). In Seneca (*fr.* xiii, 49) we find the words: "One woman appears in the streets in richer attire, another is honoured by all, but I, poor wretch, am disdained and contemned in the women's meeting." Suetonius (*Galba*, 5) also knows of the women's meeting, as an apparently permanent institution to represent women's interests. Under the emperor Heliogabalus (Ael. Lamprid. *Heliog.*, 4) an assembly room was built for the "senate of women" on the Quirinal, where the *conventus matronalis* (assembly of married women) had been accustomed to meet. Lampridius uses the words *mulierum senatus*. However, he calls its decrees "ridiculous," and says they were concerned chiefly with questions of etiquette. It was, therefore, of no political importance. Friedlander's conjecture (*History of Roman Morals*, v, 423) may be true: he thinks that these assemblies dated back to some religious union of women.

And there is no political significance in the event which Livy (xxxiv, I) describes so vividly; however, it is important for the comprehension of the character of the Roman woman, and for that reason we shall discuss it in some detail. In 215 B.C., during the terrible pressure of the Hannibalic War, the Romans introduced a law, the *lex Oppia*, which laid restrictions on the use of ornaments and carriages by women. However, after the victory of Rome, these severe measures seemed to be less needful, and the women exerted themselves to have the law removed. It was repealed in 195, in the consulship of M. Porcius Cato, although that conservative of conservatives backed it with all his influence and authority. Here is Livy's account:

> The anxieties of great wars impending or newly ended were interrupted by an affair which sounds unimportant but developed into a great and bitter struggle. M. Fundanius and L. Valerius, tribunes of the commons, proposed to the commons that the Oppian Law should be repealed. It had been enacted by C. Oppius, tribune of the commons, in the consulship of Q. Fabius and Ti. Sempronius while the Punic war was raging; it provided that no woman should possess more than half an ounce of gold, wear a garment of many colours, or ride in a carriage within Rome or a provincial town or within a mile of either of those places unless for public worship. The tribunes Marcus and Publius Janius Brutus defended the law, and said they would not allow it to be repealed; many distinguished men appeared to back it or oppose it; the Capitol was crowded with its supporters and opponents. Neither influence, nor modesty, nor their husbands' commands could keep the married women within doors. They beset all the streets in Rome and all the approaches to the forum, imploring the men who were going down to the forum that they should allow their former luxuries to be legalized, now that there was general prosperity in Rome. Every day the crowds of women grew, for they even came into the city from the provincial towns and market-boroughs. And now they dared to go to the consuls, the prætors, and the other magistrates and beseech them. But they found one of the consuls, M. Porcius Cato, quite inexorable.

Livy now describes a great duel of oratory between the chief opponents—Cato the diehard, and the liberal Valerius; he states all the grounds which they adduced for and against repealing the law. The most interesting parts of their speeches are those in which they put forward entirely opposite views on the character and the ideal position of women in law and in public life. Cato says: "Our ancestors laid down that women might carry out no business—even private business—without supervision from her guardian, and they confined them to the authority of their parents, brothers, and husbands. But we—save the mark!—are allowing them to take part in the government of the country and mingle with the men in the forum, the meetings, and the voting-assemblies. What else are they doing at this moment, in the highways and byways, except supporting a bill sponsored by the tribunes and voting for the repeal of a law? Give rein to that headstrong creature woman, that unbroken beast, and then hope that she herself will know where to stop her excesses! If you do not act, this will be one of the least of the moral and legal obligations against which women rebel. What they wish to have is freedom in all things—or, rather, if we are to tell the truth, licence in all things." Later in his speech Cato condemns especially the fact that women want freedom in order to have more luxury. "What honourable pretext can be adduced for this revolt of the women? 'We wish to be resplendent in gold and purple,' we are told, 'to ride through the city in carriages on feast-days and working days as if we were celebrating a triumph over the law which we conquered and repealed and over your votes which we captured and carried off; we wish no limits to extravagance and display.' "

The tribune Valerius meets Cato by declaring: " 'Before this, women have appeared in public—think of the Sabine women, the women who met Cariolanus, and other cases. Besides, it is right to remove laws without trouble as soon as the circumstances which called them forth have changed, as is done in other cases. . . . Shall all other ranks and kinds of men,' " he says (and here we are again quoting Livy's version), " 'feel the benefit of the country's prosperity, while only our wives are deprived of the fruits of peace and tranquillity? We men shall wear purple on our official and priestly garments; our children will wear the toga with the purple stripe; we concede the right of wearing the purple stripe to magistrates in the colonial towns and to the lowest magistrates in Rome, the overseers of the wards—not only while they are alive, but even that they be burned wearing it when dead. Shall we then forbid the women to wear purple? When you can have a purple saddle-cloth, is your wife to be forbidden to have a purple cloak? and are the trappings of your horse to be more splendid than the dress of your wife?' " He makes the point that even if this concession is made, the women will still be under the authority of their husbands and fathers. " 'As long as her kinsmen are alive, a woman is never free from her slavery; and she herself prays that she will never have the freedom brought by widowhood and bereavement. They would rather you should decide on their adornment than the laws. And you ought to keep them under your authority and guardianship, not in slavery to you, and you should prefer to be called fathers and husbands than masters. . . . In their weakness they must accept whatever decision you make. The more powerful you are the more moderately you should use your power.' "

These speeches as given in Livy may not be authentic. Still, they reproduce the atmosphere and outlook of the opposition; even in Livy's time men of the ruling classes still opposed the emancipation of women thus. We may remind our readers that after this memorable meeting of the senate the women did not rest till the law they thought obsolete was repealed. But it should not be imagined that with this success women began to exercise any important influence on the government of Rome. In principle women were, and always remained disqualified from taking part in politics.

Masters and Slaves

Roman society in the age of Cicero and Cæsar was a society of status under siege. The mobility of money and the rise of freedmen complimented patrician decline. The world of Trimalchio was also the world of Tiro, Cicero's slave and chief secretary who was able to buy a small farm after Cicero freed him. Other freedmen entered commerce, while Diochares and Theophanes were political agents of Cæsar and Pompey respectively. Horace's father was a freedman, and the poet himself grew wealthy in Roman society, moving as did others of his class in the affinities of great patricians. After a life of extreme poverty, the poet had become the darling of Roman society. By 38–33 B.C., he was himself a slave-owner, and to that fact we owe one of the great literary diatribes on Roman slavery.

In Book Two of Horace's *Satires* the slave Davus engages his master on the subject of slavery. A privileged house slave, Davus was worlds removed from agricultural labor or the slavery of galley or mine; yet he was, strictly speaking, no part of Roman "society." He was, in law, not a person but a chattel. Aristotle had called slaves "living tools." Cicero kept them, asked the governor of Illyricum to hunt down one who had stolen books, wrote of them as *pueri* (youth) fit only to carry letters or be bought or sold, and apologizes for a lack of decorum in his grief at the death of one who used to read to him. Yet apart from these things, capitalists valued slaves whose labor made their masters independent of the market. Rome's wars enslaved whole peoples, as when Caesar sold 53,000 captives into slavery in one campaign. Varro thought the skilled slave essential to the survival of government and economy. There were perhaps 200,000 of these "living tools" in Horace's Rome.

Hence the sting of Davus's conversation with Horace. The satirist raises the question: Who is the slave, the chattel or the man dependent on him? The Saturnalia (December holiday) gives Davus license to speak his mind, and he pours out a diatribe on vices and virtues, loyalty and infidelity, right reason and unruly desire. The slave becomes Rome's teacher. Abandon restraint—the Republic had given way to Caesarism and the Empire—and be enslaved! Rome's master classes are in no wise superior to their slaves. Davus sleeps with sluts, Horace with a friend's wife. Who is the base man, the real slave under the yoke of a husband's rage? Playing between the Stoic philosophy of freedom as passion restrained by reason and social and legal realities of master and slave, Horace rings all the changes on what slavery means to Rome and its decadence. Its echo in literature remains in Swift's *On the Poor Man's Contentment* or in our own time in Richard Wright and Solzhenitsyn. Men who insist, as Horace does, that certain freedoms valued by others are illusory, easily encourage their suppression. By denying the importance of freedoms guaranteed by law, he made his apology for slavery. Davus exposed with contempt the freedman's feigned indifference to privilege and passion.

From *Satires*

Horace

MY SLAVE IS FREE TO SPEAK UP FOR HIMSELF

Iamdudum ausculto et cupiens tibi dicere servus

DAVUS: I've been listening for quite sometime now,
    wanting to have
A word with you. Being a slave, though, I haven't the nerve.
HORACE: That you, Davus?
DAV: Yes, it's Davus, slave as I am.
    Loyal to my man, a pretty good fellow: *pretty* good,
    I say. I don't want you thinking I'm too good to live.
HOR: Well, come on, then. Make use of the freedom
    traditionally yours
    At the December holiday season. Speak up, sound off!
DAV: Some people *like* misbehaving: they're persistent
    and consistent.
    But the majority waver, trying at times to be good,
    At other times yielding to evil. The notorious Priscus
    Used to wear three rings at a time, and then again, none.
    He lived unevenly, changing his robes every hour.
    He issued forth from a mansion, only to dive
    Into the sort of low joint your better-class freedman
    Wouldn't want to be caught dead in. A libertine at Rome,
    At Athens a sage, he was born, and he lived, out of season.
        When Volanerius, the playboy, was racked by the gout
    In the joints of his peccant fingers (so richly deserved),
    He hired a man, by the day, to pick up the dice
    For him and put then in the box. By being consistent
    In his gambling vice, he lived a happier life
    Than the chap who tightens the reins and then lets them
        flap.
HOR: Will it take you all day to get to the bottom of this
    junk,
    You skunk?
DAV: But I'm saying, *you're* at the bottom.
HOR: How so, you stinker?
DAV: You praise the good old days, ancient fortunes, and
    manners,
    And yet, if some god were all for taking you back,
    You'd hang back, either because you don't really think
    That what you are praising to the skies is all that superior
    Or because you defend what is right with weak defenses

SOURCE: Horace, *The Satires and Epistles of Horace*, trans. Smith P. Bovie (Chicago: The University of Chicago Press, 1959), pp. 144–150.

And, vainly wanting to pull your foot from the mud,
Stick in it all the same. At Rome, you yearn
For the country, but, once in the sticks, you praise to high
    heaven
The far-off city, you nitwit. If it happens that no one
Asks you to dinner, you eulogize your comfortable meal
Of vegetables, acting as if you'd only go out
If you were dragged out in chains. You hug yourself,
Saying how glad you are not to be forced to go out
On a spree. But Maecenas *suggests*, at the very last minute,
That you be his guest: "Bring some oil for my lamp,
    somebody!
Get a move on! Is everyone deaf around here?" In a dither
And a lather, you charge out. Meanwhile, your scrounging
    guests,
Mulvius & Co., make their departure from your place
With a few descriptive remarks that won't bear repeating—
For example, Mulvius admits, "Of course, I'm fickle,
Led around by my stomach, and prone to follow my nose
To the source of a juicy aroma, weak-minded, lazy,
And, you may want to add, a gluttonous souse.
But you, every bit as bad and perhaps a bit worse,
Have the gall to wade into me, as if you were better,
And cloak your infamy in euphemism?"
                            What if you're found out
To be a bigger fool than me, the hundred-dollar slave?
Stop trying to browbeat me! Hold back your hand,
And your temper, while I tell you what Crispinus' porter
Taught me.
            Another man's wife makes you her slave.
A loose woman makes Davus hers. Of us two sinners,
Who deserves the cross more? When my passionate nature
Drives me straight into her arms, she's lovely by lamplight,
Beautifully bare, all mine to plunge into at will,
Or, turning about, she mounts and drives me to death.
And after it's over, she sends me away neither shamefaced
Nor worried that someone richer or better to look at
Will water the very same plant. But when you go out for it,
You really come in for it, don't you? Turning yourself into
The same dirty Dama you pretend to be when you take off
Your equestrian ring and your Roman robes, and change
Your respectable self, hiding your perfumed head
Under your cape?
                Scared to death, you're let in the house,
And your fear takes turns with your hope in rattling your
    bones.
What's the difference between being carted off to be scourged
and slain, in the toils of the law (as a gladiator is),

And being locked up in a miserable trunk, where the maid,
Well aware of her mistress' misconduct, has stored you
    away,
With your knees scrunched up against your head? Hasn't
    the husband
Full power over them both, and even more over the seducer?
For the wife hasn't changed her attire or her location,
And is not the uppermost sinner. You walk open-eyed
Right under the fork, handing over to a furious master
Your money, your life, your person, your good reputation.
    Let's assume that you got away: you learned your lesson,
I trust, and will be afraid from now on, and be careful?
Oh, no! You start planning how to get in trouble again,
To perish again, enslave yourself over and over.
But what wild beast is so dumb as to come back again
To the chains he has once broken loose from?
                            "But I'm no adulterer,"
You say. And I'm not a thief when I wisely pass up
Your good silver plate. But our wandering nature will leap
When the reins are removed, when the danger is taken away.
    Are you my master, you, slave to so many
Other people, so powerful a host of other things, whom no
Manumission could ever set free from craven anxiety,
Though the ritual were conducted again and again? And
    besides,
Here's something to think about: Whether a slave who's the
    slave
Of a slave is a plain fellow slave or a "subslave," as you
    masters
call him, what am I your? You, who command me,
Cravenly serve someone else and are led here and there
Like a puppet, the strings held by others.
                        Who, then, is free?
The wise man alone, who has full command of himself,
Whom poverty, death, or chains cannot terrify,
Who is strong enough to defy his passions and scorn
Prestige, who is wholly contained in himself, well rounded,
Smooth as a sphere on which nothing external can fasten,
On which fortune can do no harm except to herself.
    Now which of those traits can you recognize as one of yours?
Your woman asks you for five thousand dollars, needles you,
Shuts the door in your face and pours out cold water,
Then calls you back. Pull your neck from that yoke!
Say, "I'm free, I'm free!" Come on, say it. You can't! A
    master
Dominates your mind, and it's no mild master who lashes
You on in spite of yourself, who goads you and guides you.
    Or when you stand popeyed in front of a painting by Pausias,
You madman, are you less at fault than I am who marvel

At the posters of athletes straining their muscles in combat,
striking out, thrusting, and parrying, in red chalk and
    charcoal,
As if they were really alive and handling these weapons?
But Davis is a no-good, a dawdler, and you? Oh, MONSIEUR
Is an EXPERT, a fine CONNOISSEUR of antiques, I ASSURE
    you.
I'm just a fool to be tempted by piping-hot pancakes.
Does your strength of character and mind make much resistance
To sumptuous meals? Why is it worse for me
To supply the demands of my stomach? My back will pay
    for it,
To be sure. But do you get off any lighter, hankering
After delicate, costly food? Your endless indulgence
Turns sour in your stomach, your baffled feet won't support
Your pampered body. Is the slave at fault, who exchanges
A stolen scraper for a bunch of grapes, in the dark?
Is there nothing slavish in a man who sells his estate
To satisfy his need for rich food?
                      Now, add on these items:
1. You can't stand your own company as long as an hour;
2. You can't dispose of your leisure in a decent fashion;
3. You're a fugitive from your own ego, a vagabond soul,
Trying to outflank your cares by attacking the bottle
Or making sorties into sleep. And none of it works:
The Dark Companion rides close along by your side,
Keeps up with and keeps on pursuing the runaway slave.

HOR: "Where's a stone?"
DAV: "What use do you have for it?"
HOR: "Hand me my arrows!"
DAV: The man is either raving or satisfying his craving
    For creative writing.
HOR: If you don't clear out, instanter,
    I'll pack you off to the farm to be my ninth planter.

CHAPTER FOUR

The Classical City: Commerce and Class

The Greek Marketplace

Gustave Glotz was the most distinguished French historian of ancient Greece of the first half of this century. The editor of a large-scale classical history, author of numerous books on every phase of Greek history, Glotz was especially hailed for his *Ancient Greece at Work* (1926), an economic history of Greece from the Homeric period to the Roman conquest. The following selection on the Athenian agora and its people is from the third of four sections (Homeric, Archaic, Athenian, Hellenistic). Glotz had elsewhere developed his ideas about Greek urban institutions, in opposition to the then prevailing views of Fustel de Coulanges.* In the first stages of Athens's history, families dominated a basically collectivist society of inherited wealth. This phase gave way to a decidedly "city" phase, which Glotz called "La Cité Démocratique," in opposition to "La Cité Aristocratique." The old families were subordinated to the spirit of individual economic enterprise. The class of urban craftsmen and that of traders became dominant. Athenian culture and economic, social, and political institutions were democratized, as the working people and even the poor achieved status.

The agora, or marketplace, was the focal point of life in mercantile, democratic Athens. It was the workshop of the new classes and expressed the social ethos of those artisans and merchants who took Hermes the Thief for their special god. The agora was the place where all classes of men met. Several of the Socratic dialogues by Plato took place there or in the export district of Athens— Piraeus. So completely did it dominate the greatest period of Athenian culture that a famous German world history could take its title from the gates of the agora.† Thus Glotz's lively account of the nature of craftsmanship and the industrial and mercantile bases of Athenian life in the fifth century B.C., brings us into contact with the men who created a new mode of society in Greece—from the wealthy merchants to the peddlers and hustlers who lived on the sweepings of the streets. Whether it is true, as one critic claimed, that Glotz made Parisians of Athenians, his book exposes the commercial basis of Athenian society in the Age of Pericles.

---

* In his *La Cité Grecque* (1928), Glotz codified ideas scattered throughout his earlier works, especially his great book on Aegean civilization.
† *Die Propylaean Weltgeschichte*, from *Propylaea*, the complex structures composing the elaborate gateway to the Acropolis.

From *Ancient Greece at Work*

Gustave Glotz

INDUSTRY

The Situation of Industry

In the fifth and sixth centuries industry in Greece assumed an economic and social importance which did not escape attention. When Socrates would indicate the composition of the Athenian Assembly, before mentioning the husbandmen and the small tradesmen he enumerates the fullers, the shoemakers, the masons, and the metal-workers. The men of the crafts can form the majority; their chiefs become the masters of the commonwealth. Demos gives himself into the hands of lamp-merchants, turners, leather-dressers, and cobblers, and Aristophanes puts this statement into the mouth of a sausage-seller. Moreover, the citizens abandon the lower kinds of work, and in the accounts of public works they appear as an aristocracy of labour, lost in the multitude of Metics and slaves. Nor was Athens an exception. The little towns of the Peloponnese swarmed with craftsmen; their military contingents were almost entirely formed of men with professions. The industrial callings even attracted women. Many freedwomen and daughters of citizens reduced to need devoted themselves to the works of Athene Ergane. They wove for custom, they sold yarn, ribbons, clothes, and bonnets of their own making, or they plaited wreaths.

But, if a great part of the population lived by industry, it does not follow that it was industry on a large scale. First, we must not be misled by the concentration of many workshops in the same city or in the same quarter. We involuntarily think of the great manufacturing towns of modern times when we see the Cerameicos in Athens entirely occupied by the potters, the tanneries collected outside the city, the Peiræeus filled with workshops which manufacture imported raw materials and work for export, and Laureion inhabited by a whole people of miners and metal-workers. In order not to misinterpret this concentration, it is sufficient to recall analogous facts. There are also in Athens a Street of the Box-Merchants and a Street of the Herm-Sellers, and the craftsmen teem round the Agora. The workshops are innumerable. Some are big enough to be called factories, but none is a huge concern of the modern kind. Rival manufacturers live next door to each other; they are jealous of each other, but the struggle is not bitter, for there is work for all and the weak are not crushed by the strong. Small industry predominates; medium-sized industry plays a certain part; large industry barely makes a vague appearance.

The first cause which prevented one whole class of industries from developing indefinitely was the persistence of work in the home. At the time when the miller Nausicydes and the baker Cyrebos were each amassing a great fortune, house-

SOURCE: Gustave Glotz, *Ancient Greece at Work. An Economic History of Greece from the Homeric Period to the Roman Conquest*, trans. M. R. Dobie (London: Routledge & Kegan Paul, 1926), pp. 263–291.

wives were still employed, like their grandmothers, in pounding the corn and kneading the dough. They kept the manufacture of clothing in their hands from the moment when the fleece was brought to them to the moment when they gave their menfolk the finished chiton. The greatest ladies of Greece taught their daughters everything connected with the making of clothes. Like the wise Arete, Queen of the Phaeacians, the mother of the tyrant Jason span and wove in her palace. Everywhere the mistress of the gynaeceum held, in Plato's words, "the government of shuttles and distaffs." Indeed it was in these home work-rooms that an industry which worked for the public was born. For that it was sufficient that the output should exceed the requirements of the house and that the surplus should be sold. This might happen without fixed intention, but it was also done with the deliberate purpose of practising a profession. In this way Athens manufactured men's clothing; Megara specialized in the worker's *exomis;* Corinth put on the markets its blankets, its *kalasireis* of fine wool, and its linen cloth; Pellene made cloaks which were in great demand; Patrae was filled with women, thanks to its byssos weaving-works; Cos had a name for its bombyx silk goods; Chios, Miletos, and Cyprus sent their hangings, their embroidered garments, and their carpets far and wide; Taras grew rich on its linen stuffs; and Syracuse transformed the wool of Sicily into textiles of many colours. But the textile industry, even when it had become a special profession, produced in small quantities. For common materials, the families only turned to it for a supplement to their own output, and prefered to call in women by the day. For the finer qualities manufacture was scattered and the demand limited.

Even those industries which were entirely in the service of the public kept some traces of the family system. The son fairly often succeeded his father. In the liberal careers the case was very frequent; the schools of medicine and music were family groups, the history of sculpture and painting is that of a few houses, and the architects of Delphi were, in succession, two Agathons, then Agasicrates, the son of the second, and lastly Agathocles, the son or brother of Agasicrates. In the same way the industrial art of the potters was handed down in the family. "How long," says Plato, "the potter's son helps his father and watches him at work before he touches the wheel himself!"

In any case there was nothing like the great factory with countless hands. The largest establishment of which we know in Attica was the shield factory which the Syracusan Cephalos founded in the Peiræeus in 435 and handed down to his sons. In 404 it contained 120 slaves. After that come the two houses managed by the father of Demonsthenes. For the people of the time these "were neither of them small industries." Now the arms factory had 32 or 33 slaves, the bed factory 20. A shield factory bequeathed to Apollodoros did, it is true, produce twice the output of the armoury bequeathed to Demosthenes, and therefore may have contained twice the personnel. Even then we have only one industry which employed more than a hundred workers, and those which employed more than twenty were considered very large. The celebrated potter Duris probably had not more than a dozen men about him. The gang of shoe-makers inherited by Timarchos consisted of nine or ten slaves, and in a mime of Herondas the fashionable shoe-maker has thirteen. The mines, it is true, present a very different aspect; there slaves were hired by the hundred. But when a man needed a big staff of working miners it was because he had obtained by auction a large number of small concessions. The State only gave out small lots. The typical mine employed about thirty men underground, about the same in the washing-room, and far less in the foundry. We know of one concessionnaire who put his hand to the pick and had for total capital a sum of 4500 drachmas; with such initial assets he cannot have had more than fifteen or twenty workers under him.

Athenian industry, then, never involved a great agglomeration of workers in one undertaking. What is typical of this industry is not the factory in which Cephalos collected over a hundred hands, but rather the hovel in which the Micylos of the poet Crates cards wool with his wife "to escape starvation," or the workshop in which, according to an inscription, the helmet-maker Dionysios worked together with his wife, the gilder Atremis. And these are not accidental, isolated cases. The Athenian craftsman had no interest in increasing the number of his workers, Xenophon says. He was in the same position as the farmer, who knew exactly how many day-workers he needed, and that one man above this number was a sheer loss.

All these advantages did not, however, draw a very large number of contractors, expecially of such as had considerable means at their disposal. The State tried all manner of devices to divide up the lots and to organize competition; orders were brought within the power of the humblest workers, alone or in partnership; craftsmen were summoned from one town to another, and sometimes from great distances.

Puny as industry was, it could not always confine itself to executing orders. Sometimes it produced in advance. The retailers and exporters enabled the craftsmen to work regularly without troubling too much about the demand. The shoemaker made to measure and sold ready-made. The armourer had to provide for sudden, large demands. When the Thirty confiscated the factory of Lysias and Polemarchos, they found in stock great quantities of gold, silver, and ivory, and seven hundred finished shields. In Thebes a band of insurgents broke into the armourers' shops and fitted themselves out with lances and swords. But the craftsmen had no advantage in producing without cease and sinking capital in the shape of stock. The demand was too restricted. Even the armourer was afraid of the repercussion of political events. Demosthenes asks his guardian why his armour works have paid nothing during his minority; it is not, he says, for want of work, as is proved by the accounts of the output; then is it because the arms manufactured could not be sold? Here, certainly, we have over-production. And here we have its effects: in Aristophanes the merchants weep over the cuirasses, trumpets, crests, helmets, and javelins for which they cannot find buyers. As a rule the manufacturer made his arrangements so that production should not outstrip orders to a dangerous extent, and hired out the slaves whom he could not employ. In the fourth century he was perpetually concerned with keeping his personnel down to what was absolutely necessary. According to Xenophon, once the blacksmith or bronze-worker neglects to regulate his work by his sales, "down goes the price of his goods, and his business is ruined." If the industry of Laureion was the only one which absorbed labour indefinitely, it was because only the silver-market absorbed output indefinitely.

Workers and Wages

Let us visit the workers in their workshops and living-quarters; let us see at close range the small men who worked with their hands.

The skilled workers were called after their trades. The labourers, "those," as Plato says, "who sell the service of their arms," were called "wage-earners" (*mis-*

thotoi). Under them were the assistants, servants, and apprentices. One and all were either free men, whether citizens or foreigners, or else slaves, but the lower you go in the scale of labour the more Metics and slaves do you find. To act as a labourer for a few days was generally for the citizen a temporary means of keeping himself, one of the extremities to which sudden misfortune might reduce a man.

As a rule the hiring of service was not the occasion of a formal contract. One craftsman made an agreement with another for the execution of certain work, sometimes of a single task, and it is often difficult to say whether one of these collaborators was the subordinate or the partner of the other. Between the employer, whether he was a workshop-owner or a customer, and the employee, whether labourer or craftsman, the relationship was loose from a legal point of view, and the terms of the agreement were free. Plato, who was prepared to make regulations for everything, would have had officials to supervise the workers and to fix their wages. But the Athenian State refrained from entering on this path. The authorities never once thought of limiting the working day. Questions of payment, in case of dispute, simply went before the law-courts. A man could claim any remuneration due to him by legal proceedings, and cases between ship-owners and seamen or dock-workers were in the competence of the *Nautodikai* of the Peiræeus. In only one case do we find the authorities prescribing a salary; the *Astynomoi* did not allow women who played the flute, harp, or cithara to take more than two drachmas, and when several applicants wished to engage the same woman they assigned her by lot; but this was a police measure. Nor did the hygiene and safety of the workshop interest the city. The employment of a free child at turning a mill was prohibited on pain of death; but the very severity of the punishment proves that it was intended, not to protect the child against too heavy labour, but to preserve the son of a citizen from servitude. The mining laws treated as crimes the destruction of pit-props and the smoking of galleries, but this was in order to prevent the rapacious concessionnaire, not from killing his miners, but from destroying public property. On principle, therefore, industry enjoyed complete liberty.

Technical progress made apprenticeship necessary in almost every profession. The advantage of agriculture, to Xenophon's mind, is that, to succeed, it is sufficient to keep your eyes open and to ask questions; the other arts require a long experience before you can live decently by them. "If you want to make a man a shoe-maker, a mason, a blacksmith . . . you send him to a master who can teach him." Even the cook took lessons from a master cook. A formal contract, often in writing, fixed the amount to be paid by the apprentice's family, the length of the indentures, and the obligations on both sides. Since the sculptors and painters demanded very high premiums, poor men could only go into their studios as assistants. This was how Lysippos and Protogenes began. The learner was subjected to harsh discipline, and was not always sure of learning his trade thoroughly, for the fear of competition made the master distrustful, and he did not communicate his most precious secrets. The importance attached to professional education is attested by the apprentices' competitions. The vase-painters represent their pupils bent over ornamental details, while Athene and Victory come and crown them. On the pedestal of a monument dedicated by a potter we read these verses: "Among those who combine earth, water, and fire in their art Bacchios won the first place by his gifts, over every rival, in the judgement of the whole of Greece, and in all the competitions organized by this city he won the crown."

Men out of work used to look for engagements on the Agora, where a special place was set apart for them, the Colonos. So the "Colonites" were the unemployed. Engagements by the year generally ran from the 16th Anthesterion (March). There was a reason for this date in the country, where it marked the resumption of work after the winter; the agricultural workers passed it on to all classes of workers. The beginning of the new period was observed with joyous celebrations.

The worker's day began very early. He rose before daylight. Aristophanes amuses himself with a description of the scene. "As soon as the cock sends forth his morning song, they all jump out of bed, blacksmiths, potters, leather-dressers, shoe-makers, bathmen, flour-dealers, lyre-turners, and shield-makers; they slip on their shoes and rush off to their work in the dark." Work no doubt went on until sunset. In the mines, where it never ceased, there were successive ten-hour shifts. For night work the millers, bakers, and pastry-cooks paid wages at skilled rates.

Inside the workshops painted on the vases we often see clothes hanging on the wall. When a man was working he wanted to be comfortable. For sedentary work he bared his upper part and legs, or took off everything, wearing only a cap. In the forges and potteries, the more clothes there are hanging on the wall, the more vases hang there too. Going near the fire made a man thirsty.

In the absence of machines, and with only a moderate division of labour, the craftsman and the labourer had relatively varied occupations. For tasks done by several men together, and especially for hard or monotonous work, the time was set by music. The flute, the pipe, and the whistle governed motions and gave orders in the ship-building yards. There were old songs for every trade, and for each operation in it. The airs which Calypso and Circe sang as they span and wove were known by all the women. Others were sung when the corn was pounded or milled. Harvesters, millers, fishermen, rowers, bathmen, all had their chanty. With the cultivation of the vine the Greeks took to Egypt the Song of the Wine-Press. Like dancing and gymnastics, manual labour was made rhythmical and gay.

Shop and workshop were open to visitors and idlers. As in Hesiod's time, men liked to go into the forge and to stand peacefully watching the workmen as they handled tongs, hammer, and polisher. They went to the barber's as the Frenchman goes to his café. Young men made appointments to meet and chat at the perfumer's. Socrates was always sure of finding an audience at the statuary's or the armourer's; when he wished to meet Euthydemos he went with a crowd of friends into a saddler's shop, and it was the leather-dresser Simon who took down his sayings in a diary.

The employer did not keep a great distance between himself and his men. On feast-days they met at the sacrifice and at the sacred meal, of which he bore the cost. The *epistatai* of the Erechtheion offer a victim "together with the workmen." At Eleusis the public slaves get each a good hunk of meat and about a gallon of good ordinary wine. It is a big present.

While labour does not appear too severe in the small workshops, it presents a very different spectacle in the mines. At Laureion each shift did ten hours' work after ten hours' rest. Five hewers, followed by twenty or twenty-five carriers, went one after the other to the face of the workings. In galleries between two and three feet wide and between two feet and a little over three feet high, they had often to crawl, and always to dig on their knees, on their stomachs, or on their backs. We can guess what the ventilation was like in these narrow passages. The heat was cruel. Heaped up bodies and smoky lamps made the air unbreathable. No hygienic precautions were taken. Nevertheless, we must not apply to Laureion the melancholy descriptions which are true of the mines of Egypt and Spain. Slaves though they were, the miners of Attica were not treated like convicts. The smaller concessionnaires mixed with the hewers, so their existence must have been endurable. Naturally the owner was prevented by his own interest from taking unnecessary risks with the health of his workers; they gave a high and steady output which he could not have got from exhausted bodies. The miners of Laureion were not shut away for the rest of their lives, like the quarrymen buried in the *latomiai* of Syracuse, who married there and begot children who fled screaming at the sight of a horse or an ox. At the centre of the district, at Thoricos, there was a theatre which could seat five thousand spectators; the mass of labourers were not denied all distractions. That slaves should flee to Deceleia, when Sparta called them to freedom, was only natural; but Laureion was never the scene of a general revolt, like Messenia or Sicily.

Where there were many workers, and the employer did not wish to manage his concern himself, he placed at their head a works-manager or foreman. Nicias had the work of his mines directed by a man to whom he had paid a talent. Demosthenes' father kept in his armour factory an overseer who, after the death of the chief, had full powers of proxy. Midas managed Athenogenes' perfumery works with all the rights which to-day are conferred by signature. The nine or ten shoemakers of Timarchos were directed by a workshop foreman. At Eleusis seventeen public slaves, employed on the temple works, had one foreman, and twenty-eight free workers, brought from Megara, had two. The foremen were usually slaves, but sometimes freedmen or foreigners. They may have made a bit on the feeding of their men, since it was they who did the catering. They had a name for being very hard. "Slave," says a comic poet, "beware of serving a former slave; when the bull is resting he forgets the yoke." They had an agile arm and a ready whip. A vase painting shows us, in a pottery shop, a slave hung up by his arms and legs and lashed unmercifully. The iron rings found here and there in the galleries of Laureion speak volumes about the discipline which reigned in the mines. But such treatment was confined to slaves. The law of Athens protected the person of the free man against every chastisement and every constraint.

The return of labour varied according to the trade and according to the period. We can calculate the time taken by the marble-workers to flute the columns of the Erechtheion. The five men of the gang which was most employed needed about sixty days to do twenty-four flutings with flat ridges, 19 ft. 6 in. long. That makes about nineteen inches per man per day. It is not much. On the other hand, at Eleusis in 329–8 the gang of three bricklayers, working steadily, laid in one day 413 squares, 17¾ in. by 17¾ in. by 4 in., i.e., altogether about 300 cubic feet. In the ancient galleries of Laureion, at points where the sterile rock is of the hardest limestone, the face of the workings, which are at least two feet high and wide, is cut in at regular intervals of four or five inches; each of these cuttings shows the normal output of five hewers working one after the other, each so long as his lamp burned, i.e., two hours. Therefore each man hewed about 250 cubic inches in an hour; this result, obtained with the pick and pitching-tool, is higher than that demanded to-day on the same sites from gun-powder and dynamite.

We now have to consider the question of wages.

Certain labourers, even in the fourth century, received no other remuneration than their food for the day. Otherwise it would not have been specified in the accounts that the workers who received wages had to feed themselves "at home." But this mode of payment had almost disappeared in Attica, except in the country. We have seen above that even the public slaves received a ration-allowance of 180 drachmas a year in money, and drew nothing in kind but clothing, and that their foreman, who did not receive clothing, had in addition to his keep a salary of 100 drachmas.

Whereas the public slave was paid at every Prytany, ten times a year, the workman was paid by the day or by the job. In the last third of the fifth century the price at Athens of labour by the day was one drachma. No difference was made in view of either the social position or the trade of the worker, and the labourer was paid as much for his day's work as the craftsman. But mere assistants received only 3 obols. For agricultural labourers food was reckoned at 2 ob., and they were given 4 ob. in cash. In the fourth century wages by the day tended in general to rise and to vary. In the years 395–391 a gang of bricklayers, consisting of a master mason and two lads, was paid from 4 dr. to 4 dr. 4 ob., the master getting 2 dr. and the lads 1 dr. or 1 dr. 2 ob. each. At Delphi, about the middle of the century, the plasterers received 30 Æginetan drachmas a month or 1 dr. 2½ ob. in Attic money a day. At Eleusis, in 329–8, the old wage of one drachma was only given to assistants; the labourers received 1½ dr., and the skilled workers 2 dr. (sawyers) or 2½ dr. (bricklayers, plasterers, carpenters).

But work did not go on the whole year round. There were many days when nothing was done. The Athenian calendar contained about sixty holidays, about as many as our own, including Sundays. On working days the citizen went to the law-court or the Assembly, which brought him the allowance of two or three obols. Moreover, there was not enough work to keep all hands constantly employed, and the free workman was not disposed to devote all his time to his trade. We find a gang of thirty-three men working full strength two days out of seven; on the other days two, four, twelve, fourteen, and even twenty-three are absent. When the marble-workers execute the fluting of the columns of the Erechtheion, the three gangs which contain citizens never do more than 22 or 23 drachmas worth of work per man in thirty-six or thirty-seven days, while the three gangs of slaves, directed by a slave or a Metic, do work worth 27, 35, and even 38 drachmas. The yearly salary of the architect was calculated on a basis of 2 dr. a day, at the time when the masons were earning 2½ dr. a day; this proves with certainty that the latter were idle for at least one fifth of the year and probably much more.[1]

---

[1] The advantage of the architect and the officials over the craftsmen and labourers lies just in the fact that they escape unemployment. Their emoluments are not at a higher rate, but they are fixed by the year and paid in halves, tenths, or twelfths. In the fifth century the Athenian architect gets 360 dr., one per day. In the first half of the following century the architect at Epidauros gets ¾ more, thanks to the Æginetan drachma. In the second half, the architect at Eleusis receives 720 dr. (2 dr. per day). At Delphi the architect is paid 360 Æginetan drachmas for at least eight years, then, in 345, he gets double, and in 342 four times the amount (almost three times as much as his colleague at Eleusis). Salaries come more and more to depend on talent and reputation.

From at least the fifth century onward work was also done by the piece. In the Erechtheion accounts the sawyers are sometimes paid by the day and sometimes by the saw-cut, the carpenters receive so much for each beam shaped or each plank laid, the decorative joiners who make the panels of the ceiling fit the frames at 6 dr. apiece and fasten the mouldings at 3 dr. apiece, and the masons wall up the intervals of the columns at 10 dr. each. For the fluting of the columns 300 dr. are paid, whatever may be the strength of the gang which does the work and whatever time they take over it. Rosettes for the soffits are ordered at the fixed price of 14 dr. The marble statuettes of the frieze cost so much per subject—60 dr. for a full-sized figure, 30 dr. for a medium figure, 20 dr. for a child. Work by the measure, a variety of work by the piece, is also done. The stone-cutters have a tariff which takes both material and size into account. Plastering and painting are done by the linear foot or by the square foot. . . .

Work by the day and work by the job gradually came to be apportioned by agreement, rough work being left to plain day-labourers, and the task which required manual skill being reserved for workers who were able to put their soul into it if they took time over it. In 408 the men working on the Erechtheion were paid a drachma a day, as labourers or as craftsmen, as well as working by the job. Towards the middle of the fourth century the same kind of work was remunerated in different ways, according to the amount of finish required. At Delphi we have a good example which shows that the falling off of work by the day was in proportion to the improvement of technical processes; the same contractor does the plastering of plain stones by the day and that of worked stones at 4 dr. per face. Work by the piece does not seem to have been an economy for the customer. Here, indeed, is a case where, combined with the contract system, it costs more. Three bricklaying jobs are done at Eleusis. In two of them, which are paid by the day, the laying of a thousand bricks comes to 13 dr. 2 ob.; in the third, a contractor charges 17 dr. the thousand. He thus makes a profit of 21–22 percent for which he has to engage and supervise the workmen and to be responsible for bad work. Gradually work by the job tended to drive out work by the day, even for plain tasks. It was better suited to an age in which the works contract placed a professional craftsman between the workman and the customer, and the distinction between the skilled worker and the plain labourer was accentuated.

But when we speak of wages we must determine their real value. What standard of living could the craftsman and the labourer attain? That there was suffering in the fourth century there is no doubt. We have only to listen to the wail of the fuller in the comedy. "In our trade we have the earnest of a livelihood, and we die of hunger all the year round, for ever hoping." But perhaps unemployment was more frequent in a trade in which competition was keener than in others. Let us look at the question as a whole. First of all, we must remember the abstemiousness of the South. In old days, says Aristophanes, if you went out for the day, "you took a calabash of drink and some dry bread, a couple of onions, and three olives." Clothes were very simple, and were made for a great part at home. For dwelling there were a few small rooms in a house of sunbaked brick. Let us try to calculate the cost of living in these circumstances.

Food consisted of two elements: (1) the *sitos*, i.e., cereals in the form of bread, scone, or porridge; (2) the *opsonion*, i.e., fresh or dried vegetables, meat, which was almost always pig, fresh or salt fish, and lastly fruit, chiefly olives and figs. The usual drink was wine, very much diluted, or spring water. It is easy to calculate the cost of the *sitos*. The grown man's ration was reckoned at one chœnix of wheat (nearly two pints) or two chœnices of barley meal a day. This was the big ration, which was demanded by the Spartan soldier in the field; he regarded it as very large, since he considered that half as much was enough for his servant.

At the end of the fifth century, therefore, when wheat was at 3 dr. the medimnus, the worker's *sitos* cost him 22½ dr. a year for 7½ medimni. With 60 dr. a year, or one obol a day, he was well fed. With an additional 60 dr. he could meet his other expenses. A single man lived comfortably on 120 dr. in the time of Pericles. It was enough if he worked one day in three. Now let us suppose the typical case of the man with a wife and two children to keep. Allow three full rations for the four of them, and their food will cost 180 dr. Clothing may be reckoned at 50 dr., housing at 36 dr., and sundry expenses at 14 dr. This makes 280 dr. altogether. These figures agree quite well with the salary of the architect, who must have been able to maintain a family decently on his 360 dr. The workman who earned one drachma a day could feed his family, if it was not too large, without even being compelled to work on every working day.

In the fourth century, when the general rise in prices brought the cost of corn up to 5 dr. the medimnus, the single man's food came to 100 dr., and that of a family of four to 300 dr. But the other expenses had not increased at the same rate, since the public slave lived well on 180 dr. a year plus clothing. A worker could, therefore, keep a wife and two children on 450 dr. It is true that the architect now received 750 dr., and a middle-class townsman complained that he could only just live on 540 dr.; but in these times the upper classes were beginning to have expensive wants which were not felt to the same degree by the working classes. Therefore a plain assistant, getting one drachma a day, could, if he worked three hundred days in the year, ensure a good average standard of living for his wife; but he was obliged to practise moderation and to be content with reduced rations if he had children. The labourer who earned 1½ dr. could feed two children, provided he worked on all days but holidays. The skilled worker with 2 dr. or 2½ dr. could save 150 or 300 dr. if he was constantly employed, or else he could rest three days in eight, or even one in two. While the unskilled worker, burdened with a family, could only manage by dint of hard work and privations, the skilled worker or small craftsman could bring up several children and give them, according to the work he did, an average or high standard of living.

TRADERS

The shortage of agricultural commodities and raw materials perforce gave Greek trade an ever increasing importance. It was absolutely necessary to collect products on the distant markets on which they abounded, and to distribute them among the cities with a growing population in which they were lacking. Plato himself, the declared enemy of trade, well defined the social usefulness of the trader when he called him the agent who ensures the regular and measured distribution of the wealth produced by nature without measure or regularity. Aristotle recognized equally that it is impossible for a country to remain in isolation, without selling or buying, importing or exporting. It is true that the theorists understood the legitimacy of exchanges in their own way. They would have been glad to bring back the days in which each sold the surplus of his production direct, and transactions were confined to natural products. But the people, especially in the democratic cities, was ignorant of doctrines which it would have regarded as crazy. Life laughed at systems. Trade assumed a grand development in Athens and the Peiræeus.

Material Conditions of Trade

At this period *kapeleia*, retail trade, or, more generally, land trade, was distinguished from *emporia*, wholesale trade, which was classed together with seaborne trade. For both branches the Athenian market was highly organized.

The agricultural producer could always go to the consumer without an intermediary. The market-gardener came to town with his fruit and his vegetables, the land-owner sent in his asses laden with wood, the Acharnian brought his baskets of charcoal. But, whereas at Locri the law forbade any transaction through an intermediary, and even any written contract, and at Erythrae retailers were excluded from the wool-market by decree, in Attica it was the natural play of economic relations which caused direct sale to survive.

The small traders, men and women, proclaimed themselves in streets and squares by their cries. The pedlars walked along the roads beside their beast with great bundles loading his sides, or else sat on their ass with a huge bale on their own shoulders. The shops in Athens were wretched stalls, clustering in the neighbourhood of the Agora or in sacred precincts like that of the Theseion. In all the quarters of the town the pot-houses sold drink and food to the common people. The character of the seller of corn, porridge, and vegetables, or of garlic, wine, and bread, is the joy of Aristophanes.

But the centre of home trade is the Agora. There all day long is found the throbbing life, political, social, and economic, of the great city. At the ends of the square are the offices of the magistrates, with the official placards which attract the curious. The crowd shelters under the porticoes with their slender colonnades. It passes in front of the frescoes of the great Polygnotos and flocks "to the Herms," where business men discuss prices, and political enthusiasts argue over the agenda of the next Assembly, and gapers listen to the town criers, and idlers chatter and wave their knotty sticks about, and the young bloods make their long white cloaks sweep in the most elegant manner. All who have something to sell, slaves with cloth which they have just made, craftsmen from the Cerameicos, Melite, or Scambonidæ, peasants who left their village before daybreak, Megarians driving their pigs, fishermen from Lake Copais, pass in every direction. Through alleys planted with trees they reach the places assigned to different goods, separated by movable barriers. One after another, at the hours fixed by the regulations, the different markets open; there are markets for vegetables, fruit, cheese, fish, meat and sausages, poultry and game, wine, wood, pottery, ironmongery, and old articles. There is even a corner for books. Every merchant has his place, which he reserves by paying a fee; in the shade of an awning or an umbrella he sets out his goods on trestles, near his cart and his resting beasts. Shoppers walk about; traders call to them; porters and messengers offer their services. Shouts, oaths, and quarrels; the *Agoranomoi* do not know to whom to listen. When the open-air markets are shut the customers make for the covered hall, which is like an Eastern bazaar, with counters occupying the end.

All these retailers had a bad name. Abuse was heaped on their rowdiness and vulgarity. Women who made their living in the street or on the Agora, or owned taverns, were suspected of misconduct, and the law did not allow prosecution for adultery in the case of persons of this class. But the small dealers were chiefly reproached with habitual rapacity, dishonesty, and lying. They overcharged, they adulterated their goods, they gave short weight, they swindled over the exchange.

The hucksters of the whole of Greece swarmed to the fairs which were held in connection with the great festivals. They poured in the wake of armies, some in order to sell horses or articles of equipment, others to buy booty and prisoners cheap, and most of them as sutlers. The Athenian fleet which sailed for Sicily was accompanied by a great number of merchant vessels. Xenophon draws a lively picture of the impetus given to the trade of a town by the passage of troops. Generals were obliged to take precautions against these flocks of cormorants by granting a limited number of permissions.

The Good Life of Petty Men

Business alienated men from the virtues of political life in Athens. Attendance at the Assembly was secured in the fifth century B.C. by the payment of three obols. Even in Solon's reforms Athenians had to be commanded to participate in politics. This evidence bears on what we have already said about the decline of public spirit and morality in our remarks on Aristotle and Thucydides. And it is in turn related to the emergence of the citizens of the lower classes chronicled by Glotz. These people were the medium from which Aristophanes made his art. He was the literary patron saint of craftsmen, small traders, and the various petty-bourgeois types whose vices, virtues, and habits motivate the action of his plays. Aristophanes witnessed the decline of Athenian public life (445–385 B.C.). What traditions of comic drama he intended it is difficult to say, since his are the only old comedy plays now extant. The selection printed here is from his third play *The Acharnians* (425 B.C.).

The action of *The Acharnians* takes place in a street by the Pnyx or meeting place of the Assembly. The central figure Dicaeopolis, a country bumpkin, harangues the Athenians on the necessity of ending the great war with Sparta. His interest seems to be restoring trade and prosperity to the city, but he blames Athens for the war. His main antagonist is Lamachus to whom the war is a source of profit. When the Assembly refuses to vote peace, Dicaeopolis concludes his own and commences trade with Sparta and her allies. The play ends with Lamachus wounded in battle and Dicaeopolis in bed with a courtesan. That Dicaeopolis (that is, good government) speaks for Aristophanes seems clear, since at the crucial moment the chorus of Acharnes agree with him, speaking directly to the audience. It is thus wrong to see in *The Acharnians* the self-interest of Dicaeopolis triumphing over virtue. The opposite is true. The title of the play refers to *acharnai*, a famous old *demos* or Attic people, renowned for peace, truth, and virtue.

It is generally recognized that Aristophanes favored rustic simplicity, the old-fashioned good citizen against the imperious demagogue, and peace. His special hatred was reserved for warmongers, profiteers, and the ruthless foreign policy chronicled by Thucydides, whose *History* echoed the playwright's caustic wit. The legitimate interests of commerce against the excessive lust for gain at any cost require both peace and honest dealing. To these themes Aristophanes returned again and again, especially in *Plutus*.

From *The Acharnians*

Aristophanes

DICAEOPOLIS (*coming out of his house and marking out a square in front of it*):

These are the confines of my market-place. All Peloponnesians, Megarians, Boeotians, have the right to come and trade here, provided they sell their wares to me and not to Lamachus. As market-inspectors I appoint these three whips of Leprean leather, chosen by lot. Warned away are all informers and all men of Phasis. They are bringing me the pillar on which the treaty is inscribed and I shall erect it in the centre of the market, well in sight of all.

(*He goes back into the house just as a Megarian enters from the left, carrying a sack on his shoulder and followed by his two little daughters.*)

MEGARIAN: Hail! market of Athens, beloved of Megarians. Let Zeus, the patron of friendship, witness, I regretted you as a mother mourns her son. Come, poor little daughters of an unfortunate father, try to find something to eat; listen to me with the full heed of an empty belly. Which would you prefer? To be sold or to cry with hunger?

DAUGHTERS: To be sold, to be sold!

MEGARIAN: That is my opinion too. But who would make so sorry a deal as to buy you? Ah! I recall me a Megarian trick; I am going to disguise you as little porkers, that I am offering for sale. Fit your hands with these hoofs and take care to appear the issue of a sow of good breed, for, if I am forced to take you back to the house, by Hermes! you will suffer cruelly of hunger! Then fix on these snouts and cram yourselves into this sack. Forget not to grunt and to say wee-wee like the little pigs that are sacrificed in the Mysteries. I must summon Dicaeopolis. Where is he? (*Loudly*) Dicaeopolis, do you want to buy some nice little porkers?

DICAEOPOLIS (*coming out of his house*): Who are you? a Megarian?

MEGARIAN: I have come to your market.

DICAEOPOLIS: Well, how are things at Megara?

MEGARIAN: We are crying with hunger at our firesides.

DICAEOPOLIS: The fireside is jolly enough with a piper.[1] But what else is doing at Megara?

MEGARIAN: What else? When I left for the market, the authorities were taking steps to let us die in the quickest manner.

DICAEOPOLIS: That is the best way to get you out of all your troubles.

MEGARIAN: True.

DICAEOPOLIS: What other news of Megara? What is wheat selling at?

MEGARIAN: With us it is valued as highly as the very gods in heaven!

DICAEOPOLIS: Is it salt that you are bringing?

MEGARIAN: Aren't you the ones that are holding back the salt?[2]

SOURCE: Aristophanes, *The Acharnians*, in *The Complete Greek Drama*, vol. 2, ed. Eugene O'Neill and Whitney J. Oates (New York: Random House, 1938), pp. 452–473.

[1] Dicaeopolis has misunderstood the Megarian, taking *peinames*, "we starve," for *pinomes*, "we drink"; hence his apparently inappropriate reply.

[2] At this time the Athenians had possession of the island of Minoa off the Megarian coast; they

DICAEOPOLIS: Is it garlic then?

MEGARIAN: What garlic! do you not at every raid like mice grub up the ground with your pikes to pull out every single head?

DICAEOPOLIS: What *are* you bringing then?

MEGARIAN: Little sows, like those they immolate at the Mysteries.[3]

DICAEOPOLIS: Ah! very well, show me them.

MEGARIAN: They are very fine; feel their weight. See! how fat and fine.

DICAEOPOLIS (*feeling around in the sack*): Hey! what's *this?*

MEGARIAN: A sow.

DICAEOPOLIS: A *sow*, you say? Where from, then?

MEGARIAN: From Megara. What! isn't it a sow then?

DICAEOPOLIS (*feeling around in the sack again*): No, I don't believe it is.

MEGARIAN: This is too much! what an incredulous man! He says it's not a sow; but we will stake, if you will, a measure of salt ground up with thyme, that in good Greek this is called a sow and nothing else.

DICAEOPOLIS: But a sow of the human kind.

MEGARIAN: Without question, by Diocles! of my own breed! Well! What think you? would you like to hear them squeal?

DICAEOPOLIS: Yes, I would.

MEGARIAN: Cry quickly, wee sowlet; squeak up, hussy, or by Hermes! I take you back to the house.

DAUGHTERS: Wee-wee, wee-wee!

MEGARIAN: Is that a little sow, or not?

DICAEOPOLIS: Yes, it seems so; but let it grow up, and it will be a fine fat thing.

MEGARIAN: In five years it will be just like its mother.

DICAEOPOLIS: But it cannot be sacrificed.

MEGARIAN: And why not?

DICAEOPOLIS: It has no tail.

MEGARIAN: Because it is quite young, but in good time it will have a big one, thick and red. But if you are willing to bring it up you will have a very fine sow.

DICAEOPOLIS: The two are as like as two peas.

MEGARIAN: They are born of the same father and mother; let them be fattened, let them grow their bristles and they will be the finest sows you can offer to Aphrodité.

DICAEOPOLIS: But sows are not immolated to Aphrodité.

MEGARIAN: Not sows to Aphrodité! Why, she's the only goddess to whom they are offered! the flesh of my sows will be excellent on your spit.

DICAEOPOLIS: Can they eat alone? They no longer need their mother?

MEGARIAN: Certainly not, nor their father.

DICAEOPOLIS: What do they like most?

MEGARIAN: Whatever is given them; but ask for yourself.

DICAEOPOLIS: Speak! little sow.

---

were thus able to intercept all her maritime commerce, and they incidentally controlled her salt-works also.

3 This brilliant scene is a riotous tissue of plays on the double meaning of the Greek word *choiros*, which signifies not only "sow" but also "female genitalia." The English word "pussy" has comparable senses, but is regrettably ill-adapted to the needs of this particular scene, which must thus remain the Hellenist's delight and the translator's despair.

DAUGHTERS: Wee-wee, wee-wee!

DICAEOPOLIS: Can you eat chick-pease? [4]

DAUGHTERS: Wee-wee, wee-wee, wee-wee!

DICAEOPOLIS: And Attic figs?

DAUGHTERS: Wee-wee, wee-wee!

DICAEOPOLIS: What sharp squeaks at the name of figs. Come, let some figs be brought for these little pigs. Will they eat them? Goodness! how they munch them, what a grinding of teeth, mighty Heracles! I believe those pigs hail from the land of the Voracians.

MEGARIAN (*aside*): But they have not eaten all the figs; I took this one myself.

DICAEOPOLIS: Ah! what curious creatures! For what sum will you sell them?

MEGARIAN: I will give you one for a bunch of garlic, and the other, if you like, for a quart measure of salt.

DICAEOPOLIS: I'll buy them. Wait for me here.

(*He goes into the house.*)

MEGARIAN: The deal is done. Hermes, god of good traders, grant I may sell both my wife and my mother in the same way!

(*An* INFORMER *enters.*)

INFORMER: Hi! fellow, what country are you from?

MEGARIAN: I am a pig-merchant from Megara.

INFORMER: I shall denounce both your pigs and yourself as public enemies.

MEGARIAN: Ah! here our troubles begin afresh!

INFORMER: Let go of that sack. I'll teach you to talk Megarian!

MEGARIAN: Dicaeopolis, Dicaeopolis, they want to denounce me.

DICAEOPOLIS (*from within*): Who dares do this thing? (*He comes out of his house.*) Inspectors, drive out the informers. Ah! you offer to enlighten us without a lamp! [5]

INFORMER: What! I may not denounce our enemies?

DICAEOPOLIS (*with a threatening gesture*): Watch out for yourself, and go off pretty quick and denounce elsewhere.

(*The* INFORMER *runs away.*)

MEGARIAN: What a plague to Athens!

DICAEOPOLIS: Be reassured, Megarian. Here is the price for your two sowlets, the garlic and the salt. Farewell and much happiness!

MEGARIAN: Ah! we never have that amongst us.

DICAEOPOLIS: Oh, I'm sorry if I said the wrong thing.

MEGARIAN: Farewell, dear little sows, and seek, far from your father, to munch your bread with salt, if they give you any.

(*He departs and* DICAEOPOLIS *takes the "sows" into his house.*)

CHORUS (*singing*): Here is a man truly happy. See how everything succeeds to his wish. Peacefully seated in his market, he will earn his living; woe to Ctesias, and all other informers who dare to enter there! You will not be

---

[4] Here we find a pun similar to that on *choiros*, for the word *erebinthos* means both "chick-pea" and "penis"; the remark about figs in the next line seems also to contain such a *double entendre*.

[5] This remark is a pun on the word *phainein*, which means both "to light" and "to inform against."

cheated as to the value of wares, you will not again see Prepis wiping his big arse, nor will Cleonymus jostle you; you will take your walks, clothed in a fine tunic, without meeting Hyperbolus and his unceasing quibblings, without being accosted on the public place by any importunate fellow, neither by Cratinus, shaven in the fashion of the adulterers, nor by this musician, who plagues us with his silly improvisations, that hyper-rogue Artemo, with his arm-pits stinking as foul as a goat, like his father before him. You will not be the butt of the villainous Pauson's jeers, nor of Lysistratus, the disgrace of the Cholargian deme, who is the incarnation of all the vices, and endures cold and hunger more than thirty days in the month.

(*A* BOEOTIAN *enters, followed by his slave, who is carrying a large assortment of articles of food, and by a troop of flute players.*)

BOEOTIAN: By Heracles! my shoulder is quite black and blue. Ismenias, put the penny-royal down there very gently, and all of you, musicians from Thebes, strike up on your bone flutes "The Dog's Arse."

(*The Musicians immediately begin an atrocious rendition of a vulgar tune.*)

DICAEOPOLIS: Enough, damn you; get out of here! Rascally hornets, away with you! Whence has sprung this accursed swarm of Chaeris fellows which comes assailing my door?

(*The Musicians depart.*)

BOEOTIAN: Ah! by Iolas! Drive them off, my dear host, you will please me immensely; all the way from Thebes, they were there piping behind me and they have completely stripped my penny-royal of its blossom. But will you buy anything of me, some chickens or some locusts?

DICAEOPOLIS: Ah! good day, Boeotian, eater of good round loaves. What do *you* bring?

BOEOTIAN: All that is good in Boeotia, marjoram, penny-royal, rush-mats, lampwicks, ducks, jays, woodcocks, water-fowl, wrens, divers.

DICAEOPOLIS: A regular hail of birds is beating down on my market.

BOEOTIAN: I also bring geese, hares, foxes, moles, hedgehogs, cats, lyres, martins, otters and eels from the Copaic lake.

DICAEOPOLIS: Ah! my friend, you who bring me the most delicious of fish, let me salute your eels.

BOEOTIAN (*in tragic style*): Come, thou, the eldest of my fifty Copaic virgins, come and complete the joy of our host.

DICAEOPOLIS (*likewise*): Oh! my well-beloved, thou object of my long regrets, thou art here at last then, thou, after whom the comic poets sigh, thou, who art dear to Morychus. Slaves, hither with the stove and the bellows. Look at this charming eel, that returns to us after six long years of absence. Salute it, my children; as for myself, I will supply coal to do honour to the stranger. Take it into my house; death itself could not separate me from her, if cooked with beet leaves.

BOEOTIAN: And what will you give me in return?

DICAEOPOLIS: It will pay for your market dues. And as to the rest, what do you wish to sell me?

BOEOTIAN: Why, everything.

DICAEOPOLIS: On what terms? For ready-money or in wares from these parts?

BOEOTIAN: I would take some Athenian produce, that we have not got in Boeotia.

DICAEOPOLIS: Phaleric anchovies, pottery?

BOEOTIAN: Anchovies, pottery? But these we have. I want produce that is wanting with us and that is plentiful here.

DICAEOPOLIS: Ah! I have the very thing; take away an informer, packed up carefully as crockery-ware.

BOEOTIAN: By the twin gods! I should earn big money, if I took one; I would exhibit him as an ape full of spite.

DICAEOPOLIS (*as an informer enters*): Hah! here we have Nicarchus, who comes to denounce you.

BOEOTIAN: How small he is!

DICAEOPOLIS: But all pure evil.

NICARCHUS: Whose are these goods?

DICAEOPOLIS: Mine; they come from Boeotia, I call Zeus to witness.

NICARCHUS: I denounce them as coming from an enemy's country.

BOEOTIAN: What! you declare war against birds?

NICARCHUS: And I am going to denounce you too.

BOEOTIAN: What harm have I done you?

NICARCHUS: I will say it for the benefit of those that listen; you introduce lampwicks from an enemy's country.

DICAEOPOLIS: Then you even denounce a wick.

NICARCHUS: It needs but one to set an arsenal afire.

DICAEOPOLIS: A wick set an arsenal ablaze! But how, great gods?

NICARCHUS: Should a Boeotian attach it to an insect's wing, and, taking advantage of a violent north wind, throw it by means of a tube into the arsenal and the fire once get hold of the vessels, everything would soon be devoured by the flames.

DICAEOPOLIS: Ah! wretch! an insect and a wick devour everything!

(*He strikes him.*)

NICARCHUS (*to the* CHORUS): You will bear witness, that he mishandles me.

DICAEOPOLIS (*to the* BOEOTIAN): Shut his mouth. Give me some hay; I am going to pack him up like a vase, that he may not get broken on the road.

(*The* INFORMER *is bound and gagged and packed in hay.*)

LEADER OF THE CHORUS: Pack up your goods carefully, friend; that the stranger may not break it when taking it away.

DICAEOPOLIS: I shall take great care with it. (*He hits the* INFORMER *on the head and a stifled cry is heard.*) One would say he is cracked already; he rings with a false note, which the gods abhor.

LEADER OF THE CHORUS: But what will be done with him?

DICAEOPOLIS: This is a vase good for all purposes; it will be used as a vessel for holding all foul things, a mortar for pounding together law-suits, a lamp for spying upon accounts, and as a cup for the mixing up and poisoning of everything.

LEADER OF THE CHORUS: None could ever trust a vessel for domestic use that has such a ring about it.

DICAEOPOLIS: Oh! it is strong, my friend, and will never get broken, if care is taken to hang it head downwards.

BOEOTIAN: Well then, I will proceed to carry off my bundle.

LEADER OF THE CHORUS: Farewell, worthiest of strangers, take this informer, good for anything, and fling him where you like.

DICAEOPOLIS: Bah! this rogue has given me enough trouble to pack! Here! Boeotian, pick up your pottery.

BOEOTIAN: Stoop, Ismenias, that I may put it on your shoulder, and be very careful with it.

DICAEOPOLIS: You carry nothing worth having; however, take it, for you will profit by your bargain; the informers will bring you luck.

(*The* BOEOTIAN *and his slave depart;* DICAEOPOLIS *goes into his house; a slave comes out of* LAMACHUS' *house.*)

SLAVE: Dicaeopolis!

DICAEOPOLIS (*from within*): What's the matter? Why are you calling me?

SLAVE: Lamachus wants to keep the Feast of Cups, and I come by his order to bid you one drachma for some thrushes and three more for a Copaic eel.

DICAEOPOLIS (*coming out*): And who is this Lamachus, who demands an eel?

SLAVE (*in tragic style*): He is the terrible, indefatigable Lamachus, who is always brandishing his fearful Gorgon's head and the three plumes which o'ershadow his helmet.

DICAEOPOLIS: No, no, he will get nothing, even though he gave me his buckler. Let him eat salt fish while he shakes his plumes, and, if he comes here making any din, I shall call the inspectors. As for myself, I shall take away all these goods; (*in tragic style*) I go home on thrushes' wings and blackbirds' pinions.

(*He goes into his house.*)

FIRST SEMI-CHORUS (*singing*): You see, citizens, you see the good fortune which this man owes to his prudence, to his profound wisdom. You see how, since he has concluded peace, he buys what is useful in the household and good to eat hot. All good things flow towards him unsought. Never will I welcome the god of war in *my* house; never shall *he* sing the "Harmodius" at my table; he is a sot, who comes feasting with those who are overflowing with good things and brings all manner of mischief in his train. He overthrows, ruins, rips open; it is vain to make him a thousand offers, to say "be seated, pray, and drink this cup, proffered in all friendship"; he burns our vine-stocks and brutally spills on the ground the wine from our vineyards.

SECOND SEMI-CHORUS (*singing*): This man, on the other hand, covers his table with a thousand dishes; proud of his good fortunes, he has had these feathers cast before his door to show us how he lives. (*A woman appears, bearing the attributes of Peace.*) Oh, Peace! companion of fair Aphrodité and of the sweet Graces, how charming are thy features and yet I never knew it! Would that Eros might join me to thee, Eros crowned with roses as Zeuxis shows him to us! Do I seem somewhat old to thee? I am yet able to make thee a threefold offering; despite my age I could plant a long row of vines for you; then beside these some tender cuttings from the fig; finally a young vine-stock, loaded with fruit, and all around the field olive trees, to furnish us with oil wherewith to anoint us both at the New Moons.

(*A* HERALD *enters.*)

HERALD: Oyez, oyez! As was the custom of your forebears, empty a full pitcher of wine at the call of the trumpet; he who first sees the bottom shall get a wine-skin as round and plump as Ctesiphon's belly.

DICAEOPOLIS (*coming out of the house; to his family within*): Women, children, have you not heard? Faith! do you not heed the herald? Quick! let the hares boil and roast merrily; keep them turning; withdraw them from the flame; prepare the chaplets; reach me the skewers that I may spit the thrushes.

LEADER OF FIRST SEMI-CHORUS: I envy you your wisdom and even more your good cheer.

DICAEOPOLIS: What then will you say when you see the thrushes roasting?

LEADER OF FIRST SEMI-CHORUS: Ah! true indeed!

DICAEOPOLIS: Slave! stir up the fire.

LEADER OF FIRST SEMI-CHORUS: See, how he knows his business, what a perfect cook! How well he understands the way to prepare a good dinner!

(*A* HUSBANDMAN *enters in haste.*)

HUSBANDMAN: Ah! woe is me!

DICAEOPOLIS: Heracles! What have we here?

HUSBANDMAN: A most miserable man.

DICAEOPOLIS: Keep your misery for yourself.

HUSBANDMAN: Ah! friend! since you alone are enjoying peace, grant me a part of your truce, were it but five years.

DICAEOPOLIS: What has happened to you?

HUSBANDMAN: I am ruined; I have lost a pair of steers.

DICAEOPOLIS: How?

HUSBANDMAN: The Boeotians seized them at Phylé.

DICAEOPOLIS: Ah! poor wretch! and do you still wear white?

HUSBANDMAN: Their dung made my wealth.

DICAEOPOLIS: What can I do in the matter?

HUSBANDMAN: Crying for my beasts has lost me my eyesight. Ah! if you care for poor Dercetes of Phylé, anoint mine eyes quickly with your balm of peace.

DICAEOPOLIS: But, my poor fellow, I do not practise medicine.

HUSBANDMAN: Come, I adjure you; perhaps I shall recover my steers.

DICAEOPOLIS: Impossible; away, go and whine to the disciples of Pittalus.

HUSBANDMAN: Grant me but one drop of peace; pour it into this little reed.

DICAEOPOLIS: No, not a particle; go and weep somewhere else.

HUSBANDMAN (*as he departs*): Oh! oh! oh! my poor beasts!

LEADER OF SECOND SEMI-CHORUS: This man has discovered the sweetest enjoyment in peace; he will share it with none.

DICAEOPOLIS (*to a slave*): Pour honey over this tripe; set it before the fire to dry.

LEADER OF SECOND SEMI-CHORUS: What lofty tones he uses! Did you hear him?

DICAEOPOLIS (*to the slaves inside the house*): Get the eels on the gridiron!

LEADER OF SECOND SEMI-CHORUS: You are killing me with hunger; your smoke is choking your neighbours, and you split our ears with your bawling.

DICAEOPOLIS: Have this fried and let it be nicely browned.

(*He goes back into the house. A* WEDDING GUEST *enters, carrying a package.*)

WEDDING GUEST: Dicaeopolis! Dicaeopolis!

DICAEOPOLIS: Who are you?

WEDDING GUEST: A young bridegroom sends you these viands from the marriage feast.

DICAEOPOLIS: Whoever he be, I thank him.

WEDDING GUEST: And in return, he prays you to pour a glass of peace into this vase, that he may not have to go to the front and may stay at home to make love to his young wife.

DICAEOPOLIS: Take back, take back your viands; for a thousand drachmae I would not give a drop of peace. (*A young woman enters*) But who is she?

WEDDING GUEST: She is the matron of honour; she wants to say something to you from the bride privately.

DICAEOPOLIS: Come, what do you wish to say? (*The* MATRON OF HONOUR *whispers in his ear.*) Ah! what a ridiculous demand! The bride burns with longing to keep her husband's tool at home. Come! bring hither my truce; to her alone will I give some of it, for she is a woman, and, as such, should not suffer under the war. Here, friend, hand me your vial. And as to the manner of applying this balm, tell the bride, when a levy of soldiers is made, to rub some in bed on her husband, where most needed. (*The* MATRON OF HONOUR *and the* WEDDING GUEST *depart.*) There, slave, take away my truce! Now, quick, bring me the wine-flagon, that I may fill up the drinking bowls!

(*The slave leaves. A* HERALD *enters.*)

LEADER OF THE CHORUS (*in tragic style*): I see a man, "striding along apace, with knitted brows; he seems to us the bearer of terrible tidings."

HERALD (*in tragic style*): Oh! toils and battles and Lamachuses!

(*He knocks on* LAMACHUS' *door.*)

LAMACHUS (*from within; in tragic style*): What noise resounds around my dwelling, where shines the glint of arms.

(*He comes out of his house.*)

HERALD: The Generals order you forthwith to take your battalions and your plumes, and, despite the snow, to go and guard our borders. They have learnt that a band of Boeotians intend taking advantage of the Feast of Cups to invade our country.

LAMACHUS: Ah! the Generals! they are numerous, but not good for much! It's cruel, not to be able to enjoy the feast!

DICAEOPOLIS: Oh! warlike host of Lamachus!

LAMACHUS: Wretch! do you dare to jeer me?

DICAEOPOLIS: Do you want to fight this four-winged Geryon?

LAMACHUS: Oh! oh! what fearful tidings!

DICAEOPOLIS: Ah! ah! I see another herald running up; what news does he bring me?

(*Another* HERALD *enters.*)

HERALD: Dicaeopolis!

DICAEOPOLIS: What is the matter?

HERALD: Come quickly to the feast and bring your basket and your cup; it is the priest of Bacchus who invites you. But hasten, the guests have been waiting for you a long while. All is ready—couches, tables, cushions, chaplets, perfumes, dainties and whores to boot; biscuits, cakes, sesamé-bread, tarts, lovely dancing women, and the "Harmodius." But come with all speed.

LAMACHUS: Oh! hostile gods!

DICAEOPOLIS: This is not astounding; you have chosen this great ugly Gorgon's head for your patron. (*To a slave*) You, shut the door, and let someone get ready the meal.

LAMACHUS: Slave! slave! my knapsack!

DICAEOPOLIS: Slave! slave! a basket!

LAMACHUS: Take salt and thyme, slave, and don't forget the onions.

DICAEOPOLIS: Get some fish for me; I cannot bear onions.

LAMACHUS: Slave, wrap me up a little stale salt meat in a fig-leaf.

DICAEOPOLIS: And for me some nice fat tripe in a fig-leaf; I will have it cooked here.

LAMACHUS: Bring me the plumes for my helmet.

DICAEOPOLIS: Bring me wild pigeons and thrushes.

LAMACHUS: How white and beautiful are these ostrich feathers!

DICAEOPOLIS: How fat and well browned is the flesh of this wood-pigeon!

LAMACHUS (*to* DICAEOPOLIS): My friend, stop scoffing at my armour.

DICAEOPOLIS (*to* LAMACHUS): My friend, stop staring at my thrushes.

LAMACHUS (*to his slave*): Bring me the case for my triple plume.

DICAEOPOLIS (*to his slave*): Pass me over that dish of hare.

LAMACHUS: Alas! the moths have eaten the hair of my crest.

DICAEOPOLIS: Shall I eat my hare before dinner?

LAMACHUS: My friend, will you kindly not speak to me?

DICAEOPOLIS: I'm not speaking to you; I'm scolding my slave. (*To the slave*) Shall we wager and submit the matter to Lamachus, which of the two is the best to eat, a locust or a thrush?

LAMACHUS: Insolent hound!

DICAEOPOLIS: He much prefers the locusts.

LAMACHUS: Slave, unhook my spear and bring it to me.

DICAEOPOLIS: Slave, slave, take the sausage from the fire and bring it to me.

LAMACHUS: Come, let me draw my spear from its sheath. Hold it, slave, hold it tight.

DICAEOPOLIS: And you, slave, grip well hold of the skewer.

LAMACHUS: Slave, the bracings for my shield.

DICAEOPOLIS: Pull the loaves out of the oven and bring me these bracings of my stomach.

LAMACHUS: My round buckler with the Gorgon's head.

DICAEOPOLIS: My round cheese-cake.

LAMACHUS: What clumsy wit!

DICAEOPOLIS: What delicious cheese-cake!

LAMACHUS: Pour oil on the buckler. Hah! hah! I can see reflected there an old man who will be accused of cowardice.

DICAEOPOLIS: Pour honey on the cake. Hah! hah! I can see an old man who makes Lamachus of the Gorgon's head weep with rage.

LAMACHUS: Slave, full war armour.

DICAEOPOLIS: Slave, my beaker; that is *my* armour.

LAMACHUS: With this I hold my ground with any foe.

DICAEOPOLIS: And I with this in any drinking bout.

LAMACHUS: Fasten the strappings to the buckler.

DICAEOPOLIS: Pack the dinner well into the basket.

LAMACHUS: Personally I shall carry the knapsack.

DICAEOPOLIS: Personally I shall carry the cloak.

LAMACHUS: Slave, take up the buckler and let's be off. It is snowing! God help us! A wintry business!

DICAEOPOLIS: Take up the basket, mine's a festive business.

(*They depart in opposite directions.*)

LEADER OF THE CHORUS: We wish you both joy on your journeys, which differ so much. One goes to mount guard and freeze, while the other will drink, crowned with flowers, and then lie with a young beauty till he gets his tool all sore.

CHORUS (*singing*): I say it freely; may Zeus confound Antimachus, the poet-historian, the son of Psacas! When Choregus at the Lenaea, alas! alas! he dismissed me dinnerless. May I see him devouring with his eyes a cuttle-fish, just served, well cooked, hot and properly salted; and the moment that he stretches his hand to help himself, may a dog seize it and run off with it. Such is my first wish. I also hope for him a misfortune at night. That returning all-fevered from horse practice, he may meet an Orestes, mad with drink, who will crack him over the head; that wishing to seize a stone, he, in the dark, may pick up a fresh turd, hurl, miss him and hit Cratinus.

(*The slave of* LAMACHUS *enters.*)

SLAVE OF LAMACHUS (*knocking on the door of* LAMACHUS' *house, in tragic style*): Captives present within the house of Lamachus, water, water in a little pot! Make it warm, get ready cloths, cerate, greasy wool and bandages for his ankle. In leaping a ditch, the master has hurt himself against a stake; he has dislocated and twisted his ankle, broken his head by falling on a stone, while his Gorgon shot far away from his buckler. His mighty braggadocio plume rolled on the ground; at this sight he uttered these doleful words, "Radiant star, I gaze on thee for the last time; my eyes close to all light, I die." Having said this, he falls into the water, gets out again, meets some runaways and pursues the robbers with his spear at their backsides. But here he comes, himself. Get the door open.

(*In this final scene all the lines are sung.*)

LAMACHUS (*limping in with the help of two soldiers and singing a song of woe*): Oh! heavens! oh! heavens! What cruel pain! I faint, I tremble! Alas! I die! the foe's lance has struck me! But what would hurt me most would be for Dicaeopolis to see me wounded thus and laugh at my ill-fortune.

DICAEOPOLIS (*enters with two courtesans, singing gaily*): Oh! my god! what breasts! Swelling like quinces! Come, my treasures, give me voluptuous kisses! Glue your lips to mine. Haha! I was the first to empty my cup.

LAMACHUS: Oh! cruel fate! how I suffer! accursed wounds!

DICAEOPOLIS: Hah! hah! Hail! Lamachippus!

LAMACHUS: Woe is me!

DICAEOPOLIS (*to the one girl*): Why do you kiss me?

LAMACHUS: Ah, wretched me!

DICAEOPOLIS (*to the other girl*): And why do you bite me?

LAMACHUS: 'Twas a cruel score I was paying back!

DICAEOPOLIS: Scores are not evened at the Feast of Cups!

LAMACHUS: Oh! Oh! Paean, Paean!

DICAEOPOLIS: But to-day is not the feast of Paean.

LAMACHUS (*to the soldiers*): Oh! take hold of my leg, do; ah! hold it tenderly, my friends!

DICAEOPOLIS (*to the girls*): And you, my darlings, take hold of my tool, both of you!

LAMACHUS: This blow with the stone makes me dizzy; my sight grows dim.

DICAEOPOLIS: For myself, I want to get to bed; I've got an erection and I want to make love in the dark.

LAMACHUS: Carry me to the surgeon Pittalus. Put me in his healing hands!

DICAEOPOLIS: Take me to the judges. Where is the king of the feast? The wine-skin is mine!

LAMACHUS (*as he is being carried away*): That spear has pierced my bones; what torture I endure!

DICAEOPOLIS (*to the audience*): You see this empty cup! I triumph! I triumph!

CHORUS: Old man, I come at your bidding! You triumph! you triumph!

DICAEOPOLIS: Again I have brimmed my cup with unmixed wine and drained it at a draught!

CHORUS: You triumph then, brave champion; thine is the wine-skin!

DICAEOPOLIS: Follow me, singing "Triumph! Triumph!"

CHORUS: Aye, we will sing of thee, thee and thy sacred wine-skin, and we all, as we follow thee, will repeat in thine honour, "Triumph, Triumph!"

The New Slums of Rome

Jérôme Carcopino was the director of the École Française de Rome. His *Daily Life in Ancient Rome* (1940) was intended to describe the people and the city at the height of the Empire. For the student who wants to see city streets come alive or how people worked, what they ate, how they made money and on what they spent it, Carcopino's book is invaluable. It also provides vivid descriptions of schools, gods, what government required of Romans and what it did for them especially in the second century A.D. Taking Juvenal, Martial, and Pliny the Younger as his chief guides, Carcopino in fact leads one into the inferno of city life—*pace* Dante!

The selection printed here reveals Carcopino's special skills in archaeology as he digs out for us the crowded tenements of the poor. We see thousands of Romans ill housed in buildings doomed to collapse, plagued by poor water supplies and worse sanitation. What strikes us immediately is the poverty unrelieved by the feats of technology for which Rome was in some respects so justly famous. But, as in our own society, the wealth of nations which harnesses nature and technology to the chariots of empire and war does so by cheating its citizens.* It is Carcopino's judgment that the uses of war have too often dazzled us and blinded us to the abuses of people which flourished in imperial Rome. By illuminating archaeological and literary evidence through a witty use of modern standards, this book becomes a comparative sociology of cities and thus doubly valuable.

From *Daily Life in Ancient Rome*

Jérôme Carcopino

Houses and Streets

In contrast to the Pompeian *domus*, the Roman *insula* grew steadily in stature until under the Empire it reached a dizzy height: Height was its dominant characteristic and this height which once amazed the ancient world still astounds us by its striking resemblance to our own most daring and modern buildings. As early as the third century B.C. *insulae* of three stories (*tabulata, contabulationes, contignationes*) were so frequent that they had ceased to excite remark. In enumerating the prodigies which, in the winter of 218–217 B.C., preluded the invasion of Hannibal, Livy mentions without further comment the incident of an ox which escaped from the cattle market and scaled the stairs of a riverside *insula* to fling itself into the void from the third story amid the horrified cries of the onlookers. By the end of the republic the average height of the *insulae* indicated by this anecdote had already been exceeded. Cicero's Rome was, as it were, borne

---

SOURCE: Jérôme Carcopino, *Daily Life in Ancient Rome*, ed. Henry T. Rowell, trans. E. O. Lorimer (New Haven: Yale University Press; London: Routledge & Kegan Paul, 1940), pp. 22–44.

* Since 1945, American governments have spent $1100 billion on armaments and war, more than the total value of all fixed capital in domestic and industrial buildings combined (1970). Yet, perhaps because of this, we have the eighteenth highest infant mortality rate in the world and slums as gross as those Petronius and Carcopino describe so well.

aloft and suspended in the air on the tiers of its apartment houses: *"Romam cenaculis sublatum atque suspensam."* The Rome of Augustus towered even higher. In his day, as Vitruvius records, "The majesty of the city and the considerable increase in its population have compelled an extraordinary extension of the dwelling houses, and circumstances have constrained men to take refuge in increasing the height of the edifices." This remedy proved so perilous that the emperor, alarmed by the frequent collapse of buildings, was forced to regulate it and to forbid private individuals to erect any building more than 20 metres high. It followed that avaricious and bold owners and contractors vied with each other in exploiting to the full the freedom still left them under this decree. Proofs abound to show that during the empire period the buildings attained a height which for that epoch was almost incredible. In describing Tyre at the beginning of the Christian Era, Strabo notes with surprise that the houses of this famous oriental seaport were almost higher than those of Imperial Rome. A hundred years later, Juvenal ridicules this aerial Rome which rests only on beams as long and thin as flutes. Fifty years later Aulus Gellius complains of stiff, multiple-storied houses (*"multis arduisque tabulatis"*); and the orator Aelius Aristides calculates in all seriousness that if the dwellings of the city were all reduced to one story they would stretch as far as Hadria on the upper Adriatic. Trajan in vain renewed the restrictions imposed by Augustus and even made them more severe by imposing a limit of eighteen metres on the height of private houses. Necessity, however, knows no law: and in the fourth century the sights of the city included that giant apartment house, the Insula of Felicula, besides the Pantheon and the Column of Marcus Aurelius. It must have been erected a century and a half before, for at the beginning of the reign of Septimius Severus (193–211) its fame had already spread across the seas. When Tertullian sought to convince his African compatriots of the absurdity of the heretical Valentinians, who filled the infinite space which separates the Creator from his creatures with mediators and intermediaries, he rallied the heretics on having "transformed the universe into a large, furnished apartment house, in whose attics they have planted their god under the tiles (*"ad summas tegulas"*) and accuses them of "rearing to the sky as many stories as we see in the Insula of Felicula in Rome."

Despite the edicts of Augustus and Trajan, the audacity of the builders had redoubled and the Insula of Felicula towered above the Rome of the Antonines like a sky-scraper. Even if this particular building remained an exception, an unusually monstrous specimen, we know from the records that all around it rose buildings of five and six stories. Martial was fortunate in having to climb only to the third floor of his quarters on the Quirinal, for many other tenants of the house were worse lodged. In Martial's *insula* and in the neighbouring blocks of flats there were many dwellers perched much higher up, and in the cruel picture Juvenal paints of a fire in Rome, he seemed to be addressing one unfortunate who, like the god of the Valentinians, lived under the tiles: "Smoke is pouring out of your third floor attic, but you know nothing of it; if the alarm begins on the ground floor, the last man to burn will be he who has nothing to shelter him from the rain but the tiles, where the gentle doves lay their eggs."

There are two types of these innumerable and imposing structures, whose summits were invisible to the passer-by unless he stepped back some distance. In the more luxurious, the ground floor or most of it was let as a whole to one tenant. This floor had the prestige and the advantages of a private house and was often dignified by the name of *domus* in contrast to the flats or *cenacula* of the upper stories.

Only people of consequence with well-lined purses could indulge in the luxury of such a *domus*. We know for instance that in Caesar's day Caelius paid for his annual rent of 30,000 sesterces ($1200.00). In the humbler *insulae* the ground floor might be divided into booths and shops, the *tabernae*, which we can visualize the better because the skeletons of many have survived to this day in the Via Biberatica and at Ostia. Above the *tabernae* lowlier folk were herded. Each *taberna* opened straight onto the street by a large arched doorway extending its full width, with folding wooden leaves which were closed or drawn across the threshold every evening and firmly locked and bolted. Each represented the storehouse of some merchant, the workshop of some artisan, or the counter and show-window of some retailer. But in the corner of each *taberna* there was nearly always a stair of five or six steps of stone or brick continued by a wooden ladder. The ladder led to a sloping loft, lit directly by one long oblong window pierced above the centre of the doorway, which served as the lodging of the storekeeper, the caretakers of the shop, or the workshop hands. Whoever they might be, the tenants of *tabernae* had never more than this one room for themselves and their families; they worked, cooked, ate, and slept there, and were at least as crowded as the tenants of the upper floors. Perhaps on the whole they were even worse provided for. Certainly they frequently found genuine difficulty in meeting their obligations. To bring pressure to bear on a defaulter, the landlord might "shut up the tenant" (*percludere inquilinum*), that is, make a lien on his property to cover the amount due.

There were, then, differences between the two types of apartment house which are known by the common name of *insula*, but almost all resulted from the primary distinction between those houses where the *rez-de-chaussée* formed a *domus* and those in which it was let out in *tabernae*. The two types might be found side by side, and they obeyed the same rules in the internal arrangement and external appearance of their upper stories.

Let us for a moment consider the Rome of our own day. It is true that in the course of the last sixty years, and particularly since the parcelling out of the Villa Ludovisi, Rome has seen the separate development of "aristocratic quarters." But prior to that, an equalitarian instinct had always tended to place the most stately dwellings and the humblest side by side; and even today the stranger is sometimes surprised to turn from a street swarming with the poorest of the poor and find himself face to face with the majesty of a Palazzo Farnese. This brotherly feature of the living Rome helps us to reconstruct in imagination the Rome of the Caesars where high and low, patrician and plebeian, rubbed shoulders everywhere without coming into conflict. Haughty Pompey did not consider it beneath his dignity to remain faithful to the Carinae. Before migrating for political and religious reasons into the precincts of the Regia, the most fastidious of patricians, Julius Caesar, lived in the Subura. Later Maecenas planted his gardens in the most evilly reputed part of the Esquiline. About the same period the ultra-wealthy Asinius Pollio chose for his residence the plebeian Aventine, where Licinius Sura, vice emperor of the reign of Trajan, also elected to make his home. At the end of the first century A.D. the emperor Vespasian's nephew and a parasite poet like Martial were near neighbours on the slopes of the Quirinal, and at the end of the second century Commodus went to dwell in a gladiatorial school on the Caelian.

It is true that when they were laid waste by fire the various quarters of the city rose from their ashes more solid and more magnificent than before. Nevertheless, the incongruous juxtapositions which persist to this day were repeated with a minimum of change after each renovation, and every attempt to specialise the fourteen regions of the Urbs was foredoomed to failure. Hypersensitive people, anxious to escape the mob, were driven to move to a greater and greater distance, to take refuge on the fringes of the Campagna among the pines of the Pincian or the Janiculum, where they could find room for the parks of their suburban villas. The common people, meanwhile, driven out of the centre of the city by the presence of the court and the profusion of public buildings but nevertheless fettered to it by the business transacted there, overflowed by preference into the zones intermediate between the fora and the outskirts, the outside districts adjacent to the Republican Wall, which the reform of Augustus had with one stroke of the pen incorporated in the Urbs. The *Regionaries* record the number of *insulae* or apartment blocks in each region, and the number of *vici* or arteries serving the *insulae;* and separate averages may be obtained for the eight regions of the old city and the six regions of the new. The average for the older regions is 2965 *insulae* with 17 *vici* and for the newer 3429 *insulae* with 28 *vici*. We note that the largest number of *insulae* were massed in the new city; and that they attained the greatest size not in the old city where there were 174 *insulae* per *vicus*, but in the new where there were only 123 per *vicus*. The *Regionaries* also locate for us the Insula of Felicula, the giant sky-scraper in the ninth region, known as the region of the Circus Flaminius, in the very heart of the new city. Isolated soundings lead us to the same conclusion as do mass statistics: the successful experiments of imperial city planning caused the huge modern-style apartment blocks of ancient Rome to increase in number and grow to immoderate size.

Seen from the outside, all these monumental blocks of flats were more or less identical in appearance and presented a fairly uniform façade to the street. Piled story upon story, the large-bayed *cenacula* were superimposed one above the other; the first steps of their stone staircases cut through the line of the *tabernae* or the walls of the *domus*. Reduced to its governing essentials, the plan of these buildings is familiar. They might well be urban houses of today or yesterday. From a study of the best preserved of their ruins, the most competent experts have been able to reproduce on paper the original plan and elevation; and these drawings show such startling analogies with the buildings in which we ourselves live that at first sight we are tempted to mistrust them. A more attentive examination, however, bears witness to the conscientious accuracy of these reconstructions. M. Boethius, for instance, has brought together on one photographic plate such and such a section of Trajan's market or such and such a building at Ostia and an equivalent existing piece of building in the Via dei Cappellari at Rome or the Via dei Tribunali at Naples. By this means he has demonstrated a surprising resemblance—at moments approaching identity—between these plans, separated in time by so many centuries. If they could rise from the dead, the subjects of Trajan or of Hadrian would feel they had come home when they crossed the threshold of these modern roman *casoni;* but they might with justice complain that in external appearance their houses had lost rather than gained in the course of the ages.

Superficially compared with its descendant of the Third Italy, the *insula* of Imperial Rome displays a more delicate taste and a more studied elegance, and in truth the ancient building gives the more modern impression. Here wood and rubble are ingeniously combined in its facings and there bricks are disposed with cunning skill, all harmonised with a perfection of art which has been forgotten among us since the Norman mansions and the castles of Louis XIII. Its doorways and its windows were no less numerous and often larger. Its row of shops was usually protected and screened with the line of a portico. In the wider streets its stories were relieved by a picturesque variety either of loggias (*pergulae*) testing on the porticos or of balconies (*maeniana*). Some were of wood, and the beams that once supported them may be found still embedded in the masonry; others were brick, sometimes thrown out on pendentives whose lines of horizontal impost are the parents of the parallel extrados, sometimes based on a series of cradle-vaults supported by large travertine consoles firmly embedded in the masonry of the prolonged lateral walls.

Climbing plants clung round the pillars of the loggias and the railing of the balconies, while most of the windows boasted miniature gardens formed of pots of flowers such as the elder Pliny has described. In the most stifling corners of the great city these flowers assuaged somewhat the homesickness for the countryside which lay heavy on the humble town dweller sprung from a long line of peasant ancestors. We know that at the end of the fourth century the host of a modest inn at Ostia, like that in which Saint Augustine set his gentle and memorable discourse with Saint Monica, always surrounded his guest house with green and shady trees. The Casa dei Dipinti, considerably older still, seems to have been completely festooned with flowers, and the highly plausible reconstruction of it which MM. Calza and Gismondi have made suggests a garden city in every respect like the most attractive ones that enlightened building societies and philanthropic associations are putting up today for the workmen and lower middle classes of our great towns.

Unfortunately for this *insula*, the most luxurious of those to which archaeology has so far introduced us, its external appearance belied its comforts. The architects had indeed neglected nothing in its outward embellishment. They had paved it with tiles and mosaics. By long and costly processes they had clothed it with colours as fresh and living in their day as those of the frescoes of Pompeii, though now three parts obliterated. To these it owes the name by which learned Italians call it, Casa dei Dipinti, the House of Paintings. I dare not assert that it was equipped with *laquearia* enamelled on movable plaques of arbor vitae or carved ivory, such as wealthy upstarts like Trimalchio fixed above their dining-tables and worked by machinery to rain down flowers or perfume or tiny valuable gifts on their surprised and delighted guests, but it is not improbable that the ceilings of the rooms were covered with the gilded stucco which the elder Pliny's contemporaries admired. Be that as it may, all this luxury had its price, and the most opulent of the *insulae* suffered from the fragility of their construction, the scantiness of their furniture, insufficient light and heat, and absence of sanitation.

Archaic Aspects of the Roman House

These lofty buildings were far too lightly built. While the *domus* of Pompeii easily covered 800 to 900 square metres, the *insulae* of Ostia, though built according to the specifications which Hadrian laid down, were rarely granted such extensive foundations. As for the Roman *insulae*, the ground plans recoverable from the cadastral survey of Septimius Severus, who reproduced them, show that they usually varied between 300 and 400 square metres. Even if there were no smaller ones (which is extremely unlikely) of which all trace has been buried for ever in the upheavals of the terrain, these figures are misleading: a foundation of 300 square metres is inadequate enough to carry a structure of 18 to 20 metres high, particularly when we remember the thickness of the flooring which separated the stories from each other. We need only consider the ratio of the two figures given to feel the danger inherent in their disproportion. The lofty Roman buildings possessed no base corresponding to their height and a collapse was all the more to be feared since the builders, lured by greed of gain, tended to economise more and more at the expense of the strength of the masonry and quality of the materials. Vitruvius states that the law forbade a greater thickness for the outside walls than a foot and a half, and in order to economise space the other walls were not to exceed that. He adds that, at least from the time of Augustus, it was the custom to correct this thinness of the walls by inserting chains of bricks to strengthen the concrete. He observes with smiling philosophy that this blend of stone-course, chains of brick and layers of concrete in which the stones and pebbles were symmetrically embedded, permitted the convenient construction of buildings of great height and allowed the Roman people to create handsome dwellings for themselves with little difficulty: *"populus romanus egregias habet sine impeditione habitationes."*

Twenty years later Vitruvius would have recanted. The elegance he so admired had been attained only at the sacrifice of solidity. Even after the brick technique had been perfected in the second century, and it had become usual to cover the entire façade with bricks, the city was constantly filled with the noise of buildings collapsing or being torn down to prevent it; and the tenants of an *insula* lived in constant expectation of its coming down on their heads. We may recall the savage and gloomy tirade of Juvenal: "Who at cool Praeneste, or at Volsinii amid its leafy hills, was ever afraid of his house tumbling down? . . . But here we inhabit a city propped up for the most part by slats: for that is how the landlord patches up the crack in the old wall, bidding the inmates sleep at ease under the ruin that hangs above their heads."

The satirist has not exaggerated and many specific cases provided for in the legal code, the *Digest*, take for granted precisely the precarious state of affairs which excited Juvenal's wrath.

> Suppose, for instance, that the owner of an *insula* has leased it for a sum of 30,-000 sesterces to a principal tenant who by means of subletting draws from it a revenue of 40,000 sesterces, and that the owner presently, on the pretext that the building is about to collapse, decides to demolish it; the principal tenant is entitled to bring an action for damages. If the building was demolished of necessity, the plaintiff will be entitled to the refund of his rent and nothing further. On the other hand, if the building has been demolished only to enable the owner to erect a better and ultimately more remunerative building, the defendant must further pay to the principal tenant who has been compelled to evict his subtenants whatever sum the plaintiff has thus lost.

This text is suggestive both in itself and in its implications. The terms in which it is couched leave no doubt as to the frequency of the practices of which it speaks, and they indicate that the houses of Imperial Rome were at least as fragile as the old American tenements which not so long ago collapsed or had to be demolished in New York.

The Roman houses, moreover, caught fire as frequently as the houses of Stamboul under the Sultans. This was because, in the first place, they were unsubstantial; further, the weight of their floors involved the introduction of massive wooden beams, and the movable stoves which heated them, the candles, the smoky lamps, and the torches which lighted them at night involved perpetual risk of fire; and finally, as we shall see, water was issued to the various stories with grudging hand. All these reasons combined to increase both the number of fires and the rapidity with which they spread. The wealthy Crassus in the last century of the republic devised a scheme for increasing his immense fortune by exploiting these catastrophes. On hearing the news of an outbreak, he would run to the scene of the disaster and offer profuse sympathy to the owner, plunged in despair by the sudden destruction of his property. Then he would offer to buy on the spot—at a sum far below its real value—the parcel of ground, now nothing but a mass of smouldering ruins. Thereupon, employing one of the teams of builders whose training he had himself superintended, he erected a brand new *insula*, the income from which amply rewarded him for his capital outlay.

Even later, under the empire, after Augustus had created a corps of *vigiles* or fire-fighting night watchmen, the tactics of Crassus would have been no less successful. In spite of the attention Trajan paid to the policing of the Urbs, outbreaks of fire were an everyday occurrence in Roman life. The rich man trembled for his mansion, and in his anxiety kept a troop of slaves to guard his yellow amber, his bronzes, his pillars of Phrygian marble, his tortoise-shell inlays. The poor man was startled from his sleep by flames invading his attic and the terror of being roasted alive. Dread of fire was such an obsession among rich and poor alike that Juvenal was prepared to quit Rome to escape it: "No, no, I must live where there is no fire and the night is free from alarms!" He had hardly overstated the case. The jurists echo his satires, and Ulpian informs us that not a day passed in Imperial Rome without several outbreaks of fire: "*plurimis uno die incendiis exortis.*"

The scantiness of furniture at least reduced the gravity of each of these catastrophes. Granted that they were warned in time, the poor devils of the *cenacula* (like that imaginary Ucalegon whom Juvenal ironically saddled with the epic name of one of the Trojans of the Aeneid) were quickly able "to clear out their miserable goods and chattels." The rich had more to lose and could not, like Ucalegon, stuff all their worldly possessions into one bundle. Apart from their statues of marble and bronze, their furniture, however, was sparse enough, for wealth displayed itself not in the number of items but in their quality, the precious materials employed, and the rare shapes which bore witness to their owner's taste.

In the passage of Juvenal quoted above, the millionaire he pictures was taking precaution to save not what we nowadays would call "furniture," but his curios and *objets d'art*. For every Roman, the main item of furniture was the bed (*lectus*) on which he slept during his siesta and at night and on which he reclined by day to eat, read, write, or receive visitors. Humble people made shift with a shelf of masonry built along the wall and covered with a pallet. Those better off had handsomer and more elaborate couches in proportion to their means. Most beds were single ones (*lectuli*). There were double beds for married couples (*lecti geniales*); beds for three which graced the dining-room (*triclinia*); and those who wished to make a splash and astonish the neighbours had couches for six. Some were cast in bronze; most were simply carved in wood, either in oak or maple, terebinth or arbor vitae, or it might be in those exotic woods with undulating grain and changing lights which reflected a thousand colours like a peacock's tail (*lecti pavonini*). Some beds boasted bronze feet and a wooden frame, others again ivory feet and a frame of bronze. In some cases the woodwork was inlaid with tortoise-shell; in some the bronze was nielloed with silver and gold. There were even some, like Trimalchio's, of massive silver.

Whatever its nature and style, the bed was the major piece of furniture alike in the aristocratic *domus* and in the proletarian *insula*, and in many cases it deterred the Romans from seeking to provide themselves with anything else. Their tables (*mensae*) had little in common with ours. They never developed into sturdy tables with four legs—those were introduced late in history through the intermediary of Christian rites. When the empire was in its glory, the *mensa* was a set of little shelves in tiers, supported on one leg, and used to display for a visitor's admiration the most valuable treasures of the house (*cartibula*). Alternatively, it might be a low table of wood or bronze with three or four adjustable supports (*trapezophores*), or a simple tripod whose folding metal legs usually ended in a lion's claw. As for seats, remains of these are—not without reason—more rarely found in the excavations than tables. The armchair with back, the *thronus*, was reserved for the divinity; the chair with a more or less sloping back, the *cathedra*, was especially popular with women. Great ladies, whose indolence is a target for Juvenal's scorn, would languidly repose in them, and we have literary record of their existence in two houses: the reception hall in the palace of Augustus—Corneille's "Be seated, Cinna" is derived directly from Seneca's account—and in the room (*cubiculum*) where the younger Pliny invited his friends to come and talk with him. Also they appear in literature as the distinguishing property of the master who is teaching in a *schola* or, in connection with religious ceremonies, as the property of the *frater arvalis* of the official religion, of the head of certain esoteric pagan sects, and later of the Christian presbyter. We speak with perfect right, therefore, of the "Chair of Saint Peter" or the "chair" of a university professor.

Ordinarily the Romans were content with benches (*scamna*) or stools (*subsellia*) or *sellae* without arms or back, which they carried about with them out of doors. Even when the seat was the magistrate's ivory *sella curulis*, or made of gold like Julius Caesar's, it was never more than a folding "camp stool." The rest of the furniture, the essentials apart from beds, consisted of the covers, the cloths, the counterpanes, the cushions, which were spread over or placed on the bed, at the foot of the table, on the seat of the stool, and on the bench; and finally, the eating utensils and the jewellery. Silver table services were so common that Martial ridicules patrons who were too niggardly of their Saturnalian gifts to give their clients at least five pounds (a trifle over three pounds avoirdupois) of silverware. Only the very poor used earthenware. The rich had vessels carved by a master hand, sparkling with gold and set with precious stones. Reading some of the ancient descriptions, one seems to relive a scene out of the *Arabian Nights*, set in spacious, unencumbered rooms where wealth is revealed only by the profusion and depth of the divans, the iridescence of damask, the sparkle of jewellery and of damascened copper—and yet all the elements of that "comfort" to which the West has grown so much attached are lacking.

Even in the most luxurious Roman house, the lighting left much to be desired:

though the vast bay windows were capable of flooding it at certain hours with the light and air we moderns prize, at other times either both had to be excluded or the inhabitants were blinded and chilled beyond endurance. Neither in the Via Biberatica nor in Trajan's market nor in the Casa dei Dipinti at Ostia do we find any traces of mica or glass near the windows, therefore the windows in these places cannot have been equipped with the fine transparent sheets of *lapis specularis* with which rich families of the empire sometimes screened the alcove of a bedroom, a bathroom or garden hothouse, or even a sedan chair. Nor can they have been fitted with the thick, opaque panes which are still found in place in the skylight windows of the baths of Herculaneum and Pompeii, where they provided a hermetic closure to maintain the heat without producing complete darkness. The dwellers in a Roman house must have protected themselves, very inadequately, with hanging cloths or skins blown by wind or drenched by rain; or overwell by folding shutters of one or two leaves which, while keeping cold and rain, midsummer heat or winter wind at bay, also excluded every ray of light. In quarters armed with solid shutters of this sort the occupant, were he an ex-consul or as well known as the younger Pliny, was condemned either to freeze in daylight or to be sheltered in darkness. The proverb says that a door must be either open or shut. In the Roman *insula*, on the contrary, the tenant could be comfortable only when the windows were neither completely open nor completely shut; and it is certain that in spite of their size and number, the Romans' windows rendered them neither the service nor the pleasure that ours give us.

In the same way, the heating arrangements in the *insula* were extremely defective. As the *atrium* had been dispensed with, and the *cenacula* were piled one above the other, it was impossible for the inhabitants of an *insula* to enjoy the luxury common to the peasantry, of gathering round the fire lighted by the womenfolk in the centre of their hovels, while sparks and smoke escaped by the gaping hole purposely left in the roof. It would be a grave mistake, moreover, to imagine that the *insula* ever enjoyed the benefit of central heating with which a misuse of language and an error of fact have credited it. The furnace arrangements which are found in so many ruins never fulfilled this office. They consisted of, first, a heating apparatus (the *hypocausis*) consisting of one or two furnaces which were stoked, according to the intensity of heat desired and the length of time it was to be maintained, with wood or charcoal, faggots or dried grass; second, an exit channel through which the heat, the soot, and the smoke penetrated indiscriminately into the adjacent *hypocaustum;* third, the heat-chamber (the *hypocaustum*) characterised by piles of bricks in parallel rows, between and over which heat, soot, and smoke circulated together; and finally the heated rooms resting on, or, rather, suspended above the *hypocaustum* and known, therefore, as the *suspensurae.* Whether or not they were connected with it by the spaces within their partition walls, the *suspensurae* were separated from the *hypocaustum* by a flooring formed of a bed of bricks, a layer of clay, and a pavement of stone or marble. This compact floor was designed to exclude unwelcome or injurious exhalations and to slow down the rise of temperature. It will be noticed that in this device the heated surface of the *suspensurae* was never greater than the surface of the *hypocaustum* and its working demanded a number of *hypocauses* equal to, if not greater than, the number of *hypocausta.* It follows, therefore, that this system of furnaces had nothing to do with central heating and was not applicable to many-storied buildings. In ancient Italy it was never used to heat an entire building, unless it was one single and isolated room like the latrine excavated in 1929 at Rome between the Great Forum and the Forum of Caesar. Moreover, even in the buildings where such a furnace system existed, it never occupied more than a small fraction of the house: the bathroom in the best-equipped villas of Pompeii or the *caldarium* of the public baths. It need hardly be stressed that no traces of such a system have been found in any of the *insulae* known to us.

This was not the worst. The Roman *insula* lacked fireplaces as completely as furnaces. Only a few bakeries at Pompeii had an oven supplied with a pipe somewhat resembling our chimney; it would be too much to assume that it was identical with it, for of the two examples that can be cited, one is broken off in such a way that we cannot tell where it used to come out, and the other was not carried up to the roof but into a drying cupboard on the first floor. No such ventilation shafts have been discovered in the villas of Pompeii or Herculaneum; still less, of course, in the houses of Ostia, which reproduce in every detail the plan of the Roman *insula*. We are driven to conclude that in the houses of the Urbs bread and cakes were cooked with a fire confined in an oven, other food simmered over open stoves, and the inhabitants themselves had no remedy against the cold but what a brazier could provide. Many of these were portable or mounted on runners. Some were wrought in copper or bronze with great taste and skill. But the grace of this industrial art was scant compensation for the brazier's limited heating power and range. The haughtiest dwellings of ancient Rome were strangers alike to the gentle, equal warmth which the radiator spreads through our rooms and to the cheerfulness of our open fires. They were threatened moreover by the attack of noxious fumes and not infrequently by the escape of smoke which was not always prevented either by the thorough drying or even by the preliminary carbonisation of the fuel (*ligna coctilia, acapna*).

To make matters worse, the *insula* was as ill supplied with water as with light and heat. I admit that the opposite opinion is generally held. People forget that the conveyance of water to the city at State expense was regarded as a purely public service from which private enterprise had been excluded from the first, and which continued to function under the empire for the benefit of the collective population with little regard for the needs of private individuals. According to Frontinus, a contemporary of Trajan, eight aqueducts brought 222,237,060 gallons of water a day to the city of Rome, but very little of this immense supply found its way to private houses.

In the first place, it was not until the reign of Trajan and the opening on June 24, 109, of the aqueduct called by his name, *aqua Traiana*, that fresh spring water was brought to the quarters on the right bank of the Tiber; until then, the inhabitants had to make their wells suffice for their needs. Secondly, even on the left bank access to the distributory channels connected by permission of the princeps with the *castella* of his aqueducts was granted, on payment of a royalty, only to individual concessionaires and to ground landlords; and certainly up to the beginning of the second century these concessions were revocable and were, in fact, brutally revoked by the administration on the very day of the death of a concessionaire. Finally, and most significantly, it seems that these private water supplies were everywhere confined to the ground floor, the chosen residence of the capitalists who had their *domus* at the base of the apartment blocks.

In the colony of Ostia, for instance, which, like its neighbour Rome, possessed an aqueduct, municipal channels, and private conduits, no building that has so far been excavated reveals any trace of rising columns which might have conveyed spring water to the upper stories. All ancient texts, moreover, whatever the period in which they were written, bear conclusive witness to the absence of any such installations. Under the empire, the poet Martial complains that his town house lacks water although it is situated near an aqueduct. In the *Satires* of Juvenal the water-carriers (*aquarii*) are spoken of as the scum of the slave population. The jurists of the first half of the third century considered the water-carriers so vital to the collective life of each *insula* that they formed, as it were, a part of the building itself and, like the porters (*ostiarii*) and the sweepers (*zetarii*), were inherited with the building by the heir or legatees. The praetorian prefect Paulus, in issuing instructions to the *praefectus vigilum*, did not forget to remind the commandant of the Roman firemen that it was part of his duty to warn tenants always to keep water ready in their rooms to check an outbreak: "*ut aquam unusquisque inquilinus in cenaculo habeat iubetur admonere*."

Obviously, if the Romans of imperial times had needed only to turn a tap and let floods of water flow into a sink, this warning would have been superfluous. The mere fact that Paulus expressly formulated the warning proves that, with a few exceptions to which we shall revert later, water from the aqueducts reached only the ground floor of the *insula*. The tenants of the upper *cenacula* had to go and draw their water from the nearest fountain. The higher the flat was perched, the harder the task of carrying water to scrub the floors and walls of those crowded *contignationes*. It must be confessed that the lack of plentiful water for washing invited the tenants of many Roman *cenacula* to allow filth to accumulate, and it was inevitable that many succumbed to the temptation for lack of a water system such as never existed save in the imagination of too optimistic archaeologists.

Far be it from me to stint my well-deserved admiration for the network of sewers which conveyed the sewage of the city into the Tiber. The sewers of Rome were begun in the sixth century B.C. and continually extended and improved under the republic and under the empire. The *cloacae* were conceived, carried out, and kept up on so grandiose a scale that in certain places a waggon laden with hay could drive through them with ease; and Agrippa, who perhaps did more than any man to increase their efficiency and wholesomeness by diverting the overflow of the aqueducts into them through seven channels, had no difficulty in travelling their entire length by boat. They were so solidly constructed that the mouth of the largest, as well as the oldest of them, the *Cloaca Maxima*, the central collector for all the others from the Forum to the foot of the Aventine, can still be seen opening into the river at the level of the Ponte Rotto. Its semicircular arch, five metres in diameter, is as perfect today as in the days of the kings to whom it is attributed. Its patinstra-tufa voussoirs have triumphantly defied the passage of twenty-five hundred years. It is a masterpiece in which the enterprise and patience of the Roman people collaborated with the long experience won by the Etruscans in the drainage of their marshes; and such as it has come down to us, it does honour to antiquity. But it cannot be denied that the ancients, though they were courageous enough to undertake it, and patient enough to carry it through, were not skilful enough to utilise it as we would have done in their place. They did not turn it to full account for securing a cleanly town or ensuring the health and decency of the inhabitants.

The system served to collect the sewage of the *rez-de-chausée* and of the public latrines which stood directly along the route, but no effort was made to connect the *cloacae* with the private latrines of the separate *cenacula*. There are only a few houses in Pompeii whose upstairs latrines were so designed that they could empty into the sewer below, whether by a conduit connecting them with the sewer or by a special arrangement of pipes, and the same can be said of Ostia and Herculaneum. But since this type of drainage is lacking in the most imposing *insulae* of Ostia as in those of Rome, we may abide in general by the judgment of Abbé Thédenat, who thirty-five years ago stated unequivocally that the living quarters of the *insulae* had never at any time been linked with the *cloacae* of the Urbs. The drainage system of the Roman house is merely a myth begotten of the complacent imagination of modern times. Of all the hardships endured by the inhabitants of ancient Rome, the lack of domestic drainage is the one which would be most severely resented by the Romans of today.

The very rich escaped the inconvenience. If they lived in their own *domus*, they had nothing to do but construct a latrine on the ground level. Water from the aqueducts might reach it and at worst, if it was too far distant from one of the sewers for the refuse to be swept away, the sewage could fall into a trench beneath. These cess trenches, like the one excavated near San Pietro in 1892, were neither very deep nor proof against seepage, and the manure merchants had acquired the right—probably under Vespasian—to arrange for emptying them. If the privileged had their *domus* in an *insula*, they rented the whole of the ground floor and enjoyed the same advantage as in a private house. The poor, however, had a longer way to go. In any case they were forced to go outside their homes. If the trifling cost was not deterrent, they could pay for entry to one of the public latrines administered by the *conductores foricarum*. The great number of these establishments, which the *Regionaries* attest, is an indication of the size of their clientele. In Trajan's Rome, as today in some backward villages, the immense majority of private people had to have recourse to the public latrine. But the comparison cannot be pushed further. The latrines of ancient Rome are disconcerting on two counts; we need only recall the examples of Pompeii, of Timgad, of Ostia, and that already alluded to at Rome itself, which was heated in winter by a *hypocausis:* the *forica* at the intersection of the Forum and the Forum Iulium.

The Roman *forica* was public in the full sense of the term, like soldiers' latrines in war time. People met there, conversed, and exchanged invitations to dinner without embarrassment. And at the same time, it was equipped with superfluities which we forego and decorated with a lavishness we are not wont to spend on such a spot. All round the semicircle or rectangle which it formed, water flowed continuously in little channels, in front of which a score or so of seats were fixed. The seats were of marble, and the opening was framed by sculptured brackets in the form of dolphins, which served both as a support and as a line of demarcation. Above the seats it was not unusual to see niches containing statues of gods or heroes, as on the Palatine, or an altar to Fortune, the goddess of health and happiness, as in Ostia; and not infrequently the room was cheered by the gay sound of a playing fountain as at Timgad. Let us be honest with ourselves: we are amazed at this mixture of delicacy and coarseness, at the solemnity and grace of the decorations and the familiarity of the actors. It is like nothing but the fifteenth-century *madrasas* in Fez, where the latrines were also designed to accommodate a crowd, and decorated with exquisitely delicate stucco and covered with a lacelike ceiling of cedar wood. Suddenly Rome—where even the latrines of the imperial palace, as majestic and ornate as a sanctuary beneath its dome, contained three seats side by side—Rome at once mystic and sordid, artistic and carnal, without embarrassment and without shame seems to join hands with the distant Haghrab at the epoch of the Merinids, so far removed from us in time and space.

But the public latrines were not the resort of misers or of the very poor. These folk had no mind to enrich the *conductores foricarum* to the tune of even one as. They preferred to have recourse to the jars, skilfully chipped down for the purpose, which the fuller at the corner ranged in front of his workshop. He purchased permission for this from Vespasian, in consideration of a tax to which no odour clung, so as to secure gratis the urine necessary for his trade. Alternatively they clattered down the stairs to empty their chamber pots (*lasana*) and their commodes (*sellae pertusae*) into the vat or *dolium* placed under the well of the staircase. Or if perhaps this expedient had been forbidden by the landlord of their *insula*, they betook themselves to some neighbouring dungheap. For in Rome of the Caesars, as in a badly kept hamlet of today, more than one alley stank with the pestilential odour of a cess trench (*lacus*) such as those which Cato the Elder during his censorship paved over when he cleaned the *cloacae* and led them under the Aventine. Such malodorous trenches were extant in the days of Cicero and Caesar; Lucretius mentions them in his poem, *De rerum natura*. Two hundred years later, in the time of Trajan, they were still there and one might see unnatural mothers of the Megaera type, anxious to rid themselves of an unwanted child, surreptitiously taking advantage of a barbarous law and exposing a new-born infant there; while matrons grieving over their barrenness would hasten no less secretly to snatch the baby, hoping to palm it off on a credulous husband as their own, and thus with a supposititious heir to still the ache in his paternal heart.

There were other poor devils who found their stairs too steep and the road to these dung pits too long, and to save themselves further trouble would empty the contents of their chamber pots from their heights into the streets. So much the worse for the passer-by who happened to intercept the unwelcome gift! Fouled and sometimes even injured, as in Juvenal's satire, he had no redress save to lodge a complaint against the unknown assailant; many passages of the *Digest* indicate that Roman jurists did not disdain to take cognisance of this offence, to refer the case to the judges, to track down the offender, and assess the damages payable to the victim. Ulpian classifies the various clues by which it might be possible to trace the culprit.

If [he says] the apartment [*cenaculum*] is divided among several tenants, redress can be sought only against that one of them who lives in that part of the apartment from the level of which the liquid has been poured. If the tenant, however, while professing to have sub-let [*cenacularium exercens*], has in fact retained for himself the enjoyment of the greater part of the apartment, he shall be held solely responsible. If, on the other hand, the tenant who professes to have sub-let has in fact retained for his own use only a modest fraction of the space, he and his sub-tenants shall be jointly held responsible. The same will hold good if the vessel or the liquid has been thrown from a balcony.

But Ulpian does not exclude the culpability of an individual if the inquiry is able to fix the blame on one guilty person, and he requests the praetor to set in equity a penalty proportionate to the seriousness of the injury. For instance:

When in consequence of the fall of one of these projectiles from a house, the body of a free man shall have suffered injury, the judge shall award to the victim in addition to medical fees and other expenses incurred in his treatment and necessary to his recovery, the total of the wages of which he has been or shall in future be deprived by the inability to work which has ensued.

Wise provisions these, which might seem to have inspired our laws relating to accidents, but which we have failed to adopt in their entirety, for Ulpian ends with a notable restriction. In formulating his final paragraph he expresses with unemotional simplicity his noble conception of the dignity of man: "As for scars or disfigurement which may have resulted from such wounds, no damages can be calculated on this count, for the body of a free man is without price."

The lofty sentiment of this phrase rises like a flower above a cess pit and serves to accentuate our dismayed embarrassment at the state of affairs of which these subtle legal analyses give a glimpse. Our great cities are also shadowed by misery, stained by the uncleanness of our slums, dishonoured by the vice they harbour. But at least the disease which gnaws at them is usually localised and confined to certain blighted quarters, whereas we get the impression that slums invaded every corner of Imperial Rome. Almost everywhere throughout the Urbs the *insulae* were the property of owners who had no wish to be concerned directly in their management and who leased out the upper stories to a promoter for five-year terms—in return for a rent at least equal to that of the ground-floor *domus*. This principal tenant who set himself to exploit the sub-letting of the *cenacula* had no bed of roses. He had to keep the place in repair, obtain tenants, keep the peace between them, and collect his quarterly payments on the year's rent. Not unnaturally he sought compensation for his worries and his risks by extorting enormous profits. Ever-rising rent is a subject of eternal lamentation in Roman literature.

In 153 B.C. an exiled king had to share a flat with an artist, a painter, in order to make ends meet. In Caesar's day the humblest tenant had to pay a rent of 2000 sesterces ($80) a year. In the times of Domitian and of Trajan, one could have bought a fine estate at Sora or Frusino for the price of quarters in Rome. So intolerable was the burden of rent that the sub-tenants of the first lessee almost invariably had to sub-let in their turn every room in their *cenaculum* which they could possibly spare. Almost everywhere, the higher you went in a building, the more breathless became the overcrowding, the more sordid the promiscuity. If the *rez-de-chaussée* was divided into several *tabernae*, they were filled with artisans, shopkeepers and eating-house keepers, like those of the *insula* which Petronius describes. If it had been retained for the use of one privileged possessor, it was occupied by the retainers of the owner of the *domus*. But whatever the disposition of the ground floor, the upper stories were gradually swamped by the mob: entire families were herded together in them; dust, rubbish, and filth accumulated; and finally bugs ran riot to such a point that one of the shady characters of Petronius' *Satyricon*, hiding under his miserable pallet, was driven to press his lips against the bedding which was black with them. Whether we speak of the luxurious and elegant *domus* or of the *insulae*—caravanserais whose heterogeneous inhabitants needed an army of slaves and porters under the command of a servile steward to keep order among them—the dwelling-houses of the Urbs were seldom ranged in order along an avenue, but jostled each other in a labyrinth of steep streets and lanes, all more or less narrow, tortuous, and dark, and the marble of the "palaces" shone in the obscurity of cut-throat alleys.

Streets and Traffic

If some magic wand could have disentangled the jumble of the Roman streets and laid them end to end, they would certainly have covered a distance of 60,000 *passus*, or approximately 89 kilometres. So we learn from the calculations and measurements carried out by the censors Vespasian and Titus in A.D. 73. And the elder Pliny, moved to pride by the contemplation of this immense extent of streets, compares with it the height of the buildings they served and proclaims that there existed in all the ancient world no city whose size could be compared to that of Rome.

The size is not to be denied, but if instead of admiring the imaginary and orderly perspective which Pliny plotted in a straight line on his parchment, we consider the actual layout of Roman streets, we find them forming an inextricably tangled net, their disadvantages immensely aggravated by the vast height of the buildings which shut them in. Tacitus attributes the ease and speed with which the terrible fire of A.D. 64 spread through Rome to the anarchy of these confined streets, winding and twisting as if they had been drawn haphazard between the masses of giant *insulae*. This lesson was not lost on Nero; but if in rebuilding the burnt-out *insulae* he intended to reconstruct them on a more rational plan with better alignment and more space between, he failed on the whole to achieve his aim. Down to the end of the empire the street system of Rome as a whole represented an inorganic welter rather than a practical and efficient plan. The streets always smacked of their ancient origin and maintained the old distinctions which had prevailed at the time of their rustic development: the *itinera*, which were tracks only for men on foot, the *actus*, which permitted the passage of only one cart at a time, and finally the *viae* proper, which permitted two carts to pass each other or to drive abreast. Among all the innumerable streets of Rome, only two inside the old Republican Wall could justly claim the name of *via*. They were the Via Sacra and the Via Nova, which respectively crossed and flanked the Forum, and the insignificance of these two thoroughfares remains a perpetual surprise. Between the gates of the innermost enclosure and the outskirts of the fourteen regions, not more than a score of others deserved the title: the roads which led out of Rome to Italy, the Via Appia, the Via Latina, the Via Ostienvis, the Via Labicana, etc. They varied in width from 4.80 to 6.50 metres, a proof that they had not been greatly enlarged since the day when the Twelve Tables had prescribed a maximum width of 4.80 metres.

The majority of the other thoroughfares, the real streets, or *vici*, scarcely attained this last figure and many fell far below it, being simple passages (*angiportus*) or tracks (*semitae*) which had to be at least 2.9 metres wide to allow for projecting balconies. Their narrowness was all the more inconvenient in that they constantly zigzagged and on the Seven Hills rose and fell steeply—hence the name of *clivi* which many of them, like the Clivus Capitolinus, the Clivus Argentarius, bore of good right. They were daily defiled by the filth and refuse of the neighbouring houses, and were neither so well kept as Caesar had decreed in his law, nor always furnished with the foot-paths and paving that he had also prescribed.

Caesar's celebrated text, graven on the bronze tablet of Heraclea, is worth rereading. In comminatory words he commands the landlords whose buildings face on a public street to clean in front of the doors and walls, and orders the aediles in each quarter to make good any omission by getting the work done through a contractor for forced labour, appointed in the usual manner of state contractors, at a fee fixed by preliminary bidding, which the delinquent will be obliged forthwith to pay. The slightest delay in payment is to be visited by exaction of a double fee. The command is imperative, the punishment merciless. But ingenious as was the machinery for carrying it out, this procedure involved a delay—of ten days at least—which must have usually defeated its purpose, and it cannot be denied that gangs of sturdy sweepers and cleaners directly recruited and employed by the aediles would have disposed of the business more promptly and more satisfactorily. We have, however, no indication that this was ever done, and the idea that in this case the State should have taken the authority and responsibility off the shoulders of the private individual could not possibly have entered the head of any Roman, though he were gifted with the genius of a Julius Caesar.

It is my opinion that the Romans had been equally unsuccessful in extending to the whole city the sidewalks (*margines, crepidines*) or even the paving (*viae stratae*) with which Caesar in his day had dreamed of furnishing them. The archaeologists who differ from me in this matter cite in all seriousness the wide pavements of the Italian roads, forgetting that the paving of the Via Appia in 312 B.C. preceded by sixty-five years the paving on the Clivus Publicius inside the old republican city. Alternatively, they take refuge once again in the example of Pompeii, ignoring how treacherous is this analogy. The comparison of Roman conditions with those of Pompeii is as invalid in the matter of *vici* as in the matter of *insulae*. If the streets of Imperial Rome had been as generally paved as they suppose, the Flavian praetor of whom Martial writes would not have been obliged to "walk right through the mud" in using them nor would Juvenal in his turn have had his legs caked with mud. As for foot-paths, it is impossible that they lined the streets, which were becoming completely submerged under the rising tide of outspread merchandise until Domitian intervened with an edict forbidding the display of wares on the street. His edict is commemorated in the epigram: "Thanks to you, Germanicus, no pillar is now girt with chained flagons . . . nor does the grimy cook-shop monopolise the public way. Barber, tavern-keeper, cook, and butcher keep within their own threshold. Now Rome exists, which so recently was one vast shop."

Had the above-mentioned edict any permanent effect? We may be permitted to doubt it. The retreat of the hucksters may have been secured, or not, by day at the will of a despotic emperor; it certainly took place of its own accord at night. This is in fact one of the characteristics which most markedly distinguishes Imperial Rome from contemporary cities: when there was no moon its streets were plunged in impenetrable darkness. No oil lamps lighted them, no candles were affixed to the walls; no lanterns were hung over the lintel of the doors, save on festive occasions when Rome was resplendent with exceptional illuminations to demonstrate her collective joy, as when Cicero rid her of the Catilinarian plague. In normal times night fell over the city like the shadow of a great danger, diffused, sinister, and menacing. Everyone fled to his home, shut himself in, and barricaded the entrance. The shops fell silent, safety chains were drawn across behind the leaves of the doors; the shutters of the flats were closed and the pots of flowers withdrawn from the windows they had adorned.

If the rich had to sally forth, they were accompanied by slaves who carried torches to light and protect them on their way. Other folk placed no undue reliance on the night watchmen (*tebaciarii*), squads of whom, torch in hand, patrolled the sector—too vast to be completely guarded. Each of the seven cohorts of *vigiles* was theoretically responsible for the policing of two regions. No ordinary person ventured abroad without vague apprehension and a certain reluctance. Juvenal sighs that to go out to supper without having made your will was to expose yourself to the reproach of carelessness; and if his satire goes too far in contending that the Rome of his day was more dangerous than the forest of Gallinaria or the Pontine marshes, we need only to turn the leaves of the *Digest* and note the passages which render liable to prosecution by the *praefectus vigilum* the murderers (*sicarii*), the housebreakers (*effractores*), the footpads of every kind (*raptores*) who abounded in the city, in order to admit that "many misadventures were to be feared" in her pitch-dark *vici*, where in Sulla's day Roscius of Ameria met his death. Not all night adventures were tragic, though the belated wanderer exposed himself to death or at least to the danger of pollution "whenever windows opened above his head behind which someone was not yet asleep." The least serious kind of mishap was that which overtook the sorry heroes of Petronius' story, who leaving Trimalchio's table very late and slightly "merry," lost their way for lack of a lantern in the rabbit warren of unnamed, unnumbered, unlit streets, and reached home barely before daybreak.

All communications in the city were dominated by this contrast between night and day. By day there reigned intense animation, a breathless jostle, an infernal din. The *tabernae* were crowded as soon as they opened and spread their displays into the street. Here barbers shaved their customers in the middle of the fairway. There the hawkers from Transtiberina passed along, bartering their packets of sulphur matches for glass trinkets. Elsewhere, the owner of a cook-shop, hoarse with calling to deaf ears, displayed his sausages piping hot in their saucepan. Schoolmasters and their pupils shouted themselves hoarse in the open air. On the one hand, a money-changer rang his coins with the image of Nero on a dirty table, on another a beater of gold dust pounded with his shining mallet on his well-worn stone. At the cross-roads a circle of idlers gaped round a viper tamer; everywhere tinkers' hammers resounded and the quavering voices of beggars invoked the name of Bellona or rehearsed their adventures and misfortunes to touch the hearts of the passers-by. The flow of pedestrians was unceasing and the obstacles to their progress did not prevent the stream soon becoming a torrent. In sun or shade a whole world of people came and went, shouted, squeezed, and thrust through narrow lanes unworthy of a country village; and fifteen centuries before Boileau sharpened his wit on the *Embarras de Paris*, the traffic jams of ancient Rome provided a target for the shafts of Juvenal.

It might have been hoped that night would put an end to the din with fear-filled silence and sepulchral peace. Not so; it was merely replaced by another sort of noise. Ordinary men had by now sought sanctuary in their homes, but the human stream was, by Caesar's decree, succeeded by a procession of beasts of burden, carts, their drivers, and their escorts. The great dictator had realised that in alleyways so steep, so narrow, and so traffic-ridden as the *vici* of Rome the circulation by day of vehicles serving the needs of a population of so many hundreds of thousands caused an immediate congestion and constituted a permanent danger. He therefore took the radical and decisive step which his law proclaimed. From sunrise until nearly dusk no transport cart was henceforward to be allowed within the precincts of the Urbs. Those which had entered during the night and had been overtaken by the dawn must halt and stand empty. To this inflexible rule four exceptions alone were permitted: on days of solemn ceremony, the chariots of the Vestals, of the Rex Sacrorum, and of the Flamines; on days of triumph, the chariots necessary to the triumphal procession; on days of public games, those which the official celebration required. Lastly one perpetual exception was made for every day of the year in favour of the carts of the contractors who were engaged in wrecking a building to reconstruct it on better and hygienic lines. Apart from these few clearly defined cases, no daytime traffic was allowed in ancient Rome except for pedestrians, horsemen, litters, and carrying chairs. Whether it was a pauper funeral setting forth at nightfall or majestic obsequies gorgeously carried out in full daylight, whether or not the funeral procession was preceded by flute-players and horn-blowers or followed by a long cortège of relations, friends, and hired mourners (*praeficae*), the dead, enshrined in a costly coffin (*capulum*) or laid on a hired bier (*sandapila*), made their last journey to the funeral pyre or the tomb on a simple handbarrow borne by the vespali *lones*.

On the other hand, the approach of night brought with it the legitimate commotion of wheeled carts of every sort which filled the city with their racket. For it must not be imagined that Caesar's legislation died with him and that to serve their own customs or convenience individuals sooner or later made his Draconian regulations a dead letter. The iron hand of the dictator held its sway through the centuries, and his heirs, the emperors, never released the Roman citizens from the restraints which Caesar had ruthlessly imposed on them in the interests of the public welfare. On the contrary, the emperors in turn consecrated and strengthened them. Claudius extended them to the municipalities of Italy; Marcus Aurelius to every city of the empire without regard to its own municipal statutes; Hadrian limited the teams and the loads of the carts allowed to enter the city; and at the end of the first century and the beginning of the second we find the writers of the day reflecting the image of a Rome still definitely governed by the decrees of Julius Caesar.

According to Juvenal the incessant night traffic and the hum of noise condemned the Roman to everlasting insomnia. "What sleep is possible in a lodging?" he asks. "The crossing of waggons in the narrow, winding streets, the swearing of drovers brought to a standstill would snatch sleep from a sea-calf or the emperor Claudius himself." Amid the intolerable thronging of the day against which the poet inveighs immediately after, we detect above the hurly-burly of folk on foot only the swaying of a Liburnian litter. The herd of people which sweeps the poet along proceeds on foot through a scrimmage that is constantly renewed. The crowd ahead impedes his hasty progress, the crowd behind threatens to crush his loins. One man jostles him with his elbow, another with a beam he is carrying, a third bangs his head with a wine-cask. A mighty boot tramps on his foot, a military nail embeds itself in his toe, and his newly mended tunic is torn. Then of a sudden panic ensues: a waggon appears, on top of which a huge log is swaying, another follows loaded with a whole pine tree, yet a third carrying a cargo of Ligurian marble. "If the axle breaks and pours its contents on the crowd, what will be left of their bodies?"

Thus, under the Flavians and under Trajan, just as a century and a half earlier after the publication of Caesar's edict, the only vehicles circulating by day in Rome were the carts of the building contractors. The great man's law had survived its author's death, and this continuity is symptom of the quality which guarantees to Imperial Rome a unique position among the cities of all time and every place. With effortless ease Rome harmonised the most incongruous features, assimilated the most diverse forms of past and present, and while challenging the remotest comparisons, she remains essentially and for all time incomparable. We have seen her arrogant and fragile sky-scrapers rise to heights which her engineering could scarcely justify, we have seen the most modern refinements of extravagant luxury existing side by side with preposterous discomfort and mediaeval barbarity, and now we are faced with the disconcerting traffic problems of her streets. The scenes they witness seem borrowed from the *suqs* of an oriental bazaar. They are thronged by motley crowds, seething and noisy, such as might jostle us in the square Jama' Alfna of Marrakesh, and filled with a confusion that seems to us incompatible with the very idea of civilisation. And suddenly in the twinkling of an eye they are transformed by a logical and imperious decree, swiftly imposed and maintained generation after generation, symbol of that social discipline which among the Romans compensated for their lack of techniques, and which the West today, oppressed by a multiplicity of discoveries and the complexities of progress is for its salvation striving to imitate.

The Rivalries of Plebes and Proles

We have seen how the historical developments of the Republic created in Rome a complex industrial society on top of the older layers of patrician foundations. The history of the city during the Empire became that of an economy based on slave labor in agriculture and industry and a concomitant depression of free labor and lesser craftsmen. The steady extension of human slave labor impinged on the free classes in every way. It impoverished nearly all laborers and degraded most. Small-scale enterprises were as uncompetitive in manufacture and trade as they were in agriculture. Enforced idleness and poverty created the dole for many, while the Horaces, Trimalchios, and Columellas of the Empire gave themselves to the cultivation of ruinous tastes and institutions.

Ludwig Friedländer studied Roman craftsmen, industry, professions, and proletarians in a vast history of imperial society. His four-volume work (*Darstellung aus der Sittengeschichte Roms*), *Roman Life and Manners Under the Early Empire* (1907–1913), was put through seven German editions in his lifetime. Its status as a classic of descriptive social history is sometimes missed in the translation which is often marred by choppiness. But the authority of Friedländer was great in everything pertaining to society under the Empire. His outline of the contours of social classes stresses the division of labor, the decadence of city life, and the depressive effect of slavery for some on the freedom of all. This he does especially well by the simple device of contrast, as when he described a poor man's wealth—his blanket—or his account of how Juvenal's slaves cleared the way for the poet's entrance to the Circus, or yet again in the story of the raven's funeral.

From *Roman Life and Manners Under the Early Empire*

Ludwig Friedländer

THE THIRD ESTATE

Riches and Poverty

Rome was mainly peopled by the "third estate";[1] in this class the proletariate formed the majority; it lived on the "*panem et circenses*," the generous distribution of which was ever inflating its numbers. The great distributing of corn gave only the majority of the male freemen the barest livelihood, and thus in the over-big over-rich city, there was also poverty and need. The poor man, says Martial, may well be a Stoic, and despise his life; it was no merit of his. Their dark rooms, two hundred steps up, were not as high as a man's stature. Their hearth was cold, a jug with a broken handle, a mat, a heap of straw, an empty bedstead was their furniture, a short *toga* by day and night their only protection against the cold; vinegar-wine and black bread their food. Bread, beans and turnips (the work-

Source: Ludwig Friedländer, *Roman Life and Manners Under the Early Empire*, trans. J. H. Freese and Leonard A. Magnus (London: Routledge & Kegan Paul, 1908), pp. 144–167.

[1] The author here is using an expression from another historic period to mean the Roman lower classes; it should not be seen in any other way.

man's lunch), lentils, onions, garlic, peas (one *as* bought a good meal), and fish, were their diet; leeks and a boiled sheep's head, or a smoked pig's head, was luxury. On July 1, the usual day for moving, many poor families might be seen, driven out by the estate-agent, after he had taken all their best property in distraint; with what was left they were sent into the street, "a first of July disgrace," says Martial. A pale-faced man, exhausted with frost and hunger, the "Irus of his day," and three women more like megaeras, dragging a bedstead one leg short, and a table two legs short, and other rubbish, a horn lamp and lantern, broken crockery, a rusted coalbox, a pot stinking of fish, an old wreath of black fleabane (esteemed a cure and hung up in bedrooms), a piece of Toulouse cheese, string to support the absent leek and garlic, a pot filled with a cheap depilatory. Why should they seek a dwelling? the poet asked; they could live free on the bridge. Bridges, steps, thresholds, inclines, were, as in modern Rome, the beggars' resorts: there, and in the *fora*, their picture of woe, their rags, and their maims and wounds (blind men led by a dog), sought to awaken pity: their hoarse voices intoned petitions for alms. Their refuge in the cold rain of December might be an open archway; their dog their sole friend, and their food dog's bread (bran-bread); their wealth a staff, a blanket or a mat, and a knapsack; their salvation solitary death.

Not all the poor were so poor. Some might be well off, by some vicissitude, such as raised slaves. Clesippus, a hunchbacked ugly slave, who had learnt fulling, was bought by Gegania at auction as a part of a lot with a Corinthian candelabrum; he became his mistress' lover, and her heir. When a rich man, he worshipped, as his god, his saviour candelabrum. Such careers were not rare under the Emperors. Licinius Sura, the friend of Trajan, had a love-boy Philostorgos, whose wealth irritated a friend of Epictetus; the philosopher rejoined that Fate was not to blame, but he would not purchase wealth at that price. Juvenal had to see his former barber become owner of many country-houses, and be as rich as a lord; and Martial look on at a freedman showmaker, who used to pull up old skins with his teeth, living in luxury on his former patron's Praenestine estates. Former horn-players in gladiatorial touring companies often themselves came to give gladiatorial games, as also did, under Domitian, a shoemaker at Bologna, and a fuller at Modena. (Modena was a centre for fulling and wool-making). Such pieces of luck were few: but often did dirty dealers or auctioneers make more than advocates. 24,000 sesterces seems to have sufficed for a modest livelihood for one. Juvenal's Naevolus wishes for 20,000, and a little silver, not finely wrought, but not too simple, two strong Moesian slaves, to clear his way to the Circus, and two skilful artisans, to earn him something. This poverty would be bearable.

Industry: Crafts and Trade

There was no unemployment at Rome for the willing worker, however poor. There was no export trade, and little manufacture, except perhaps of some military articles: a union of such catapult-makers is known of. The imports were enormous, and Rome was the principal exchange of the world. Transporters of wares on the river, which Pliny calls "gentlest bearer of the world's traffic," stevedores warehouses and their staff, the many middlemen, employed thousands as sailors, divers, weighers, clerks, agents, commissioners and porters; and in the exchange, besides the great bankers, there were many little money-lenders. The usurer flourished in Rome. Ambrosius' description may well literally be ante-dated: in him, usurers systematically fleeced rich young men; either by loans, or by inducing them to purchase valueless goods: they would lend on old family estates, and dun hard, if necessary, exacting newer and severer terms.

The underlings in trades and crafts were mostly slaves and freedmen working for their master, possibly on commission. The rich supplied many of their own needs through their slaves. A Yorkshire inscription in a tavern runs: "Hail, oh god of this place; slave (*servule*), be zealous at your goldsmith's work in this inn." But the need of providing the huge population with the necessary living wage, and the various and exigent luxuries of the rich, kept many independent handicraftsmen and dealers at work. But most of these even must have been freedmen, as the slaves would live for themselves on what they had learnt for their masters, and as the poor free contemned such employment.

Scanty as the sources are, the crafts and trades of Rome may be reconstructed. The gazetteers of the early fourth century mention 254 bakeries (fifteen to twenty per *regio*, and twenty-four in one) and 2300 oil-shops. Foodstuffs gave employment to thousands: and there were special markets for cattle, pigs, corn, vegetables, fish and delicacies; clothing-trades, dwellings, furniture were also amply catered for.

The complex division of labour incident to highly developed industrialism en-tailed a large populace engaged in the trades and the crafts. Many such workers of single articles formed guilds. From the ancient guild of the shoemakers, dat-ing back to Numa, there had seceded the bootmakers' guild, which Alexander Severus reorganized. Sandal-makers there also were sufficient to name a street, slipper-makers, ladies shoemakers and other specialists in this craft, all similarly organized. Trajan reorganized the bakers' union, and, besides this, many special pastries had their proper guild. The copper-smelters were specialized into pot-ters, makers of candelabra, lanterns, weights, helmets and shields: the ironmak-ers into fabricants of locks, knives, axes and hatchets, scythes and swords. Under Augustus a guild of ladder and stepmakers (*scalarii*) had been formed. The restoration of works of art in metal busied modellers, founders, polishers, gilders, sculptors, chisellers and machine-workers; jewelry workers in pearls and dia-monds; besides the guilds of gold and silversmiths, there were ringmakers', gold-beaters' and gilders' unions. St. Augustine compares the lesser gods, with their very limited power, to the craftsmen in the Streets of the Silversmiths, where every completed work has passed through many hands, where the mastery of the whole is hard to learn, and the special part easy. As in the nineteenth century, colonies of similar workers congregated at Rome and elsewhere in Italy in special districts.

The division of labour was most complex in art: the enormous demand for the unique products resulted in a large manufacture of them. The mural decorations of Pompeii, as also at Rome and elsewhere point to decorators' guilds, in which painters of houses, of arabesques, flowers, animals and scenery worked together. In the sculptors' yards of Rome (mostly in the ninth *regio*, between the *Porticus Europae*, the *Circus Agonalis* and the *Via Recta;* also perhaps near the marble-wharves) statues were often transformed, even new heads put on: in the Pan-dects it is observed that a legacy of a statue held good, though an arm had been substituted before the testator's death. Tombstones were a special branch of work: in Petronius Habinnas is a stoneworker, of whom Trimalchio orders his monuments. In the Pandects a partnership for restoring gravestones is mentioned, one member giving the capital, and the other his ability. Some workers specialized on *genii*, their yards stood behind the Temple of Castor; and others in eyes of some coloured material. This manner of production, together with the frequent employment of slaves, led to a great cheapness of statues, portraits ranging up from 3000 sesterces. Similarly, expanding trade caused specialization: thus, lupine-sellers were a special branch of vegetable-dealers: in drugs, colours, salts, essences, toilette accessories, there were many special guilds; and so, too, amongst the tailors for cloaks, mantles, summer garments, etc.

These very heterogeneous industries were carried on in the most frequented parts: at the end of the first century the nuisance of the shops which projected from the house-fronts had to be inhibited. Streets were named after their occupants, e.g., the streets of the corn-merchants, harness-makers, sandal-makers, wood-traders, glaziers, salve-dealers, scythe-makers. Such congregation was no doubt involved by the specialized labour, as St. Augustine observed. The *Via Sacra*, one of Rome's principal highways, was the Bond Street of Rome, containing, according to inscriptions, besides purveyors of luxury, jewellers, metal founders, and sculptors, a dealer in colours, a flute-maker, and a writing-teacher. Ivory dice might there be bought, and "Caietan" cord (really from Gaul), crystal balls, peacock feather fans, and other ladies' articles, fruit for dessert and crowns for the topers at banquets. Rome's best shops, however, at about A.D. 100 were in the markets, surrounding the *Saepta* of the Campus Martius. There fine slaves were to be had, large citron-wood table-tops, ivory, banquet couches inlaid with tortoise-shell, ancient bronze statues, crystal and *murrha* vessels, silver beakers of antique design, collars of gold set with emeralds, big pearl earrings, as well as less expensive wares. Other shops for luxury were in the *regio Tusca*, and in the arcades surrounding the Circus.

The shops and taverns in all their breadth opened on to the street, and were closed in with linen curtains, covered with notices or paintings: they also had signboards. Some tablets in relief have been preserved, which were either signs, or they may have decorated the shopkeepers' tombstones. On a ham-dealer's sign, there are five hams ranged arow. Two of these advertise two branches of a tailor's firm for men and for women; the design is a man or woman and their retinue examining proffered articles. One such sign represents a hare, two wild boars, and several large birds hanging on the wall; and a young girl bargaining with the mistress of the shop; both figures' costume, bearing and execution show the refining influence of Greece, like other examples, such as funeral monuments, which testify "how much more general than in modern days, even in later antiquity and the lower ranks, was the need of beautifying life by art, and leaving posterity some record and remembrance." Thus workmen would have tombstones depicted with scenes from their lives; there is one relief, a well-known group of the three Graces, and a matron fully clothed, her dress drawn over her head: the inscription is *"to the four sisters"*; it must have been the sign of a shop or inn, or, maybe, bawdy-house. A Greek-Latin book of dialogues contains the following: "I am going to the tailor. How much does this pair cost? One hundred *denarii*. How much is the waterproof? Two hundred *denarii*. That is too dear; take a hundred. Impossible; I have to pay so much for it wholesale. What shall I then give? What you like. [To the slave or attendant] Give him 125 *denarii*. Shall we go to the linen-merchant?" Bargaining was customary, as Juvenal witnesses: a schoolteacher must expect a deduction from his salary, like a dealer on his rugs and bed-linen.

It is almost only from such reliefs and inscriptions that we get some insight into the lives of these many workers. They remind us how little we know; one inscription at Hierapolis in Great Phrygia of the second or third century, records a society of purple-workers founded for mutual support, another in Sardes, A.D. 459, a quarrel—such as frequently arose among the Mauri in the Eastern Empire —between men and masters (ἐργοδόται καὶ ἐργολάβοι): the former had discontinued a building-operation. The roman *mimi* and *atellanae*, which borrowed their scene from these social strata, are unfortunately lost, and Petronius' *bourgeois* are South Italian only. The extant literature is entirely from a sphere which had more contempt than interest for the lower orders, the men who, day by day, tucked in their tunics behind the counter, or stood in apron and cap by their bench in the workshop, where nothing noble could be made, only the daily bread earned—where cheap goods were sold at 50 percent profit, be they hides or fragrances; for them profit was the best of scents. People of higher standing were no more connected with even the uncleanest business (e.g., the letting of houses and property for brothels) than are Russian grandees with the cheap spirit-stills; the transaction of such affairs by slaves and freedmen left them untainted: little tradesmen were taunted for their innocent cupidity. Pliny says that tailors watched the sinking of the Pleiades, on the 11th November, to see if it were cloudy—a sign of a rainy winter—or bright—a sign of a rough winter; and to judge whether to raise the price of cloaks or underclothing: he sees in this a cheating disposition. Workmen received, if satisfactory, a small extra wage called *corollarium*.

Workmen and small tradesmen were the most conservative class, as every revolution, riot, or civil war spelt their ruin. Cicero, enouncing a sempiternal truth, says: "Most of the innkeepers desire peace. Their livelihood depends on their custom, which involves calm: every time they are closed, they lose, and what if they are burnt down?" This was what happened in street fights. In the combat between the people and the Praetorians, 237–8, the latter, pelted from the roofs, set fire to the closed doors of inns and houses. It was usual to see in inns, shops, workshops and offices badly executed busts, almost caricatures, of the reigning Emperors. On the imperial birthdays and festivals in the Emperor's honour, and other occasions, taverns were decked with laurels and lighted up: on days of imperial mourning they were closed.

The medieval guilds had their patron saints; the Roman their tutelary gods with their holy days. Most general was the 19th March, the foundation day of the Temple of Minerva, protectress of all craft and art guilds, on the Aventine: later on, the festival was prolonged till the evening of the 23rd March. According to Ovid, spinstresses, weavers, fullers, painters, shoemakers, sculptors, doctors, schoolmaster (and their pupils had holidays) took part in it. The 9th of June (the Day of Vesta) was celebrated by millers and bakers; donkeys were wreathed with flowers and loaves, and mills were garlanded. The guild of musicians (especially flautists), who played at acts of public worship, had their feast in the Temple of Jupiter on the Capitol, and, on the 13th June, crossed Rome in masks and disguised (generally in female garb), intoxicated, singing parodies set to ancient tunes. At such industrial festivals processions were common: one picture at Pompeii represents a procession of carpenters, and figures of men sawing, and others similar were carried about by the younger men. At Rome and elsewhere the guilds had a regular part in great processions, such as triumphs and imperial arrivals, and flaunted their flags bravely: no doubt they had their own banner bearers. A marble tablet found at Ostia contains a monthly list of members of an association (one a woman), and their contributions, the interest on which, at 12 percent went towards a common celebration of all their birthdays. One general festival of the common people (March 15) was the day of Anna Perenna, a goddess of the year, held in an orchard at the first mile-stone on the *Via Flaminia* (not far from the Porta del Popolo). The girls sang very free satires of ancient origin; men and women lay down together on the grassy shores of the Tiber, in the open, or in huts, or improvised tents of reeds covered with their *togas*. There they had drinking-bouts, and prayed for as many years as they drew spoons from the bran-pie; sang theatrical ditties, capered about, and returned home, riotously supporting one another, the mockery of all who came their way.

The guilds also provided for funerals: but most of the poor, who could not raise the money for their own burial, contributed monthly to death-funds, which admitted free men, freedmen and slaves, and assured the members proper exsequies, generally in the *Columbaria;* these were huge vaulted buildings with rows of pigeon holes. Such associations also had their standing festivals, and celebrated the birthday of their patron deity (the dedication-day of his idol) with a banquet. Amongst the extant statutes of such associations, those of the "Adorers of Diana and Antinous" at Lanuvium (Città Lavigna) of A.D. 133 give an interesting glimpse into the nature of these mortuary guilds, and an idea of their festivals. By way of preface is the warning: "Do thou, novice, first read the statutes, and have heed not to encumber thyself or thy heirs." New members paid an entrance fee of 100 sesterces, and an *amphora* of good wine: the annual contribution of fifteen sesterces was paid in monthly sums of five *asses*. Three hundred sesterces were allowed for individual funerals, except for suicides, fifty being reserved for the procession and distributed at the pyre. Grievances were to be dealt with at general meetings, so as "on feast-days to carouse at ease." Carousals were organized by four members appointed for every year; they had to provide covers or pillows for the couches, hot water and dishes and four *amphorae* of good wine, and a loaf of two *asses* and four sardines for every member. The expenses of the meals were probably met out of the interest of a capital sum given by some benefactor: wine could scarcely have been insufficient, as besides initial contributions, slaves on manumission had to give an *amphora*. There were six drinking-bouts a year; at the two principal ones, the birthdays of Diana and Antinous, the president, who was elected for five years, and received double portions, had to distribute oil at the public bath before the banquet. He also had to offer wine and incense at all feast days. At the end of his term, if he had administered well, he received one and a half times his share. Any one, who, at a feast, quarrelled and left his seat, was sconced four sesterces; any one guilty of abusive language, twelve; any one who insulted the president, twenty. When Christianity made inroads among these small tradesmen, many would not immediately renounce the benefits of these associations. In the letters of Bishop Cyprianus of Carthage, one Martialis is accused of "frequenting the disgusting banquets of an association and by its means burying his children in heathen fashion in unhallowed graves."

Their manners, no doubt, left something to be desired. Hucksters in salt-fish were said, by a Graecism, to snuffle with their elbows, and their boys were unusually vulgar. Masters not unfrequently abused their disciplinary powers against their apprentices, even if free; one such clumsy pupil had his eye knocked out with a last. In the civic life of Pompeii, apprentices were not insignificant. Amongst the city election placards posted on to the walls, there is one written by the apprentices, and another by one Saturninus, "*cum discentes*" [*sic*]. Some inscriptions on graves faithfully record the good qualities of the deceased. One freedman goldsmith, "a master-maker of Clodian jars," is eulogized by his patron, as one "who insulted no one, never thwarted his patron. He always had a mass of gold and silver and was never avaricious." The tombstone of a freedman pearl-dealer on the *Via Sacra* begs the passer-by not to injure the grave where the bones of a man rest who was good, merciful, and a friend of the poor. One Lucius Nerusius Mithres, according to the hexameters on his tombstones, which are an acrostic of his name, was well known in the Holy City for his conscientious business habits as a goat-skin dealer, and for having, as a contractor, rendered all his dues to the fisc, and made equitable bargains. He had prospered, built himself a marble house, helped the indigent: and his great desert he deemed his building a burial-place for all his freedmen and freedwomen and their heirs.

Workmen often made an extra income by rearing and training birds, though necessarily in competition with professional rearers. Manilius mentions trainers who traversed the town with caged birds, all their possessions being one sparrow. In these accounts, few as they are, workmen are several times mentioned as owners. A poor shoemaker had trained a raven to congratulate Augustus: the purchase was declined, as Augustus had sycophants enough already: but the bird repeated at the right time his master's complaint: "All my trouble's gone for nothing," and was bought at a high price. A barber on the Forum had a magpie, which imitated musical sounds and men's and animals' tones: one day a stately funeral procession stopped in front of the barber's house, and the *tuba* blowers played a long piece. The magpie was silent awhile; envious magic was suspected: then she sang the whole of the funeral music from beginning to end. In the reign of Tiberius a raven flew out of a nest in the Temple of Castor into a shoemaker's opposite, and the master taught it speech. The raven learned to fly every morning to the rostrum, to address Tiberius, Drusus and Germanicus by name, to greet other passers, and was the wonder of all Rome. The owner of a shop near by killed it (on the pretext of anger, as it had dirtied a pair of new shoes), and the enraged people drove the murderer out of the district and afterwards murdered him. The bird was solemnly borne by two Moors to a pyre on the Via Appia, followed by a multitude with wreaths. Pliny, on the authority of the *Acta diurna*, dates this on the 28th March, 35.

Miscellaneous Crafts, Arts and Pursuits

General

Not only handwork and shopkeeping, but many other, often profitable, businesses were considered ungentlemanly. The poor free-born man of liberal education spoke contemptuously of men who grew rich by the undertaking of funerals, bakeries, bathhouses, the farming of river and harbour dues, public works, the clearing of the *cloacae*, or as auctioneers of rubbish or priceless wares, and through other businesses.

Two employments were so vile as to disqualify (by a law of Caesar's) for election to city offices: those of undertakers and public criers (*praecones*). *Praecones* were used for all purposes of publicity, e.g., to proclaim lost articles, runaway slaves; but their main function was auctioneering, and the close connexion of this business and that of the public jesters may have caused their disrepute. But this fact made it very profitable: for public auction in Rome included, in part, our commission-agencies, which effected the sale of the superfluities of an inheritance, or the raising of a sudden loan. In the stead of the business-man, a professional middle-man came in, the *coactor argentarius* or *exactionum*, so called from his having to dun for single demands: a profession in general disesteem. For his trouble and risk he received 1 percent on the price: and the quantity of the work explains the lowness of the rate; it might be more in especially troublesome business. Horace's father had been a *coactor*, and profits were then as low as those of a public crier; but by about A.D. 50, in consequence of greater facility and promptitude in transfers, both made larger profits. Strabo says houses were constantly changing hands. Arruntius Euarestus, an auctioneer who angled successfully in the troubled waters after the murder of Caligula, was (according to Josephus) as rich as the richest, and powerful then and after. In Martial a maiden is wooed by ten poets, seven lawyers, four tribunes and two praetors, and, without further thought, espoused to an auctioneer. Nor foolishly. A boy, if he is to make his way, ought not to study or poetize; rather should he play the cither or flute, or, if long-headed, the hammer or architect's rule.

Architecture Cicero collocates with medicine as a useful art, and of all arts the most respectable: it was also the most profitable. Under Augustus (according to Vitruvius) even the profession was so overcrowded that they had to advertise themselves, and many were incompetent. The demand increased owing to the numbers of erections, public and private, necessitated in part by fire and collapse, in part by the building mania of the rich: the smallness of the city was compensated by the endless possibilities of the country.

Excepting for some famous and highly paid artists, little is known of the despised sculptors and painters. Musicians might make much: the economical Vespasian paid Terpnus and Diodorus the citharists 200,000 for their performances at the rededication of the Theatre of Marcellus. Teaching was profitable: and the salaries of distinguished singers and citharists awoke the envy of the learned. "Learn," says Juvenal to the rhetorician, "the fee Chrysogonus and Pollio make for teaching rich children the cither, and you will tear up Theodorus' manual." Nero and Domitian patronized music more than any others. Under Domitian Martial wrote as above, and that he was going to return from Forum Cornelii to Rome, as a citharist. Art, too, like music, paid well, if popular, especially acting and dancing, and also fencing and circus-driving.

The learned professions were relegated, like the arts and their appliances, to the third estate, or to slaves. Galen selects, as noble departments, medicine, rhetoric, music, geometry, arithmetic, accountantship, astronomy, grammar, and law; to which sculpture and painting may be added. The last two, as employments, brought in the quickest returns: hence Lucian was apprenticed to his uncle, the sculptor. The learned professions required laborious and unremunerative preparation for years, and "hard it was, for men in poor circumstances to rise" (Juvenal). There is some information as to their position in life.

During the first centuries, the teaching profession had no assured position, nor the esteem of an official security. Education was in the Early Empire not regarded as a matter of State, and, in the second century, to a very slight extent, being, if anything, municipalized. Up to then education was a private concern, and only fostered by the immunity of teachers from town rates. . . .

Most of the teachers followed no inner call, but only the need of bread, as may be gathered from the fact that, of the most famous and learned grammarians or philologists of Rome in the first century, several (on the authority of Suetonius) only chanced on this occupation, or came to it as a makeshift. Many built up their knowledge as slaves or freedmen in the service of erudites, or as pedagogues accompanying a boy to school. The famous Orbilius had been servant to a magistrate, and then served in the foot and the horse. Marcus Valerius Probus of Berytus, who was even more renowned, only took to linguistics after losing every chance of a subaltern's post. A third had begun as a prize-fighter, a fourth as a clown. Pertinax, the future Emperor, however, the son of a freedman wood-dealer, exchanged this profession, unsuitable to him, for military service.

It was universally held that a teacher's bread was hardly won. Ausonius says that no grammarian was ever happy, and that an exception to this rule would be an offender against grammar. The work and conditions were severe, the gain scanty; few were those consolable by the feeling that it was a lofty and royal calling "to instruct the witless in good manners and holy knowledge." The lessons began at or before daybreak: the teacher had to be up and about, long before the smith or the weaver, to breathe air sullied by the boys' lamps, which begrimed the busts of Horace and Virgil in the schoolroom. The number of school hours varied, but, with the authority of Ausonius, was usually six. Galen tells a tale of a grammarian Diodorus, who was subject to epileptic fits, unless he took nourishment during this time, and, on his advice, took, at the third or fourth hour, bread soaked in wine, which restored him. But, in Greek–Latin schools, the pupils at midday went home for their early meal, changed and came back for the afternoon. The usual entering age, according to Paul of Aegina, whose rules more or less define the practice in the early centuries and in the West, was six or seven; geometry, grammar and gymnastics begun at twelve; the years from fourteen to twenty given to higher mathematics and philosophy and the severer exercises. The teacher's worst task was to keep discipline among the excitable and fidgety boys, whose moral education he had to supervise like a father.

Bread and Circuses

In the imperial city of Rome the plebeians became paupers. The relationship of this development to others—slavery, the freedman classes, war, and the growth of a corrupt, bureaucratic state—we have already touched upon elsewhere. What remains now is to show the consequences. Roman civilization was engulfed by barbarism long before the invasions which ended the Empire. It was not so much killed by Goths as by its own decadence. Suicide, not murder, is perhaps the proper coroner's verdict on the society of the dole, bread, and circuses. State-supported spectacles to drain the frustrations of the urban masses and pacify discontents were the solution for poverty and slums engineered by the Empire. In reality, bread and circuses were a form of social control over people whose deprivation worried Rome's rulers. The plebeian class was made wholly dependent on the state for its very existence.

This fact of Roman life was already clearly established under Augustus Caesar, as we learn from Tacitus and other reliable writers. But it is in the pages of Suetonius (Gaius Suetonius Tranquillus), a second-rate historian and literary biographer, that we find the right blend of event and observer. As the witness of the revival of government under Trajan and Hadrian, Suetonius drew on official materials for his *The Twelve Caesars*. He was Hadrian's private secretary. He had conversed with men who knew Tiberius, Caligula, Claudius, and Nero. Pliny the Younger was his friend. In the following selection from the life of Augustus, we see the emperor in his relationship to the plebes. Yet we do not see the society of bread and circuses, only the relationship between imperial policy and the necessity of diversions for the masses. The economic policies of Augustus helped destroy the small farmers and guild craftsmen whose poverty he then used in building his own power. The Empire made the urban poor destitute and then completed their degradation under the guise of social welfare.

From *The Twelve Caesars*

Suetonius

AUGUSTUS

23. He suffered only two heavy and disgraceful defeats, both in Germany, the generals concerned being Lollius and Varus. Lollius's defeat was ignominious rather than of strategic importance; but Varus's nearly wrecked the Empire, since three legions with all their officers and auxiliary forces, and the general staff, were massacred to a man. When the news reached Rome, Augustus ordered the Guards to patrol the City at night and prevent any rising; then prolonged the terms of the provincial governors, so that the allies should have men of experience, whom they trusted, to confirm their allegiance. He also vowed to celebrate Games in honour of Juppiter Greatest and Best as soon as the political situation

SOURCE: Suetonius, *The Twelve Caesars*, trans. Robert Graves (Baltimore: Penguin Books; London: A. P. Watt & Son, 1957), pp. 62–63, 67, 68, 70–71, 73–78.

improved; similar vows had been made during the Cimbrian and Marsian Wars. Indeed, it is said that he took the disaster so deeply to heart that he left his hair and beard untrimmed for months; he would often beat his head on a door, shouting: "Quinctilius Varus, give me back my legions!" and always kept the anniversary as a day of deep mourning.

24. Augustus introduced many reforms into the Army, besides reviving certain obsolete practices, and exacted the strictest discipline. He grudged even his generals home-leave, and granted this only during the winter. When a Roman knight cut off the thumbs of his two young sons to incapacitate them for Army service, Augustus had him and his property publicly auctioned; but, realizing that a group of tax-collectors were bidding for the man, knocked him down to an imperial freedman—with instructions that he should be sent away and allowed a free existence in some country place. He gave the entire Tenth Legion an ignominious discharge because of their insolent behaviour, and when some other legions also demanded their discharge in a similarly riotous manner, he disbanded them, withholding the bounty which they would have earned had they continued loyal. If a company broke in battle, Augustus ordered the survivors to draw lots, then executed every tenth man, and fed the remainder on barley bread instead of the customary wheat ration. Company commanders found absent from their posts were sentenced to death, like other ranks, and any lesser dereliction of duty earned them one of several degrading punishments—such as being made to stand all day long in front of general headquarters, sometimes wearing tunics without sword-belts, sometimes carrying ten-foot poles, or even sods of turf—as though they had been private soldiers whose task it was to measure out and build the camp ramparts.

25. When the Civil Wars were over, Augustus no longer addressed the troops as "Comrades," but as "Men"; and had his sons and step-sons follow suit. He thought "Comrades" too flattering a term: consonant neither with military discipline, nor with peacetime service, nor with the respect due to himself and his family. Apart from the City fire brigades, and militia companies raised to keep order during food shortages, he enlisted freedmen in the Army only on two occasions. The first was when the veteran colonies on the borders of Illyricum needed protection; the second, when the Roman bank of the Rhine had to be held in force. These soldiers were recruited, as slaves, from the households of well-to-do men and women, and then immediately freed; but he kept them segregated in their original companies, not allowing them either to mess with men of free birth or to carry arms of standard pattern. . . .

30. Augustus divided the City into districts and wards; placing the districts under the control of magistrates annually chosen by lot, and the wards under supervisors locally elected. He organized stations of night-watchmen to alarm the fire brigades; and, as a precaution against floods, cleared the Tiber channel which had been choked with an accumulation of rubbish and narrowed by projecting houses. Also, he improved the approaches to the City: repaving the Flaminian Way as far as Ariminium, at his own expense, and calling upon men who had won triumphs to spend their prize money on putting the other main roads into good condition. . . .

32. Many of the anti-social practices that endangered public peace were a legacy of lawlessness from the Civil Wars; but some were of more recent origin. For example, bandit parties infested the roads armed with swords, supposedly

worn in self-defence, which they used to overawe travellers—whether free-born or not—and force them into slave-barracks built by the landowners. Numerous so-called "workmen's guilds," in reality organizations for committing every sort of crime, had also been formed. Augustus now stationed armed police in bandit-ridden districts, had the slave-barracks inspected, and dissolved all workmen's guilds except those that had been established for some time and were carrying on legitimate business. Since the records of old debts to the Public Treasury had become by far the most profitable means of blackmail, Augustus burned them; also granting title-deeds to the occupants of City sites wherever the State's claim to ownership was disputable. When persons had long been awaiting trial on charges that were not pressed, and therefore continued to wear mourning in pub-lic—with advantage to nobody, except their gleeful enemies—Augustus struck the cases off the lists and forbade any such charge to be renewed unless the plain-tiff agreed to suffer the same penalty, if he lost the case, as the defendant would have done. To prevent actions for damages, or business claims, from either not being heard or being prorogued, he increased the legal term by another thirty days—a period hitherto devoted to public games in honour of distinguished citi-zens. . . .

34. The existing laws that Augustus revised, and the new ones that he en-acted, dealt, among other matters, with extravagance, adultery, unchastity, brib-ery, and the encouragement of marriage in the Senatorial and Equestrian Orders. His marriage law being more rigorously framed than the others, he found him-self unable to make it effective because of an open revolt against several of its clauses. He was therefore obliged to withdraw or amend certain penalties exacted for a failure to marry; to increase the rewards he offered for large families; and to allow a widow, or widower, three years' grace before having to marry again. Even this did not satisfy the knights, who demonstrated against the law at a public entertainment, demanding its repeal; where upon Augustus sent for the children whom his grand-daughter Agrippina had borne to Germanicus, and publicly displayed them, some sitting on his own knee, the rest on their father's —and made it quite clear by his affectionate looks and gestures that it would not be at all a bad thing if the knights imitated that young man's example. When he then discovered that bachelors were getting betrothed to little girls, which meant postponing the responsibilities of fatherhood, and that married men were fre-quently changing their wives, he dealt with these evasions of the law by shorten-ing the permissible period between betrothal and marriage, and by limiting the number of lawful divorces.

35. The Senatorial Order now numbered more than 1000 persons, some of whom were popularly known as the "Orcus Men." This was really a name for ex-slaves freed in the masters' wills, but had come to describe senators who had bribed or otherwise influenced Mark Antony to enrol them in the Order on a pretence that Julius Caesar, before he died, had chosen them for this honour. The sight of this sad rabble, wholly unworthy of office, decided Augustus to restore the Order to its former size and repute by two new acts of enrolment. First, each member was allowed to nominate one other; then Augustus and Agrippa together reviewed the list and announced their own choice. When Augustus presided on this second occasion he is said to have worn a sword and a steel corselet beneath his tunic, with ten burly senatorial friends crowding around him. According to Cremutius Cordus, the senators were not even then permitted to approach Au-

gustus's chair, except singly and after the folds of their robes had been carefully searched. Though shaming some of them into resignation, he did not deny them the right to wear senatorial dress, or to watch the Games from the Orchestra seats, or to attend the Order's public banquets. He then encouraged those selected for service to a more conscientious (and less inconvenient) discharge of their duties, by ruling that each member should offer incense and wine at the altar of whatever temple had been selected for a meeting; that such meetings should not be held more than twice a month—at the beginning and in the middle—and that, during September and October, no member need attend apart from the few whose names were drawn by lot to provide a quorum for the passing of decrees. . . .

Augustus revised the roll of citizens, ward by ward; and tried to obviate the frequent interruptions of their trades or businesses which the public grain-distribution entailed, by handing out tickets, three times a year, valid for a four months' supply; but was implored to resume the former custom of monthly distributions, and consented. He also revived the traditional privilege of electing all the City magistrates, not merely half of them (he himself had been nominating the remainder), and attempted to suppress bribery by the imposition of various penalties; besides distributing on Election Day a bounty of ten gold pieces from the Privy Purse to every member both of the Fabian tribe—the Octavian family were Fabians—and of the Scaptian tribe, which included the Julians. His object was to protect the candidates against demands for further emoluments.

Augustus thought it most important not to let the native Roman stock be tainted with foreign or servile blood, and was therefore very unwilling to create new Roman citizens, or to permit the manumission of more than a limited number of slaves. Once, when Tiberius[1] requested that a Greek dependent of his should be granted the citizenship, Augustus wrote back that he could not assent unless the man put in a personal appearance and convinced him that he was worthy of the honour. When Livia[2] made the same request for a Gaul from a tributary province, Augustus turned it down, saying that he would do no more than exempt the fellow from tribute— "I would far rather forfeit whatever he may owe the Privy Purse than cheapen the value of the Roman citizenship." Not only did he make it extremely difficult for slaves to be freed, and still more difficult for them to attain full independence, by strictly regulating the number, condition, and status of freedmen; but he ruled that no slave who had ever been in irons or subjected to torture could become a citizen, even after the most honourable form of manumission.

Augustus set himself to revive the ancient Roman dress and once, on seeing a group of men in dark cloaks among the crowd, quoted Virgil indignantly:

Behold them, conquerors of the world, all clad in Roman gowns!

and instructed the aediles that no one should ever again be admitted to the Forum, or its environs, unless he wore a gown and no cloak. . . .

41. His generosity to all classes was displayed on many occasions. For instance, when he brought the treasures of the Ptolemies to Rome at his Alexandrian triumph, so much cash passed into private hands that the interest rate on loans dropped sharply, while real estate values soared. Later, he made it a rule that whenever estates were confiscated and the funds realized by their sale ex-

---

1 Tiberius was Augustus's sucessor.
2 Livia was the wife of Augustus.

ceeded his requirements, he would grant interest-fee loans for fixed periods to anyone who could offer security for twice the amount. The property qualification for senators was now increased from 8000 to 12,000 gold pieces, and if any member of the Order found that the value of his estate fell short of this, Augustus would make up the deficit from the Privy Purse. His awards of largesse to the people were frequent, but differed in size: sometimes it was four gold pieces a head, sometimes three, sometimes two and a half; and even little boys benefited, though hitherto eleven years had been the minimum age for a recipient. In times of food shortage he often sold grain to every man on the citizens' list at a very cheap rate; occasionally he supplied it free; and doubled the number of free money-coupons.

42. However, to show that he did all this not to win popularity but to improve public health, he once sharply reminded the people, when they complained of the scarcity and high price of wine, that: "Marcus Agrippa, my son-in-law, has made adequate provision for thirsty citizens by building several aqueducts." Again, he replied to a demand for largesse which he had, in fact, promised: "I always keep my word." But when they demanded largesse for which no such promise had been given, he issued a proclamation in which he called them a pack of shameless rascals, and added that though he had intended to make them a money present, he would now tighten his purse-strings. Augustus showed equal dignity and strength of character on another occasion when, after announcing a distribution of largesse, he found that the list of citizens had been swelled by a considerable number of recently freed slaves. He gave out that those to whom he had promised nothing were entitled to nothing, and that he refused to increase the total sum; thus the original beneficiaries must be content with less. In one period of exceptional scarcity he found it impossible to cope with the public distress except by expelling every useless mouth from the City, such as the slaves in the slave-market, all members of gladiatorial schools, all foreign residents with the exception of physicians and teachers, and a huge crowd of household-slaves. He writes that when at last the grain supply improved: "I had a good mind to discontinue permanently the supply of grain to the City, reliance on which had discouraged Italian agriculture; but refrained because some politician would be bound one day to revive the dole as a means of ingratiating himself with the people." Nevertheless, in his handling of the food problem he now began to consider the interests of farmers and corn merchants as much as the needs of city dwellers.

43. None of Augustus's predecessors had ever provided so many, so different, or such splendid public shows. He records the presentation of four Games in his own name and twenty-three in the names of other City magistrates who were either absent or could not afford the expense. Sometimes plays were shown in all the various City districts, and on several stages, the actors speaking the appropriate local language; and gladiators fought not only in the Forum or the Amphitheatre, but in the Circus and Enclosure as well; or the show might, on the contrary, be limited to a single wild-beast hunt. He also held athletic competitions in the Campus Martius, for which he put up tiers of wooden seats; and dug an artificial lake beside the Tiber, where the present Caesarean Grove stands, for a mock sea-battle. On these occasions he posted guards in different parts of the City to prevent ruffians from turning the emptiness of the streets to their own advantage. Chariot races and foot races took place in the Circus, and among those who hunted the wild beasts were several volunteers of distinguished family. Augustus

also ordered frequent performances of the Troy Game by two troops, of older and younger boys; it was an admirable tradition, he held, that the scions of noble houses should make their public début in this way. When little Gaius Nonius Asprenas fell from his horse at one performance and broke a leg, Augustus comforted him with a golden torque and the hereditary surname of "Torquatus." Soon afterwards, however, he discontinued the Troy Game, because Asinius Pollio the orator attacked it bitterly in the House; his grandson, Aeserninus, having broken a leg too.

Even Roman knights sometimes took part in stage plays and gladiatorial shows until a Senatorial decree put an end to the practice. After this, no person of good family appeared in any show, with the exception of a young man named Lycius; he was a dwarf, less than two feet tall and weighing only 17 lb. but had a tremendous voice. At one of the Games Augustus allowed the people a sight of the first group of Parthian hostages ever sent to Rome by leading them down the middle of the arena and seating them two rows behind himself. And whenever a strange or remarkable animal was brought to the City, he used to exhibit it in some convenient place on days when no public shows were being given: for instance, a rhinoceros in the Enclosure; a tiger on the stage of the Theatre; and a serpent nearly ninety feet long in front of the Comitium, where popular assemblies were held.

Once Augustus happened to be ill on the day that he had vowed to hold Games in the Circus, and was obliged to lead the sacred procession lying in a litter; and when he opened the Games celebrating the dedication of Marcellus's Theatre, and sat down in his chair of state, it gave way and sent him sprawling on his back. A panic started in the Theatre during a public performance in honour of Gaius and Lucius; the audience feared that the walls might collapse. Augustus, finding that he could do nothing else to pacify or reassure them, left his own box and sat in what seemed to be the most threatened part of the auditorium.

44. He issued special regulations to prevent the disorderly and haphazard system by which spectators secured seats for these shows; having been outraged by the insult to a senator who, on entering the crowded theatre at Puteoli, was not offered a seat by a single member of the audience. The consequent Senatorial decree provided that at every public performance, wherever held, the front row of stalls must be reserved for senators. At Rome, Augustus would not admit the ambassadors of independent or allied kingdoms to seats in the orchestra, on learning that some were mere freedmen. Other rules of his included the separation of soldiers from civilians; the assignment of special seats to married commoners, to boys not yet come of age, and, close by, to their tutors; and a ban on the wearing of dark cloaks, except in the back rows. Also, whereas men and women had hitherto always sat together, Augustus confined women to the back rows even at gladiatorial shows: the only ones exempt from this rule being the Vestal Virgins, for whom separate accommodation was provided, facing the praetor's tribunal. No women at all were allowed to witness the athletic contests; indeed, when the audience clamoured at the Games for a special boxing match to celebrate his appointment as Chief Pontiff, Augustus postponed this until early the next morning, and issued a proclamation to the effect that it was the Chief Pontiff's desire that women should not attend the Theatre before ten o'clock. . . .

To be brief: Augustus honoured all sorts of professional entertainers by his friendly interest in them; maintained, and even increased, the privileges enjoyed by athletes; banned gladiatorial contests if the defeated fighter were forbidden to plead for mercy; and amended an ancient law empowering magistrates to punish stage-players wherever and whenever they pleased—so that they were now competent to deal only with misdemeanors committed at games or theatrical performances. Nevertheless, he insisted on a meticulous observance of regulations during wrestling matches and gladiatorial contests; and was exceedingly strict in checking the licentious behaviour of stage-players. When he heard that Stephanio, a Roman actor, went about attended by a page-boy who was really a married woman with her hair cropped, he had him flogged through all the three theatres—those of Pompey, Balbus, and Marcellus—and then exiled. Acting on a praetor's complaint, he had a comedian named Hylas publicly scourged in the hall of his own residence; and expelled Pylades not only from Rome, but from Italy too, because when a spectator started to hiss, he called the attention of the whole audience to him with an obscene movement of his middle finger. . . .

46. After thus improving and reorganizing Rome, Augustus increased the population of Italy by personally founding twenty-eight veteran colonies. He also supplied country towns with municipal buildings and revenues; and even gave them, to some degree at least, privileges and honours equalling those enjoyed by the City of Rome. This was done by granting the members of each local senate the right to vote for candidates in the City Elections; their ballots were to be placed in sealed containers and counted at Rome on polling day. To maintain the number of knights he allowed any township to nominate men capable of taking up such senior Army commands as were reserved for the Equestrian Order; and, to encourage the birth-rate of the Roman commons, offered a bounty of ten gold pieces for every legitimate son or daughter whom a citizen could produce, on his tours of the City wards.

Part 2

The Middle Ages and the Renaissance

To cover the thousand years between the fall of the Roman Empire in the West and the rise of the new empires overseas with the word *medieval* is a convenience. The alternate term *Middle Ages* connotes something between antiquity and modernity. This usage stems from humanists of Renaissance Europe, men who embraced the classical part and eschewed the barbarism of the *Gothic* interlude. They rejected what was medieval because they felt remote from it while living in it. Yet the society of medieval Europe was more diverse and attractive than humanist images of it; it preserved more of the classical world than they admitted, and it created much that is alive today—universities, towns, monarchies, family patterns, languages, religions, modes of painting, and patterns of agriculture. Through a span of years as great as the combined empires of Greece and Rome, and over twice as many centuries as anything we may call modern Western civilization, medieval civilization truly mediated between the classical world and our own. Yet its modes of thought and expresssion, its methods of getting and exchanging wealth, and its political and social institutions seem but distant cousins to our own. Perhaps we will find in the pages of this part something of kinship with the humanists and also the shock of recognition of what of the Middle Ages is alive today.

CHAPTER FIVE

Land and Labor: Material Bases
of a Traditional Society

On the Two Feudal Ages

Marc Bloch died in 1943, a victim of Nazi executioners. Before his capture, he had served his country in the Resistance and the world of scholarship through the breadth of an unfettered, humane, historical imagination. His great works on French and European social and economic history were well known by 1940, when his *Feudal Society* (*La Société Féodale*) came to print from "the author on active service." In the following selection from this work, we can appreciate Bloch's concern for the limiting conditions of medieval existence—soil, topography, climate, technique, population—and the institutions and thought created within them. For Bloch, feudalism was a system of human relations, including those of production, which in combination constituted the social order of medieval Europe. Lords, peasants, fiefs, manors, towns, and vills were constituent parts. But the spread of people on the land was more variable than the other limiting conditions. And so Bloch began his study of feudal society by writing about population and its impact on the material bases of life in Europe. In Chapter Seven we will again present a selection by Bloch, which discusses how man's environment limits his thought and feeling.

From *Feudal Society*

Marc Bloch

The First Feudal Age: Density of Population

It is and always will be impossible for us to calculate, even approximately, the population of Western countries during the first feudal age. Moreover, there undoubtedly existed marked regional variations, constantly intensified by the spasms of social disorder. Compared with the veritable desert of the Iberian plateaux, which gave the frontier regions of Christendom and Islam the desolate appearance of a vast "no man's land"—desolate even in comparison with early Germany, where the destruction wrought by the migrations of the previous age was being slowly made good—the country districts of Flanders and Lombardy seemed relatively favoured regions. But whatever the importance of these con-

SOURCE: Marc Bloch, *Feudal Society*, trans. L. A. Manyon (Chicago: University of Chicago Press, 1968), pp. 60–71.

trasts and whatever their effect on all the aspects of civilization, the fundamental characteristic remains the great and universal decline in population. Over the whole of Europe, the population was immeasurably smaller than it has been since the eighteenth century or even since the twelfth. Even in the provinces formerly under Roman rule, human beings were much scarcer than they had been in the heyday of the Empire. The most important towns had no more than a few thousand inhabitants, and waste land, gardens, even fields and pastures encroached on all sides amongst the houses.

This lack of density was further aggravated by very unequal distribution. Doubtless physical conditions, as well as social habits, conspired to maintain in the country districts profound differences between systems of settlement. In some districts the families, or at least some of them, took up their residence a considerable distance apart, each in the middle of its own farmland, as was the case, for example, in Limousin. In others on the contrary, like the Île-de-France, they mostly crowded together in villages. On the whole, however, both the pressure of the chiefs and, above all, the concern for security militated against too wide dispersal. The disorders of the early Middle Ages had in many cases induced men to draw nearer to each other, but these aggregations in which people lived cheek by jowl were separated by empty spaces. The arable land from which the village derived its sustenance was necessarily much larger in proportion to the number of inhabitants than it is today. For agriculture was a great devourer of space. In the tilled fields, incompletely ploughed and almost always inadequately manured, the ears of corn grew neither very heavy nor very dense. Above all, the harvests never covered the whole area of cultivation at once. The most advanced systems of crop-rotation known to the age required that every year half or a third of the cultivated soil should lie fallow. Often indeed, fallow and crops followed each other in irregular alternation, which always allowed more time for the growth of weeds than for that of the cultivated produce; the fields, in such cases, represented hardly more than a provisional and short-lived conquest of the waste land, and even in the heart of the agricultural regions nature tended constantly to regain the upper hand. Beyond them, enveloping them, thrusting into them, spread forests, scrub and dunes—immense wildernesses, seldom entirely uninhabited by man, though whoever dwelt there as charcoal-burner, shepherd, hermit or outlaw did so only at the cost of a long separation from his fellow men.

The First Feudal Age: Intercommunication

Among these sparsely scattered human groups the obstacles to communication were many. The collapse of the Carolingian empire had destroyed the last power sufficiently intelligent to concern itself with public works, sufficiently strong to get some of them carried out. Even the old Roman roads, less solidly constructed than has sometimes been imagined, went to rack and ruin for want of maintenance. Worse still, bridges were no longer kept in repair and were lacking at a great number of river-crossings. Added to this was the general state of insecurity, increased by the depopulation to which it had itself in part contributed. Great was the surprise and relief at the court of Charles the Bald, when in the year 841 that prince witnessed the arrival at Troyes of the messengers bringing him the crown jewels from Aquitaine: how wonderful that such a small number of men, entrusted with such precious baggage, should traverse without accident those vast areas infested on all sides by robbers! The Anglo-Saxon Chronicle shows much less surprise when relating how, in 1061, one of the greatest nobles of England, Earl Tostig, was captured and held to ransom by a handful of bandits at the gates of Rome.

Compared with what the world offers us today, the speed of travel in that age seems extremely slow. It was not, however, appreciably slower than it was at the end of the Middle Ages, or even the beginning of the eighteenth century. By contrast with today, travel was much faster by sea than by land. From 60 to 90 miles a day was not an exceptional record for a ship: provided (it goes without saying) that the winds were not too unfavourable. On land, the normal distance covered in one day amounted, it seems, to between 19 and 25 miles—for travellers who were in no hurry, that is: say a caravan of merchants, a great nobleman moving round from castle to castle or from abbey to abbey, or an army with its baggage. A courier or a handful of resolute men could by making a special effort travel at least twice as fast. A letter written by Gregory VII at Rome on the 8th December 1075 arrived at Goslar, at the foot of the Harz, on the 1st of January following; its bearer had covered about 29 miles a day as the crow flies—in reality, of course, much more. To travel without too much fatigue and not too slowly it was necessary to be mounted or in a carriage. Horses and mules not only go faster than men; they adapt themselves better to boggy ground. This explains the seasonal interruption of many communications; it was due less to bad weather than to lack of forage. The Carolingian *missi* had earlier made a point of not beginning their tours till the grass had grown. However, as at present in Africa, an experienced foot-traveller could cover astoundingly long distances in a few days and he could doubtless overcome certain obstacles more quickly than a horseman. When Charles the Bald organized his second Italian expedition he arranged to keep in touch with Gaul across the Alps partly by means of runners.

Though poor and unsafe, the roads or tracks were in constant use. Where transport is difficult, man goes to something he wants more easily than he makes it come to him. In particular, no institution or method could take the place of personal contact between human beings. It would have been impossible to govern the state from inside a palace: to control a country, there was no other means than to ride through it incessantly in all directions. The kings of the first feudal age positively killed themselves by travel. For example, in the course of a year which was in no way exceptional, the emperor Conrad II in 1033 is known to have journeyed in turn from Burgundy to the Polish frontier and thence to Champagne, to return eventually to Lusatia. The nobleman with his entourage moved round constantly from one of his estates to another; and not only in order to supervise them more effectively. It was necessary for him to consume the produce on the spot, for to transport it to a common centre would have been both inconvenient and expensive. Similarly with the merchant. Without representatives to whom he could delegate the task of buying and selling, fairly certain in any case of never finding enough customers assembled in one place to assure him a profit, every merchant was a pedlar, a "dusty foot" (*pied poudreux*), plying his trade up hill and down dale. The cleric, eager for learning or the ascetic life, was obliged to wander over Europe in search of the master of his choice: Gerbert of Aurillac studied mathematics in Spain and philosophy at Rheims; the Englishman Stephen Harding, the ideal monachism in the Burgundian abbey of Molesmes. Before him, St. Odo, the future abbot of Cluny, had travelled through

France in the hope of finding a monastery whose members lived strictly according to the rule.

Moreover, in spite of the old hostility of the Benedictine rule to the *gyrovagi*, the bad monks who ceaselessly "vagabonded about," everything in contemporary clerical life favoured this nomadism: the international character of the Church; the use of Latin as a common language among educated priests and monks; the affiliations between monasteries; the wide dispersal of their territorial patrimonies; and finally the "reforms" which periodically convulsed this great ecclesiastical body and made the places first affected by the new spirit at once courts of appeal (to which people came from all parts to seek the good rule) and mission centres whence the zealots were despatched for the conquest of the Catholic world. How many foreign visitors came to Cluny in this way! How many Cluniacs journeyed forth to foreign lands! Under William the Conqueror almost all the dioceses and great abbeys of Normandy, which the first waves of the "Gregorian" revival were beginning to reach, had at their head Italians or Lorrainers; the archbishop of Rouen, Maurille, was a man from Rheims who, before occupying his Neustrian see, had studied at Liège, taught in Saxony and lived as a hermit in Tuscany.

Humble folk, too, passed along the highways of the West: refugees, driven by war or famine; adventurers, half-soldiers, half-bandits; peasants seeking a more prosperous life and hoping to find, far from their native land, a few fields to cultivate. Finally, there were pilgrims. For religious devotion itself fostered travel and more than one good Christian, rich or poor, cleric or layman, believed that he could purchase salvation of body and soul only at the price of a long journey.

As has often been remarked, it is in the nature of good roads to create a vacuum around them—to their own profit. In the feudal age, when all roads were bad, scarcely any of them was capable of monopolizing the traffic in this way. Undoubtedly such factors as the restrictions of the terrain, tradition, the presence of a market here or a sanctuary there, worked to the advantage of certain routes, although far less decisively than the historians of literary or artistic influences have sometimes believed. A fortuitous event—a physical accident, the exactions of a lord in need of money—sufficed to divert the flow, sometimes permanently. The building of a castle on the old Roman road, occupied by a race of robber knights—the lords of Méréville—and the establishment some distance away of the St. Denis priory of Toury, where merchants and pilgrims found by contrast a pleasant reception, were sufficient to divert the traffic from the Beauce section of the road from Paris to Orleans permanently westward, so that the ancient roadway was abandoned from that time on. Moreover from the beginning of his journey to the end, the traveller had almost always the choice of several itineraries, of which none was absolutely obligatory. Traffic, in short, was not canalized in a few great arteries; it spread capriciously through a multitude of little blood-vessels. There was no castle, burg, or monastery, however far from the beaten track, that could not expect to be visited occasionally by wanderers, living links with the outer world, although the places where such visits were of regular occurrence were few.

Thus the obstacles and dangers of the road in no way prevented travel. But they made each journey an expedition, almost an adventure. If men, under pressure of need, did not fear to undertake fairly long-journeys (they feared it less, perhaps, than in centuries nearer to our own) they shrank from those repeated comings and goings within a narrow radius which in other civilizations form the texture of daily life; and this was especially so in the case of humble folk of settled occupations. The result was an ordering of the scheme of human relations quite different from anything we know today. There was scarcely any remote little place which had not some contacts intermittently through that sort of continuous yet irregular "Brownian movement" which affected the whole of society. On the other hand, between two inhabited centres quite close to each other the connections were much rarer, the isolation of their inhabitants infinitely greater than would be the case in our own day. If, according to the angle from which it is viewed, the civilization of feudal Europe appears sometimes remarkably universalist, sometimes particularist in the extreme, the principal source of this contradiction lay in the conditions of communication: conditions which favoured the distant propagation of very general currents of influence as much as they discouraged, in any particular place, the standardizing effects of neighbourly intercourse.

The only more or less regular letter-mail service which functioned during the whole of the feudal era was that which linked Venice to Constantinople. Such a thing was practically unknown in the West. The last attempts to maintain a royal posting-service, on the model left by the Roman government, had disappeared with the Carolingian empire. It is significant of the general disorganization that the German monarchs themselves, the true heirs of that empire and its ambitions, should have lacked either the authority or the intelligence necessary to secure the revival of an institution clearly so indispensable to the control of vast territories. Sovereigns, nobles, prelates were obliged to entrust their correspondence to special couriers, otherwise—as was usual among persons of lesser rank—the transport of letters was simply left to the kindness of passing travellers; as, for instance, the pilgrims on their way to St. James of Galicia. The relative slowness of the messengers, the mishaps that at every stage threatened their progress, meant that the only effective authority was the one on the spot. Forced constantly to take the gravest steps—the history of the papal legates is in this respect very instructive—every local representative of a great potentate tended only too naturally to act for his personal advantage and thus finally to transform himself into an independent ruler.

As for knowledge of distant events, everyone, whatever his rank, was obliged to rely on chance encounters. The picture of the contemporary world which the best-informed men carried in their minds presented many lacunae; we can form an idea of them from the unavoidable omissions even from the best of these monastic annals which are as it were the written reports of medieval news-hawks. Moreover, it was seldom exact as to time. It is, for example, remarkable to find a person so well placed for acquiring information as Bishop Fulbert of Chartres showing astonishment on receiving gifts for his church from Cnut the Great: for he admits that he believed this prince to be still a heathen, although in fact he had been baptized in infancy. The monk Lambert of Hersfeld is quite well informed about German affairs, but when he goes on to describe the grave events which occurred in his time in Flanders (a region bordering on the Empire and in part an imperial fief), he soon makes a series of the strangest blunders. Such an imperfect state of knowledge was a poor foundation for any large political designs.

The First Feudal Age: Trade and Currency

The life of the Europe of the first feudal age was not entirely self-contained. There was more than one current of exchange between it and the neighbouring civilizations, and probably the most active was that which linked it to Moslem Spain, as witnessed by the numerous Arab gold pieces which, by this route, penetrated north of the Pyrenees and were there sufficiently sought after to become the object of frequent imitations. In the western Mediterranean, on the other hand, long-distance navigation was now practically unknown. The principal lines of communication with the East were elsewhere. One of them, a sea-route, passed through the Adriatic, at the head of which lay Venice, to all appearance a fragment of Byzantium, set in a world apart. On land the Danube route, for a long time severed by the Hungarians, was almost deserted. But farther north, on the trails which joined Bavaria to the great market of Prague and thence, by the terraces on the northern flank of the Carpathians, continued to the Dnieper, caravans passed back and forth, laden on the return journey with products of Constantinople or of Asia. At Kiev they met the great transversal which, running across the plains and from river to river, linked the riparian countries of the Baltic with the Black Sea, the Caspian or the oases of Turkestan. For the West had missed its chance of being the intermediary between the north and north-east of the continent and the eastern Mediterranean, and had nothing to offer on its own soil to compare with the mighty comings and goings of merchandise which made the prosperity of Kievian Russia.

Not only was this trade restricted to very few routes; it was also extremely small in volume. What is worse, the balance of trade seems to have been distinctly unfavourable—at any rate with the East. From the eastern countries the West received almost nothing except a few luxury articles whose value—very high in relation to their weight—was such as to take no account of the expense and risks of transport. In exchange it had scarcely anything to offer except slaves. Moreover, it seems that most of the human cattle rounded up on the Slav and Lettish territories beyond the Elbe or acquired from the slave-traders of Britain took the road to Islamic Spain; the eastern Mediterranean was too abundantly provided with this commodity from its own sources to have any need to import it on a large scale. The profits of the slave-trade, in general fairly small, were not sufficient to pay for the purchase of precious goods and spices in the markets of the Byzantine world, of Egypt or of nearer Asia. The result was a slow drain of silver and above all of gold. If a few merchants unquestionably owed their prosperity to these remote transactions, society as a whole owed scarcely anything to them except one more reason for being short of specie.

However, money was never wholly absent from business transactions in feudal Europe, even among the peasant classes, and it never ceased to be employed as a standard of exchange. Payments were often made in produce; but the produce was normally valued item by item in such a way that the total of these reckonings corresponded with a stipulated price in pounds, shillings and pence. Let us therefore avoid the expression "natural economy," which is too summary and too vague. It is better to speak simply of shortage of currency. This shortage was further aggravated by the anarchic state of minting, another result of the subdivision of political authority and the difficulty of communication: for each important market, faced with the threat of shortage, had to have its local mint. Except for the imitation of exotic coinages and apart from certain insignificant little pieces, the only coins now produced were *denarii*, which were rather debased silver pieces. Gold circulated only in the shape of Arab and Byzantine coins or imitations of them. The *libra* and the *solidus* were only arithmetical multiples of the *denarius*, without a material basis of their own. But the various coins called *denarii* had a different metallic value according to their origin. Worse still, even in one and the same area almost every issue involved variations in the weight or the alloy. Not only was money generally scarce, and inconvenient on account of its unreliability, but it circulated too slowly and too irregularly for people ever to feel certain of being able to procure it in case of need. That was the situation, in the absence of a sufficiently active commerce.

But here again, let us beware of too facile a formula—the "closed economy." It would not even apply exactly to the small farming operations of the peasants. We know that markets existed where the rustics certainly sold some of the produce of their fields or their farmyards to the towns-folk, to the clergy, to the men-at-arms. It was thus that they procured the *denarii* to pay their dues. And poor indeed was the man who never bought a few ounces of salt or a bit of iron. As to the "autarky" of the great manors, this would have meant that their masters had gone without arms or jewels, had never drunk wine (unless their estates produced it), and for clothes had been content with crude materials woven by the wives of tenants. Moreover, even the inadequacies of agricultural technique, the disturbed state of society, and finally the inclemency of the weather contributed to maintain a certain amount of internal commerce: for when the harvest failed, although many people literally died of starvation, the whole population was not reduced to this extremity, and we know that there was a traffic in corn from the more favoured districts to those afflicted by dearth, which lent itself readily to speculation. Trade, therefore, was not nonexistent, but it was irregular in the extreme. The society of this age was certainly not unacquainted with either buying or selling. But it did not, like our own, live by buying and selling.

Moreover, commerce, even in the form of barter, was not the only or perhaps even the most important channel by which at that time goods circulated through the various classes of society. A great number of products passed from hand to hand as dues paid to a chief in return for his protection or simply in recognition of his power. It was the same in the case of that other commodity, human labour: the *corvée* furnished more labourers than hire. In short, exchange, in the strict sense, certainly played a smaller part in economic life than payment in kind; and because exchange was thus a rare thing, while at the same time only the poorest could resign themselves to living wholly on their own produce, wealth and well-being seemed inseparable from authority.

Nevertheless, in an economy so constituted the means of acquisition at the disposal even of the powerful were, on the whole, singularly restricted. When we speak of money we mean the possibility of laying by reserves, the ability to wait, the "anticipation of future values"—everything that, conversely, the shortage of money particularly impedes. It is true that people tried to hoard wealth in other forms. The nobles and kings accumulated in their coffers gold or silver vessels and precious stones; the churches amassed liturgical plate. Should the need arise for an unexpected disbursement, you sold or pawned the crown, the goblet or the crucifix; or you even sent them to be melted down at the local mint. But such liquidation of assets, from the very fact of the slowing down of exchange which made it necessary, was never easy nor was it always profitable; and the hoarded treasure itself did not after all constitute a very large amount. The great as well as the humble lived from hand to mouth, obliged to be content with the resources of the moment and mostly compelled to spend them at once.

The weakness of trade and of monetary circulation had a further consequence of the gravest kind. It reduced to insignificance the social function of wages. The latter requires that the employer should have at his disposal an adequate currency, the source of which is not in danger of drying up at any moment; on the side of the wage-earner it requires the certainty of being able to employ the money thus received in procuring for himself the necessities of life. Both these conditions were absent in the first feudal age. In all grades of the hierarchy, whether it was a question of the king's making sure of the services of a great official, or of the small landlord's retaining those of an armed follower or a farm-hand, it was necessary to have recourse to a method of remuneration which was not based on the periodic payment of a sum of money. Two alternatives offered: one was to take the man into one's household, to feed and clothe him, to provide him with "prebend," as the phrase went; the other was to grant him in return for his services an estate which, if exploited directly or in the form of dues levied on the cultivators of the soil, would enable him to provide for himself.

Now both these methods tended, though in opposite ways, to create human ties very different from those based on wages. Between the prebend-holder and the master under whose roof he lived the bond must surely have been much more intimate than that between an employer and a wage-earner, who is free, once his job is finished, to go off with his money in his pocket. On the other hand, the bond was almost inevitably loosened as soon as the subordinate was settled on a piece of land, which by a natural process he tended increasingly to regard as his own, while trying to reduce the burden of service. Moreover, in a time when the inadequacy of communications and the insufficiency of trade rendered it difficult to maintain large households in relative abundance, the "prebend" system was on the whole capable of a much smaller extension than the system of remuneration based on land. If feudal society perpetually oscillated between these two poles, the narrow relationship of man and man and the looser tie of land tenure, the responsibility for this belongs in large part to the economic regime which, to begin with at least, made wage-earning impracticable.

The Economic Revolution of the Second Feudal Age

We shall endeavour, in another work, to describe the intensive movement of repopulation which, from approximately 1050 to 1250, transformed the face of Europe: on the confines of the Western world, the colonization of the Iberian plateaux and of the great plain beyond the Elbe; in the heart of the old territories, the incessant gnawing of the plough at forest and wasteland; in the glades opened amidst the trees or the brushwood, completely new villages clutching at the virgin soil; elsewhere, round sites inhabited for centuries, the extension of the agricultural lands through the exertions of the assarters. It will be advisable then to distinguish between the stages of the process and to describe the regional variations. For the moment, we are concerned only with the phenomenon itself and its principal effects.

The most immediately apparent of these was undoubtedly the closer association of the human groups. Between the different settlements, except in some particularly neglected regions, the vast empty spaces thenceforth disappeared. Such distances as still separated the settlements became, in any case, easier to traverse. For powers now arose or were consolidated—their rise being favoured by current demographic trends—whose enlarged horizons brought them new responsibilities. Such were the urban middle classes, which owed everything to trade. Such also were the kings and princes; they too were interested in the prosperity of commerce because they derived large sums of money from it in the form of duties and tolls; moreover they were aware—much more so than in the past—of the vital importance to them of the free transmission of orders and the free movement of armies. The activity of the Capetians towards that decisive turning-point marked by the reign of Louis VI, their aggressions, their domanial policy, their part in the organization of the movement of repopulation, were in large measure the reflection of considerations of this kind—the need to retain control of communications between the two capitals, Paris and Orleans, and beyond the Loire or the Seine to maintain contact with Berry or with the valleys of the Oise and the Aisne. It would seem that while the security of the roads had increased, there was no very notable improvement in their condition; but at least the provision of bridges had been carried much farther. In the course of the twelfth century, how many were thrown over all the rivers of Europe! Finally, a fortunate advance in harnessing methods had the effect, about the same time, of increasing very substantially the efficiency of horse-transport.

The links with neighbouring civilizations underwent a similar transformation. Ships in ever greater numbers ploughed the Tyrrhenian Sea, and its ports, from the rock of Amalfi to Catalonia, rose to the rank of great commercial centres; the sphere of Venetian trade continually expanded; the heavy wagons of the merchant caravans now followed the route of the Danubian plains. These advances were important enough. But relations with the East had not only become easier and more intimate. The most important fact is that they had changed their character. Formerly almost exclusively an importer, the West had become a great supplier of manufactured goods. The merchandise which it thus shipped in quantity to the Byzantine world, to the Latin or Islamic Levant and even—though in smaller amounts—to the Maghreb, belonged to very diverse categories. One commodity, however, easily dominated all the rest. In the expansion of the European economy in the Middle Ages, cloth played the same vital rôle as did metal and cotton goods in that of nineteenth-century England. If in Flanders, in Picardy, at Bourges, in Languedoc, in Lombardy, and yet other places—for the cloth centres were to be found almost everywhere—the noise of the looms and the throbbing of the fullers' mills resounded, it was at least as much for the sake of foreign markets as for local requirements. And undoubtedly this revolution, which saw our Western countries embarking on the economic conquest of the world by way of the East, is to be explained by a multiplicity of causes and by looking—as far as possible—towards the East as well as towards the West. It is none the less true that it could not have occurred without the demographic changes mentioned above. If the population had not been more numerous than before and the cultivated area more extensive; if the fields—their quality improved by augmented manpower and in particular by more intensive ploughing

—had not become capable of yielding bigger and more frequent harvests, how could so many weavers, dyers or cloth-shearers have been brought together in the towns and provided with a livelihood?

The North was conquered, like the East. From the end of the eleventh century Flemish cloth was sold at Novgorod. Little by little, the route of the Russian plains became hazardous and was finally closed. Thenceforward Scandinavia and the Baltic countries turned towards the West. The process of change which was thus set in motion was completed when, in the course of the twelfth century, German merchants took over the Baltic. From that time onwards the ports of the Low Countries, especially Bruges, became the centres where northern products were exchanged not only for those of the West itself but also for merchandise from the East. Strong international links united the two frontiers of feudal Europe by way of Germany and especially through the fairs of Champagne.

Such a well-balanced external trade could not fail to bring a flow of coin and precious metals into Europe and so add substantially to its monetary resources. This relative easing of the currency situation was reinforced—and its effects multiplied—by the accelerated rhythm of circulation. For in the very heart of the West the progress of repopulation, the greater ease of communications, the cessation of the invasions which had spread such an atmosphere of confusion and panic over the Western world, and still other causes which it would take too long to examine here, had led to a revival of commerce.

Let us avoid exaggeration, however. The picture would have to be carefully shaded—by regions and by classes. To live on their own resources remained for long centuries the ideal—though one that was rarely attained—of many peasants and most villages. Moreover, the profound transformations of the economy took place only very gradually. It is significant that of the two essential developments in the sphere of currency, one, the minting of larger pieces of silver much heavier than the *denarius*, appeared only at the beginning of the thirteenth century (and even at that date in Italy alone) and the other, the resumption of the minting of gold coins of an indigenous type, was delayed till the second half of the same century. In many respects, what the second feudal age witnessed was less the disappearance of earlier conditions than their modification. This observation applies to the part played by distance as well as to commerce. But the fact that the kings, the great noble, and the manorial lords should have been able to begin once more to amass substantial wealth, that wage-earning, sometimes under legal forms clumsily adapted from ancient practices, should have increasingly supplanted other methods of remunerating services—these signs of an economy in process of revival affected in their turn, from the twelfth century onwards, the whole fabric of human relations.

Furthermore, the evolution of the economy involved a genuine revision of social values. There had always been artisans and merchants; individuals belonging to the latter class had even been able, here and there, to play an important rôle, though collectively neither group counted for much. But from the end of the eleventh century the artisan class and the merchant class, having become much more numerous and much more indispensable to the life of the community, made themselves felt more and more vigorously in the urban setting. This applies especially to the merchant class, for the medieval economy, after the great revival of these decisive years, was always dominated, not by the producer, but by the trader. It was not for the latter class that the legal machinery of the previous age—founded on an economic system in which they occupied only an inferior place—had been set up. But now their practical needs and their mental attitude were bound to imbue it with a new spirit. Born in the midst of a very loosely knit society, in which commerce was insignificant and money a rarity, European feudalism underwent a fundamental change as soon as the meshes of the human network had been drawn closer together and the circulation of goods and coin intensified.

Dialogue on Economic Life

The social order of the Middle Ages was always more complex in reality than would appear from the comments of contemporaries. Yet there is in Aelfric's tenth-century teaching of Latin something of value to us in understanding his society. The monk reveals the world of ordinary labor with its hierarchies of occupations and obligations stretching from the simplest plowman to the most august "counselor." The elements of authority and subordination emerge clearly in these pages of this conversational grammar by the Abbot of Eynsham (born 955). Although we do not see the whole body of society and the entire range of duties, dues, and responsibilities, we do see clearly enough the difficult search for security in this life and salvation in that to come. Lamentation is the natural mode of peasant expression, no matter what his special craft or skill. The merchant, by contrast, seems self-assured and perhaps pitiless. Above all else there is insistence that men be content with their lot. The order of human work is itself God's providence.

From *Work*

Aelfric of Eynsham

PUPIL: We children beg you, teacher, to teach us how to speak Latin correctly, for we are very ignorant and make mistakes in our speech.

TEACHER: What do you want to talk about?

PUPIL: What do we care what the subject is, provided the language be correct, and the discourse be useful, not idle and base?

TEACHER: Do you desire to be flogged in your learning?

PUPIL: We had rather be flogged for learning's sake than be ignorant; but we know that you are kind and will not inflict blows upon us unless we force you to do so.

TEACHER: I ask an answer to this: What is your work at present?

PUPIL: I am a monk by profession and I sing every day the seven services of the hours with my brethren and am occupied with reading and singing, but nevertheless I should like, between times, to learn Latin.

TEACHER: What do these your comrades know?

PUPIL: Some are plowmen, some shepherds, some oxherds; and some are hunters, some fishermen, some fowlers, some merchants, some shoemakers, some salters, and some bakers.

TEACHER: Plowman, what can you say for yourself? How do you do your work?

PLOWMAN: O, dear master, I work very hard; I go out at daybreak, drive the oxen to the field and yoke them to the plow. Never is winter weather so severe that I dare to remain at home; for I fear my master. But when the oxen are

---

SOURCE: Aelfric of Eynsham, *Work*, in *English Literature from Widsith to the Death of Chaucer*, trans. Allen R. Benham (New Haven: Yale University Press, 1916), pp. 26–33.

yoked to the plow and the share and coulter fastened on, every day I must plow a full acre or more.

TEACHER: Have you any one to help you?

PLOWMAN: I have a boy who urges on the oxen with a goad. He is now hoarse from cold and shouting.

TEACHER: Do you do anything else in the course of a day?

PLOWMAN: I do a great deal more. I have to fill the bins of the oxen with hay and water them and clean their stalls.

TEACHER: Oh! Oh! that is hard work!

PLOWMAN: The labor is indeed great, because I am not free.

TEACHER: What is your work, shepherd, have you anything to do?

SHEPHERD: Yes indeed, master, I have. In the early morning I drive my sheep to their pasture and stand over them in heat or cold with dogs lest wolves devour them. I lead them back to their folds and milk them twice a day. In addition I move their folds, make cheese and butter and am faithful to my master.

TEACHER: Well, oxherd, what is your work?

OXHERD: O my master, my work is very hard. When the plowman unyokes the oxen, I lead them to pasture and all night I stand over them and watch for thieves. Then in the early morning I turn them over to the plowman after I have fed and watered them.

TEACHER: Is this one of your friends?

OXHERD: Yes, he is.

TEACHER: Can you do anything?

HUNTER: I know one craft.

TEACHER: What is it?

HUNTER: I am a hunter.

TEACHER: Whose?

HUNTER: The king's.

TEACHER: How do you carry on your work?

HUNTER: I weave my nets and put them in a suitable place, and train my dogs to follow the wild beasts until they come unexpectedly to the nets and are entrapped. Then I kill them in the nets.

TEACHER: Can't you hunt without nets?

HUNTER: Yes, I can hunt without them.

TEACHER: How?

HUNTER: I chase wild beasts with swift dogs.

TEACHER: What wild beasts do you catch?

HUNTER: Harts, boars, does, goats and sometimes hares.

TEACHER: Did you go out to-day?

HUNTER: No, because it is Sunday; but I was out yesterday.

TEACHER: What luck did you have?

HUNTER: I got two harts and a boar.

TEACHER: How did you catch them?

HUNTER: The harts I took in a net and the boar I slew.

TEACHER: How did you dare to kill a boar?

HUNTER: The dogs drove him to me, and I, standing opposite to him, slew him suddenly.

TEACHER: You were very brave.

HUNTER: A hunter should not be afraid; for many kinds of wild beasts live in the woods.

TEACHER: What do you do with your game?

HUNTER: I give the king what I take because I am his hunter.

TEACHER: What does he give you?

HUNTER: He clothes me well and feeds me. Occasionally he gives me a horse or a ring that I may pursue my craft more willingly.

TEACHER: What craft do you follow?

FISHERMAN: I am a fisherman.

TEACHER: What do you gain by your craft?

FISHERMAN: Food and clothes and money.

TEACHER: How do you catch your fish?

FISHERMAN: I go out in my boat, throw my net in the river, cast in my hook baited and take in my creel whatever comes to me.

TEACHER: What if they are unclean fish?

FISHERMAN: I throw the unclean ones back and keep the clean for meat.

TEACHER: Where do you sell your fish?

FISHERMAN: In the city.

TEACHER: Who buys them?

FISHERMAN: The citizens; I do not catch as many as I could sell.

TEACHER: What sorts of fish do you catch?

FISHERMAN: Eels and pike, minnows and turbots, trout and lampreys; in short, whatever swims in running water.

TEACHER: Why don't you fish in the sea?

FISHERMAN: Sometimes I do, but seldom; because a large boat is needed for sea-fishing.

TEACHER: What do you catch in the sea?

FISHERMAN: Herring and salmon, dolphins and sturgeons, oysters and crabs, mussels, periwinkles, cockles, flounders, sole, lobsters and many others.

TEACHER: Wouldn't you like to catch a whale?

FISHERMAN: No.

TEACHER: Why not?

FISHERMAN: Because it is a dangerous thing to catch a whale. It is safer for me to go to the river with my boat than to go with many ships to hunt whales.

TEACHER: Why so?

FISHERMAN: Because I prefer to take a fish that I can kill than one that with a single blow can swallow not only me but my companions also.

TEACHER: Yet, many catch whales without danger and get a good price for them.

FISHERMAN: I know it, but I do not dare; for I am very timid.

TEACHER: What have you to say, fowler? How do you catch the birds?

FOWLER: I entice them in many ways, sometimes with nets, sometimes with nooses, sometimes with lime, sometimes by whistling, sometimes with a hawk and sometimes with traps.

TEACHER: Have you a hawk?

FOWLER: Yes.

TEACHER: Can you tame it?

FOWLER: Yes; what good would it be to me, if I could not tame it?

HUNTER: Give me a hawk.

FOWLER: I will gladly, if you will give me a swift dog. Which hawk do you prefer, the larger or the smaller?

HUNTER: Give me the larger one.

TEACHER: How do you feed your hawks?

FOWLER: They feed themselves and me in the winter, and in the spring I let them fly in the woods. In the autumn I take the young birds and tame them.

TEACHER: And why do you let the tame ones go?

FOWLER: Because I don't want to feed them in the summer, since they eat a good deal.

TEACHER: Many people feed those that they have tamed, even through the summer, that they may have them ready again.

FOWLER: Yes, so they do; but I do not take so much trouble for them, because I can get others, not one only, but many more.

TEACHER: What can you say, merchant?

MERCHANT: I say that I am useful to the king and to the magistrates and to the wealthy and to all the people.

TEACHER: How is that?

MERCHANT: I go aboard my ship with my goods and row over parts of the sea, sell my things and buy precious treasures that are not produced in this country. These latter I bring here with great peril from the sea. Sometimes I suffer shipwreck and lose all my wares, hardly escaping with my life.

TEACHER: What do you bring us?

MERCHANT: Purple goods and silk, precious gems and gold, strange raiment and spices, wine and oil, ivory and brass, copper and tin, sulphur and glass, and the like.

TEACHER: Do you sell your goods for the same price for which you bought them?

MERCHANT: No; what profit would I then have from my labor? But I sell them dearer than I bought them, that I may make a profit. Thus I feed myself, my wife and my son.

TEACHER: And you, shoemaker, what do you do that is useful to us?

SHOEMAKER: My craft is a cunning one and very useful to you.

TEACHER: How?

SHOEMAKER: I buy hides and skins and prepare them by my art and make of them various kinds of footwear—slippers, shoes and gaiters; bottles, reins and trappings; flasks and leathern vessels; spurstraps and halters; purses and bags. None of you could pass a winter without the aid of my craft.

TEACHER: Salter, how is your craft useful to us?

SALTER: Who of you would relish his food without the savor of salt? Who could fill either his cellar or his storeroom without the aid of my craft? behold, all butter and cheese would you lose, nor would you enjoy even your vegetables, without me.

TEACHER: And what do you say, baker? Does any one need your craft, or could we live without you?

BAKER: Life might be sustained for a while without my craft, but not long nor well. Truly, without my skill, every table would be empty. Without bread all food would cause sickness. I strengthen the heart of man. I am the strength of men and few would like to do without me.

TEACHER: What shall we say of the cook? Do we need his skill for anything?

THE COOK SAYS: If you should send me away from your midst, you would be compelled to eat your vegetables green and your meat uncooked, and you could have no nourishing broth without my skill.

TEACHER: We do not need your skill, nor is it necessary to us; for we ourselves could cook the things which should be cooked and roast the things that should be roasted.

THE COOK SAYS: If you send me away, that is what you will have to do. Nevertheless, without my skill, you cannot eat.

TEACHER: Monk, you who are talking with me, I have persuaded myself that you have good comrades and that they are very necessary. Now, who are these?

PUPIL: I have smiths—a blacksmith, a goldsmith, a silversmith, a coppersmith, a carpenter and many other workers at various trades.

TEACHER: Have you any wise counselor?

PUPIL: I certainly have. How could our community be ruled without a counselor?

TEACHER: What would you say, wise man? Among these crafts which seems to you the greatest?

COUNSELOR: I tell you that among all these occupations the service of God seems to me to hold the first place; for thus it is written in the Gospels: "Seek ye first the kingdom of God and his righteousness and all these things shall be added to you."

TEACHER: And among the worldly crafts which seems to you to be first?

COUNSELOR: Agriculture, because the farmer feeds us all.

THE BLACKSMITH SAYS: Where would the farmer get his plowshare, or mend his coulter when it has lost its point, without my craft? Where would the fisherman get his hook, or the shoemaker his awl, or the tailor his needle, if it were not for my work?

THE COUNSELOR RESPONDS: Verily, you speak the truth; but we prefer to live with the farmer rather than with you; for the farmer gives us food and drink. What you give us in your shop is sparks, noise of hammers and blowing of bellows.

THE CARPENTER SPEAKS: How could you spare my skill in building houses, in the use of various tools, in building ships and in all the things I make?

THE COUNSELOR SAYS: O comrades and good workmen, let us quickly settle these disputes, and let there be peace and harmony among us. Let each one benefit the others with his craft and agree always with the farmer who feeds us and from whom we get fodder for our horses. And this advice I give to all workers, that each one shall follow his own craft diligently, for he who forsakes his craft shall be himself forsaken by his craft. Whoever you are, priest or monk or layman or soldier, exercise yourself in this. Be satisfied with your office; for it is a great disgrace for a man to be unwilling to be what he is, and what it is his duty to be.

TEACHER: Well, children, how have you enjoyed this conversation?

PUPIL: Pretty well, but you speak profoundly and beyond our age. Speak to us according to our intelligence that we may understand what you say.

TEACHER: Here is a simple question for you: why are you so eager to learn?

PUPIL: Because we do not wish to be like stupid animals that do not know anything but grass and water.

TEACHER: And what is your wish?

PUPIL: We wish to be wise.

TEACHER: In what wisdom? Do you wish to be crafty or to assume a thousand shapes, skillful in deceiving, astute in speaking, graceful, speaking good and thinking evil, using soft words, feeding fraud within, like a whited sepulcher, beautiful without, but full of corruption?

PUPIL: We do not wish for this kind of wisdom; for he is not wise who deceives himself with pretenses.

TEACHER: But how would you be wise?

PUPIL: We wish to be simple without hypocrisy, and wise that we may turn from evil and do good. . . .

* * *

Wondrously wrought and fair its wall of stone,[1]
shattered by Fate! The castles rend asunder,
the work of giants moldereth away.
Its roofs are breaking and falling; its towers crumble
in ruin. Plundered those walls with grated doors—
their mortar white with frost. Its battered ramparts
are shorn away and ruined, all undermined
by eating age. The mighty men that built it,
departed hence, undone by death, are held
fast in the earth's embrace. Tight is the clutch
of the grave, while overhead for living men
a hundred generations pass away.
    Long this red wall, now mossy gray, withstood
while kingdom followed kingdom in the land,
unshaken 'neath the storms of heaven—yet now
its towering gate hath fallen. . . .

---

1 Poem, included in an eleventh-century manuscript collection of earlier Anglo-Saxon literature, has been tentatively ascribed to the tenth century. But all attempts to date it precisely or to identify the "ruined city" with Roman Bath or another town have failed.

The Rise of Dependent Cultivation

One of the most important elements in the shaping of popular life and culture in the Middle Ages was the rise of dependent cultivation of the land. This feature of the mature seignorial society of the West subordinated peasant cultivators to aristocratic possessors of the soil. Land became the social cement which bound warrior lords to kings and kinglets. It also became the bond between lord and peasant in the wake of the Roman imperial collapse and the defeat of tribal institutions between the fifth and tenth centuries of the Christian era. Prosper Boissonnade's account of these developments stresses the triumph of petty aristocrats and their growing monopoly of social, economic, legal, and political privilege. It also accepts the "Germanic" thesis, that private property in land grew out of the decay of collectivism and the defeat of tribal communities. These developments had the blessing of the Church, which benefited from them, and helped pave the way for institutions of government powerful enough to halt the marauding of which Bloch wrote. Dependence, however, was the price of stability.

From *Life and Work in Medieval Europe*
Prosper Boissonnade

Colonization and the relative progress of agricultural production turned primarily to the advantages of the upper ranks of Western society. On the other hand, the free communities, which owned the soil in common in certain parts of Christendom, were little by little eliminated either wholly or partially from this ancient possession.

From the seventh to the tenth centuries the collective property of tribes and village communities, even more than that of family communities, was subjected to a series of severe shocks. Tribal property, the primitive form of agrarian collectivism, survived only in the Celtic countries, Ireland, Wales, and Scotland. The soil of Ireland still belonged in the seventh century to 184 tribes or clans, each of which possessed enough territory to pasture 3000 to 9000 cows, and which were subdivided into 552 districts called *carrows* of about 525 to 1050 acres each, each carrow being in turn subdivided into four quarters, and each quarter containing four family estates. As in Scotland and Wales the clan held the land in common, and was one in peace as in war. It had its petty kings, its chiefs and nobles and clients, but no man possessed any individual property save his household goods, and each held only a right of usufruct over his strip of the tribal domain. Heaths, forests, and pasturelands were kept for common use, and ploughlands were redivided periodically for a term among the family groups. There even persisted traces of the ancient co-ownership of cattle to the profit of the clan, although by the seventh century herds had become private property. In each district of Ireland the free population lived communistically in immense wooden buildings, protected by earthworks and divided into three galleries. The

SOURCE: Prosper Boissonnade, *Life and Work in Medieval Europe* (New York: Harper & Row, Publishers; London: Routledge & Kegan Paul, 1949), pp. 78–90.

people lived and fed there in common, seated upon benches, and all the free families of the district slept there upon beds of reeds. But even in these regions, where isolation had caused the perpetuation of primitive forms of civilization, tribal property was soon shaken by the formation of great domains by the tribal chieftains (*pencenedl*), and nobles (*uchelwrs* or *machtiern*), as well as by the building up of the wide lands of the Celtic Church and the organization of family property.

Everywhere else in the Christianized Germanic West, the decline of the collective property of the tribe was infinitely more rapid. In England, as Vinogradoff has shown, there is no trace of property owned collectively by the Anglo-Saxon tribes, now formed into states, or of the collective property of the shire or district. At most there are a few vestiges of common lands in the little division of the hundred, and a few survivals of a tribal régime in regions influenced by the Celts. But everywhere the old Germanic institution of the common property of the township survived. Forests, pasturelands, heaths, and bogs remained undivided among the members of the village community, who possessed equal rights of property and usage over them. Meadows and ploughlands were divided into lots, the former enclosed for part of the year, the latter lying in open fields. Each free member of the village community had the right to a certain number of long strips or "furlongs" of about an acre, separated from the others by bands of turf known as balks, so that each family enjoyed an equal share in the cultivated lands and those under fallow. After the corn had been harvested and the hay cut, these fields were thrown open to the cattle of all the family groups, as the common pastures were all the year round. The land was cultivated by the same methods and in common by the members of the rural community, who yoked teams of eight or twelve oxen to plough the soil. The collective property of the Anglo-Saxon village grew rapidly less in proportion as there were organized royal and seigniorial manors, which claimed forests and common lands for themselves, and left the village community with nothing but its old methods of co-operative cultivation and its periodical distributions of the fields.

Germany, breaking away more completely still from the régime of agrarian collectivism, knew tribal property no more, and was little by little giving up the system of collective village property called the mark. Diminished, on the one hand, by legal appropriations resulting on the breaking up of waste lands, and, on the other, by voluntary alienations agreed to by the village communities, and by the usurpations of princes and great landowners, the mark broke up in all the Germanic lands from the Elbe to the Rhine and the Scheldt. It maintained itself henceforth only in the form of numerous commons (*allmends*), over which, moreover, the village community often preserved only rights of usage. In the regions of Gaul, Spain, and Italy, where the mark system had succeeded in establishing itself, it was unable to survive, and left no trace save in the commons, in certain rights of usage such as common of shack over the open fields, and sometimes in the system of co-operative ploughing. The Roman peasants continued to be acquainted only with the public property of the state, which passed to the kings, and the communal property of the townships of freemen, the *vici*, which became less and less important as time went on.

Without suffering the same profound decay, family property in its primitive forms was obliged to undergo a modification under the action of the individualistic conceptions of Roman Law, and the influence of economic necessities, working in favour of individual private property. But at the same time family ownership regained at the expense of the collective ownership of tribe and village a part of the ground which it lost to individual ownership, so that its power was less severely shaken. Family property, at first limited to the possession of household goods, cattle, garden and house, and to a temporary usufruct in the arable holdings (Celtic *tate*, Anglo-Saxon *hide*, and Germanic *hufe*) of about 40 to 120 acres redivided at more or less regular intervals, was in the end extended to these holdings, which began to be held in permanent occupation. Family property was also increased by lands recovered from the waste by the family community, which thereupon became private property. At the same time, however, the old family property (the Anglo-Saxon *ethel*, the *terra aviatica*, *salica*), indivisible, inalienable, belonging to the whole group of relatives, reserved for male members only to the exclusion of women, cultivated and enjoyed in common, this ancient property which still existed in the Celtic and Germanic countries in the sixth century, suffered a series of shocks under the influence of the individualistic tendencies of Roman civilization. Between the seventh and ninth centuries, in these regions, the father received the right to name an heir, to divide up the land, and to make grants thereof. Wills became general. Women and girls were admitted to a share in the inheritance even of land. Alienations of the family domain were allowed within certain limits.

Thus the principle of private and individual property was built up and spread among the new races. Rising out of partition or succession, it was increased by the fruit of personal labour, of what were known as "aquisitions," and especially of assarts, or land recovered from the waste by the labour of pioneers. It received barbarian names, *bookland* in England, *alod* in Germany and Gaul, but it was fundamentally the same as the old Roman property, the *possessio* or *sors*, over which the individual has full rights, which was now triumphing over the primitive conception long surviving among the Celts and Germans.

This movement, which tended to transfer the ownership of the soil, almost the sole source of wealth, from collective groups, whether tribe, hundred, village, or family, to individuals, worked chiefly in favour of the classes which were then in possession of political and social power. It was not the small but the large property which benefited by the disappearance of communal ownership. The possession of the soil became the appanage of those who, in the division of social labour, had seized upon the functions of government and upon material and spiritual power. First of all the new rulers of states, kings and kinglets of diverse origin, built up great domains for themselves. In the Celtic countries they accumulated them in the shape of disinherited lands, the third share in all booty, and tribal lands. Elsewhere in the Germanic countries they acquired a large part of the soil by conquest, by judicial sentences, by the partial appropriation of the lands of village communities and marks, by seizure of the public lands of the Roman Treasury. Much the same thing happened on the territories of the old Roman Empire. Often the new powers thus got into their possession a third of the land, and in Lombard Italy, for a short time, they appropriated as much as a half. Everywhere, it is true, they squandered this treasure, but magnificent vestiges still remained in the ninth century. In the Italian peninsula a ninth part of the soil then formed part of the royal domain, and the Carolingian princes owned at this period 100 domains in Lombardy, 205 in Piedmont, and 320 in Alamannia, Bavaria, Thuringia, and Ostmark. Hundreds of thousands of square miles were in royal hands in the West. They were mainly composed of forests and waste lands, but there was also a good deal of cultivated land among them.

Side by side with this great princely property, the great Church property likewise extended from century to century, slowly built up, a piece here and a piece there, out of the munificence of kings and of the upper classes, even of humbler folk, and by dint also of reclamation and agricultural colonization. There are plain indications that about a third of the soil of Western Christendom belonged to the Church in the ninth century, although it suffered from the policy of secularization pursued by the first Carolingians, and from the frequent usurpations of powerful laymen. In England men saw kings endow forty abbeys at a stroke, and bestow upon them the tenth of their royal domains. In Christian Germany abbeys and bishoprics were overwhelmed with gifts. Prüm had 2000 "manses," 119 villages, and two great forests; Fulda, 15,000 carucates of land; Tegernsee, 12,000; St. Gall, 160,000 arpents. To the Abbey of Lorsch belonged 2000 "manses" and two large forests; to Gandersheim, 11,000 carucates of land. Hersfeld had 1702 domains. Many monasteries had no less than 1000 to 2000 carucates of property, and the bishoprics of Augsberg, Salzburg, and Freisingen owned from 1000 to 1600 "manses." In the words of a Merovingian king: "All wealth has been handed over to the churches." There have come down 72 grants by Charlemagne in favour of the churches of Germany; Louis the Pious made 600 grants to them, and the two first Ottos, 1541. The churches and monasteries in the old imperial lands were no less favoured. One Bishop of Langres owned a whole county. The Abbey of Saint-Remi of Rheims held 693 domains, and Saint-Germain-des-Prés had 1727, covering an area of 150,000 hectares. Saint-Wandrille, near Rouen, had 1727 "manses" and 10,000 subjects at the end of the seventh century, and 4824 domains in the ninth. Luxeuil numbered 15,000. The Abbey of Saint-Martin of Tours ruled over 20,000 serfs. In Italy 2000 "manses" were in the hands of the Bishopric of Bologna and the patrimony of the Papacy was a sort of vast state, the possessions of which were sown all over a part of the West.

The territorial power of the lay aristocracy was due to a combination of usurpation committed at the expense of communal property, violence employed against small proprietors, pressure exercised on the rulers of the states, and colonization. Already growing under the Roman Empire, it overran all bounds during the Dark Ages. It triumphed everywhere in the West from the Celtic lands, in which the *uchelwers* and *machtiern* created vast domains for themselves out of tribal and family lands, to the Roman countries, where the old and new nobility amalgamated and grew rich by despoiling short-sighted kings and the enfeebled ranks of the small proprietors. In England the nobles, earls, thanes, and ealdormen built up opulent manors with such success that in the eleventh century a few great families had succeeded in gaining possession of two-thirds of the soil of England. In Germany a Duke of Bavaria, in the eighth century, possessed 276 "manses" in one district and 100 in another. In the same country the head of the great Welf family, in the tenth century, owned 4000. It has been calculated that instead of 900 hectares, which was the average extent of a great Roman estate in Gaul, the great Merovingian estate was as large as 1800 to 2600. Moreover, the same person usually possessed several such estates. In the Carolingian period the lands of the aristocracy were often scattered in different regions. Nobles of the first or second rank, Gallo-Roman, Visigothic, and Lombard, thus succeeded in completing to their own advantage the process of concentration of landed property, which had been begun in the preceding period.

The possession of land assured to the bodies, families, and individuals who managed thus to monopolize it so great a power that it led to the formation of a new nobility, distinct both from that of Germanic and that of Roman society, but built up of elements borrowed from both and combined with more recent institutions.

Almost everywhere in the West, from the Celtic to the Germanic and Roman lands, the nobility of race or birth grew less, but a landed aristocracy grew greater under a variety of names, and began to coalesce with the aristocracy of service, formed of persons, sometimes of quite humble rank, who were attached to the king's service and with the aristocracy of high officials, to whom was delegated the exercise of public authority, and who gradually transformed their revocable function into a hereditary office. Thus the Irish, Welsh, and Armorican *uchelwrs*, *pencenedls*, *cinnidls*, *machtiern*, *baires* (cow-owners), the Anglo-Saxon thanes, earls, and ealdormen, the Frankish and Gallo-Roman *edelings*, *antrustions*, *nobiles*, *proceres*, *optimates*, dukes and counts, the Visigothic and Romano-Spanish *nobiliores*, *gardings*, *judices*, dukes and counts, the Lombard and Italian *gasindes*, *gastaldes*, dukes and counts—all these went to form a single class of lords or great men (*seniores*, *optimates*, *proceres*, *potentes*), which replaced that of the Roman senators and the ancient Germanic and Celtic nobility. Below them and in their service were grouped their agents, familiars, and men-at-arms (*familia*, *maisnie*, *comitatus*, *truste*), whom their patronage ennobled to such a point, indeed, that in Germany, the Low Countries, and Italy, and even for a short time in Gaul, simple serving men, the *ministeriales*, often of servile condition, were promoted to the rank of nobles. Sometimes freely, sometimes by force, sometimes in spite of the kings, and sometimes with their concurrence, the aristocracy, assuming to itself the protective rôle which, in civilized society, the state claims to exercise over individuals, subordinated landowners of small or medium estate by granting them their patronage, by spreading the custom of commendation among them and by enrolling them as vassals. Thus was created a whole hierarchy of domains and of free dependants, under the protection of the great landowners, on condition of a mutual exchange of services and the concession to the vassals of estates known as *beneficia* or *precaria*, which, although granted at first on revocable or temporary titles, very soon became hereditary. Finally, a new stage was reached in a part of the West when the state despoiled itself of the attributes of public power (administration, police, justice, the right to levy taxes and raise troops) by the concession of immunities, and the great landowner became not only a high official, but also a sort of sovereign in his own domain. Then it was that on the top of the seigniorial régime, which placed men in economic and social dependence upon the landed aristocracy, there was superimposed the feudal régime, which, in the ninth, and especially in the tenth, century, finally conferred political sovereignty upon the aristocratic class, at least in France and Northern Spain, if not in Germany, Italy, and England.

Already masters of the greater part of the land, the landed aristocracy thus became, in the course of 400 years, masters of the men who dwelt thereon. The great domain was indeed the solid basis upon which their power was founded. Preserving its fundamental integrity, despite the partitions which detached fragments from it, and bearing in its very name the memory of its chief owner, the *villa*, or *massa*, or *curtis*, or *saltus*, or *sala*, or *fundus*, or *manor*, as it was called, was a little kingdom, almost a little world, governed by a master, the lord, who was invested with absolute authority, and had in his possession all the elements necessary to economic existence. In it an organization, founded on a hierarchy of functions and divisions of labour, secured the satisfaction of all the needs of masters and subjects alike. At its centre rose the seigniorial dwelling-place (*palatium*, *fronhof*, *salhof*, *castellum*, hall), half castle and half farmhouse, often surrounded by a wall or palisade, rough enough in the Celtic and Germanic countries, but already more elegant and comfortable in the Romance lands. Everything was arranged for the sojourn of the master and his household. Farm buildings, stables, storehouses, cellars, barns, workshops, were all grouped round the hall. Everything was there, even a chapel for the life of the spirit, which was no more forgotten than the life of the body.

Economic organization reached the highest pitch of perfection on the great monastic domains, and then on the great imperial or royal domains. In the former, wherein a sort of communistic ideal prevailed, where the rule restrained individualism, and where all the monks were equals in the distribution of labour and of its produce, there reigned an inflexible discipline, which assigned to each his task and his reward, according to the principles of a sort of co-operative society for both production and consumption. The administration of each of the economic services was presided over by a cellarer, or else by a provost or dean. Each monk had his function, just as each subject of the monastery had his; gardeners, labourers, fishermen, foresters, swineherds, oxherds, or shepherds worked under the order of the foremen of each occupation. A strict economy ruled in the distribution and preservation of produce. Everything—harvests, instruments of tillage, and iron tools, even old habits and old shoes—was looked after with the greatest care. In the great royal domains, such as those of Aquitaine, to which Charlemagne's famous capitulary *De Villis* applied, the same organization was applied with less rigidity. Agents (*judices*, *majores*) or stewards there governed one domain or several, having under their orders special agents at the head of each service, and the whole population of subjects. The Lombard household, with its *gastald*, its *massarius*, its *domesticus*, and the Germanic *hof* with its *meier* or *vogt* present the same spectacle. Everywhere in the West identity of needs gave rise to similar organs in the great domain.

The lands of the great domain were divided into a collection of holdings worked by tenants, to whom their cultivation was confided (*terra indominicata*), and a section reserved by the lord to be farmed directly by himself. The latter was called the lord's demesne (*dominicum*, *terra dominicata*). It comprised not only the central nucleus of lands, on which stood the lord's dwelling-place, but also other lands scattered about over more or less distant parts of the estate. It was cultivated by the labour of serfs grouped round the central hall, or else by that of *coloni* and serfs, who were provided with separate holdings. The Abbey of Saint-Germain-des-Prés, for example, which reserved 6471 hectares as its own demesne, and divided up 17,112 hectares among its *coloni* and serfs, farmed the former by means of three days of labour exacted weekly from the tenants settled on the latter. The demesne usually comprised arable lands, meadows, and forests. That of Verrières contained 300 hectares of ploughland, 95 arpents of vines, and 60 of meadows, besides a large wood, and the demesne of Vitry-en-Auxerrois was on the same scale. It was in this way that the great landowner secured the very considerable quantity of the necessities of life which he needed. At Bobbio the demesne of the abbey provided the monks in the ninth century with 2100 hogsheads of corn, 2800 pounds of oil, 1600 cartloads of hay, and a quantity of cheese, salt, chestnuts, and fish. It maintained a large number of cattle, especially swine. All the food products—cereals, meat, oil, milk, wine—all the raw materials necessary to life—wool, flax, wood—and the greater part of the necessary manufactures came from the demesne, and from the rents imposed upon the tenants of land allotted by the lord. The latter and his household got food and clothing and everything which they needed from the demesne. Charlemagne himself lived with his folk in this fashion.

Finally, it was the great domain which was the centre of the social life of the upper classes. Great men dwelt there, sometimes in wooden habitations, rough and devoid of art, as in England and Ireland, sometimes, as in the Roman lands, in stone houses, where the old traditions of luxury lived again, more or less remodelled by Germanic barbarism. These great men there led a life of violent physical exercise, of hunting, and of great meals, mingled with more or less refined amusements, the characteristic life of all aristocracies at half-civilized epochs. Only an élite rose during the Carolingian period to the conception of intellectual pleasures. For the other members of the aristocratic classes, the sensual life of pure materialism, in its different aspects, remained the sole ideal to which they were capable of rising.

This violent, fierce, and ambitious rural aristocracy, which disciplined labour and reduced it to serfdom in the great domain, pursued with tenacity the destruction of the small free properties, which hindered its expansion, just as the independence of the small owners gave umbrage to its authority. Small free proprietors remained for a long time sufficiently numerous in the West to represent a social power, with which kings and great men had to count. Gallic *cymrys* and Irish *fines*, Armorican *boni viri*, Anglo-Saxon *ceorls*, Germanic *frilingen*, Burgundian *minofledes*, Gallo-Roman *ingenui* (free owners), Visigothic *allodiales*, Lombard *ahrimanns*, Italian *primi homines*, *bozadores*, they all had many traits in common. First of all there was the moderate size of their domains, no more than 120 acres (the *hide* or *hufe*) in England and Germany, and often less in the Gallo-Roman manor. But the small owner was absolute master therein, as absolute as the great man in his *villa* or his *curtis*. He had the right to enjoy all the possessions of the Germanic village community or the commons of the Roman townships. His property and his person were inviolable, and under the special protection of custom and law. He had the right to bear arms; he sat on the judgment seat with his peers; he was summoned to assemblies. He administered the affairs of his village or township, forming with his neighbours the common council, which is even found in Roman lands. But he was a butt for the galling jealousy of the great landowner, because his lands were intermingled with those of the latter and prevented him from rounding off his estate. In a single canton of the district of Salzburg, in the eighth century, 237 small estates were thus entangled among 21 ecclesiastical domains, 17 lands of vassals, and 12 great ducal properties. When all things else bent before the great territorial lord, the freeman with his independent nature, living proudly with his family in his little village, on his isolated holding, or in his Roman township, dared to raise his head and look his powerful neighbour in the face. Therefore, a duel to the death was waged throughout the West between the small proprietor and the great. Decimated by the wars in which it was obliged to take part, submitted to absorbing and onerous public charges, ill-defended by the royal power, which ought in self-interest to have leant upon it, and which only intermittently recognized that this was so, this middle class defended itself vigorously. Already in decay in Italy and Gaul, in the fifth and sixth centuries, it succeeded in building itself up again in the eighth, and became numerous and influential in the Carolingian Empire. The abdication of the central power during the following period; the weakness of Charlemagne's successors, which obliged freemen to commend themselves to the great lords by the Edict of Mersen (847), together with the shock of the last invasions, brought an almost complete victory to the landed capitalism of the time, represented by the great proprietors. It was not obtained without a struggle. We read of freemen grouping themselves together in associations or unions for mutual defence (gilds), organizing revolts at various points, rising in insurrection against the landed aristocracy, and outlawed by princes, who imagined themselves to be defending the social order. The small free properties, usurped by the aristocracy or alienated in its favour, were at last absorbed in the great domains, or else transformed either into benefices or into *precaria* and placed in dependence upon them. Nevertheless, little islands of free landowners managed to maintain themselves everywhere where physical nature and the power of tradition stayed the movement for the concentration of land, which was going on in favour of the aristocratic classes. In Lower Saxony, Frisia, the German Marches, maritime Flanders, in the regions of the Alps and Pyrenees, in Aquitaine and Southern Gaul, in the northern and eastern counties of England, in the high March of Spain, and in a few parts of Italy the population of small owners remained free and proud, a powerless minority in the midst of thousands of men who remained in subjection, or had been cast back into it.

The Organization of the Manor

The account of Boissonnade prepares us for the following selection—a description of the manor as the chief social and economic unit of agrarian society. The class of documents of which this is an example had its origins in Carolingian practices of administration. But the great age of the manor was the thirteenth century. Under the pressure of a rapidly growing population, seignorial elements expanded the demesne worked by peasants in most manors. Lords managed their tenants—who ranged in stature from abject bondsmen to prosperous peasants, free smallholders leasing estates—through the skills of managerial classes. Seignorial rights and peasant dues were fixed in rentals and estate survey subject to enforcement in the customary court of the manor. There the lord might sometimes take cognizance of cases himself, with the aid of a jury from the manorial tenants. Often his steward or bailiff (seneschal) who governed the demesne for him held the tenants to their obligations. This document tells us something of the routine of labor among ordinary people in that society. It also reveals the existence of the town (vill) as a self-governing community of peasants in many matters.

From *Seneschaucie*

Anonymous

The Office of Seneschal

The seneschal of lands ought to be prudent and faithful and profitable, and he ought to know the law of the realm to protect his lord's business and to instruct and give assurance to the bailiffs who are beneath him in their difficulties. He ought two or three times a year to make his rounds and visit the manors of his stewardship, and then he ought to inquire about the rents, services, and customs, hidden or withdrawn, and about franchises of courts, lands, woods, meadows, pastures, waters, mills, and other things which belong to the manor and are done away with without warrant, by whom, and how: and if he be able let him amend these things in the right way without doing wrong to any, and if he be not, let him show it to his lord, that he may deal with it if he wish to maintain his right.

The seneschal ought, at his first coming to the manors, to cause all the demesne lands of each to be measured by true men, and he ought to know by the perch of the country how many acres there are in each field, and thereby he can know how much wheat, rye, barley, oats, peas, beans, and dredge one ought by right to sow in each acre, and thereby can one see if the provost or the hayward account for more seed than is right, and thereby can he see how many ploughs are required on the manor, for each plough ought by right to plough nine score acres, that is to say: sixty for winter seed, sixty for spring seed, and sixty in fallow. Also he can see how many acres ought to be ploughed yearly by boon or custom, and

SOURCE: E. Lamond, trans., *Seneschaucie* (London, New York: Longmans Green, 1890).

how many acres remain to be tilled by the ploughs of the manor. And further he can see how many acres ought to be reaped by boon and custom, and how many for money. And if there be any cheating in the sowing, or ploughing, or reaping, he shall easily see it. And he must cause all the meadows and several pastures to be measured by acres, and thereby can one know the cost, and how much hay is necessary every year for the sustenance of the manor, and how much stock can be kept on the several pastures, and how much on the common. . . .

The Office of Bailiff

The bailiff ought to be faithful and profitable, and a good husbandman, and also prudent, that he need not send to his lord or superior seneschal to have advice and instruction about everything connected with his baillie, unless it be an extraordinary matter, or of great danger; for a bailiff is worth little in time of need who knows nothing, and has nothing in himself without the instruction of another. The bailiff ought to rise every morning and survey the woods, corn, meadows, and pastures, and see what damage may have been done. And he ought to see that the ploughs are yoked in the morning, and unyoked at the right time, so that they may do their proper ploughing every day, as much as they can and ought to do by the measured perch. And he must cause the land to be marled, folded, manured, improved, and amended as his knowledge may approve, for the good and bettering of the manor. He ought to see how many measured acres the boon-tenants and customary-tenants ought to plough yearly, and how many the ploughs of the manor ought to till, and so he may lessen the surplus of the cost. And he ought to see and know how many acres of meadow the customary-tenants ought to mow and make, and how many acres of corn the boon-tenants and customary-tenants ought to reap and carry, and thereby he can see how many acres of meadow remain to be mowed, and how many acres of corn remain to be reaped for money, so that nothing shall be wrongfully paid for. And he ought to forbid any provost or bedel or hayward or any other servant of the manor to ride on, or lend, or ill-treat the cart-horses or others. And he ought to see that the horses and oxen and all the stock are well kept, and that no other animals gaze in or eat their pasture. . . .

The Office of Provost

The provost ought to be elected and presented by the common consent of the township, as the best husbandman and the best approver among them. And he must see that all the servants of the court rise in the morning to do their work, and that the ploughs be yoked in time, and the lands well ploughed and cropped, and turned over, and sown with good and clean seed, as much as they can stand. And he ought to see that there be a good fold of wooden hurdles on the demesne, strewed within every night to improve the land. . . .

Let no provost remain over a year as provost, if he be not proved most profitable and faithful in his doings, and a good husbandman. Each provost ought every year to account with his bailiff, and tally the works and customs commuted in the manor, whereby he can surely answer in money for the surplus in the account, for the money for customs is worth as much as rent. . . .

The Office of Hayward

The hayward ought to be an active and sharp man, for he must, early and late, look after and go round and keep the woods, corn, and meadows and other things belonging to his office, and he ought to make attachments and approvements faithfully, and make the delivery by pledge before the provost, and deliver them to the bailiff to be heard. And he ought to sow the lands, and be over the ploughers and harrowers at the time of each sowing. And he ought to make all the boon-tenants and customary-tenants who are bound and accustomed to come, do so, to do the work they ought to do. And in haytime he ought to be over the mowers, the making, and the carrying, and in August assemble the reapers and the boon-tenants and the labourers and see that the corn be properly and cleanly gathered; and early and late watch so that nothing be stolen or eaten by beasts or spoilt. . . .

The Office of Ploughmen

The ploughmen ought to be men of intelligence, and ought to know how to sow, and how to repair and mend broken ploughs and harrows, and to till the land well, and crop it rightly; and they ought to know also how to yoke and drive the oxen, without beating or hurting them, and they ought to forage them well, and look well after the forage that it be not stolen nor carried off; and they ought to keep them safely in meadows and several pastures, and other beasts which are found therein they ought to impound. . . .

The Office of Waggoners

The waggoner ought to know his trade, to keep the horses and curry them, and to load and carry without danger to his horses, that they may not be overloaded or overworked, or overdriven, or hurt, and he must know how to mend his harness and the gear of the waggon. And the bailiff and provost ought to see and know how many times the waggoners can go in a day to carry marl or manure, or hay or corn, or timber or firewood, without great stress; and as many times as they can go in a day, the waggoners must answer for each day at the end of the week. . . .

The Office of Cowherd

The cowherd ought to be skilful, knowing his business and keeping his cows well, and foster the calves well from time of weaning. And he must see that he has fine bulls and large and of good breed pastured with the cows, to mate when they will. And that no cow be milked or suckle her calf after Michaelmas, to make cheese of rewain; for this milking and this rewain make the cows lose flesh and become weak, and will make them mate later another year, and the milk is better and the cow poorer. . . .

And every night the cowherd shall put the cows and other beasts in the fold during the season, and let the fold be well strewed with litter or fern, as is said above, and he himself shall lie each night with his cows. . . .

The Office of Swineherd

The swineherd ought to be on those manors where swine can be sustained and kept in the forest, or in woods, or waste, or in marshes, without sustenance from the grange; and if the swine can be kept with little sustenance from the grange during hard frost, then must a pigsty be made in a marsh or wood, where the swine may be night and day. . . .

The Office of Shepherd

Each shepherd ought to find good pledges to answer for his doings and for good and faithful service, although he be companion to the miller. And he must cover his fold and enclose it with hurdles and mend it within and without, and repair the hurdles and make them. And he ought to sleep in the fold, he and his dog; and he ought to pasture his sheep well, and keep them in forage, and watch them well, so that they be not killed or destroyed by dogs or stolen or lost or changed, nor let them pasture in moors or dry places or bogs, to get sickness and disease for lack of guard. No shepherd ought to leave his sheep to go to fairs, or markets, or wrestling matches, or wakes, or to the tavern, without taking leave or asking it, or without putting a good keeper in his place to keep the sheep, that no harm may arise from his fault. . . .

The Office of Dairymaid

The dairymaid ought to be faithful and of good repute, and keep herself clean, and ought to know her business and all that belongs to it. She ought not to allow any under-dairymaid or another to take or carry away milk, or butter, or cream, by which the cheese shall be less and the dairy impoverished. And she ought to know well how to make cheese and salt cheese, and she ought to save and keep the vessels of the dairy, that it need not be necessary to buy new ones every year. . . .

The dairymaid ought to help to winnow the corn when she can be present, and she ought to take care of the geese and hens and answer for the returns and keep and cover the fire, that no harm arise from lack of guard.

The Peasant Family
in Thirteenth-Century England

Neither technical innovations nor changes in the systems of cultivation provided much relief from labor and the rigors of subsistence in a heavily populated and land-poor agrarian society. The peasant farmer might be a prosperous free peasant or a quarter-virgater working from seven to ten acres. He might be Chaucer's Franklin whose table rained meat and drink, or a man hardly able to live on the margin of subsistence or even below it. Whatever his condition, he had to work his own allotment, yield dues in kind or money (or both), provide his lord with body services or agreed-upon substitutes, and use the labor of his family in order to live. When we speak of peasant farming, we have in mind chiefly the manorial society which thrived in the great open-field country of the European plains and low valleys. This "champion" area flourished especially in the English Midland counties—an area studied intensively by George C. Homans and described in his important book on English peasant villages of the thirteenth century. Homans's work is neither a history of agriculture nor one of rural society. It is rather a collective biography of a real village world inhabited by real people, focused on the family unit and its right of inheritance, and thus in itself a contribution to the sociology of "champion" England.

From *English Villagers of the Thirteenth Century*

George C. Homans

THE FAMILY IN CHAMPION COUNTRY

The last chapter in this study of the families of villagers in the Middle Ages must be a miscellany. It must fill in details; it must summarize; it must generalize. In the first place, the family cannot be seen apart from its setting: the house and the household. Besides his acres of tillage scattered over the open fields, a villager would have a messuage beside those of his fellows in the village proper. A messuage had at least enough room for a house and yard, outbuildings, and a garden. Houses were poor things, of a timbered framework filled in with wattle and daub—the ancestors of the English half-timbered houses. They were easily built and easily moved, since the posts and beams were the only materials which were costly enough to be worth moving. They were easily destroyed: we read of a burglar "digging through" the wall of a house in order to make a theft.

But the smallness and wretchedness of the houses of villagers can be exaggerated. For instance, the specifications of a house are given in the court rolls of the year 1281 of Halesowen, Worcs. It was to be thirty feet long between the walls and fourteen feet broad, with corner posts, three new doors and two windows. Such a house was not large, but it was a house of a peculiar sort. It was to be built by a man for his mother when she retired and he took over her land. In

SOURCE: George C. Homans, *English Villagers of the Thirteenth Century* (Cambridge, Mass.: Harvard University Press, 1941), pp. 208–219.

short, it was a "dower house," a dependance. The main house on the holding may have been much larger.

Houses varied, in the thirteenth century as today, with the wealth of their owners. In fact, the members of the village social classes often took their names from their houses. The villagers of the lowest class were called cotters after their cotes, which may have been not only smaller and poorer than the houses of the more substantial villagers but also different in plan. In some places the villagers of the middle class were called husbands, that is, they were the bonds who had houses rather than cotes. Even in the house of a yardling in a prosperous village, most of the members of the household may have eaten and slept together in the room which was dominated by the hearth. But such houses must have had other rooms. We have seen settlements made before manorial courts in which the holder of a tenement would agree that his aged father and mother should have a room at the end of his house in which to dwell in their old age. The holder himself and his wife must often have had a room of their own.

After looking at the house, we must look at the household. The household, rather than the family, was the actual working unit. The person who held the tenement, the holder, the *husbond*, was the head of the household and directed the husbandry of the tenement. He must have been responsible, after his father's death or retirement, for marrying off his sisters and providing his brothers with proper portions out of the profits of the tenement. Anyone who lived on the tenement, who took meals there, was his *mainpast*. That is, the husbond was held responsible for the good behavior of the members of his household. If they did evil, he was bound to produce them in court and even answer for the damages they had done.

Next in consequence after the husbond came the housewife, the head of the women's work in the household, and after her the husbond's father and mother, if they were still alive. If any of his brothers and sisters had decided to remain on the holding and unmarried, the husbond was bound to find them sustenance and must also have put them to work. Of the husbond's children, the most important was the son who stayed on the hearth, the son and heir. As he grew older, he must have become more and more closely associated with his father in the management of the holding, until at last his father was ready to retire and leave him in entire control. The other children fell into two groups, more distinct in men's minds then than they are today: those who stayed at home and had their sustenance from their father, and those who were making their own living. In widely separated parts of England, the latter were called by a special name, as if they were recognized as a separate class in the community. They were called *selfodes* or *sulfodes*. The derivation of this word is unknown—it may have come from such form as *selfhood*—but its meaning is made plain in the following passage from a statement of the customs of Cirencester, Glos., dating from about 1209:

> Villeins of Cirencester, that is, of the lord King, while they are under the rod and power of their fathers, and at mainpast of their fathers and mothers, their parents will acquit by the bidreaps which the latter do for the lord King or his farmer. But as soon as the same villeins have their own free power, and live by their own labor, and are made *sulfodes*, then each of them ought to do three bidreaps for the lord King.

These bidreaps (*precarie*) were the days in harvest when everyone in the manor was bound to turn out to reap the lord's corn.

The customs of the bidreaps, as they are recorded for different manors, have much to tell about the parts played by the various members of a village household in the economy of the holding. For one thing, some members of a household were released from the duty of coming to the bidreaps. Presumably their other work was so important that it could not be neglected. Foremost among them was the housewife. Another person often exempted was the family shepherd. Some villages supported common shepherds, but in others each holding must have had its own. Other persons sometimes exempted were the "nurse" and the "marriageable daughter." Perhaps these two were in fact the same person: that is, one of the older daughters of the family had the duty of taking care of the younger children. Still another was the "master servant." These exemptions give some hint of the size of households and the number of duties to be performed on a large holding.

Besides the children and other kinsmen of the holder, two classes of servants, or *hewes*, to give them their old English name, might be members of a village household. One class was made up of those servants who slept or took their meals in the main house of the holding. Whether men-servants or maids, they were likely to be unmarried. We must remember the *anilepimen* and *anilepiwymen*, the single men and single women who are mentioned in the custumals as hired farm hands. These house servants would be allowed food, clothing, and lodging, besides perhaps being paid something in coin for their work.

The other class was made up of those servants who were given dependent cottages on the messuage. They might well be married and keeping house for themselves. They were commonly called *undersettles*, that is, they were settled under the holder of the tenement. Such a name shows how medieval people thought of society as a hierarchy of classes one above the other, the members of the lower class holding their lands of members of the one next higher. This scheme of thought was most elaborately worked out in the customs of the members of the military classes and is now called the feudal system, but the feudal system in its main lines was the scheme of thought of all members of society.

The undersettles were sometimes called coterells, lesser cotter. Such men had to do a share of the works owed to the lord from the holding. In a custumal of Aldingbourne, Sussex, drawn up in 1257, the statement of the services due from a yardling at the lord's works in harvest specifies that:

> He shall send to the oat boonwork all his hands, and anyone who holds a cottage of him shall come with his hands to the boonwork.

Again, at the harvest works of Haddenham and Cuddington, Bucks.:

> All the villeins ought to come with their entire households except their wives and shepherds. And if the tenant has two men, he ought not work; if he does not have two, he will work. And if any one of them has an undersettle, let him come to the first bidreap.

That is, if the tenant had two able-bodied men in his household, he was exempted from working himself. At Brandon, Suffolk, the arrangements were similar, according to a passage in the Ely custumal of 1277:

> It is to be known that every undersettle or anilepiman or anilepiwyman, holding a house or a *bord*, no matter of whom he holds it, will find one man at each of the three boonworks of harvest, at the lord's food.

The *bord* of this entry links these men with the *bordarii* of Domesday Book, where it is hard to distinguish between the *bordarii* and the cotters (*cottarii*). A *bord* was a small house, a cottage. These special words for houses suggest that there were various traditional kinds of villagers' houses, differing in plan as well as in size.

It is possible that the position of an undersettle resembled that of a Scotch cotter in later centuries. If it was, the holder of the tenement on which the undersettle lived gave him a cottage and an acre or two of land and helped him in the tillage. In return the undersettle worked for the holder as a farm hand. But often an undersettle was simply a man who leased a house or land from the holder of a tenement. So he was in 1326 at Littleport, Cambs., in the Fen country, when the following entry was made in the court rolls of the village:

> The jurors present that outsiders who come in and hire their houses of various people and hold nothing of the lord common in the fen with their beasts and take other benefits in common, and these are called undersettles.

These undersettles were outsiders, holding no land directly of the lord of the manor. Yet they enjoyed rights of common which ought to have belonged only to true villagers, and the court wanted to know by what warrant they did so. It appeared that each of them reaped in harvest half an acre of the lord's corn and did other light services in return for his privileges. Any person who occupied a small part of a tenement held by another man and rendered his share of the rents owed to the lord from the tenement may have been called an undersettle. Under this definition, the heir's brother or sister on whom he had settled a few acres of the family holding would have been an undersettle. Such sub-tenants are often mentioned in manorial records. For instance, at Pavenham, Beds., about 1270, John, son of Robert Cerne, held one yardland and had as tenants William West, Simon West, William Reyn', and others. How could the yardland support all these men, besides the chief holder and his family? We do not know, but we know that this yardland was nothing exceptional. The Hundred Rolls show that many another holding of that size or even smaller was crowded with minor sub-tenants.

Four classes of persons, then, might have lived on a village holding in champion country. There was the immediate family of the holder: himself, his wife, and their children. There were others of his kin: his mother and father, his unmarried brothers and sisters. There were the serving-men and maids boarding in the house. And there were undersettles. On any particular holding, one or more of these classes might not be represented, and if a holding were small its population would be reduced. But many a yardling must have kept a numerous household.

One way of summarizing the traditional organization of families in the champion country of England during the Middle Ages will be to compare it with a similar traditional organization in another time and country. For this purpose a good description exists of peasant families in the Luneburg district of northwestern Germany as they were in the middle of the last century. A typical family of prosperous peasants in the Luneburg possessed a compact holding of land with a house and farm buildings upon it, the *hof*. Such a holding was supposed in custom to be indivisible, and some of the *höfe* had in fact remained undivided and in the hands of the same families for many centuries.

It was through the working of the customs of inheritance of the country that the *hof* remained in the family which had always held it. The customs provided that the *hof* should descend at the death of its last holder to one of his sons, and in default of sons to one of his daughters. In some places, the eldest son inherited, in some the youngest; in some the peasant himself selected the son who was to have the *hof* after his death. Commonly the eldest son succeeded.

If he had the right to choose, the peasant selected his son and heir in good time and associated this son with him in the management of the farm. Often the choice of an heir and the heir's marriage went together: the two were parts of the same event. To understand what happened, we must bear in mind the arrangement of the rooms of the *hof*, which was traditional and determined by the family customs. The door of the house opened into a large room which served as kitchen and living room for the whole household. On the left of the kitchen was the "grandfather's room," the room occupied by the former head of the household, and another room where prayers were said and family meals were taken. On the right were the rooms of the peasant and his children and those where the servants slept. When the heir married, his father moved from his former quarters into the "grandfather's room," and the heir and his wife took over the rooms his father had left. From that day on, the father took a smaller and smaller part in the management of the farm and left more and more to his heir. Thus each generation in its appointed time took a definite place both in the house itself and in the household.

Families were large. When the brothers of the heir, those who were not to inherit the land, came to man's estate, they could do one of two things. They could leave the *hof* to seek their fortunes or they could remain in the *hof*, but if they took the latter course they had to remain unmarried. For all those who left the *hof*, the peasant felt bound to provide portions out of the savings of the farm. In the same way he provided dowers for the daughters of the house when they married.

There were many likenesses between this family and household of the Luneburg of north-western Germany in the nineteenth century and a typical villager's family and household in the champion country of England in the thirteenth. Some of the phrases of the English court rolls even suggest that in that country as in Germany, when the father turned the holding over to his heir, he moved from the room which he had formerly occupied into another one traditionally reserved for the old folk. The chief difference between the Luneburg family and that of medieval England is that the land belonging to the one was a compact area, while the land belonging to the other was made up of strips scattered all over the fields of a champion village. The custom of both families was that the holding was not to be divided or alienated. These likenesses are only what we should have expected to find. From north-western Germany came the Angles and Saxons who settled the champion country of England. From north-western Germany they brought their language and their social order. This order was still nearly intact in the England of the thirteenth century. In later times it changed considerably, while in the German homeland, in a district which remained somewhat isolated and entirely given over to husbandry, it was preserved much longer. Nor were Germany and England the only places where families of this sort were to be found. They were at the heart of the ancient social order in many parts of north-western Europe.

This traditional family organization had great virtues. There is a danger of considering the rules of inheritance, life-tenancies, and alienation simply as rules of law in abstraction from the conditions under which men live together. When the rules are studied as they were followed in village life, they are seen to fall into a consistent system of custom, according to which the rights of every member of a family in the means of subsistence possessed by the family were established from birth to death and from generation to generation. Every child knew what he had to expect and knew that if he were once given the means of making his living he was secure in holding them. Some certainty and security for the future are necessary to men if they are to be useful members of society.

But presumably there is more than one traditional family organization—indeed we know there is—which provides this certainty and security. The particular family organization which we are studying had virtues of its own. Adapted to a society in which land was the paramount source of wealth, it insured that in every generation of a family at least one man, his wife, and their children had a decent subsistence from every established holding of land. But it retarded any further increase of population which would press upon the means of subsistence: the sons who were not to inherit a holding had to get out or remain unmarried. It preserved a stable social order at home. We have already seen how the intricate dispositions of a champion village were preserved in part by its customs of inheritance. If all went by rule, the same holdings, whatever their rank, would remain in the same families generation after generation. At the same time it supplied plenty of men for the rough work at the frontiers of society. The great French sociologist Frédéric LePlay felt that families of this particular type, which he called stem- or root-families (*familles-souches*) because like the root of a vine they were continually sending forth new shoots, which were continually cut back —he felt that these families had been one of Europe's great sources of strength. LePlay observed that these families steadily provided, for colonization, for trade, for war, landless men with their fortunes to make. And so far as we can tell from its history or judge from similar instances, this admirable organization was not thought out by some primitive Lycurgus and imposed by him on his countrymen, but must have been the product of a long train of interactions between the interests and sentiments of men and the external conditions of their lives.

In some societies the family relationships are extended and generalized until a man considers himself in some way kin of every other member of the community. If a stranger cannot show that he is a kinsman, he is regarded as an enemy and treated as such. Thus in ancient China a village was also a clan. The English place-names in -*ing* suggest that English villages may once have been of this sort: Reading means Red's people or descendants. And the court rolls sometimes speak of "the blood of the village." But upon the whole there is little evidence that the inhabitants of an English village in the thirteenth century thought of themselves as a body of kinsmen. Indeed the stem-family organization may have emphasized the distinction between the family and the community. When the marriage of the heir customarily coincided with his taking over the management of the holding, the history of the blood which held the tenement became the history of a single sequence of small families, each pulsing in its time, each dying away. Since in every generation the sons who were not to inherit land either left the holding or did not marry, the number of actual kinsmen could not increase rapidly. In inheritance the immediate family was favored at the expense of what later became the aristocratic tradition of descent in the strict male line: in default of sons, a tenement was often divided among the daughters of the last holder, and only if the last holder left no children did it go undivided to his brother.

The terms used to reckon and indicate kinship often reflect the facts of family organization. Accordingly in north-western Europe they supported the emphasis put upon the small family group consisting of a man, his wife, and their children. In reckoning closeness of blood relationship, men reckoned by degrees of descent from an original mated couple, a man and a woman. In like manner, the terms for distant kinsmen were made merely by adding prefixes to the terms for persons closer to the small family. There were then, as there are now, great-grandfathers and grandfathers, great-uncles, grandsons, and cousins of different degrees. The point is that the small family was the unit to which the terms for more distant kinsmen were referred. This method is not the one used in many societies.

There was some extension of terms of kinship to unrelated or distantly related members of the community. Friends of the family are still called uncles, aunts, or cousins by courtesy, and in medieval England a pleasant old fellow was probably the uncle of every boy in the village. But this process was carried less far in the stem-family organization than it is in some societies. A further circumstance which may have weakened in the thirteenth century a man's consciousness that he had a large body of kinsmen was the instability of family names. A man was by no means sure of bearing the same last name as his father and his brothers, but was likely to be called Hugh's son, after his father, or Reeve, after his office, or Marden, after his home village, or Atwell, after the place of his house in the village, or Turnpenny, after some personal characteristic, without regard for the names his ancestors had borne. It is true that in that century and later ones the use of family names became more and more settled. Indeed many of the present English family names date from the thirteenth century.

In some societies, then, the larger social groups, the village, the clan, the tribe, are extensions of the family group. They are in fact, or are held to be, groups of kinsmen. In the society of the champion country, family and village tended to be two bodies different in kind. Better than many another society could have prepared them, this society prepared western Europeans and their descendants for the social isolation in which many families live in the great urban agglomerations of today.

The distinctions which were made by English lawyers in the Middle Ages between the different tenures and between common law and local custom have obscured the fact that most men lived under rules which were alike in essentials. If Kent, parts of East Anglia, and the West be left out of account, the different systems of family law were all much alike. Any tenement, a barony or a burgage as well as a yardland, descended to one son of the last holder. In default of sons, it was divided among the daughters. It was subject to the life-tenancies of free bench and courtesy. In their main lines, the rules of family law were the same for the earl and the husbandman, and for the burgess too, though he was somewhat more free than the others to sell or devise his holding. Where the rules differed, it was only in non-essentials, such as the particular fraction of the holding the widow was to enjoy as her bench. In important matters they were the same, and in particular they all held to what we have seen as the central principle of the organization of medieval families: the permanent association of a given blood line with an established tenement unit.

Any member of society, bond or free, belonged to a family much like all other families in its organization. In what concerned his family's tenement, whether that was a cotland or an earldom, he felt the same sentiments as other Englishmen and thought in terms of the same schemes. How strong a force this uniformity was in holding together the classes of society, in maintaining society as an organism functioning harmoniously, we cannot estimate. But certainly the constitutional historians have not been sufficiently concerned with the fact that in ancient and stable societies the organization of the government is often a reproduction, with a difference, of the organization of the common family. Thus in Norway, where the custom of much of the country was that land was held in common by the sons of the last holder, the kingship also in ancient times was held in common by all the sons of the last king. Naturally the system engendered periodic warfare. In England, on the contrary, the realm descended to one of the king's sons and one only, just as any man's tenement descended to one of his sons. Clearly this factor will not sufficiently account for the establishment in England and other countries of the principle of primogeniture in the descent of the kingship, but it must at least be taken into consideration. The constitutional historians speak of the anxiety engendered by the king's alienation of some of his rights or a part of the demesne of the crown as if it were simply a fear of the taxation which might become necessary when the king could not live of his own. There was more force in the anxiety than this theory accounts for. When the king made an alienation, the subject felt a little of the irritation he would have felt if his father had alienated a part of his family's tenement. A king of England, then as now, had to take care to avoid doing violence, not only to the persons and possessions of his subjects, but also to their moral sentiments.

We have been concerned here with the rules governing the disposition of family possessions. These matters are the only ones which the surviving records— for the most part legal records—can reveal in detail. As a result, we understand the main lines of family organization, but much remains unknown. Court rolls have little to say about the day-by-day life of the family, about the play of the children, the work of the men and women, the company around the hearth, the behavior of the family in the births, marriages, and deaths of its members. They have little to say about the language people used in addressing their kinsmen and in talking about family affairs. They have little to say about the relations between the family and the rites of the parish church, although the central object of worship in the Middle Ages was a holy family: a compassionate mother, a stern but just father, and their son. This kind of knowledge is likely to be forever barred to the student of a dead society.

Technology, Water Mills, and the Burdens of Labor

We have already noted Marc Bloch's central concern for the relationships of land, labor, and wealth as they are molded by the environment. But Bloch also understood how these raw materials were subject to human inventiveness; hence his abiding interest in technology, amply illustrated in his studies on animal traction, the stirrup, and the application of waterpower to do work once done by men or beasts. Innovations in technology had far-reaching effects on the nature of work and social organization. Changes in the plow and harness expanded food production and enhanced the material culture, even the forms of village life. The development of the stirrup revolutionized warfare and helped bring on the dominance of warrior-aristocrats. Naturally, technological change often created human stress and dislocation. This we know well enough in our own lives. Historians of the use of the water mill in making cloth in England have shown how the need for rapid streams caused the relocation of the industry, from the low-lying southeast to the mountain stream-courses of the northern and western counties. Bloch explored the impact on peasant life of milling grain by waterpower, emphasizing how new tools created new trades and the specialization of labor. He also deals explicitly with the social conflicts produced by the triumph of one mode of production over another, as the mill became a standing symbol of the lords' monopolistic powers over peasants.

From *Land and Work in Medieval Europe*

Marc Bloch

THE ADVENT AND TRIUMPH OF THE WATER MILL

When the first water-wheels began to turn in the rivers the art of grinding cereals, in Europe and in the civilised lands of the Mediterranean, was already more than a thousand years old. In the beginning, the procedure must have been rudimentary in the extreme, the grain being simply crushed with rough stones. But in prehistoric times—we are not here concerned with the exact moment or place—the invention of real tools brought a decisive advance. Then came the pestle and mortar, or a stone roller moving to and fro on a long flat surface, as depicted in Egyptian statuettes, worked by women, usually in a kneeling position. Next came the revolving millstone. Invented in the Mediterranean basin—perhaps in Italy—in the course of the two or three centuries before the Christian era, it had found its way into Gaul shortly before the Roman Conquest. It too could be worked by manpower, and often was in fact. Although if Samson, whom the Bible depicts as grinding corn for his masters the Philistines, certainly never turned a millstone because it was still unknown in Palestine at the period when the story of this worthy Strong Man was written, there were later on in the Ro-

SOURCE: Marc Bloch, *Land and Work in Medieval Europe*, trans. J. E. Anderson (London: Routledge & Kegan Paul, 1967), pp. 136–160.

man world innumerable slaves—and even some free men, such as Plautus in his penniless youth—who braced their muscles to this monotonous task. But the new invention made it possible for the first time to substitute animal for human effort in the grinding of corn—usually the horse or the ass. When Caligula on one occasion requisitioned all the horses in Rome, bread became scarce because there was no means of turning corn into flour. But the same invention made possible another and much greater advance. The simplicity and regularity of rotary motion compared with the complicated movements required by the older method opened up the way for the use of a force which, although more blind than any animal-power, naturally moves in one constant direction—namely the power of running water. Without the *mola versatilis* there would never have been a water mill.

These two stages did in fact follow in fairly quick succession. There was a water mill about the year 18 B.C. at Cabira in Pontus among the outlying buildings of the palace formerly built by Mithradates, and no doubt contemporary with the buildings as a whole. In that case this would be the earliest specimen of precise date—from 120 to 63 B.C. A Greek epigram generally attributed to the Augustan Age features some nymphs who are grinding corn, and the expressions used in it make it quite clear that the water goddesses have only recently been obliged to submit to this enslavement. About the same period, the Latin author Vitruvius describes the apparatus in detail; a little later Pliny notes mill-wheels in the rivers of Italy. Although, as usual, these texts do not throw any light on the actual moment of birth of the invention, it cannot be mere chance that they are massed together in such a short space of time. All the indications point to a narrowly restricted period—the last century before the Christian era—and in all probability to the Eastern Mediterranean as the place of origin. It is significant that Vitruvius should only know the new machine by its Greek name, *hydroletes:* and from Greece it must rapidly have spread to Italy.

These conclusions are supported by what we know of its history in the rest of Europe. On the rivers of Gaul, the first mills to which our documents vouchsafe a reference are the ones turned by a small tributary of the Moselle in the 3rd century. In southern Germany, these contrivances spread sufficiently quickly and widely after invasions to claim the attention of the Alaman and Bavarian laws from the first half of the 8th century onwards. In the north, in regions less open to Gallic and Roman influence, they spread more slowly. Our documents clearly show the broad lines that they followed. There were slaves, like the Bavarian who was taken captive by the Thuringians about the year 770 and built his master a mill; there were colonists, like those Frankish warriors whose village, founded in 775 on the Unstrut, was given the evocative name of Mühlhausen; there were monks and nuns, like the inmates of Tauberbishofsheim, who settled in the great Odenwald forest about the year 732. The construction of water-wheels went ahead as fast as immigrants came in with the technical skills of their own countries. In Great Britain, there is no known example before 838. In Ireland, the legal collection of Senchus Mor mentions water mills in the 9th and 10th centuries. According to a legend which was probably not far from the truth, the oldest of these was said to be the work of a foreigner summoned for that express purpose "from beyond the seas." Among the Slavs of Bohemia and on the shores of the Baltic, this invention, though familiar for more than a thousand years to the waterside populations of the Mediterranean, does not seem to have penetrated before the 12th century. There, too, it had made its way from west to east, following the immigration routes. We have evidence of this in an incident reported in a chronicle by a priest of Holstein. Some Saxon peasants had settled during the 10th century in Schleswig and Wagria. Their settlements were later destroyed in a Slav offensive. When almost two hundred years later another group of Germans arrived and re-settled in the country, they came upon traces of the previous occupation, amongst which were the embankments that had been raised to form ponds for working water mills. From this point onwards, then, these were among the most characteristic material signs of Western civilisation in any territory that came to be colonised. And lastly as regards Scandinavia, water mills were introduced into Denmark in the second half of the 12th century, and into Iceland round about 1200; but they hardly became general in Nordic societies until the 14th century. There must of course be a good deal of uncertainty about this information. We would never dare to assert that in such and such a year among this particular people the mill-wheel began for the first time to be turned by water-power. But the few landmarks upon which we are forced to rely do nevertheless provide us with a sufficiently lively image of events. They provide an almost regular series of isochromes which radiate without the slightest doubt from the Mediterranean world. Moreover linguistics come to the aid of history at this point. In the Germanic and Celtic languages, and even in certain Slav languages, the word for water mill was directly or indirectly borrowed from the Latin.

There is, of course, good reason to be astonished at first sight by the undoubted Mediterranean origin of this great technical improvement. For the uneven flow usually characteristic of rivers in this kind of climate would not seem to make them suitable for the production of motive power. On the other hand it is equally true that they did not suffer from the disadvantages of frosts and ice which often interrupted the supply of flour under more northerly skies, once the water mill had come into almost general use. Nevertheless the apparent anomaly still remains. But it may perhaps be possible to resolve it. Nothing is more certain than that the revolving millstone was the creation of Mediterranean civilisations. Now this first invention—for which no one will be tempted to find an explanation in geographical determinism—did, as we have seen, condition the second. But there is more to it still. A water-wheel driven by a current is adaptable to many other uses besides the turning of a millstone. More particularly, if fitted with scoops attached to its rim, it can actually pick up water and deliver it into any given basin or irrigation channel. This is still the practice, as it was at the beginning of our era, among the Mediterranean peoples. Nay, it goes back to a very remote past. Strabo, to whom we owe the mention of the oldest known water mill at Cabira, was the first writer to make express reference to these water-lifting wheels. He saw them at work in Egypt, where, although unknown in the time of the Pharaohs, they seem in fact to have spread widely under Roman rule. A little before this date, however, in a passage that is unfortunately obscure, Lucretius appears to allude to them. Let us however go back to Vitruvius. He classifies the mill among the machines for drawing up water—rather a peculiar method of classification, which one is tempted to explain in terms of some historical reminiscence. In more precise terms, he describes it as follows: a wheel with scoops attached to the rim, simply worked by a man's foot; then a wheel with scoops attached, but also having water-vanes, so that it can be turned by the water-power of the river; and finally the evolution into a water mill. This order might well represent the relationship between the three contrivances. In other words we should probably see in the water mill the further development after a short period of an invention which had been contrived in early times to facilitate irrigation, and which was naturally at home in regions where agriculture was always one long struggle against the drought of summer. It must be freely admitted that this is no more than a hypothesis of a most conjectural kind; but it will serve at any rate to give at least a provisionally satisfactory answer to one of those "why's" that are both the torment and the delight of the historian's craft.

The first and most obvious effect on the social order of this technical advance was a new step forward in the specialisation of skills among artisans. The tool created the trade. In the days when a Greek poet described the villages wakening at dawn to the sound of grain being crushed by the pestle, and later on when the revolving millstone had been introduced, the preparation of flour in the country-side was the domestic task of the slave or the housewife; in the large towns, it was one of the tasks of the baker. *Pistor*—"the grinder": this remained to the very end the Roman baker's name. The millstone, worked by hand or by horse, figured among his familiar attributes on monuments, and in his shop, along with the oven, as one of the instruments of his trade. For water mills on the other hand millers were needed. Their build, clearly differentiated from all others, appears at Rome for the first time on an inscription in 448. It is not part of our present purpose to trace the history of this profession, whose nature varied moreover to a large degree according to the time and the place concerned. At first, as at Rome, the miller was a member of a guild; then generally he became a *sergent* or sei-gnorial farmer; finally he became a self-employed craftsman who was his own master, for the miller in ancient Europe enjoyed great variety of status. He pro-voked a good deal of hostility too, often expressed in the small talk of an idle moment. "An art or a science? Is the miller's an honest calling?"—this was the question that occupied the first pages of Hans Hering's *Traité singulier des moulins*, which the "philosopher of Oldenburg" wrote in 1663. The German proverb was no more in doubt about the matter than Chaucer: "Why do storks never nest on a mill? Because they are afraid the miller will steal their eggs." But let us turn aside from the echoes of these ancient village spites, however instructive they may be in their dogged persistence, their overtones still continuing to sound in the *cahiers* of 1789. In any analysis of former rural societies, as of the middle classes who had so often risen from the ranks of the small craftsmen peasant, the miller is always seen to have his allotted place alongside the innkeeper or the cattle-dealer. And all this, thanks to the ingenious mind which first entrusted the millstone to the "water-sprites."

But it is more especially in the history of technology that the initiative taken by this nameless pioneer constitutes a date of real importance.

The generations immediately before ours, as well as our own, have witnessed a tremendous revolution in transport, animal traction giving place to purely me-chanical forms of energy. Not very different was the revolution that took place in another sphere with the coming of the water mill. But in the course of all this progress towards relieving the animate world of physical effort, the history of which more or less summarises the essentials of technical evolution—iron replac-ing wood, coal supplanting charcoal, chemical dyes taking the place of cochineal and indigo—in the course of this increasingly direct control exercised by man over elemental natural forces, without recourse to the power of animals, the ad-vances made shortly before the birth of Christ were, in one sense, the most decisive of all. For the force thereby harnessed was one of the most familiar and most easily used, as well as one of the most powerful; it is the self-same force that our turbines aim at harnessing today. Its importance lay moreover in the fact that the creature who stood to gain by saving his muscle-power was man just as much as the animal. Finally, it was important because it was the first such step, and was in effect to be the only one up to the time when the steam engine was invented. For a wheel fitted with vanes could with only slight modifications transmit its motion to a great many other contrivances as well as millstones. Olive-presses and tanning-mills were no more than simple applications of the process of crushing and grinding by stones. But it was only a short while before the invention spread widely. The hydraulic saw dates from at least the 3rd century. By the 11th century the first fulling-mills recorded in our texts were already echoing with their heavy and urgent throb through some of the alpine valleys where the last survivors of the race were still at work in the 20th century. The bellows and the smithy's trip-hammer do not seem to have been much later in their appearance by the riverside. Then other multifarious and novel uses were devised, so much so that the earliest manufactories of the 17th and 18th centuries, whose machines used a system of interconnected wheels roughly similar to what Vitruvius describes, and were driven by water-power, were really nothing more than descendants of the ancient water mill; and in England they long continued to be called "mills."

But that is not the end of the story. For the internal mechanism of the water mill also represented a step forward in the equipment of humanity whose signifi-cance far surpassed the fairly modest history of milling. This was not, of course, true of all water mills. Up to quite recent times one could see working, in various regions where mechanisation was still fairly primitive, mills with horizontal mill-wheels, fixed at water level, and connected by means of a simple rigid beam to the moving millstone placed immediately above it. The existence of this singularly rudimentary type raises some vexing problems. Scattered as it appears to be from one end of our world to the other, in regions as remote from one another as Syria, Roumania, Norway and the Shetlands, it is hardly possible to ascribe its invention to any specific civilisation. Moreover it is totally different from the apparatus described in the clearest of the ancient texts like Vitruvius. Hence we are led to wonder whether this is not a genuine example of a technical regression such as might well have occurred among peoples accustomed to a very crude level of material existence. It may well have seemed easier to them to imitate the action of a force like that of water, well known to everyone, rather than reproduce mechan-ical contrivances that had already reached some degree of complexity. Whatever the truth may be about this hypothesis, the question is clearly still an open one, and would be worth going into more fully. But there is no doubt at all that the Greco-Roman mill contained a vertical wheel. It seems originally to have been quite often driven from underneath, "by the river as it flowed on its course," as Pliny puts it. This is also how Vitruvius describes it. But quite early on—for this is the picture we already get in the epigram in the *Anthology*—a system of simple canalisation made it possible, when required, to let the water flow on to the vanes "towards the top." Now in one way or another this arrangement con-fronted the constructors with a mechanical difficulty, the original magnitude of which tends to be concealed from us today by the common spectacle of an all-too-cunningly mechanised world. The difficulty was that the movement, in passing from the vertical millwheel to the necessarily horizontal millstone, had to change its plane. An arrangement of cogwheels made this possible, thus introducing a principle with an immense future ahead of it, of which the mill was one of the very first examples.

About the beginning of the Christian era, the Greco-Roman civilisation, whose consumption of flour was enormous, had therefore at its disposal for the production of this essential foodstuff a piece of machinery that had already been brought to a remarkable degree of perfection. It was in fact the first machine whose use seemed capable of ameliorating the lives of countless numbers of human beings. The astonishing thing is that, having it at their disposal, they were so slow at bringing it into general use.

For—let us make no mistake about it—although the invention of the water mill took place in ancient times, its real expansion did not come about until the Middle Ages. It is possible to collect significant evidence of this fact relating to Gaul: one reference in the 3rd century; another round about the year 500, showing the machine to be still exceptional; five to my knowledge in the 6th century (one of which is contained in the Salic Law, which suggests the existence of a great number of mills); and finally a great many references during the rest of the Frankish period. The disproportion between the amount of documentary material —quotations included—of the Roman period and of Merovingian age is not so great as to be considered merely fortuitous. But the most telling example is provided by Rome itself. Under Caligula, as we have seen, flour was supplied to the city by horse-driven millstones. In order to find references to water mills, we must —despite the relative abundance of texts—pursue the matter as far as the middle of the 4th century. At that point there is a reference to mills at the Janiculum, fed by a stream from Trajan's aqueduct, which continued to appear in documents of all kinds right up to the 7th century. From then on they were considered indispensable to the life of the population: witness the care with which emperors and Gothic kings vied with each other in ensuring that no water from the mill-races should be channelled off for other uses; witness above all the predicament in which Belisarius found himself during the siege of Rome by Totila through the destruction of the water conduits. The only defensive measure he could take was to set up improvised millwheels in boats floating on the river Tiber.

Now this failure to develop to the full the possibilities of technical devices that lay ready to hand was not an isolated instance in the ancient world. "Rome," writes Gautier, "did not exercise over the forces of nature any dominion comparable with the development of her political organisation." I fully agree; but one may be allowed to wonder whether Rome did in fact really desire any such dominion. And if we should reach the conclusion that she was not particularly keen to exercise such a dominion, it may not be altogether impossible to understand the reason why.

Suetonius relates that when Vespasian was rebuilding the Capitol which had been burnt down during the last of the civil wars, an artisan put before him proposals for a machine that would have allowed the columns to be cheaply transported to the top of the slope. The prince rewarded the inventor but declined the invention. "Let me still be allowed," he said, "to give the people a livelihood." This anecdote is instructive on more than one count. Greco-Roman civilisations were well enough endowed with quick eyes and ready minds, and it is unlikely that they lacked the quality of technical inventiveness; consider for example the ingenuity displayed in their siege-engines or their methods of house-warming. Nor were the generations contemporary with the first millwheels stupid enough not to realise that all progress in mechanisation would economise in human muscle-power. "Spare your hands, which have been long familiar with the mill-stone, you maidens who used to crush the grain. Henceforth you shall sleep long, oblivious of the crowing cocks who greet the dawn. For what was once your task, Demeter has now handed on to the Nymphs." This epigram from the *Anthology*, which we have already invoked as a witness, might well deserve to be selected as the motto of a society in a position to make machinery a source of joy and dignity for mankind. There is the same note of rejoicing, though expressed in less poetic terms, in the work of an agronomist like Palladius. Much later on, and under very different skies, Irish legend traced the origin of the first water mill to a king's love for a beautiful captive, telling how he would fain spare his mistress the fatigue of turning the millstone when she was great with child. But the ancient world hardly seemed to feel the need to spare human effort because, in relation to its agricultural capacity, it was at the beginning of the Christian era very thickly populated. Moreover, although these rough tasks could have been accomplished by an impersonal natural force, it was the custom to hand them over to a body of labour that was among the cheapest and the commonest of any of the resources of that age. For it must of course be realised that the case of building works for the Capitol was exceptional. Rome, hypertrophied in its economic functions, saw its streets seething with a hungry proletariat whom the governing classes were only too glad to help find a living by employing them in public works. People like this would not have been willing to push the millstone, which was no doubt why the majority of the City's mills were worked by horses. Elsewhere, on the great estates, the millstones were not usually turned by paid workers nor by horses or asses, nor—as Pliny tells us in a reminiscence about even more primitive methods—did people pound the grain in antiquated mortars. This hard labour belonged to the slaves, sometimes men, more often women, the sisters of those menials to whom the poet in the *Anthology* so mercifully promised repose. The lords of the great *latifundia*, who were much less sympathetic to the burdens of humble people, had no reason to instal expensive machinery when their markets and their very houses were overflowing with human cattle. As for more modest households and for bakers, who would in any case have been unable to afford such heavy expenditure, many of them were quite well enough off to have their own domestic slaves; or else they did their own work themselves. In the large towns like Rome, no doubt, water mills would have been of the greatest service. But as is common knowledge, an invention rarely spreads until it is strongly felt to be a social necessity, if only for the reason that its construction then becomes a matter of routine.

Now this necessity was just becoming apparent towards the end of the Empire. In general the population was declining, there were in particular difficulties in the supply of slave labour; there was a tendency for the great gangs of slaves formerly fed directly by the master to be broken up and their members dispersed among holdings that were separated from the domain—though this is not the place to enquire into the reasons for these phenomena. It is enough for us to accept them as we find them—solid facts, among the most incontrovertible of all those that have dominated the evolution of European societies in the period between Antiquity and the Middle Ages. The proof that men were then beginning to run short of manpower for the millstones lies in the fact that they thought of supplementing the supply of slaves by using condemned criminals. Up till then, the latter had only been put to work in the mines. Constantine was the first to add another alternative to this ancient penalty, that of forced labour in the public mills. As a palliative it was obviously of limited use, and, even in its own field, clearly insufficient. It was better to turn to a machine that had long ago been thought out, but had only been very incompletely exploited. We know what was in fact done. Perhaps the invention owed its birth to some individual flash of genius. But effective progress lay in transforming the idea into practical reality and this only took place under the pressure of social forces. Just because these two stages seem to have been so sharply marked off from one another, the history of the water mill, being part of the general history of technology, has the peculiar merit of being a spontaneous event; and it illustrates features of development which are more or less universal.

However, we must beware of imagining that this was a victory achieved at one stroke. Before the ancient processes of grinding by animal or human power finally retreated in face of the water mill, and later on the windmill, and even in face of steam flour-milling—for in some places the struggle continued as late as this—they were to have a further lengthy span of life, marked by bitter social conflicts. Unfortunately this story, which we shall attempt to sketch, is veiled in considerable obscurity.

Hand mills, grain crushed by the hands of men—these terms which frequently occur in our texts are ambiguous. Leaving on one side the stone roller, which does not seem to have had a very long period of use in the West, we are left with the mortar and the revolving millstone. How are we to tell which of these two methods, one still very rudimentary and the other already much more sophisticated, is being referred to on each occasion in the texts? To set and maintain a heavy stone in motion is such hard work that one might be tempted to rule it out in cases where a woman is shown grinding corn. But nothing of the sort: for there are various accounts of an exceptionally clear nature which describe slaves or housewives "pushing" or "turning" the millstone. Moreover the *molae trusatiles* of early Roman times, which could only be heavy because they were at first confined to the homes of the rich, where they were worked by slaves or horses, were soon replaced even in classical times. As rotary millstones made their way into the humblest dwellings, so the *molae trusatiles* were replaced by smaller, less powerful and lighter models for domestic use. Taken in their strict sense, expressions like *manumolae* suggest the idea of a millstone rather than a mortar. But how can we have much confidence in the precision of a vocabulary which we know on other counts to have been so hazy? One can hardly say more than that the language of the texts, reports by certain travellers at periods not far removed from our own day, and the testimony of some rare objects that have unfortunately never been inventoried, give the impression that the millstone gradually supplanted the old prehistoric device which had held the field so long. It is better frankly to admit our ignorance of the early periods. It is an ignorance that furnishes a good example of the difficulties lurking in the path of the historian of science as soon as he attempts to come to grips with the facts. Moreover, the old types of mill handed down from classical times to the Middle Ages can surely not have remained without improvement in the course of so many centuries. And this holds good whatever the motive-power may have been. By a strange piece of application of the most recent to the most ancient device, hand mills or horse-mills seem sometimes to have been provided with a system of cog-wheels. But here, too, our sources nearly always leave us in the dark. We shall therefore be forced in what follows to adopt a necessary though inconvenient schematisation, and simply compare and contrast the different types of apparatus according to the nature of their motive-power.

There was one important and elementary reason which for a long time slowed down the victory of the water mill. Throughout the world there are regions which lack rivers and streams. As transport difficulties made it impossible to rely upon a supply of flour from mills situated at any distance, people lacking water power had no alternative but to content themselves with the ancient methods, at any rate until the day when their problems were solved by an even newer invention—the windmill. It was probably borrowed from the Arab world, coming into the West about the end of the 12th century, and thereafter making rapid progress, at least in northern France, during the next ten or twelve decades. When we find that a land-register at Orsonville in 1360, in the dry district of the Beauce, notes wind-mills and horse-driven mills side by side, it does not need a very bold imagination to conjecture that the latter had preceded the former. We must add to this the fact that not all water-courses were equally suitable for turning water-wheels; moreover even the best of them could not escape being frozen over, or flooded, or even becoming dried out. The Abbot of Saint-Alban was a wise man when he repaired the monastic water mills in the 13th century, and replaced one of them, whose feed-channel had dried up, by "a very fine horse-driven mill." Even in 1741, although the suburbs of Paris had an abundance of water mills on their rivers, and windmills on their hills, the Controller General, remembering both the great frosts of the previous winter and the floods of the year before, invited the city to equip itself also with hand mills.

But it was doubtless not only the vagaries of nature that prompted these ministerial recommendations. Traditionally it was considered prudent to guard against siege. There was not a single fortress in the Middle Ages that did not have its hand mills. Philip Augustus was careful to see that the castles on which he lavished so much attention were provided with them. The inhabitants of Nîmes, when they were putting their defences in order in the year of the battle of Poitiers, placed "ten or twelve" of these contrivances within the walled city. And it was not a question of guarding against an imaginary danger: the inhabitants of Parma, when besieged for months on end by the Emperor Frederick II, had all their water-courses and canals cut off, and would have succumbed to famine if they had not had hand- and horse-driven mills at their disposal. Thus it came about that warfare, upsetting as usual all normal economic conditions, constantly forced men back upon ancient and rudimentary techniques.

Finally the need for mobility seemed for a long time to make portable machines advisable. It was only natural that mills—evidently hand-worked—should have been loaded on to the Carolingian army waggons at a period when enormous tracts of territory—particularly in Germany—existed in ignorance of the water mill. It is stranger to see the Norman merchants as late as the 13th century providing themselves with this kind of equipment in their peregrinations. No doubt this was due to other than strictly technical considerations. One of these was of an economic nature: in many places, bread and flour were only supplied one day at a time, and the traveller often had no other course open to him but to buy and even to import unground corn. The other reason is connected with a social institution whose importance will become clear later on: the privately owned mill enabled people to avoid the dues for grinding as they went on their travels. These seignorial rights (*banalités*) could be very burdensome.

But apart from these admittedly exceptional cases, the fact remains that even where water was plentiful, and even where there was no risk of warfare, the old contrivances continued to do duty for a long period. We must no doubt take into account during the first centuries of the Middle Ages the slowness with which any innovation was likely to spread, the continued existence of considerable bands of slaves in the service of important people, and finally the habits brought with them from their original homes by the barbarian conquerors. The slave who ground the corn was looked down on with peculiar contempt, and was always less effectively protected in her life and in her honour than her sisters employed in household duties; and she remained for a long time a familiar figure in the home of the German chieftain—witness the Northern sagas and the ancient laws of Frisia and of Kent. In the royal *villa* of Marlenheim in Alsace, quite close to the clear-flowing waters of the Mossig, hand mills were still in use at the end of the 6th century, worked by female servants. But fairly rapidly, at least in ancient *Romania* and the adjacent districts of Germania, wherever natural conditions were not unfavourable to the change, hand- and horse-operated mills disappeared from the great domains. It was an accomplished fact in Gaul at the time of the Carolingian *Polyptyques* and the *Capitulare de villis*, and in England at the time of Domesday Book, all of which, for those who have ears to hear, are loud with the music of the millwheel. But there was one out-of-the-way place left where the old-fashioned methods were still carried on—namely the homes of the peasants.

Let us imagine the various conditions that would be required for the establishment of a water-driven mill. Not only did it require the legal right to draw on the water; but the costs of construction and repair prevented its being profitable unless there was a sufficiently large quantity of grain to grind. It is a striking fact that among the earliest-mentioned mills in our documents, many of them—from the 4th century onwards in Rome, from the 6th at Dijon and Geneva—were intended for supplying urban populations. The Roman ones were managed by a corporation under very strict State control: the later Empire had no intention of entrusting the capital's food-supply to private initiative. We do not know the status of those at Dijon and Geneva. But there is clear proof that there was no difficulty in making them a paying concern. In the country, there may well have been collective bodies administered by the village communities. This system may perhaps have existed in Ireland, where the ancient tribal structure of society was favourable to group effort. In the barbarian kingdoms, on the other hand, there is no documentary evidence of any such system. If Bavarian law held mills to be public places, this was not on the grounds that they were common property. Even if this latter description could have applied to some of them, there were certainly others to which it was not applicable, at any rate those constructed by the monasteries. Now the law lumped them all together indiscriminately under the same label; they were deemed "public," and therefore enjoyed a special "peace," granted by whatever master they may have been dependent on, because they embodied a common effort by a number of men for a purpose that was worthy of protection, in the same way as a market, for example. Even where—as in Frisia —the community was exceptional in managing to avoid being stifled by seignorial authority, the peasants only took advantage of their liberty to remain obstinately faithful to their own individual mills. They were not prepared to come to a friendly agreement with one another and adapt technical progress to their own requirements.

All the mills whose history we can more or less follow were in fact seignorial in origin. Many of them were dependent upon monasteries, which were already fairly numerous. In addition they were obliged to feed an equal and often larger number of servants, domestic vassals and passing travellers. Hence their consumption of flour was considerable—about 2000 *muids* a year (more than 9 tons at Corbie in the 9th century, according to Abbé Alard's estimate, which left out of account the provision of food for guests). We can be quite certain that no possible economy in manpower would have escaped their attention, even though under the strict Rule the monks were themselves in duty bound to carry out the heaviest tasks. Ascetics like Germain d'Auxerre and Radegonde might well take upon themselves as a mortification the heavy and menial work of grinding corn; but the wise abbot of Loches preferred a water mill, which "allowed a single brother to do the work of several," thus setting free a whole group of pious souls, probably for the work of prayer. About this same time, Cassiodorus boasted, among the advantages of the site he had chosen for his model foundation of a *Vivarium*, the fact that it had a river suitable for watermills. There can be no doubt that these monastic constructions—as the story of the Saint of Loches bears witness—often served as an example to lay lords. They, too, maintained on their estates imposing numbers of armed retainers and agricultural servants. In order to be able to feed such large numbers, all manors, whether they were ecclesiastically owned or not, possessed some demesne land, cultivated directly by the lord, as well as dues paid by the tenants, largely in the form of agricultural produce. Once it was harvest-time there would be growing piles of corn, ready for the mill. It is probable moreover that from this point onwards an appreciable part of the revenue of the seignorial mill came from the dues levied on the peasants of the surrounding district, whether or not they were tenants, who found it convenient to have their own corn ground at the central mill. Perhaps local despots were already trying to transform this permission into an obligation. But as yet custom did not support their efforts. This would be the explanation, for example, of the tenants of the Saint-Bertin monks in the 9th century, the serfs of Saint-Denis at Concevreux in the 10th century, and no doubt many others like them, whose humble story is not mentioned in any document, continuing to grind their corn at home, more often than not with their own hands.

From the 10th century onwards, however, a profound change took place in the economic and legal framework of rural life. Using their power of command— which was called the "*ban*"—and fortified by their right to deal out justice, a right whose growth was at that time facilitated by the absence of any adequate machinery of justice in the State, the lords—or at least a large number of them— succeeded in setting up certain monopolies very much to their own advantage, monopolies concerning the use of the baking-oven, the wine-press, the breeding-boar or bull, the sale of wine or beer, at any rate during certain months; monopolies in the supply of horses for treading out corn, when this custom prevailed; and lastly—probably the most ancient and certainly the most widespread of all—a monopoly over the mill. Now society at this period was on principle inclined to confuse the ideas of what was just with what was customary. Novel claims were soon converted into custom, and the *banalités* soon became an integral part of seignorial right, and remained so as long as the seignory survived. (In Canada, where the French social system had been imported under the Bourbons, they lasted as late as 1854.) From this point onwards the lord's mill was the only one where tenants of land on which it was erected were allowed to grind their corn, subject, of course, to a respectable payment to the lord of the mill and the mill-stream. Sometimes—since rights regularly overlapped—this obligation would even extend to villagers living in neighbouring lordships whose lords were too feeble or too unskilful to succeed in winning this privilege on their own account. For French legal theory in the 13th century with its tendency to schematize was inclined to regard this as one of the highest judicial rights (*haute justice*)—a right that belonged to none but these. Thus it came about that when from the 11th and 12th centuries onwards the great demesnes began to crumble away, and a little later on when money payments gradually began to replace the payment of dues in kind, the seignorial mills, which under the old free system would have risked standing idle, were certain of a long and useful life since they were guaranteed a steady supply of customers bound by hard custom to bring them their corn.

As may be guessed, such compulsory rights did not prevail without a struggle. It was no very difficult matter to prevent the construction of other water mills, windmills or even horse-driven mills on the lord's land. A little vigilant policing and timely agreements were quite enough to prevent the peasants from taking their corn to any local competitors. But a serious obstacle of another kind arose from the multiplicity of domestic mills, which had gone on steadily working down the centuries in almost every cottage. The lords decided to declare war upon them.

It will unfortunately never be possible to give a detailed account of this long dispute as far as the greater part of France, and perhaps also Germany, is concerned. This is at any rate true as far as I have been able to ascertain from a somewhat incomplete enquiry made more incomplete by one of the most unfortunate gaps in the technical equipment of historians in France. I refer to the regrettable—let us bluntly call it the ridiculous—habit which allows editors of charters to deprive their readers of any subject index. As if these collections only existed to provide the genealogists with tables of proper names with which to do their gymnastics! The absence of testimony in any quantity seems however quite certain. Moreover for France in particular the silence of the documents is easily explicable. This country revelled in seignorial rights (*banalités*). They not only extended over a greater number of activities than elsewhere, but they scored their greatest successes at a remarkably early date. Now by reason of this very precocity the period of their establishment, covering roughly the 10th and 11th centuries, happens to coincide with a time when documents are more scarce than at any other period of the Middle Ages. When source-material once more became abundant, the decisive stage of the struggle was already over. A very great piece of luck enables us to see the monks of Jumièges, in an agreement dated 1207, breaking up any hand mills that might still exist on the lands of Viville. The reason is no doubt that this little fief, carved out of a monastic estate for the benefit of some high-ranking *sergent* of the abbot, had in fact escaped for a long period the payment of seignorial dues. The scenes that took place in this corner of the Norman countryside under Philip Augustus must have had many precedents in the days of the last Carolingians or the first Capetians. But they escape the meshes of the historian's net.

The victory was not however complete by the end of the Middle Ages, for many old hand-operated mills still existed here and there, in more or less intermittent use. If they had not still been a familiar object in the 14th century, we should surely not have seen the Ecorcheurs (brigands) stationed by the Dauphin in Alsace, forcing their prisoners, as the slavemaster had forced his charges in former times, to turn the millstone. In the towns, people whose rank put them outside seignorial jurisdiction did not disdain on occasions to go in for a little hand-milling at home. An example of this was a canon of Montpezat in Quercy, in the years before 1380. But the most important reason of all is that in the countryside seignorial authority, harassing though it was, was very poorly served. It was therefore often incapable of acting with that continuity which alone would have made it possible to reduce the peasants, past masters in the art of passive resistance, to complete submission. At least in certain regions it was left with a great deal to do when, during later centuries, new forces came into play, destroying the routine of country life.

In Germany, the Sovereigns of the territorial states had gained possession of a considerable number of seignorial rights which had always been less split up there than in France, and had, perhaps from the very start, been recognised as royal privileges. They threw all their energies into the business of supervision, as did the Prussian State for example in the 18th and at the beginning of the 19th century, in Westphalia, Pomerania and Eastern Prussia. But in this last province hand mills were such an inseparable part of the equipment traditionally dear to the heart of the Slav population that it was sometimes only possible to forbid their use to German settlers, and for the rest of the inhabitants simply to limit their number.

In France, the struggle was taken up again with more powerful weapons during the period of feudal reaction in the 17th and 18th centuries. An effective ally was found in the great judicial bodies that were such citadels of privilege. One after the other the *Parlements* of Dijon and Rouen pursued the hand mills. The struggle was specially bitter in Brittany, where country life remained for a long while—and even down to our own times—very little mechanised and still peculiarly primitive. Attachment to the ancient methods of grinding seems moreover to have been much less widespread in western Brittany, entirely Celtic in language, than in the eastern cantons, largely Gallic, and at that period probably the most impoverished. The villagers would seem by their own accounts to have fallen back upon these practices only in times of drought or when the common mill was not working properly, or for milling buckwheat, which it appears that millers were often in the habit of refusing to grind. There can however really be no doubt that the competition of so many "hand mills" hidden away in the cottages must have made serious inroads on the profits from the legal monopoly. For that matter, the lords did not so much claim to suppress them as to make the use of them subject to the payment of a due. The very multiplicity of the successive decrees passed by the *Parlement* of Rennes shows how dogged the resistance must have been. Among the various forms of "feudal tyranny," this particular one still roused some of the most lively of all the registered protests in the Breton *cahiers* of 1789.

But it is in England especially that the war of wind and water against human muscle is seen in its clearest light.

Manorial rights were not an institution native to England. The Norman conquerors had imported them for the continent as one of the principal elements in the "manorial" system which after the almost total dispossession of Saxon aristocracy they methodically established, by superimposing it on what remained of a much looser form of dependence. It is true that in England the system of seignorial monopolies always remained less complete than on the other shore of the Channel. But manorial rights over the mill were generally introduced, though not without resistance. Opposition was all the more passionate in this country because by reason of its remoteness from Mediterranean influences, and the strong impress of German and Scandinavian civilisation, the water mill, though familiar from the end of the 11th century on the great estates, only won its way very slowly among the middle classes. It is characteristic that among the privileges granted to English burgesses there should frequently figure this clause, totally unknown in French and German urban charters, allowing the use of the hand mill, or more rarely the horse-driven mill. This was the state of affairs at Newcastle, Cardiff and Tewkesbury during the 12th century, and in London even in the middle of the 14th century. But the pressure of rich and powerful communities was needed to overcome this tolerance. "The men shall not be allowed to possess any hand mills"—such was the clause inserted by the canons of Embsay in Yorkshire between 1120 and 1151, in a charter in which a noble lady made over to them a certain water mill. It expressed their own attitude and that of all lords of rivers and manors. There were occasions when milling stones were seized by the lord's officials in the very houses of the owners and broken in pieces; there were insurrections on the part of housewives; there were law-suits which grimly pursued their endless and fruitless course, leaving the tenants always the losers. The chronicles and monastic cartularies of the 13th and 14th centuries are full of the noise of these quarrels. At St. Albans they assumed the scale of a veritable milling epic.

In this small Hertfordshire town, to which the monks who were lords of the place obstinately refused to give any privileges, the example of neighbouring citizens stirred up—in the words of the monastic chronicler—a particularly "indomitable tenantry." They were a collection of artisans rather than peasants, and it was not only the dues for milling grain or malt and the miller's exactions that they sought to avoid by milling at home. The drapers among them also claimed, in defiance of the lord's fulling-mill, the right to set up their own fulling stocks for the pressing of material, at least for the coarser kinds of cloth, for it was generally considered at this period that fine materials must be fulled under foot. The first quarrel broke out in 1274, accompanied by the usual incidents. Millstones and lengths of cloth were confiscated; there was mutual violence perpetrated by the lord's officials and by tenants; a league was formed among the inhabitants, who clubbed together to establish a common purse to maintain their cause at law, whilst monks, barefoot before the High Altar, were chanting penitential psalms; attempts were made by the women to win the queen over to their side, but the abbot had taken the precaution of having her smuggled into the monastery by a secret entrance. Finally there were lengthy proceedings before the royal court, ending inevitably in the defeat of the recalcitrant party, who sought to appease their offended lord by the gift of five fine barrels of wine. Another incident took place in 1314. Then in 1326 the citizens demanded a charter which should contain among other clauses the right to domestic milling. This led to an open insurrection, in which the monastery was twice besieged. Final agreement was only reached under pressure from the king, but it left the problem of the lord's monopolies unresolved. Taking advantage of this uncertainty, the inhabitants soon had anything up to eighty hand mills working in their homes. But in 1331 a new abbot—Richard II, the terrible leprous abbot—entered the lists. He won the day by going to law. From all over the town the millstones were brought in to the monastery, and the monks paved their parlours with them, like so many trophies. But when in 1381 the great insurrection of the common people broke out in England and Wat Tyler and John Ball emerged as leaders, the people of St. Albans were infected by the same fever and attacked the abbey. They destroyed the notorious paved floor, the monument to their former humiliation, and as the stones were doubtless no longer any use for grinding, they broke them up and each took fragments of them as a sign of victory and solidarity, "as the faithful do on Sundays with the holy bread." The deed of liberation which they extorted from the monks recognised their freedom to maintain "hand mills" in every home. The insurrection however proved to be like a blaze of straw that soon burns itself out. When it had collapsed all over England, the charter of St. Albans and all the other extorted privileges were annulled by royal statute. But was this the end of a struggle that had lasted more than a century? Far from it. The chronicler, as he draws to the close of his story, has to admit that for malting at any rate the detestable hand mills have come into action again and have been again forbidden.

They were destined to give humble service throughout the length and breadth of England for a long time to come. It is true that the narratives we possess hardly make any further mention of them. But here and there a manorial "record" allows us to lift the veil for a moment. The Tudors had long ago succeeded the Plantagenets when in 1547 the people of the royal manor of Kingsthorpe obtained recognition for their right to grind at least a certain quantity of their grain at home. At Bury in Lancashire, it was not until the restoration of the Stuarts that the lord of the manor succeeded in suppressing competition with his own mill; and even so the obstinacy of those representing the parish made the suit drag on till 1713. This was only seventy-three years before the first large-scale flour-mill was opened in London.

In short, when the iron and coal age opened, the ancient prehistoric tools had nowhere yielded altogether to the "engines" which for so many centuries had also relied upon the inanimate forces of wind and water. There is therefore nothing surprising about the survival in working order of hand mills in Ireland, Scotland, the Shetlands, Norway, East Prussia, and nearly everywhere in Slav territory, right up to the end of the 19th century; and even perhaps in our own day they have not altogether ceased to function. For these regions, situated as they are on the fringe of the West, had in all respects long been faithfully wedded to a fairly rudimentary degree of mechanisation. Prussian villagers were still grinding grain in 1896 according to the elementary methods of their ancestors, and felt obliged, like them, to hide from strangers as they did so—as though the lord's monopolies were still in existence. In these actions we can recognise, not only the dim potency of tradition, but also the fact that in the North winter frosts were not very favourable to the use of running water-power; moreover, in the Shetlands, Norway, Scotland, and even Ireland, there was no seignorial authority comparable to that prevailing in France. But even in the heart of our own civilisation, a more searching enquiry would no doubt reveal more than one scattered example of a similar survival. The Breton hand mills, whose history might well tempt a scholar more familiar with the province than I am, were surely given something of a new lease of life by the suppression of the seignorial regime. In the second half of the 19th century Lamprecht observed near the tributaries of the Moselle, whose waters had turned some of the earliest millwheels in Gaul, material traces of the relatively recent practice by which, with the aid of human power, corn was crushed between two revolving stones.

These anomalies should not however be allowed to lead us astray. When the steam engine arrived to put the finishing touches to the defeat of the hand mill and mortar, by far the greater part of the flour consumed in the countryside and in the towns of the West had for centuries been milled by wind- or water-power. No doubt the peasants, if left to their own devices, would have clung still longer to the old ancestral ways. And by imposing heavy milling dues seignorial lords, owners of the manorial mills, may sometimes have encouraged unintentionally a fidelity to the past; but in the end they destroyed it through the use of force. In more than one respect these seignorial undertakings were not unlike some of our great present-day commercial enterprises. First of all they saw themselves constrained by the shortage of manpower to bring about this great improvement in human equipment; then they harshly imposed the system on all around them. Thus technical progress resulted from two constraining forces; and this, we may be quite sure, is not the only example.

A Cahier of 1484 and Kirchmair's Story of Geismayr's Dream: The Poor and Peasants as Petitioners and Rebels

In *1984* George Orwell began his story of oppression by stating a paradox: "Until they are conscious, they cannot rebel; until they rebel, they cannot be conscious." Centuries of life among the proletarians of town and country and even among the petty possessors of medieval society taught the same lesson. Medieval people were not only bound to an Ixion's wheel of economic necessity, they were also harnessed to a world view which stressed subordination, passivity in the face of oppression, and a heavily moralized view of the special providence which shows in the fall of a sparrow. Yet they did protest and, when driven beyond the limits of human endurance, they broke the cake of customary obedience—they rebelled. The two documents grouped together here show these twin strains of populist action.

The petition of the Third Estate in the Estates-General of 1484 was drawn up from the many *cahiers* sent up by local communities. The people suffered from royal exactions and seignorial exploitation. Both the "spider king," Louis XI (1461–1493), and the French nobility grasped that money was the sinew of power. And so they taxed. Clergy and nobility secured exemptions, while on the Third Estate the full burdens fell: taille, gabelle, and others too numerous to mention. Thus the maturing of royal authority after a century of civil war and domestic discord restored peace but not freedom from oppression to France's people. The army became a symbol of Valois dynastic ambitions in Italy. French power and French diplomacy were both costly. Only the expanded system of taxation paid the bills. But for most Frenchmen glory came at too high a price—in hardship and often in humiliation at the hands of aristocrats. It is this double sense of grievance we find reflected in the people as petitioners.

While the French poor appealed to their king in representative assemblies, the German peasants found such courses blocked. Throughout the fifteenth century the power of the bishops waxed. By the early sixteenth century a history of German peasant insurrections against petty tyrants could be written: in Bohemia in the 1440s; in the manifold risings of the Bundschuh between 1502 and 1517; and in smaller, sporadic campaigns throughout southwestern Germany between 1450 and 1490. Elsewhere in Europe great risings had occurred: in Flanders in the 1320s; in the Ile-de-France in the 1350s; and in the Peasants' Rebellion of 1381 in England.

Protest and this record of risings gave witness to peasant consciousness of oppression. The revolts—in and out of Germany—make for interesting comparative studies. Spontaneity was one general characteristic. Great brutality among rebels and repressors was another. Lack of sophisticated political programs was a third. Perhaps the most interesting common facet, however, is that of ideology. Norman Kohn's *Pursuit of the Millennium* drew attention to the consistent peasant appeal from the justice of this world to what was just in God's sight. Peasant movements were thus often linked to heresy, especially heresies which promised the Second Coming now, or some literal realization of Paradise on Earth. This use of the language of religion should not surprise us. Nor ought we to be shocked by the excesses of peasants acting in what they thought was God's cause —not if we recall the Crusades and the ritual slaughter of Jews in times of crisis.

Before the Peasants' War (1524–1525), the millennial expectation was prominent in peasant ballads, popular revolutionary sermons, and rebel catechisms. In the account of Michael Geismayr and his "plan of reform" given by George Kirchmair, we recognize the peasant longing for freedom and a return to some fancied golden age of justice and equality. The notion of the spiritual freedom of Christians in peasant hands became more worldly.

From *Journal des états généraux de France tenus à Tours en 1484*

THE PLIGHT OF THE FRENCH POOR

■ 1484 ■

For the third and common estate, the people of the three estates declare that this kingdom is at present like a body which has been drained of its blood by various wounds, to such an extent that all its members are empty. And just as the blood is the nourishment of the corporal life, so the finances of the kingdom are the nourishment of the commonwealth. The members are the clergy, the nobles, and people of the third estate, who are drained and denuded of resources; and there is no longer a bit of gold or silver in the said members except for those who have been near the king and have shared in his benefactions. And to understand the cause of this extreme poverty of the kingdom, it should be known that for eighty or a hundred years this poor French body has been drained almost continuously in various and pitiable ways. . . .

As for the little people, one could not imagine the persecution, poverty, and misery they have suffered and still suffer in many ways.

First of all, since that time [1461, the death of Charles VII] no region has been free from the continual going and coming of armed men, living off the poor people, now the standing companies, now the feudal levies of nobles, now the free archers, sometimes the halberdiers and at other times the Swiss and pikemen all of whom have done infinite harm to the people.

And one should note and consider with pity the injustice and iniquity suffered by this poor people, for the men of arms are hired to defend them from oppression and yet it is they who oppress them the most. It is necessary for the poor labourer to pay and hire those who beat him, dislodge him from his house, make him sleep on the ground, deprive him of his substance; and yet securities are granted to the men of arms to preserve and defend them, and to protect their goods!

And the iniquity of this practice is clear enough. For when the poor labourer has worked all day long in weariness and sweat of his body, and has gathered the fruit of his labour, from which he expects to live, they come to take from him part of the fruit of his labour, to hand over to someone who will perhaps beat him

SOURCE: James Bruce Ross and Mary Martin McLaughlin, eds., "The Plight of the French Poor," in *The Portable Renaissance Reader* (New York: The Viking Press, 1953), pp. 214–218.

before the end of the month, and will come to remove the horses who have tilled the land which bore the fruit with which the man of war is paid. And when the poor labouring man has paid with great difficulty the quota he owes as tallage, for the hire of the men of arms, and when he takes comfort in what is left to him, hoping it will be enough to live on for the year, or to sow, there suddenly come men of arms who eat up or waste this little reserve which the poor man has saved to live on.

And there is still worse. For the man of war is not satisfied with the goods which he finds in the hut of the labourer but forces him by blows with stick or spear to go to town to get wine, white bread, fish, spices, and other luxuries. And, in truth, if God did not counsel the poor and give them patience, they would succumb in despair. And if in times past there were many evils, it has been even worse since the death of the king. And if the people had not felt the hope of some relief on the accession of the new king, they would have abandoned their labours.

And as for the intolerable burden of tallage and taxes which the poor people of this kingdom have not carried, to be sure, for that would have been impossible, but under which they have died and perished from hunger and poverty, the mere description of the grievousness of these imposts would cause infinite sadness and woe, tears of pity, great sighs and groans from sorrowing hearts; not to mention the enormity of the evils which followed and the injustice, violence, and extortion with which these taxes were imposed and seized.

And to consider these burdens which we may call not only intolerable but even deadly and pestiferous, who would ever have thought or dreamed of seeing this poor people so badly treated, who were formerly called free [françoys]. Now we may call them a people in worse condition than serfs, for serfs are nourished and this people has been crushed by intolerable burdens, such as securities, duties, impositions, and excessive tallage. While in the time of King Charles VII the quotas of the tallage imposed by the parish officials were counted only in twenties, such as twenty, forty, sixty pounds, after his death they began to be levied by hundreds, and since, they have grown from hundreds to thousands. And in many parishes where in the time of the late King Charles only forty or sixty pounds of tallage a year were levied, up to a thousand pounds were imposed in the year of the death of the last king [Louis XI]. And in the time of King Charles, in the duchies, Normandy, Languedoc, and others, the tallage was only in thousands but now it is in millions. And in the province of Normandy . . . there have followed many great and pitiable consequences, for some have fled and sought shelter in England, Brittany, and elsewhere, others in great numbers have died of hunger, others in despair have killed their wives and children and themselves, seeing they had nothing left to live on. And many men, women, and children, having no animals, are forced to work yoked to the plough, and others labour at night out of fear that in daylight they will be seized and apprehended for the said tallage. And as a result parts of the land have remained unploughed, and all because they have been subject to the will of those who wish to enrich themselves with the substance of the people, and without the consent and deliberation of the three estates. . . .

And as to the manner of raising these tallages and taxes, great pillage and robbery have been committed which everyone knows about. Among great abuses and injustices, all notorious, it has happened that the individuals of a parish who had already paid their quota and share have been imprisoned to pay what their neighbours owe, and even more than the other parishioners owed. And they were not through after paying the quota and share of the others but must also pay the sergeant, jailer, and clerk, or else suffer harm and the loss of their earnings. These things considered, it seems to the said estates that the king should take pity on his poor people, and relieve them of the said tallage and taxes, as he has proclaimed, in order that they may be able to live under him. And this they beg of him very humbly. . . .

And may it please my lords who take pensions to content themselves with the income from their own lords, without taking any extraordinary pensions or sums of money. Or at least if some receive them, let the pensions be reasonable, moderate, and bearable, out of regard for the afflictions and miseries of the poor people. For these pensions and monies are not taken from the domain of the king, nor could he supply them, but they are taken entirely from the third estate; and it is only the poor labourer who contributes to paying the said pensions. And thus it often happens that the poor labourer and his children die of hunger, for the substance on which he was to live was taken for the said pensions. And there is no doubt that in the payment of these there is sometimes a piece of money which has come out from the purse of a labourer whose chidren beg at the gates of those who receive the said pensions. And often the dogs are fed with bread bought with the pennies of the poor labourer on which he was to live.

From *Social Reform and the Reformation*

J. S. *Schapiro*

THE PEASANTS' WAR IN TYROL

■ 1525 ■

There arose in this country a cruel, terrible, and inhuman insurrection of the common peasant folk; I was there at the time and beheld strange and wondrous things. Certain factious and noisy people had undertaken to rescue by force from the judge a condemned rebel who had done wrong and who had justly been sentenced to punishment. After they had done this on a Wednesday, on Whitsunday the peasants, young and old, flocked together from all the mountains and valleys, although they did not know what they would do. Then when a great crowd of them had gathered together in the Mühland meadow in the Eisack valley, they concluded that they would deliver themselves of their burdens. A noble lord, Sigmund Brandisser, who was bailiff in Rodenegg, went to the assembled peasants and pointed out to them all the danger, mockery, damage, trouble, and care that would ensue. Although they promised him not to take action, but to present their grievances before their rightful prince, who was then at Innsbruck, they did not keep their promise, and on Whitsunday night they attacked Brixen, and in defiance of God and right they plundered and robbed all the priests, canons, and chaplains. Then they assembled before the bishop's courtyard and drove away all his councillors and servants with great violence, and in such an inhuman way that it cannot be described.

The people of Brixen forgot their duty to their Bishop Sebastian no less

SOURCE: J. S. Schapiro, *Social Reform and the Reformation* (New York: Columbia University Press, 1909), pp. 234–240.

quickly than the peasants of Neustift had forgotten theirs to their lord, the Provost Augustine. In sum, no one thought of duty, loyalty, promises, or anything else. The people of Brixen and the peasants were of one mind. Each group had its leaders. Without any notice, without any reason, these leaders with five thousand men marched on the monastery of Neustift, and fell on the church on Friday, May 12, 1525. Of the wantonness of which they were guilty there one could write a whole book. The Provost Augustine, a pious man, was driven out and pursued, and the priests were so insulted, mocked, and tortured that each must have been made ashamed of the name and sign of priest. The peasants did more than twenty-five thousand florins' worth of damage to the church, in destroying the building and in looting silver, ornaments, furnishings and vessels, documents and books. The insolence, drunkenness, blasphemy, and sacrilege with which the house of God was desecrated at this time no one can describe. It would also have been burned, but God would not suffer this to happen.

On Saturday, May 13, the peasants chose a leader, Michael Geismayr, a squire's son from Sterzing, a malicious, evil, rebellious, but crafty man. As soon as he was chosen their leader, the plundering of priests went on through the whole land. There was no priest in the land so poor but that he must lose all he possessed. Afterwards they fell upon many of the nobility, and destroyed many of them, for no one was able to arm himself for defence. Even the Archduke Ferdinand and his excellent wife knew that they were safe nowhere. For in the whole country, in the valley of the inn and on the Etsch, in the towns and among the peasants, there was such rioting, such an uproar and tumult, that an honest man might hardly walk in the streets. Robbing, plundering, and thieving were so common that even many pious men were tempted, who afterwards bitterly repented. And yet, to tell the truth, no one grew rich from the robbing, plundering, and stealing.

Michael Geismayr's Plan of Reform

■ 1526 ■

At the very outset you must pledge your lives and property, not to desert each other but to cooperate at all times; always to act advisedly and to be faithful and obedient to your chosen leaders. You must seek in all things, not your own welfare, but the glory of God and the commonweal, so that the Almighty, as is promised to those who obey Him, may give us His blessing and help. To Him we entrust ourselves entirely because He is incorruptible and betrays no one.

All those godless men who persecute the Eternal Word of God, who oppress the poor and who hinder the common welfare, shall be extirpated.

The true Christian doctrines founded on the Holy Word of God shall be proclaimed, and you must zealously pledge yourselves to them.

All privileges shall be done away with, as they are contrary to the Word of God, and distort the law which declares that no one shall suffer for the misdeeds of others.

All city walls, castles, and fortresses shall be demolished. From now on cities shall cease to exist and all shall live in villages. From cities result differences in station in the sense that one deems himself higher and more important than another. From cities come dissension, pride, and disturbances; whereas in the country absolute equality reigns.

All pictures, images, and chapels that are not parish churches (which are a horror unto God and entirely un-Christian) shall be totally abolished throughout the land.

The Word of God is to be at all times faithfully preached in the empire, and all sophistry and legal trickery shall be uprooted and all books containing such evil writings burned.

The judges, as well as the priests in the land, shall be paid only when they are employed, in order that their services may be obtained at the least expense.

Every year each community shall choose a judge and eight sworn jurors who shall administer the law during that year.

Court shall be held every Monday, and all cases shall be brought to an end within two days. The judges, sworn scribes, advocates, court attendants, and messengers shall not accept money from those concerned in the lawsuit, but they shall be paid by the community. Every Monday all litigants shall appear before the court, present their cases, and await decision.

There shall be only one government in the land, which should be located at Brixen as the most suitable place, because it is in the center of the empire, and contains many monasteries and other places of importance. Hither shall come the officials from all parts of the land, including several representatives from the mines, who shall be chosen for that purpose.

Appeals shall be taken immediately to this body and never to Meran, where it is useless to go. The administration at Meran shall be forthwith abolished.

At the seat of government there shall be established a university wherein the Word of God alone shall be taught. Three learned members of this university, well versed in Holy Scriptures (from which alone the righteousness of God can be taught), shall be appointed members of the government. They shall judge all matters according to the commands of God, as is proper among a Christian people.

Each province shall, after consulting with the others, decide whether the taxes are to be abolished from now on or whether a "free year" shall be established as is ordained in the Bible. In the meanwhile taxes should be collected for public purposes. We must remember that the empire will need money for carrying on war.

It is in the general interest to abolish customs tariffs in the interior but to permit them at the frontiers; this will establish the principle of taxing imports and not exports.

Every man shall pay the tithe according to the Word of God; it shall be spent in the following manner: In each parish there shall be a priest to preach the Scriptures, and he shall be supported from the tithe in a respectable fashion. The rest of the tithe shall be given to the poor; but such regulation shall be made as will do away with house-to-house begging, so that idle loafers may no longer be permitted to collect charity.

The monasteries and houses of the Teutonic Knights shall be turned into asylums. In some of these only sick people shall be housed; and they must be well cared for with food and medicines. In others old people who can no longer work shall be maintained; and in some, poor uneducated children shall be respectably brought up. The poor who remain at home shall be assisted on the advice of the district judge, since he is best informed. Such people shall be provided for, according to their needs, from the tithe or by charity. If the tithe be not enough for the support of the priests and the poor, then let each man loyally give charity according to his ability, and any shortage shall be made up from the public treasury. One official shall do nothing else except look after the asylums and the poor. Every judge, each in his own district, shall, by the means of the tithe, charity, and public appeals, be helpful to the poor at their homes. They shall be provided not only with meat and drink, but with clothing and other necessities as well, so that good morals prevail in the land. . . .

No one shall engage in business, and so avoid being contaminated with the sin of usury. Good regulations shall be enacted to prevent scarcity as well as to prohibit overcharging and cheating; so that all things may be sold at an honest and fair price. Let some place in the land be fixed upon (Trent, for example, on account of its central location) where all the manufactured articles shall be made. Silk, cloth, velvet, and shoes shall be produced there under the supervision of an official. Whatever cannot be grown in our country, as spices, shall be imported; shops shall be opened in several appointed places where all sorts of things shall be sold. No profit is to be made, as all things are to be sold at cost. By such means will all deceit and trickery be prevented and all things be bought at their proper value. Money will remain in the country, and this will be for the benefit of the common man. The official and his assistants, charged with the duty of enforcing these regulations, shall be paid fixed salaries. . . .

All smelting houses and mines of tin, silver, copper, and other metals found in the country, which belong to the nobles or to associations of foreign merchants, such as the Fuggers, Hochstätters, Baumgartners, and others like them, shall be confiscated and given over to public ownership; in all justice, they have forfeited them as they have acquired the mines by unjust and cruel means. The workmen were paid their wages in bad wares and bad money, though in appearance they were given more in amount than their earnings. The prices of spices and other wares rose because of bad currency. All coiners of money who bought silver of these monopolists had to pay their arbitrary prices. This indirectly resulted to the disadvantage of the poor man, who found that the rewards of his labour had decreased. All the merchants through whose hands the bad coins passed demanded still higher prices. As a result the whole world was entangled in this un-Christian usury. In such manner were the princely fortunes made, which, in all fairness, should be forfeited.

There shall be a superintendent over all the mines in the country who must be resworn every year. He shall have power to supervise every transaction and shall permit no smelting to be done except by the government. The metals shall be bought when prices are low. The miners shall be paid their wages in cash and not in goods, in order that peace and satisfaction may exist among the workers. If the mines are worked in an orderly and systematic manner there will be enough profit from them to pay the running expenses of the government. If the income is not sufficient for this purpose, a penny tax shall be laid on all to equalize the burden. Every effort should be made, however, to get the most out of the mines. The profits of one mine should be used to open another, because, through mining, the country can get the largest income with the least labour.

This is Geismayr's constitution when he dreams in his chimney corner and imagines himself a prince.

CHAPTER SIX

Towns and Commerce: Urban Institutions in Traditional Society

Commerce and the Revival of Cities

Medieval society was a traditional, agrarian society, even at the close of the fifteenth century. Spain, which had perhaps the largest urban population in Europe, contained scarcely 15 percent of its people in towns. But there is something misleading in our initial statement. The role of money and commerce in the medieval economy was never wholly negligible in the centuries before the "commercial revolution" and oversea expansion of the Renaissance. Yet many historians believed for decades the thesis of Pirenne, that between roughly A.D. 450 and the early eleventh century commerce had been overcome by barbarism and cities abandoned under the pressure of the Moslem closure of the Mediterranean Sea. To complement his thesis about the failure of commerce and hence of cities —true urban life for Pirenne was always based on trade and industry—the great Belgian historian developed another in a series of articles and in some lectures given in America in 1922.*

Urban life in Europe was discontinuous. Its revival turned on the resurrection of commerce alone. Medieval cities owed nothing to religion, political traditions, or other impulses. Despite the successful revision of Pirenne's thesis by historians (especially with respect to Gaul, Germany, and England), we may well open our consideration of the urban life of the Middle Ages with his exploration of the relationship between commerce and towns.

---

* "Les Origines des Constitutions Urbaines au Moyen-Âge," *Revue Historique*, vol. 53 (1893), pp. 52ff., and vol. 57 (1895), pp. 57ff. The lectures were given in English and first published in that language by Princeton University Press (1925). A French translation followed.

From *Medieval Cities: Their Origins and the Revival of Trade*

Henri Pirenne

THE REVIVAL OF COMMERCE

The end of the ninth century was the moment when the economic development of Western Europe that followed the closing of the Mediterranean was at its lowest ebb. It was also the moment when the social disorganization caused by the raids of the barbarians and the accompanying political anarchy reached a maximum.

The tenth century, if not an era of recovery, was at least an era of stabilization and relative peace. The surrender of Normandy to Rollo (912) marked in the West the end of the great Scandinavian invasions, while in the east Henry the Fowler and Otto I checked and held the Slavs along the Elbe and the Hungarians in the valley of the Danube (934, 955). At the same time the feudal system, which had definitely displaced the monarchy, was established in France on the débris of the old Carolingian order. In Germany, on the contrary, the somewhat later development of society enabled the princes of the House of Saxony to resist the encroachments of the lay aristocracy. On their side they had the powerful influence of the bishops and used it to restore the ascendancy of the monarchy. In assuming the title of Roman Emperor, they laid claim to the universal authority which Charlemagne had exercised.

If all this was not accomplished without bitter conflicts, nevertheless it was decidedly productive of good. Europe ceased to be overrun by ruthless hordes. She recovered confidence in the future, and, with that confidence, courage and ambition. The date of the renewal of a cooperative activity on the part of the people might well be ascribed to the tenth century. At that date, likewise, the social authorities began once more to acquit themselves in the rôle which it was their place to play. From now on, in feudal as well as in episcopal principalities, the first traces could be seen of an organized effort to better the condition of the people. The prime need of that era, hardly rising above anarchy, was the need of peace, the most fundamental and the most essential of all the needs of society.

The first Truce of God was proclaimed in 989. Private wars, the greatest of the plagues that harassed those troubled times, were energetically combated by the territorial counts in France and by the prelates of the imperial Church in Germany.

Dark though the prospect still was, the tenth century nevertheless saw in outline the picture which the eleventh century presents. The famous legend of the terrors of the year 1000 is not devoid, in this respect, of symbolic significance. It is doubtless untrue that men expected the end of the world in the year 1000. Yet the century which came in at that date is characterized, in contrast with the preceding one, by a recrudescence of activity so marked that it could pass for the vigorous and joyful awakening of a society long oppressed by a nightmare of anguish. In every demesne was to be seen the same burst of energy and, for that

SOURCE: Henri Pirenne, *Medieval Cities: Their Origins and the Revival of Trade*, trans. Frank D. Halsey (Princeton: Princeton University Press, 1952), pp. 56–76.

matter of optimism. The Church, revivified by the Clunisian reform, undertook to purify herself of the abuses which had crept into her discipline and to shake off the bondage in which the emperors held her. A mystic enthusiasm of which she was the inspiration, animated her congregations and launched them upon the heroic and grandiose enterprise of the Crusades which brought Western Christianity to grips with Islam. The military spirit of feudalism led her to initiate and to succeed in epic undertakings. Norman knights went to battle with Byzantines and Moslems in southern Italy, and founded there the principalities out of which was later to arise the Kingdom of Sicily; other Normans, with whom were associated Flemings and Frenchmen from the north, conquered England under the leadership of Duke William. South of the Pyrenees the Christians drove before them the Saracens of Spain; Toledo and Valencia fell to their hands (1072–1109).

Such undertakings testify not only to energy and vigor of spirit; they testify also to the health of society. They would have obviously been impossible without that native strength which is one of the characteristics of the eleventh century. The fecundity of families seemed, at this date, to be as general among the nobility as among the peasants. Younger sons abounded everywhere, feeling themselves crowded for room on their natal soil and eager to try their fortunes abroad. Everywhere were to be met adventurers in search of money or work. The armies were full of mercenaries, "Coterelli" or "Brabantiones," letting their services to whoever wished to employ them. From Flanders and Holland bands of peasants were setting out, by the beginning of the twelfth century, to drain the *Mooren* on the banks of the Elbe. In every part of Europe labor was offered in superabundant quantity and this is undoubtedly the explanation of the increasing number, from then on, of great reclamation projects in clearing land and diking streams.

It does not appear that, from the Roman era to the eleventh century, the area of cultivated land had been perceptibly increased. Save in the Germanic countries, the monasteries had hardly altered, in this respect, the existing situation. They were almost always established on old estates and did nothing to decrease the extent of the woods, the heaths and the marshes contained within their demesnes. But it was quite a different matter when once the increase of population permitted these unproductive terrains to be put to good use. Just about the year 1000 there began a period of reclamation which was to continue, with steady increase, up to the end of the twelfth century. Europe "colonized" herself, thanks to the increase of her inhabitants. The princes and the great proprietors turned to the founding of new towns, where flocked the "younger sons" in quest of lands to cultivate. The great forests began to be cleared. In Flanders appeared, about 1150, the first *polders*. (A "polder" is diked land, reclaimed from the sea.) The Order of the Cistercians, founded in 1098, gave itself over at once to reclamation projects and the clearing of the land.

It is easy to see that the increase in population and the burst of renewed general activity of which it was both cause and effect, operated from the very first to the benefit of an agricultural economy. But this condition should, before long, have had its effect upon trade as well. The eleventh century, in fact, brings us face to face with a real commercial revival. This revival received its impetus from two centers of activity, one located in the south and the other in the north: Venice on one side and the Flemish coast on the other. And this is merely another way of saying that it was the result of an external stimulus. The contact with foreign trade, maintained at these two points, first caused it to appear and spread. Quite likely it could have come about in some other way. Commercial activity might have been revived by virtue of the trend of general economic life. The fact is, however, that this was not the case. Just as the trade of the West disappeared with the shutting off of its foreign markets, just so it was renewed when these markets were reopened.

Venice, whose influence was felt from the very first, has a well recognized and singular place in the economic history of Europe. Like Tyre, Venice shows an exclusively commercial character. Her first inhabitants, fleeing before the approach of the Huns, the Goths and the Lombards, had sought (in the fifth and sixth centuries) a refuge on the barren islets of the lagoons at Rialto, at Olivolo, at Spinalunga, at Dorsoduro. To exist in these marshes they had to tax their ingenuity and to fight against Nature herself. Everything was wanting: even drinking water was lacking. But the sea was enough for the existence of a folk who knew how to manage things. Fishing and the preparation of salt supplied an immediate means of livelihood to the Venetians. They were able to procure wheat by exchanging their products with the inhabitants of the neighboring shores.

Trade was thus forced upon them by the very conditions under which they lived. And they had the energy and the genius to turn to profit the unlimited possibilities which trade offered them. By the eighth century the group of islets they occupied was already thickly populated enough to become the see of a special diocese.

At the date when the city was founded, all Italy still belonged to the Byzantine Empire. Thanks to her insular situation, the conquerors who successively overran the peninsula—first the Lombards, then Charlemagne, and finally, still later, the German emperors—were not successful in their attempts to gain possession. She remained, therefore, under the sovereignty of Constantinople, thus forming at the upper end of the Adriatic and at the foot of the Alps an isolated outpost of Byzantine civilization. While Western Europe was detaching herself from the East, she continued to be part of it. And this circumstance is of capital importance. The consequence was that Venice did not cease to gravitate in the orbit of Constantinople. Across the waters, she was subject to the attraction of that great city and herself grew great under its influence.

Constantinople, even in the eleventh century, appears not only as a great city, but as the greatest city of the whole Mediterranean basin. Her population was not far from reaching the figure of a million inhabitants, and that population was singularly active. She was not content, as had been the population of Rome under the Republic and the Empire, to consume without producing. She gave herself over, with a zeal which the fiscal system shackled but did not choke, not only to trading but to industry. For Constantinople was a great port and a first-rate manufacturing center as well as a political capital. Here were to be found every manner of life and every form of social activity. Alone, in the Christian world, she presented a picture analogous to that of great modern cities with all the complexities, all the defects but also with all the refinements of an essentially urban civilization. An uninterrupted shipping kept her in touch with the coasts of the Black Sea, Asia Minor, southern Italy, and the shores of the Adriatic. Her war fleets secured to her the mastery of the sea, without which she would not have been able to live. As long as she remained powerful, she was able to maintain, in the face of Islam, her dominion over all the waters of the eastern Mediterranean.

It is easy to understand how Venice profited by her alliance with a world so different from the European west. To it she not only owed the prosperity of her commerce, but from it she first learned those higher forms of civilization, that perfected technique, that business enterprise, and that political and administrative organization which gave her a place apart in the Europe of the Middle Ages. By the eighth century she was devoting herself with greater and greater success to the provisioning of Constantinople. Her ships transported thither the products of the countries which were contiguous to her on the east and the west: wheat and wine from Italy, wood from Dalmatia, salt from the lagoons, and, in spite of the prohibitions of the Pope and the Emperor himself, slaves which she easily secured among the Slavic peoples of the shores of the Adriatic. Thence they brought back, in return, the precious fabrics of Byzantine manufacture, as well as spices which Asia furnished to Constantinople. By the tenth century the activity of the port had already attained extraordinary proportions. And with the extension of trade, the love of gain became irresistible. No scruple had any weight with the Venetians. Their religion was a religion of business men. It mattered little to them that the Moslems were the enemies of Christ, if business with them was profitable. After the ninth century they began more and more to frequent Aleppo, Cairo, Damascus, Kairwan, Palermo. Treaties of commerce assured their merchants a privileged status in the markets of Islam.

By the start of the eleventh century, the power of Venice was making as marvellous progress as her wealth. Under the Doge Pietro II Orseolo, she cleared the Adriatic of the Slavic pirates, subjected Istria and had at Zara, Veglia, Arbe, Trau, Spalato, Curzola, and Lagosta settlements or military establishments. John the Deacon extols the splendor and the glory of *Venetia Aurea*, and William of Apuleia vaunts the city "rich in money, rich in men," and declares that "no people in the world are more valorous in naval warfare, more skilful in the art of guiding ships on the sea."

It was inevitable that the powerful economic movement, of which Venice was the center, should be communicated to the countries of Italy from which she was separated only by the lagoons. There she obtained the wheat and wine which she either consumed herself or exported, and she naturally sought to create there a market for the Eastern merchandise which her mariners unloaded in greater and greater quantity on the quays by the Po. She entered into relations with Pavia, which was not long in being animated by her infectious activity. She obtained from the German emperors the right to trade freely first with the nearby cities and then with all Italy, as well as the shipping monopoly for all goods arriving in her port.

In the course of the tenth century Lombardy was inspired, by her example, with commercial life. Trade rapidly spread from Pavia to the neighboring cities. All of them made haste to share in the traffic of which Venice had given them the outstanding example and which it was to her interest to stimulate among them. The spirit of enterprise developed in one place after another.

It was not only products of the soil which kept the commercial relations with Venice flourishing. Industry was already commencing to appear. Early in the eleventh century, for example, Lucca turned to the manufacture of cloths and kept at it until much later. Probably a great many more details would be known about the beginnings of this economic revival in Lombardy if our sources of information were not so deplorably meager.

Preponderant as the Venetian influence had been in Italy, it did not make itself felt there exclusively. The south of the peninsula beyond Spoleto and Benevento was still, and so remained until the arrival of the Normans in the eleventh century, under the power of the Byzantine Empire. Bari, Tarentum, Naples and above all Amalfi, kept up relations with Constantinople similar to those of Venice. They were very active centers of trade and, no more than Venice, did not hesitate to traffic with Moslem ports.

Their shipping was, naturally, fated to find competitors sooner or later among the inhabitants of the coastal towns situated further to the north. And, in fact, after the beginning of the eleventh century we see first Genoa, then Pisa soon after, turning their attention to the sea. In 935 the Saracen pirates had again pillaged Genoa. But the moment was approaching when she was in her turn to take the offensive. There could be no question of her concluding commercial arrangements, as had Venice or Amalfi, with the enemies of her Faith. The mystic, excessive scrupulousness of the West in religious matters did not permit it, and too many hates had accumulated in the course of the centuries. The sea could be opened up only by force of arms.

In 1015–1016 an expedition was undertaken by Genoa, in cooperation with Pisa, against Sardinia. Twenty years later, in 1034, they got possession for a time of Bona on the coast of Africa; the Pisans, on their part, victoriously entered the port of Palermo in 1062 and destroyed its arsenal.

In 1087 the fleets of the two cities, encouraged by Pope Victor III, attacked Mehdia. All these expeditions were due as much to religious enthusiasm as to the spirit of adventure. With a quite different viewpoint from that of the Venetians, the Genoese and the Pisans considered themselves soldiers of Christ and of the Church, opponents of Islam. They believed they saw the Archangel Gabriel and St. Peter leading them into battle with the Infidels, and it was only after having massacred the "priests of Mahomet" and pillaged the mosque of Mehdia that they signed an advantageous treaty of commerce. The Cathedral of Pisa, built after this triumph, admirably symbolized both the mysticism of the conquerors and the wealth which their shipping was beginning to bring to them. Pillars and precious marble brought from Africa served to decorate it—it seems as if they had wished to attest by its splendor the revenge of Christianity upon the Saracens whose opulence was a thing of scandal and of envy. Those, at least, are the sentiments which an enthusiastic contemporary poem expresses:

> *Unde tua in aeternum splendebit ecclesia*
> *Auro, gemnis, margaritis et palliis splendida*[1]

Before the counter-attack of Christianity, Islam thus gave way little by little. The launching of the First Crusade (1096) marked its definite recoil. In 1097 a Genoese fleet sailed towards Antioch, bringing to the Crusaders reinforcements and supplies. Two years later Pisa sent out vessels "under the orders of the Pope" to deliver Jerusalem. From that time on the whole Mediterranean was opened, or

---

1 "Thy church will be resplendent for eternity,
  Dazzling with gold, with gems, with pearls and precious cloths."

rather reopened, to Western shipping. As in the Roman era, communications were reestablished from one end to the other of that essentially European sea.

The Empire of Islam, insofar as the sea was concerned, came to an end. To be sure, the political and religious results of the Crusade were ephemeral. The kingdom of Jerusalem and the principalities of Edessa and Antioch were reconquered by the Moslems in the twelfth century. But the sea remained in the hands of the Christians. They were the ones who held undisputed economic mastery over it. All the shipping in the ports of the Levant came gradually under their control. Their commercial establishments multiplied with surprising rapidity in the ports of Syria, Egypt and the isles of the Ionian Sea. The conquest of Corsica (1091), of Sardinia (1022) and of Sicily (1058–1090) took away from the Saracens the bases of operations which, since the ninth century, had enabled them to keep the West in a state of blockade. The ships of Genoa and Pisa kept the sea routes open. They patronized the markets of the East, whither came the products of Asia, both by caravan and by the ships of the Red Sea and the Persian Gulf, and frequented in their turn the great port of Byzantium. The capture of Amalfi by the Normans (1073), in putting an end to the commerce of that city, freed them from her rivalry.

But their progress immediately aroused the jealousy of Venice. She could not bear to share with these newcomers a trade in which she laid claim to a monopoly. It was of no moment that she professed the same Faith, belonged to the same people and spoke the same language; since they had become rivals she saw in them only enemies. In the spring of the year 1100 a Venetian squadron, lying in wait before Rhodes for the return of the fleet which Pisa had sent to Jerusalem, fell upon it unawares and ruthlessly sank a large number of vessels. So began between the maritime cities a conflict which was to last as long as their prosperity. The Mediterranean was no more to know that Roman peace which the Empire of the Caesars had once enjoined upon her. The divergence of interests was hereafter to sustain on the sea of hostility, sometimes secret and sometimes openly declared, between the rivals who contested for supremacy. The quarrels of the Italian republics of the Middle Ages are still duplicated in modern times by the continued wrangling of the States whose coasts the Mediterranean washes.

In developing, maritime commerce must naturally have become more generalized. By the beginning of the twelfth century it had reached the shores of France and Spain. After the long stagnation into which the city had fallen at the end of the Merovingian period, the old port of Marseilles took on new life. In Catalonia, Barcelona, out of which the kings of Aragon had driven the Moslems, profited in turn by the opening up of the sea. However, Italy undoubtedly kept the upper hand in that first economic revival. Lombardy, where from Venice on the east and Pisa and Genoa on the west all the commercial movements of the Mediterranean flowed and were blended into one, flourished with an extraordinary exuberance. On that wonderful plain cities bloomed with the same vigor as the harvests. The fertility of the soil made possible for them an unlimited expansion, and at the same time the ease of obtaining markets favored both the importation of raw materials and the exportation of manufactured products. There, commerce gave rise to industry, and as it developed, Bergamo, Cremona, Lodi, Verona, and all the old towns, all the old Roman *municipia*, took on new life, far more vigorous than that which had animated them in antiquity. Soon their surplus production and their fresh energy were seeking to expand abroad. In the south Tuscany was won. In the north new routes were laid out across the Alps. By the passes of the Splügen, St. Bernard and the Brenner, their merchants were to bring to the continent of Europe that same healthy stimulus which had come to them from the sea. They followed those natural routes marked by river courses—the Danube to the east, the Rhine to the north, and the Rhône to the west. In 1074 Italian merchants, undoubtedly Lombards, are made mention of at Paris; and at the beginning of the twelfth century the fairs of Flanders were already drawing a considerable number of their compatriots.

Nothing could be more natural than this appearance of southerners on the Flemish coast. It was a consequence of the attraction which trade spontaneously exerts upon trade.

It has already been shown that, during the Carolingian era, the Netherlands had given evidence of a commercial activity not to be found anywhere else. This is easily explained by the great number of rivers which flow through that country and which there unite their waters before emptying into the sea: the Rhine, the Meuse and the Scheldt. England and the Scandinavian countries were so near that land of large and deep estuaries that their mariners naturally frequented it at an early date. It was to them, as we have seen above, that the ports of Duurstede and Quentovic owed their importance. But this importance was ephemeral. It could not survive the period of the Norseman invasions. The easier access was to a country, the more it lured the invaders and the more it had to suffer from their devastations. The geographical situation which, at Venice, had safeguarded commercial prosperity was, here, naturally due to contribute to its destruction.

The invasions of the Norsemen had been only the first manifestation of the need of expansion felt by the Scandinavian peoples. Their overflowing energy had driven them forth, towards Western Europe and Russia simultaneously, upon adventures of pillage and conquest. They were not mere pirates. They aspired, as had the Germanic tribes before them with regard to the Roman Empire, to settle in countries more rich and fertile than was their homeland, and there to create colonies for the surplus population which their own country could no longer support. In this undertaking they eventually succeeded. To the east, the Swedes set foot along those natural routes which led from the Baltic to the Black Sea by way of the Neva, Lake Ladoga, the Lovat, the Volchof, the Dvina and the Dnieper. To the west, the Danes and the Norwegians colonized the Anglo-Saxon kingdoms north of the Humber. In France, they had ceded to them by Charles the Simple the country on the Channel which took from them the name of Normandy.

These successes had for their result the orientation in a new direction of the activity of the Scandinavians. Starting at the beginning of the tenth century, they turned away from war to devote themselves to trade. Their ships plowed all the seas of the north and they had nothing to fear from rivals since they alone, among the peoples whose shores those seas bathed, were navigators. It is enough to peruse the delightful tales of the sagas to get an idea of the hardihood and the skill of the barbarian mariners whose adventures and exploits they recount.

Each spring, once the water was open, they put out to sea. They were to be met in Iceland, in Ireland, in England, in Flanders, at the mouths of the Elbe, the Weser, and the Vistula, on the islands of the Baltic Sea, at the head of the Gulf of Bothnia and the Gulf of Finland. They had settlements at Dublin, at Hamburg, at Schwerin, on the island of Gotland; thanks to them, the current of trade, which starting from Byzantium and Bagdad crossed Russia by way of Kiev and Novgorod, was extended up to the shores of the North Sea and there made felt its beneficent influence. In all history there is hardly a more curious phenomenon than that effect wrought on northern Europe by the superior civilizations of the Greek and Arab Empires, and of which the Scandinavians were the intermediaries. In this respect their role, despite the differences of climate, society and culture, seems quite analogous to that which Venice played in the south of Europe. Like her, they renewed the contact between the East and the West. And just as the commercial activity of Venice did not long delay in involving Lombardy in the movement, so likewise Scandinavian shipping brought about the economic awakening of the coast of Flanders.

The geographical situation of Flanders, indeed, put her in a splendid position to become the western focus for the commerce of the seas of the north. It formed the natural terminus of the voyage for ships arriving from northern England or which, having crossed the Sound after coming out of the Baltic, were on their way to the south. As has already been stated, the ports of Quentovic and Duurstede had been frequented by the Norsemen before the period of the invasions. First one and then the other disappeared before the storm. Quentovic did not rise again from her ruins and it was Bruges, whose situation at the head of the Gulf of Zwyn was the better one, that became her heritor. As for Duurstede, Scandinavian mariners reappeared there at the beginning of the tenth century. Yet her prosperity did not last very long. As commerce flourished, it was concentrated more and more about Bruges, nearer to France and kept in a more stable condition of peace by the Counts of Flanders, whereas the neighborhood of Duurstede was too exposed to the incursions of the still half-barbaric Friesians to enjoy security. Be that as it may, it is certain that Bruges attracted to her port, more and more, the trade of the north, and that the disappearance of Duurstede, in the course of the eleventh century, definitely assured her future. The fact that coins of the Counts of Flanders, Arnold II and Baldwin IV (956–1035), have been discovered in considerable numbers in Denmark, in Prussia, and even in Russia, would attest, in the lack of written information, to the relations with those countries which Flanders kept up after this date with the help of Scandinavian mariners.

Communication with the nearby English coast was to become still more active. It was at Bruges, for example, that the Anglo-Saxon Queen Emma, expelled from England, settled about 1030. In 991–1002 the list of market-tolls at London makes mention of the Flemings as if they were the most important group of foreigners carrying on business in that city.

Among the causes of the commercial importance which so early characterized Flanders, should be pointed out the existence in that country of an indigenous industry able to supply the vessels that landed there with a valuable return cargo. From the Roman era and probably even before that, the Morini and the Menapii had been making woollen cloths. This primitive industry was due to be perfected under the influence of the technical improvements introduced by the Roman conquest. The peculiar fineness of the fleece of the sheep raised on the humid meadows of the coast was the final factor needed to insure success. The tunics (*saga*) and the cloaks (*birri*) which it produced were exported as far as beyond the Alps and there even was at Tournai, in the last days of the Empire, a factory for military clothing. The Germanic invasion did not put an end to this industry. The Franks who invaded Flanders in the fifth century continued to carry it on as had the older inhabitants before them—there is no doubt but that the Friesian cloaks of which the ninth-century historiographer speaks were made in Flanders. They seem to be the only manufactured products which furnished, in the Carolingian era, the substance of a regular trade. The Friesians transported them along the Scheldt, the Meuse and the Rhine, and when Charlemagne wanted to reply with gifts to the compliments of the Caliph Harun-al-Rashid, he found nothing better to offer him than "*pallia fresonica*." It is to be supposed that these cloths, as remarkable for their beautiful colors as for their softness, must have immediately attracted the attention of the Scandinavian navigators of the tenth century. Nowhere, in the north of Europe, were found more valuable products, and they undoubtedly had a place, side by side with the furs of the north and the Arab and Byzantine silk fabrics, among the most sought-after export goods. According to every indication, the cloths which were made mention of in the London market about the year 1000 were cloths from Flanders. And the new markets which shipping was now offering to them could not have failed to give a fresh impulse to their manufacture.

Thus commerce and industry, the latter carried on locally and the former originating abroad, joined in giving Flanders, after the tenth century, an economic activity that was to continue developing. In the eleventh century the advances made were already surprising. Thenceforth Flanders traded with the north of France the wines of which she exchangd for her cloths. The conquest of England by William of Normandy bound to the Continent that country which heretofore had gravitated in the orbit of Denmark, and multiplied the relations which Bruges had already been maintaining with London. By the side of Bruges, other mercantile centers appeared: Ghent, Ypres, Lille, Douai, Arras, Tournai. Fairs were instituted by the Counts of Thourout at Messines, Lille and Ypres.

Flanders was not alone in experiencing the salutary effects of the shipping of the north. The repercussion made itself felt along the rivers which end in the Netherlands. Cambrai and Valenciennes on the Scheldt, Liége, Huy and Dinant on the Meuse had already, in the tenth century, been mentioned as centers of trade. This was true also of Cologne and Mainz, on the Rhine. The shores of the Channel and of the Atlantic, further removed from the seat of activity of the North Sea, do not seem to have had the same importance. Hardly any mention was made of them, with the exception of Rouen, naturally in close contact with England, and, further south, Bordeaux and Bayonne whose development was much slower. As for the interior of France and Germany, they were affected only very slightly by the economic movement which little by little spread in that direction, either coming up from Italy or coming down from the Netherlands.

It was only in the twelfth century that, gradually but definitely, Western Europe was transformed. The economic development freed her from the traditional immobility to which a social organization, depending solely on the relations of man to the soil, had condemned her. Commerce and industry did not merely find a place alongside of agriculture; they reacted upon it. Its products no longer served solely for the consumption of the landed proprietors and the tillers of the soil; they were brought into general circulation, as objects of barter or as raw material. The rigid confines of the demesnial system, which had up to now hemmed in all economic activity, were broken down and the whole social order was patterned along more flexible, more active and more varied lines. As in antiquity, the country oriented itself afresh on the city. Under the influence of trade the old Roman cities took on new life and were repopulated, or mercantile groups formed round about the military burgs and established themselves along the sea coasts, on river banks, at confluences, at the junction points of the natural routes of communication. Each of them constituted a market which exercised an attraction, proportionate to its importance, on the surrounding country or made itself felt afar.

Large or small, they were to be met everywhere; one was to be found, on the average, in every twenty-five square leagues of land. They had, in fact, become indispensable to society. They had introduced into it a division of labor which it could no longer do without. Between them and the country was established a reciprocal exchange of services. An increasingly intimate solidarity bound them together, the country attending to the provisioning of the towns, and the towns supplying, in return, articles of commerce and manufactured goods. The physical life of the burgher depended upon the peasant, but the social life of the peasant depended upon the burgher. For the burgher disclosed to him a more comfortable sort of existence, a more refined sort, and one which, in arousing his desires, multiplied his needs and raised his standard of living. And it was not only in this respect that the rise of cities strongly stimulated social progress. It made no less a contribution in spreading throughout the world a new conception of labor. Before this it had been serf; now it became free, and the consequences of this fact, to which we shall return, were incalculable. Let it be added, finally, that the economic revival of which the twelfth century saw the flowering revealed the power of capital, and enough will have been said to show that possibly no period in all history had a more profound effect upon humanity.

Invigorated, transformed and launched upon the route of progress, the new Europe resembled, in short, more the ancient Europe than the Europe of Carolingian times. For it was out of antiquity that she regained that essential characteristic of being a region of cities. And if, in the political organization, the rôle of cities had been greater in antiquity than it was in the Middle Ages, in return their economic influence in the latter era greatly exceeded what it had ever been before. Generally speaking, great mercantile cities were relatively rare in the western provinces of the Roman Empire. Aside from Rome herself, there were scarcely any at all except Naples, Milan, Marseilles and Lyons. Nothing of the sort was then in existence which might be comparable to what were, at the beginning of the tenth century, ports like Venice, Pisa, Genoa or Bruges, or centers of industry such as Milan, Florence, Ypres, and Ghent. In Gaul, in fact, the important place held in the twelfth century by ancient cities such as Orleans, Bordeaux, Cologne, Nantes, Rouen, and others, was much superior to what they had enjoyed under the emperors. Finally, the extension of the economic development of medieval Europe went well beyond the limits it had reached in Roman Europe. Instead of halting along the Rhine and the Danube, it overflowed widely in Germany and reached as far as the Vistula. Regions which had been travelled over, at the beginning of the Christian era, only by infrequent traders in amber and furs and which seemed as inhospitable as the heart of Africa might have seemed to our ancestors, now burgeoned with cities. The Sound, which no Roman trading vessel had ever crossed, was animated by the continual passage of ships. They sailed the Baltic and the North Sea as they had sailed the Mediterranean. There were almost as many ports on the shores of the one as on the shores of the other.

From two quarters, trade made use of the resources which Nature had placed at its disposal. It dominated the two inland seas which between them bounded the admirably indented coast line of the continent of Europe. Just as the Italian cities had driven back the Moslems from the Mediterranean, so in the course of the twelfth century the German cities drove back the Scandinavians from the North Sea and the Baltic, on which hereafter were spread the sails of the Hanse Towns.

Thus the commercial expansion which first made its appearance at the two points at which Europe came in contact with it—by Venice with the world of the east, by Flanders with the Russo-Scandinavian world—spread like a beneficent epidemic over the whole Continent. In reaching inland, the movement from the north and the movement from the south finally met each other. The contact between them was effected at the midpoint of the natural route which led from Bruges to Venice—on the plain of Champagne, where in the twelfth century were instituted the famous fairs of Troyes, Lagny, Provins and Bar-sur-Aube, which up to the end of the thirteenth century fulfilled, in medieval Europe, the functions of an exchange and of a clearing house.

Abu al-Fadl and Benedetto Cotrugli: Two Treatises on Commerce

No matter how we account for the existence of town life in medieval Europe, the social institutions and people who traded commodities and manipulated capital were a breed apart from those found in agrarian society. Dealing in money left its mark on the mental furniture of the men who did it. The profit motive and the substitution of artificial values of exchange for the natural value (or utility) of things molded a common consciousness in Arab merchants of the ninth century and Italian experts of the fifteenth. This is apparent in the writings of men who instructed others in the beauties of commerce and the perfection of mercantile technique. Especially striking is the struggle of merchants against the restrictions placed on commerce by religious teaching and the growth of a sense of "professional" identity. It is perhaps more important to notice and describe these things than to see whether one can account for them within the framework of particular ideologies.* For the role of money in the medieval economy was that of a solvent and catalyst, and the men who dealt in it were agents of complication in social life.

From *Medieval Trade in the Mediterranean World*

Robert S. Lopez and Irving W. Raymond

Necessary Pieces of Advice for Merchants—with the Dispensation of God, the Exalted and All Powerful

If the merchant deals in bulky articles, he must have reliable associates and sufficient assistants to help him with buying, packing, loading, transporting, and selling. For if he has to take care of everything by himself, his mind and body come to grief; moreover the camel drivers, carriers, and boatmen as well as all others whose help he needs in transporting the goods are always eager to steal his property.

The best [solution for] a merchant who is thrown entirely upon his own resources will be to concern himself only with light things which he can handle by himself.

The foundation of all trade in relation to selling and buying consists in buying from a man who does not care for the article or whom need compels to accept the price [offered] and in selling to a man who is eager to acquire the article or who is under necessity to buy. For that is the surest way to fare well with the merchandise bought and to obtain a rich profit.

The merchant must be no more pessimist than optimist, since pessimism induces him to hold back his capital, but optimism induces him to take such risks that he has more to fear than to hope.

SOURCE: Robert S. Lopez and Irving W. Raymond, eds., *Medieval Trade in the Mediterranean World* (New York: Columbia University Press, 1955), pp. 410–418.

* Especially illustrative of this point is Max Weber's unsuccessful effort to link capitalism and bourgeois life to the Protestant theologies of Calvinism.

And let him also know that disappointment comes easily with too much greed in seeking advantage, and excessive pursuit of gain is the road to loss. The explanation consists in this: that between the purchase of one who has a wild desire to buy and the purchase of another who has a faint desire and who heals his soul of the madness of greed and keeps it free from the slavery of passion, [between these] there is a wide chasm and a great difference. Not otherwise is it in trade. Whoever desires too strongly becomes so blind that he cannot see the right way and loses the right insight, inclines to passion, and strays from the right judgment of reason. The best things are always those which are happy in the present and reach a beautiful ending in the future.

And whenever a merchant realizes that a certain branch of trade brings him good fortune, he ought to devote himself to it—with the exception of those enterprises which bring with them obvious dangers and which do not insure him against disappointment, because in that particular branch a man's good fortune is frequently determined by fate. It is told in the tradition of the Prophet that once a man came to him and told him that his way of earning a living was commerce, but that he had no luck in it; he bought no goods that did not become hard to sell or spoil on his hands. Then [the Prophet] said to him, "Have you ever had profits to your satisfaction on something you have purchased and on which you have taken some risk?" [The man] answered: "I do not remember that this ever occurred except in the *qard* business." [1] At that the Prophet said, "Then throw yourself into the *qard* business." The man did so and acquired riches and became well-to-do and lived in fine circumstances. When the Prophet learned this, he said, "Whoever is blessed with fortune in one enterprise should devote himself to it."

Moreover, the merchant must grant deductions in selling, for this is one of the essential means of the profession by which he earns his daily bread. It consists in this: that a merchant ought to make it definitely clear to himself that out of a profit of a dinar one half goes to deductions, either at the weighing, or at the paying, or as a gift to the middleman, or as rebate if the buyer asks for it. But if the merchant is greedy and thinks, "I have been too hasty in the bargain so that I contented myself with a profit of one dinar; and had I exerted myself, [the buyer] would have allowed me a profit of one dinar and a quarter, since he was very anxious to buy; therefore the best thing to do now is in the weighing to take over-full measure and through overweighing to recover the loss, and to exact the currency which I deem best; also I shall give nothing to any broker or middleman"—if he has said that to himself and has acted accordingly, then discord will follow. For a disruptive element thrusts itself between the thoughts of the two; the buyer turns away and [the seller] loses everything. Now the latter would like the former to come back, and he has exchanged the sparrow in the hand for the pigeon on the roof. From the Prophet the saying is transmitted, "Deductions are profit," and "God grants blessings to a man who grants deductions when paying and charging, when buying and selling." And a popular proverb says, "The oil sells the cake."

---

1 A branch of business connected with lending, not to be confused with *qired or muqarada*, the Muslim contract resembling the Western *commenda*. The story told by Abu al-Fadl is certainly untrue; the Koran condemns interest bearing loans. But it indicates that Muslim merchants like their Christian colleagues, sought for release from religious prohibitions.

BENEDETTO COTRUGLI, *ON COMMERCE AND THE PERFECT MERCHANT*

■ NAPLES, 1458 ■

On the Universal Manner and Order of Business

Inasmuch as all things in the world have been made with a certain order, in like manner they must be managed—and most particularly those which are of the greatest importance, such as the business of merchants, which, as we have said, is ordered for the preservation of the human race. Hence a merchant must manage himself and his merchandise with a certain order tending to his purpose, which is [the attainment of] wealth.

Nonetheless the order must be different according to the different substance and capital which a man happens to have. A very rich man must manage one way, a rich man another, and a man who has a small capital still another. This is why some people have knowledge and ability to manage large sums, others to manage small ones. For those who are rich and have the management of many weighty matters ought to maintain their intelligence on a high [plane], and to investigate lofty matters in a rational way, as the saying goes, "The greater the ship, the greater the labor." And we must not undertake great things trusting the advice of sailors or of certain frivolous men and passers-by. Because a sailor deals with gross things and has a gross intelligence, when he drinks in a tavern or buys bread in the [public] square people would believe that he is an important man and that he brings you valuable advice on wine and bread by saying that whoever carries it to some place would make out well with it. A moderate merchant, and especially one who has great things in his care, must not engage in purchases of grain and wine on the advice of such men. But he must strive to obtain advice from merchants and to search and probe his own mind, always remembering that excellent saying of Lactantius. . . .

A great merchant, then, must plan his business and apportion it in a [certain] order. And he must not keep all money together, but he should distribute it in various solid businesses. And this method, in my opinion, is very efficiently followed by the Florentines more than by any other people; this I say as a generalization, for many others also follow it. That is to say, [suppose] that I am a great and rich merchant in Florence. I acquire an interest (*intravengo*) in the Altoviti [partnership], whose management is in Venice, and I invest 2000 ducats of my own in that partnership (*compagnia*); and I draw one fourth of the profit. . . . And I enter into another partnership in Rome, and invest 1000 ducats in it; into another in Avignon, and invest 1000 ducats in it; in a shop of the Art of Silk, 1000 ducats. And according to my lot and substance I keep within my own management 6000 ducats with which I do business in my own name and in such merchandise as seems best to me at the moment. And since I have a finger in many places, in solid and planned [investments], I cannot but make out well, for the one makes up for the other; whereas if I had all money gathered together I would have grounds for fear—for I would always have extra money, and I would wish to catch every bird, and deal with very bad payers; or else, wishing to embrace [too] much, I would lose [my own] and take a bad fall. But in this way, if I distribute my own [capital], every partnership has its managers who are restricted and under orders. And [the managers] do not put out too much beyond the little capital (*corpo*) that they have, both because they do not always receive a commission and because they do not have too much extra money. Therefore this is a sound, solid, and healthy management for those who are very rich.

Those who have a medium amount of money, such as 4000 ducats, must manage another way. That is, they must not divide that capital of their own, but keep it solidly tied in one body, except that they may sometimes—yet seldom—make *commenda* contracts of 400 to 500 ducats, and get [them] back, and frequently see the accounting and clear the profits; so that your money may often come back into your hands. And our own people of Ragusa are very skilled in this [kind of] management; I should praise them at leisure in this respect if I did not think that a critic would ascribe this to my love of fatherland. [This happens] both because they employ wares which are quickly [disposed of], such as silver, gold, lead, copper, wax, crimson, leather, and the like, and because of the dexterity of mind which they have —[or rather, which they would have] if they did not blunder. For as they begin to increase their capital they begin to build or to overturn stones in making gardens, vineyards, and in other pursuits outside rather than inside the ground; so that they have made such a great and beautiful show of palaces that it is a wonderful thing to see. To them I shall say with St. Paul, "I give praise for everything, but I do not give praise for this. . . ."

We shall now speak of those who have little money, up to some 500 ducats. They must get personally busy with said money, and make no *commenda* contracts nor anything else; and [they must] not spread it over several businesses. And they must help the money with personal effort, for if you wish to stand still with so little money you will consume it. For usually the profits which stand still are limited and small; and with little money you cannot save.

Those who do not have anything must strive to engage in any personal exertion without being ashamed of adapting themselves to the circumstances, even as the Tragedian warns when he cries out, "It is fit to adapt oneself to the circumstance." One should not be ashamed of being with others and serving—witness the same Seneca: "Nor do I regard as shameful anything that fortune may command to the unfortunate"—and of making any low and mean exertion, provided it is honest, to enable oneself to begin to have [something]. To be with others we do not regard as mean, indeed we consider it as necessary for a merchant. . . . The experienced Genoese, Florentines, and Venetians are wont to do it today, and our own country also was accustomed to it not long ago, during my early youth. And I saw many gentlemen entrust to their [fellow] citizens their own sons, [to be] trained and placed in some good position, so that from childhood they could learn their art, of which they were much more avid than they are now; wherefore our income has grown and [our] soul has enlarged. I also saw [sons of gentlemen] exert themselves in their profession, not only in the services belonging to their required tasks but even in sweeping the shop; nor were they ashamed. But this custom has been well preserved by the Florentines—both to place [their sons] with others and to engage in any other honest endeavor, mean though it may be.

I have seen great men who, being impoverished, were not ashamed of lending horses to carters, and of engaging in brokerage, innkeeping, and any other business of the kind. And I have seen some of them return rich in a short time, with 10,000 ducats; for the sake of honesty I do not wish to name them or to extol them in praise—I should not wish to make them proud or humiliated in their glory. And it will be noted that usually, when a Genoese is impoverished through some accident of adverse fortune, he becomes a pirate, and [so do] some Catalans; the Florentines [become] brokers or artisans in some craft, and they exert themselves and help themselves with industry. . . .

Women in Bourgeois Society

In the charming vignette of late–fourteenth-century Parisian life that follows we have a chance to observe the wife of an apparently prosperous merchant in terms of his concept of her station. Presented in the form of a letter, this selection is in reality a discourse of wifely duties and of the ideal attainments of a loving and virtuous woman. It also provides some detail about the patterns of consumption possible for a Parisian bourgeois: good linen, a well-lit house with window glass, a smokeless fire, bread to spare for catching fleas, and a board groaning with exotic foods!

The life of a prosperous merchant's wife was less onerous than that of a peasant woman. She was probably freer to develop facets of a personality closed to women in classical society. But there was a crushing weight of social convention established by male dominance, of hierarchy in social life, of dependence.

From *The Goodman of Paris*

Anonymous

THE GOOD WIFE

Dear Sister,

You being the age of fifteen years and, in the week that you and I were wed, did pray me to be indulgent to your youth and to your small and ignorant service, until you had seen and learned more; to this end you promised me to give all heed and to set all care and diligence to keep my peace and my love, as you spoke full wisely, and, as I well believe, with other wisdom than your own, beseeching me humbly in our bed, as I remember, for the love of God not to correct you harshly before strangers nor before our own folk, but rather each night, or from day to day, in our chamber, to remind you of the unseemly or foolish things done in the day or days past, and chastise you, if it pleased me, and then you would strive to amend yourself according to my teaching and correction, and to serve my will in all things, as you said. And your words were pleasing to me, and won my praise and thanks, and I have often remembered them since. And know, dear sister, that all that I know you have done since we were wed until now, and all that you shall do hereafter with good intent, was and is to my liking, pleaseth me, and has well pleased me, and will please me. For your youth excuses your unwisdom and will still excuse you in all things as long as all you do is with good intent and not displeasing to me. And know that I am pleased rather than displeased that you tend rose-trees, and care for violets, and make chaplets, and dance, and sing: nor would I have you cease to do so among our friends and equals, and it is but good and seemly so to pass the time of your youth, so long as you neither seek nor try to go to the feasts and dances of lords of too high rank, for that does not become you, nor does it sort with your estate, nor mine. And as for the greater service that you say you would willingly do for me, if you were able and I taught it you,

SOURCE: E. Power, trans., *The Goodman of Paris* (London: Routledge & Kegan Paul, 1928).

know, dear sister, that I am well content that you should do me such service as your good neighbours of like estate do for their husbands, and as your kinswomen do unto their husbands. . . . And lastly, meseems that if your love is as it has appeared in your good words, it can be accomplished in this way, namely in a general instruction that I will write for you and present to you, in three sections containing nineteen principal articles. . . .

Care of a Husband

The seventh article of the first section showeth how you should be careful and thoughtful of your husband's person. Wherefore, fair sister, if you have another husband after me, know that you should think much of his person, for after that a woman has lost her first husband and marriage, she commonly findeth it hard to find a second to her liking, according to her estate, and she remaineth long while all lonely and disconsolate and the more so still if she lose the second. Wherefore love your husband's person carefully, and I pray you keep him in clean linen, for that is your business, and because the trouble and care of outside affairs lieth with men, so must husbands take heed, and go and come, and journey hither and thither, in rain and wind, in snow and hail, now drenched, now dry, now sweating, now shivering, ill-fed, ill-lodged, ill-warmed, and ill-bedded. And naught harmeth him, because he is upheld by the hope that he hath of the care which his wife will take of him on his return, and of the ease, the joys, and the pleasures which she will do him, or cause to be done to him in her presence; to be unshod before a good fire, to have his feet washed and fresh shoes and hose, to be given good food and drink, to be well served and well looked after, well bedded in white sheets and nightcaps, well covered with good furs, and assuaged with other joys and desports, privities, loves, and secrets whereof I am silent. And the next day fresh shirts and garments.

Certes, fair sister, such services make a man love and desire to return to his home and to see his goodwife, and to be distant with others. Wherefore I counsel you to make such cheer to your husband at all his comings and stayings, and to persevere therein; and also be peaceable with him, and remember the rustic proverb, which saith that there be three things which drive the goodman from home, to wit, a leaking roof, a smoky chimney, and a scolding woman. And therefore, fair sister, I beseech you that you keep yourself in the love and good favour of your husband, you be unto him gentle, and amiable, and debonair. Do unto him what the good simple women of our country say hath been done to their sons, when these have set their love elsewhere and their mothers cannot wean them therefrom.

Wherefore, dear sister, I beseech you thus to bewitch and bewitch again your husband that shall be, and beware of roofless house and of smoky fire, and scold him not, but be unto him gentle and amiable and peaceable. Have a care that in winter he have a good fire and smokeless and let him rest well and be well covered between your breasts, and thus bewitch him. And in summer take heed that there be no fleas in your chamber, nor in your bed, the which you may do in six ways, as I have heard tell. For I have heard from several that if the room be strewn with alder leaves, the fleas will be caught thereon. Item, I have heard tell that if you have at night one or two trenchers [of bread] slimed with glue or turpentine and set about the room, with a lighted candle in the midst of each trencher, they will come and be stuck thereto. The other way that I have tried and 'tis true: take a rough cloth and spread it about your room and over your bed, and all the fleas that shall hop thereon will be caught, so that you may carry them away with the cloth wheresoe'er you will. Item, sheepskins. Item, I have seen blanchets [of white wool] set on the straw and on the bed, and when the black fleas hopped thereon, they were the sooner found upon the white, and killed. But the best way is to guard oneself against those that be within the coverlets and the furs, and the stuff of the dresses wherewith one is covered. For know that I have tried this, and when the coverlets, furs, or dresses, wherein there be fleas, be folded and shut tightly up, as in the chest tightly corded with straps, or in a bag well tied up and pressed, or otherwise put and pressed so that the aforesaid fleas be without light and air and kept imprisoned, then will they perish forthwith and die. Item, I have sometimes seen in divers chambers, that when one had gone to bed they were full of mosquitoes, which at the smoke of the breath came to sit on the faces of those that slept, and stung them so hard, that they were fain to get up and light a fire of hay, in order to make a smoke so that they had to fly away or die, and this may be done by day if they be suspected, and likewise he that hath a mosquito net may protect himself therewith.

And if you have a chamber or a passage where there is great resort of flies, take little sprigs of fern and tie them to threads like to tassels, and hang them up and all the flies will settle on them at eventide; then take down the tassels and throw them out. Item, shut up your chamber closely in the evening, but let there be a little opening in the wall towards the east, and as soon as the dawn breaketh, all the flies will go forth through this opening, and then let it be stopped up. . . .

Item, have whisks wherewith to slay them by hand. Item, have little twigs covered with glue on a basin of water. Item, have your windows shut full tight with oiled or other cloth, or with parchment or something else, so tightly that no fly may enter, and let the flies that be within be slain with the whisk or otherwise as above, and no others will come in. Item, have a string hanging soaked in honey, and the flies will come and settle thereon and at eventide let them be taken in a bag. Finally meseemeth that flies will not stop in a room wherein there be no standing tables, forms, dressers or other things whereon they can settle and rest, for if they have naught but straight walls whereon to settle and cling, they will not settle, nor will they in a shady or damp place. Wherefore meseemeth that if the room be well watered and well closed and shut up, and if naught be left lying on the floor, no fly will settle there.

And thus shall you preserve and keep your husband from all discomforts and give him all the comforts whereof you can bethink you, and serve him and have him served in your house, and you shall look to him for outside things, for if he be good he will take even more pains and labour therein than you wish, and by doing what I have said, you will cause him ever to miss you and have his heart with you and your loving service and he will shun all other houses, all other women, all other services and households.

Menus and Recipes

Dinner for a Meat Day Served in Thirty-one Dishes and Six Courses

First course. [Wine of] Grenache and roasts, veal pasties, pimpernel pasties, black-puddings and sausages.

Second course. Hares in civey and cutlets, pea soup [*lit.*, strained peas], salt meat and great joints (*grosse char*), a soringue of eels and other fish.

Third course. Roast: coneys, partridges, capons, etc., luce, bar, carp, and a quartered pottage.

Fourth course. River fish à la dodine, savoury rice, a bourrey with hot sauce and eels reversed.

Fifth course. Lark pasties, rissoles, larded milk, sugared flawns.

Sixth course. Pears and comfits, medlars and peeled nuts. Hippocras and wafers. . . .

A Fish Dinner

First course. Pea soup, herring, salt eels, a black civey of oysters, an almond brewet, a tile, a broth of broach and eels, a cretonnée, a green brewet of eels, silver pasties.

Second course. Salt and freshwater fish, bream and salmon pasties, eels reversed, and a brown herbolace, tench with a larded broth, a blankmanger, crisps, lettuces, losenges, orillettes, and Norwegian pasties, stuffed luce and salmon.

Third course. Porpoise frumenty, glazed pommeaulx, Spanish puffs and chastelettes, roast fish, jelly, lampreys, congers and turbot with green sauce, breams with verjuice, leches fried, darioles and entremet. Then Dessert, Issue, and Sally-Forth. . . .

Rosée of Young Rabbits, Larks and Small Birds or Chickens

Let the rabbits be skinned, cut up, parboiled, done again in cold water and larded; let the chickens be scalded for plucking, then done again, cut up and larded, and let larks and little birds be plucked only for parboiling in sewe of meat; then have bacon lard cut up into little squares and put them into a frying pan and take away the lumps but leave the fat, and therein fry your meat, or set your meat to boil on the coal, often turning it, in a pot with fat. And while you do this, have peeled almonds and moisten them with beef broth and run it through the strainer, then have ginger, a head of clove, cedar otherwise hight *alexander* [red cedar], make some gravy and strain it and when the meat is cooked set it in a pot with the broth and plenty of sugar; then serve in bowls with glazed spices thereon.

Eel Reversed

Take a large eel and steam it, then slice it along the back the length of the bone on both sides, in such manner that you draw out the bone, tail and head all together, then wash and turn it inside out, to wit the flesh outwards, and let it be tied from place to place; and set it to boil in red wine. Then take it out and cut the thread with a knife or scissors, and set it to cool on a towel. Then take ginger, cinnamon, cloves, flour of cinnamon, grain [of Paradise], nutmegs, and bray them and set them aside. Then take bread toasted and well brayed, and let it not be strained, but moistened with wine wherein the eel hath been cooked and boil all together in an iron pan and put in verjuice, wine, and vinegar and cast them on the eel.

The Universities
and the Origins of Professions

The universities of Europe emerged from schools conducted by the clergy of cathedral chapters. By the end of the twelfth century, universities conformed to one of two distinct types. Paris was a corporation of masters (men licensed to teach) of the arts. Bologna was a union of students who hired whatever teachers suited their fancy. In the pages of Heer's *The Medieval World*, the rise of these corporations is put in the context of the often conflicting needs of monarchy and papacy for learned servants. Hence, quite apart from the intriguing questions of the nature of university life, of town-gown squabbles, of student culture as a medieval "counter-culture" (see pp. 379–391), universities expressed the diversification of professional and urban life in European society after 1100. The growth of government required men learned in the law. The expansion of urban life and its complexities of business put a new premium on more than mere literacy. The conflicts of secular and church power augmented the ability of universities to play the one against the other, in search of new privileges and more august patrons. Masters and students came chiefly from the ranks of aristocratic and mercantile families. But the university was an urban institution, and the thrust of its development gave impetus to classes of professionals whose life style was less and less tied to that of lord-based aristocracies. The university was also a secularizing force, despite the fact its members were in law clerics. That is why we introduce it in the context of towns, commerce, and burgher life.

From *The Medieval World: Europe 1100–1350*

Friedrich Heer

INTELLECTUALISM AND THE UNIVERSITIES

The medieval University of Paris was the battlefield of all the most significant intellectual conflicts of the age; it was also the place where thinkers of the later Middle Ages started to lay some of the important foundations of modern scientific thought. As an institution, it throve on conflict; perpetually at odds with officialdom (episcopal, papal, and royal), for a long time obdurate in its resistance to the Mendicant Orders, during the thirteenth and fourteenth centuries the University of Paris became a power in Europe of an entirely different order of magnitude from the universities of today. Medieval Paris was a world within a world; its interior life will later be explored from the inside; we are concerned here only with its outward appearance and structure.

It will be recalled that toward the end of the eleventh century or the beginning of the twelfth the business of teaching was passing from the monks to the secular clergy; this was a distinct revolution in intellectual matters. Abelard, although he

SOURCE: Friedrich Heer, *The Medieval World: Europe 1100–1350*, trans. Janet Sondheimer (Cleveland, Ohio: The World Publishing Company; London: George Weidenfeld & Nicolson 1962), pp. 199–203, 209–211.

knew nothing of universities, was the first teacher to attract large crowds of students to Paris from every country in Europe. There were three important schools, cells from which the future university developed: the school of the collegiate church of Ste. Geneviève, the monastic school of St. Victor, and the cathedral school of Notre Dame. The university grew up between 1150 and 1170, with the Chancellor of the cathedral church of Paris, Notre Dame, as its "ecclesiastical superintendent," empowered to confer the *licentia docendi* on all eligible applicants. The Chancellor also claimed to be the *judex ordinarius* of the scholars. For a long time the most important teachers at Paris were foreigners; it was their guild which made up the university. It is important to notice that during the decisive years of conflict between the university and the Chancellor appointed by the bishop, the Pope sided with the university (1200–20). Rome was anxious to win Paris for itself as a centre of theological studies, and this was the reason why in 1219 Honorius III forbade the teaching of Roman law at Paris. The Pope was afraid that the young theologians studying at Paris might be seduced from their proper studies by the attractions of careerism, the prospect of advancement in the service of Church and State available to those who had studied the law. This prohibition also suited the French kings, who feared Roman law because of its associations with the Empire.

By far the largest faculty in the university was that of arts, the "philosophical" faculty. The curriculum was based on the seven liberal arts, grammar, rhetoric and dialectics, which made up the *trivium*, and music, arithmetic, geometry and astronomy, which were the *quadrivium*. In the thirteenth century interest was concentrated on the subjects of the *trivium*, that is on philosophy and its ancillary disciplines; by the fourteenth century it had shifted to the more "scientific" subjects of the *quadrivium*, largely in consequence of ecclesiastical sanctions against the free-thinking philosophy of the "artists."

Study in the faculty of arts was really intended as a prologue to the study of theology. The only "full" professors at Paris were the doctors of theology, who had completed sixteen years of study, of which the years spent in the faculty of arts were only a necessary preliminary, to be looked back on with contempt in after years. Nevertheless, the theologians tended to regard the "artists" with disquiet and displeasure, often not untinged with envy. This is understandable when it is realized how greatly the artists increased in numbers in the second half of the twelfth century, so that the young masters of arts came to constitute a university on their own. By 1362 there were 441 masters in the faculty of arts as compared with 25 in theology, 25 in medicine and 11 in cannon law.

At Paris the masters of arts were the permanent element of intellectual unrest and the driving force of intellectual revolutions. The quality and historical importance of the "full" professors, the theologians, can be measured by their response to the perpetual challenge of the artists: whether, that is, they stood their ground on the battlefield of disputation or removed themselves in cowardly flight. Thomas Aquinas and his teacher Albertus Magnus both stood their ground.

The intellectual restlessness of the artists was closely connected with the uncertainty of their status. On the one hand they were teachers; all the students had to pass through their hands. But they were students as well, engaged in further studies leading to doctorates in theology and professorships. Just because they were so numerous, many hundreds of them were denied any hope of a chair, which only added to their unrest. Quite a few, including some of the best, had no wish to become theologians: they preferred the greater intellectual freedom and mobility afforded by philosophy. With a foot in both camps, at once teachers and students, the masters of arts came to occupy a key position in the university. The crisis came about the year 1220. Before 1200 the Chancellor of Notre Dame had restricted all lecturing to the Île de la Cité, where it would be under the close control of the bishop, whose representative he was. But about 1210 the masters of arts withdrew themselves from the Chancellor's jurisdiction and settled on the left bank of the Seine, in the Rue du Fouarre, "the street of straw," so-called because of the straw-covered entries to the schools. The Latin Quarter was now in being. Jurisdiction over the arts faculty was assumed by the Abbot of Ste. Geneviève, a rival of the Chancellor and the cathedral school.

A few minutes' walk from the Île de la Cité brings one to the tiny group of houses which is all that remains of the old Rue du Fouarre. They are a picturesque, typically Parisian jumble of buildings, a mixture of the styles of the last few centuries, cheek by jowl with the more venerable survivals from the Paris of the early Middle Ages which stand nearby. This maze of streets and houses, a great favourite with painters, is the monument to a past age of intellectual heroism: in this place the intellectual life of medieval Europe found its greatest freedom, here the love of controversy was extended to the limit.

Dante, in the *Divine Comedy*, puts his eulogy of Siger of Brabant, chief of the artists, into the mouth of Thomas Aquinas, chief of the theologians and Siger's great adversary:

> . . . the eternal light of Sigebert
> Who escaped not envy, when of truth he argued,
> Reading in the straw-littered street.

Condemned by the Church, Siger died wretchedly in prison. Dante, however, makes Aquinas sing his praises in Paradise, an illuminating comment on the undercurrents of medieval intellectual life.

From the beginning life among the arts students at Paris was a turbulent affair. In the early years of the thirteenth century a cell of "pantheists" and "freethinkers" was uncovered, and several priests and clerks from the schools were burned or imprisoned in consequence. Their heresy had been inspired by Amaury of Bêne and David of Dinant; in 1215 the papal legate forbade all lecturing on Aristotle's books of natural philosophy (i.e., the *Physics* and *Metaphysics*), and ordered all arts students to abstain on oath from reading the works of David of Dinant and other heretics. By 1255, however, practically all the known works of Aristotle had become required reading for students in the arts faculty. In the meantime the artists had gained control of the university, with their Rector as its head, and the oath of obedience to the Rector became the foundation of the whole academic structure.

The artists reached this position of strength as the leaders of the "Nations," the students. There were four "Nations," usually mutually antagonistic, three French (the French, Norman, and Picard) and the English, which included all students from Germany and other northern countries (in the fifteenth century it came to be called the "German" Nation). From the beginning of the thirteenth century there were also the three faculties, of arts, theology, and canon law (medicine belonged to the arts faculty).

From the early thirteenth century the university as a whole (i.e., both teachers and students) was in conflict with the Chancellor and the city of Paris. The first and largest migration of students and teachers from Paris (in 1228–29) was occasioned by carnival riots. The emigrants dispersed to Oxford, Cambridge, Angers, Toulouse, Orleans, and Rheims, there to escape the vigilance of Parisian officials, ecclesiastical and lay. (The city and its citizens had no doubt suffered much from the university, with its numerous special corporations and large crowds of students.) Rome was alarmed at the exodus. Gregory IX, whose attempt to bring the wanderers back was only partially successful, in 1231 promulgated the Bull *Parens Scientiarum*, the "Magna Carta" of the university of Paris: Paris was recognized as the mother of the sciences, and the jurisdiction over it of the Chancellor and the Bishop of Paris was severely curtailed.

After the dispersion of 1228–29 there was a partial truce between the artists and the theologians, brought about by their common opposition to the Mendicant Orders, who had taken advantage of the upheaval to extend their teaching activities. The secular clerks regarded the Dominicans and Franciscans, bound as they were by the strictest obedience to their own Orders and the Papacy, as undisguised enemies of the university. In 1253 the university imposed an oath on all masters, binding them to observe the university statutes. This was something the friars could not and would not tolerate. In 1255 a Papal Bull threatened the university with excommunication if it failed to admit friars as doctors. Thereupon the university formally dissolved itself. Such a dissolution, together with secession and cessation (i.e., suspension of all teaching), was the universities' most formidable weapon, and it could be used with impunity. A university possessed no buildings of its own; it could keep its "open house" anywhere, and its teachers and students would find a welcome everywhere in Europe.

In the course of this conflict each side denounced the other as "heretics," "atheists," and so forth. The progressive overclouding of the intellectual climate, which became more ominous as the thirteenth century went on its way, is closely connected with this quarrel. A temporary truce was reached in 1261 through the good offices of Pope Urban VI, who had studied canon law at Paris: the university as a whole agreed to admit the friars, but they were to be excluded from membership of the faculty of arts, which was preserved as an area of freedom. In addition, no religious college was to have more than one doctor acting as a regent of the university (a member of the governing body, normally composed of fully qualified Masters and Doctors), with the exception of the Dominicans, who were allowed to appoint two. This was really a victory for the university, whose corporate spirit had by now become well established; the struggle with the friars had made the university set its own house in order. In 1318 the oath of obedience was at length imposed on the friars.

As a political power the University of Paris reached its maturity in the fourteenth and early fifteenth centuries. The "eldest daughter of the French monarchy," it prided itself on being above all secular jurisdiction, including that of the *Parlement*, the king's court. In 1398 Gerson became Rector, and during his time the university soared even higher, to become judge over Popes and anti-Popes. The Council of Constance was organized into four Nations (French, English, German and Italian), a device aimed at reducing the dominant influence of the Italians and modelled on the organisation of the University of Paris.

But now the strain of so much self-confident intellectualism and striving after power began to tell. The university revealed its inner weaknesses in its posthumous condemnation of Joan of Arc as a heretic and a witch, two years after her death. It was finally brought low after a prolonged struggle with the French monarchy during the reigns of Charles VII, Louis XI and Louis XII. In the last year of the fifteenth century Louis XII rode armed into the university precincts, revoked the right of cessation, and thus finally brought the university to heel. By his action Louis had brought to an end a great variety of liberties, some of which we have failed to recover. . . .

The medieval universities trained up and molded a new class, and what might almost be described as a new type of man: the academic and the intellectual. The two terms are not automatically interchangeable. Some members of the new class spent from fourteen to sixteen years as students. The minimum age for entry was usually fourteen, and most of the students coming to the universities were between thirteen and sixteen years old. A very large percentage had their studies paid for by someone else, a wealthy relative or patron, an important cleric or some student foundation. Many were very poor. Their miserable lodgings were poorly heated and ill-lit. Their life was turbulent and often dissipated; brawls with the townsfolk and artisans, assassinations and excesses of all kind were a regular part of it. Discipline in the colleges was often harsh and gloomy, distinctly parsonical in tone, and accompanied by a highly developed system of informing. At the biennial meetings of the chapter the students were obliged to denounce one another's faults. This system affected men in different ways, according to their disposition: while an Erasmus or a Rabelais found it intolerable, Calvin adopted it as part of his own spiritual discipline.

Lectures started at a very early hour, six o'clock in winter, in summer even earlier, and often lasted for three hours. Famous teachers attracted large crowds, which the antiquated rooms serving as lecture-halls could barely accommodate. There were very few of the size and beauty of the late medieval lecture-hall at Salamanca which still survives for our admiration.

Contemporaries seem to have found the throng of students so over-whelming that they were driven to making wild estimates of their numbers. For example, it is said that in 1287 Paris had 30,000 students and Bologna 10,000. In fact Paris probably averaged about 2500, rising to 6000 or 7000 in its heyday. At the end of the thirteenth century Oxford may have had between 1500 and 2000 students.

It is highly significant that the medieval universities gave their students no spiritual training or instruction. This was something only introduced later by teachers whose personal religion had a strong emotional backing, the reformers Wyclif and Hus for instance; when it came, it marked the dawn of a new age. Strong religious and national feelings—the two were often combined—later came to affect both students and teachers, particularly in Eastern Europe.

Clerks trained in the university law schools became indispensable functionaries of the newly developing and rising states of the high Middle Ages, particularly France and England. Kings, Popes, princes, prelates, towns and corporations all competed for the services of this new class. The *clerici* were the equivalent of our "managerial class." They were administrators and bureaucrats, often the only people capable of manipulating the levers of power. Their steady rise from the twelfth century onwards, which now excites such interest, at the time went almost unobserved. Chroniclers could hardly be expected to notice it: their record is of battles and of the endless quarrels of Popes, kings, and barons.

But in the background the men of the new class were doggedly and systematically building up and enlarging the structure of the state, in France, in England and in the territorial principalities of Germany; no less important, they were turning the Church itself into a bureaucracy. They went serenely on their way, unperturbed by the misfortunes of their masters, unshaken by the frailties of kings or the *hubris* of Popes. Even their own setbacks failed to dismay them: if one of their number fell from grace, plenty were left to carry on with the work of bringing reason to bear on the exercise of power. The ablest of them were completely absorbed in their mission. Church and State should be fashioned into works of art, creations constructed with the same intellectual care as went to the making of a Gothic cathedral. The State, the Church, and the administrative apparatus were ends in themselves; for those who served them the old boundaries dissolved and became meaningless.

The education of this new class had a legalistic and intellectual bias. The grandeurs and miseries of the European intellectual have their origin in the medieval university, with its almost complete concentration on the education of men's reason, a task it admittedly performed to admiration. But there were whole areas it left untouched: it failed to reach the affections and emotions, failed to instill manners of a deeper sort, failed to mold the elemental forces of personality, failed to provide the nourishment of true religion. Medieval intellectuals often show symptoms of a split personality: their intellects might be highly developed, but their manners were vicious and uncouth, their personalities spiritually immature. They also had to cope with the occupational disability of their calling, homosexuality; it was not merely this, however, but the defectiveness of their personalities as a whole which so aroused the indignation of sensitive and genuinely devout men. The Paris of the intellectuals was Babylon to Jacques de Vitry, as it had been to Bernard of Clairvaux. Bernard exhorted teachers and students to flee from it and save their souls. To other people, however, Paris was paradise itself. The goliards and wandering scholars, intellectuals after their own fashion, had already sung its praises: "paradise on earth, the rose of the world, the balm of the universe." The hypercritical intellectuals of Europe made it their capital. They criticized everything, each other, the bureaucratic Church, and above all the monks, whom they stigmatized as slothful, gluttonous and lewd. The later attacks on monasticism coming from Valla and Erasmus belong to this acrimonious tradition. Typically urban, the intellectuals of Paris looked down on everything rustic and provincial, including the peasantry. The aristocracy they regarded as at least their equals if not their inferiors: what counted was the "inner nobility" of the intellect. One could fill a long catalogue with these shortcomings and vices. Their vocation made these men very vulnerable. If they were remorseless, it was in the nature of their calling. Robert de Sorbon asserted that "nothing is fully known until it has been chewed to shreds in argument." The pursuit of knowledge by way of logic imposed an endless process of dissection, analysis, categorization, clarification, distinction and elucidation. Not a few of them lived in the confident hope (and illusion) of becoming capable of all things: of being able to understand and comprehend everything under a precise formulation, logically arranged. They had their own version of Faust's dream: the world and all reality might be controlled if only man could arrive at a proper grasp of them, that is, if everything could be formulated in the correct way.

Jews in Christian Cities

In dealing with cities and commerce in medieval Europe, we must confront the question of the Jews—the "outsiders" in a Christian society. Paradoxically the Jew was also an "insider." The West European Jewish communities between ca. 700 and the twelfth century had a history fundamentally at odds with that subsequent to the Crusades. The Sephardic Jews of Spain, Provençal, Italy, and North Africa lived in a world of Moslem-Christian conflict. But in general they enjoyed the tolerance which had early been shown them in Moslem civilization. A much smaller community of Jews appears beyond the Alps (Ashkenazim), chiefly in Germany, only after the year A.D. 1000. With the expansion of commerce, European Christians and Jews found themselves cooperating with Arabic-speaking Jews in the Levant trade but in mutual competition. Fortunately, a large Roman Jewish population was in regular contact with the papacy, which since the time of Gregory the Great had insisted on "toleration until the end of time." This in practice meant no forced conversions. Hence Jewish centers of intellectual life and economic activity developed readily. Kings extended special protections for Jews and their property. Pope Calixtus II promulgated constitutions for Jews (*constitutiones pro Judaeis*) in Rome in 1119. But from the time of the first Lateran Council (1123), and especially under the impact of crusading zeal, Jewish life and property became less secure. Nonetheless, popes renewed their protective legislation. The full range of these *constitutiones* appears in that of Gregory X. Valued for their learning and experience in commerce, Jews were excluded from officeholding and eventually from landownership. Protection for European Jews came to mean security of person and property coupled with disfranchisement.

Yet in times of acute stress within the Christian host communities, Jews found themselves the scapegoats of Christian society—in the wrath of Innocent II, in the slaughters of Ashkenazim in Germany which followed abortive crusades, and in the wake of the Black Death. Anti-Jewish feeling easily spread. The third Lateran Council published anti-Jewish decrees. Fear, hatred, repressed desires, and envy often focused on the Jewish "outsider."

The need of victims is perhaps a universal trait in men; if so, Jews in Western society were an ideal fulfillment: a community set apart by law, occupation, political-social deprivation, and religion. In the thirteenth century especially, when Christian rulers made wide use of Jewish servants in economic and fiscal matters, and in the aftermath of the Death, the issue was a burning one. Then, great pogroms (persecutions and murders) took place in the Rhineland cities. Jacob von Konigshöfen preserves the memory of what took place in 1348 in Strasbourg.

From *Papal Protection of the Jews*

Pope Gregory X

■ 1272 ■

Gregory, bishop, servant of the servants of God, extends greetings and the apostolic benediction to the beloved sons in Christ, the faithful Christians, to those here now and those in the future. Even as it is not allowed to the Jews in their assemblies presumptuously to undertake for themselves more than that which is permitted them by law, even so they ought not to suffer any disadvantage in those [privileges] which have been granted them. [This sentence, first written by Gregory I in 598, embodies the attitude of the Church to the Jew.] Although they prefer to persist in their stubbornness rather than to recognize the words of their prophets and the mysteries of the Scriptures [which, according to the Church, foretold the coming of Jesus], and thus to arrive at a knowledge of Christian faith and salvation; nevertheless, inasmuch as they have made an appeal for our protection and help, we therefore admit their petition and offer them the shield of our protection through the clemency of Christian piety. In so doing we follow in the footsteps of our predecessors of blessed memory, the popes of Rome—Calixtus, Eugene, Alexander, Clement, Celestine, Innocent, and Honorius.

We decree moreover that no Christian shall compel them or any one of their group to come to baptism unwillingly. But if any one of them shall take refuge of his own accord with Christians, because of conviction, then after his intention will have been manifest, he shall be made a Christian without any intrigue. For, indeed, that person who is known to have come to Christian baptism not freely, but unwillingly, is not believed to possess the Christian faith. [The Church, in principle, never approved of compulsory baptism of Jews.]

Moreover no Christian shall presume to seize, imprison, wound, torture, mutilate, kill, or inflict violence on them; furthermore no one shall presume, except by judicial action of the authorities of the country, to change the good customs in the land where they live for the purpose of taking their money or goods from them or from others.

In addition, no one shall disturb them in any way during the celebration of their festivals, whether by day or by night, with clubs or stones or anything else. Also no one shall exact any compulsory service of them unless it be that which they have been accustomed to render in previous times. [Up to this point Gregory X has merely repeated the bulls of his predecessors.]

Inasmuch as the Jews are not able to bear witness against the Christians, we decree furthermore that the testimony of Christians against Jews shall not be valid unless there is among these Christians some Jew who is there for the purpose of offering testimony.

[The church council at Carthage, as early as 419, had forbidden Jews to bear witness against Christians; Justinian's law of 531 repeats this prohibition. Gregory X here—in accordance with the medieval legal principle that every man has the right to be judged by his peers—insists that Jews can only be condemned if

SOURCE: Pope Gregory X, *Papal Protection of the Jews*, in *The Jew in the Medieval World*, ed. J. R. Marcus (Cincinnati: Sinai Press, 1938).

there are Jewish as well as Christian witnesses against them. A similar law to protect Jews was issued before 825 by Louis the Pious (814–840) of the Frankish Empire.]

Since it happens occasionally that some Christians lose their Christian children, the Jews are accused by their enemies of secretly carrying off and killing these same Christian children and of making sacrifices of the heart and blood of these very children. It happens, too, that the parents of these children, or some other Christian enemies of these Jews, secretly hide these very children in order that they may be able to injure these Jews, and in order that they may be able to extort from them a certain amount of money by redeeming them from their straits. [Following the lead of Innocent IV, 1247, Gregory attacks the ritual murder charge at length.]

And most falsely do these Christians claim that the Jews have secretly and furtively carried away these children and killed them, and that the Jews offer sacrifice from the heart and the blood of these children, since their law in this matter precisely and expressly forbids Jews to sacrifice, eat, or drink the blood, or to eat the flesh of animals having claws. This has been demonstrated many times at our court by Jews converted to the Christian faith: nevertheless very many Jews are often seized and detained unjustly because of this.

We decree, therefore, that Christians need not be obeyed against Jews in a case or situation of this type, and we order that Jews seized under such a silly pretext be freed from imprisonment, and that they shall not be arrested henceforth on such a miserable pretext, unless—which we do not believe—they be caught in the commission of the crime. We decree that no Christian shall stir up anything new against them, but that they should be maintained in that status and position in which they were in the time of our predecessors, from antiquity till now.

We decree, in order to stop the wickedness and avarice of bad men, that no one shall dare to devastate or to destroy a cemetery of the Jews or to dig up human bodies for the sake of getting money. [The Jews had to pay a ransom before the bodies of their dead were restored to them.] Moreover, if any one, after having known the content of this decree, should—which we hope will not happen—attempt audaciously to act contrary to it, then let him suffer punishment in his rank and position, or let him be punished by the penalty of excommunication, unless he makes amends for his boldness by proper recompense. Moreover, we wish that only those Jews who have not attempted to contrive anything toward the destruction of the Christian faith be fortified by the support of such protection. . . .

Given at Orvieto by the hand of the Magister John Lectator, vice-chancellor of the Holy Roman Church, on the 7th of October, in the first indiction [cycle of fifteen years], in the year 1272 of the divine incarnation, in the first year of the pontificate of our master, the Pope Gregory X.

From *The Cremation of the Strasbourg Jewry*

Jacob von Königshofen

■ 1349 ■

In the year 1349 there occurred the greatest epidemic that ever happened. Death went from one end of the earth to the other, on that side and this side of the sea, and it was greater among the Saracens than among the Christians. In some lands everyone died so that no one was left. Ships were also found on the sea laden with wares; the crew had all died and no one guided the ship. The bishop of Marseilles and priests and monks and more than half of all the people there died with them. In other kingdoms and cities so many people perished that it would be horrible to describe. The pope at Avignon stopped all sessions of court, locked himself in a room, allowed no one to approach him and had a fire burning before him all the time. [This last was probably intended as some sort of disinfectant.] And from what this epidemic came, all wise teachers and physicians could only say that it was God's will. And as the plague was now here, so was it in other places, and lasted more than a whole year. This epidemic also came to Strasbourg in the summer of the above-mentioned year, and it is estimated that about sixteen thousand people died.

In the matter of this plague the Jews throughout the world were reviled and accused in all lands of having caused it through the poison which they are said to have put into the water and the wells—that is what they were accused of—and for this reason the Jews were burnt all the way from the Mediterranean into Germany, but not in Avignon, for the pope protected them there.

Nevertheless they tortured a number of Jews in Berne and Zofingen [Switzerland] who then admitted that they had put poison into many wells, and they also found the poison in the wells. Thereupon they burnt the Jews in many towns and wrote of this affair to Strasbourg, Freiburg, and Basel in order that they too should burn their Jews. But the leaders in these three cities in whose hands the government lay did not believe that anything ought to be done to the Jews. However in Basel the citizens marched to the city hall and compelled the council to take an oath that they would burn the Jews, and that they would allow no Jew to enter the city for the next two hundred years. Thereupon the Jews were arrested in all these places and a conference was arranged to meet at Benfeld [Alsace, February 8, 1349]. The bishop of Strasbourg [Berthold II], all the feudal lords of Alsace, and representatives of the three above-mentioned cities came there. The deputies of the city of Strasbourg were asked what they were going to do with their Jews. They answered and said that they knew no evil of them. Then they asked the Strasbourgers why they had closed the wells and put away the buckets, and there was a great indignation and clamour against the deputies from Strasbourg. So finally the bishop and the lords and the Imperial Cities agreed to do away with the Jews. The result was that they were burnt in many cities, and wherever they were expelled they were caught by the peasants and stabbed to death or drowned. . . .

[The town-council of Strasbourg which wanted to save the Jews was deposed

SOURCE: Jacob von Königshofen, *The Cremation of Strasbourg Jewry*, in *The Jew in the Medieval World*, ed. J. R. Marcus (Cincinnati: Sinai Press, 1938).

on the 9th/10th of February, and the new council gave in to the mob, who then arrested the Jews on Friday, the 13th.]

On Saturday—that was St. Valentine's Day they burnt the Jews on a wooden platform in their cemetery. There were about two thousand people of them. Those who wanted to baptize themselves were spared. [Some say that about a thousand accepted baptism.] Many small children were taken out of the fire and baptized against the will of their fathers and mothers. And everything that was owed to the Jews was cancelled, and the Jews had to surrender all pledges and notes that they had taken for debts. The council, however, took the cash that the Jews possessed and divided it among the working-men proportionately. The money was indeed the thing that killed the Jews. If they had been poor and if the feudal lords had not been in debt to them, they would not have been burnt. After this wealth was divided among the artisans some gave their share to the cathedral or to the Church on the advice of their confessors.

Thus were the Jews burnt at Strasbourg, and in the same year in all the cities of the Rhine, whether Free Cities or Imperial Cities or cities belonging to the lords. In some towns they burnt the Jews after a trial, in others, without a trial. In some cities the Jews themselves set fire to their houses and cremated themselves.

It was decided in Strasbourg that no Jew should enter the city for a hundred years, but before twenty years had passed, the council and magistrates agreed that they ought to admit the Jews again into the city for twenty years. And so the Jews came back again to Strasbourg in the year 1368 after the birth of our Lord.

Capital, Labor, and the Florentine Merchant Princes

In no parts of Europe were the domination of land and the power based on it less evident than in the Italian and Flemish towns. Money and office were the sources of urban power. And in the process of getting and using both things, the great merchants and guild-masters sought to convert them into the bases of oligarchic control. The transition from open communes to oligarchic republics was to be in many places a step toward despotism and the extinction of republicanism. Florence has long stood as a classic case of this sociopolitical development. Hence the significance of this excerpt from Dr. Richards's book, which originally formed part of her Introduction to a collection of economic documents, The Selfridge Manuscripts of Harvard University. Drawing on accounts and letters left by a younger branch of the Medici family, especially letters of the Constantinople factor, Giovanni di Francesco Maringhi, Dr. Richards joined the documents to an account of Florentine economics and politics and their interrelatedness. Here we can see the role of the great guilds, of the strife between *grosso* and *minuto* (the big and little people), the relentless concern for artisan freedom, and the oppression of ordinary labor. Whether or not Florence owed its existence to Fiesolans who founded it for trading purposes, as Villani claimed, her culture rested on industry and commerce. When her economy declined and European trade became Atlantic-oriented, Florence ceased to be a synonym for splendor and civic virtue.

From *Florentine Merchants in the Age of the Medici*

Gertrude R. B. Richards

FLORENCE UNDER THE MEDICI

According to Villani († 1348), Florence was founded by the Fiesolans for the purpose of trade; if this be true—and no theory as to the origin of the city has yet been advanced which seems more probable—then the history of Florence is also the history of commerce along the valley of the Arno. Certainly from Roman days on she was generally known as a market-place. Through the age of the communes she might easily have lost her economic prestige to her more fortunate rivals, Venice, Genoa, and Amalfi, all sea-ports, had it not been for the fact that she was so fortunately situated where the two meandering highways, one from Pisa to Ancona, and the other from Bologna to Rome, crossed, thus securing for her the overland trade which not even the sea-ports were able to challenge.

Fifteenth-century Florence was still a city of craftsmen, and of merchants organized into guilds or *Arti*, a fundamental and permanent element of Florentine life. Probably neither the exact date of the formation of these guilds nor the order of their establishment will ever be known accurately; but certainly by the twelfth

SOURCE: Gertrude R. B. Richards, ed., *Florentine Merchants in the Age of the Medici* (Cambridge, Mass.: Harvard University Press, 1932), pp. 35–53.

century they were functioning, and, moreover, their members were trading in the markets of northern Europe. By the statute of 1415 the guilds were listed in order of their respective importance, the seven major and fourteen minor *Arti*—which number remained unaltered until the downfall of the Republic.

I. *Le Arti Maggiori*
1. *L'Arte dei Giudici e Notai*—Judges and Notaries.
2. *L'Arte di Calimala*—Merchants of Foreign Cloth.
3. *L'Arte di Lana*—Wool-manufacturers.
4. *L'Arte de' Cambiatori*—Bankers and Money-changers.
5. *L'Arte di Seta*—Silk-manufacturers.
6. *L'Arte de' Medici e Speziali*—Doctors and Apothecaries.
7. *L'Arte de' Pellicciai e Vaiai*—Skinners and Furriers.

II. *Le Arti Minori*
1. *L'Arte de' Beccai*—Cattle-dealers and Butchers.
2. *L'Arte de' Fabbri*—Blacksmiths.
3. *L'Arte de' Calzolai*—Shoemakers.
4. *L'Arte de' Maestri di Pietre e di Legnami*—Master Stone-masons and Wood-carvers.
5. *L'Arte de' Rigattieri e de' Linaiuoli*—Retail Dealers and Linen Merchants.
6. *L'Arte de Vinattieri*—Wine-merchants.
7. *L'Arte degli Albergatori*—Inn-keepers.
8. *L'Arte de' Galigai*—Tanners.
9. *L'Arte degli Oliandoli*—Oil-merchants.
10. *L'Arte de' Correggiai*—Saddlers.
11. *L'Arte de' Chiavaiuoli*—Locksmiths.
12. *L'Arte de' Corazzai*—Armorers.
13. *L'Arte de' Legnaiuoli*—Carpenters.
14. *L'Arte de' Fornai*—Bakers.

Each one of the guilds was to a large extent self-governing and independent, although, in times of stress, they were quick to form alliances for mutual benefit. While each was empowered to define its own terms of entrance and the conditions of labor for its members, as well as standards of production, there was little marked variation in the organization. Each required every member to be a native Florentine, to furnish sponsors for his character, as well as evidence of freedom from misdemeanor, and to provide both caution-money and an entrance fee. Members of the mercantile guilds were either *maestri* (full members) or *garzoni* (apprentices); while the craft guilds had an intermediary group, *lavoranti* (laborers). Each guild had its magistrates, priors, rectors, or consuls; each had its own notaries, its financial officials, and its *provvisioni* or statutes; and each sponsored some particular church or charity which it endowed with funds and decorated with masterpieces of art. The little group of merchants whose activities form the substance of these letter-books were connected with the *Arte di Calimala*, the *Arte del Cambio*, the *Arte di Lana*, and the *Arte di Seta*.

The *Arte di Calimala* had to do with the redressing, dyeing, and finishing of woolen cloth, foreign at first and later both domestic and foreign. As early as the reign of Henry II, Florentines were at the English fairs buying cloth to be refinished; by 1182 they were carrying on a brisk trade in that commodity with southern France. As early as the thirteenth century, agencies of the *Arte di Calimala* were established in Paris, St. Denis, Provins, Lagny, Troyes, Marseilles, Arles, Avignon, Perpignan, and Toulouse. Each of these agencies had its resident agent receiving and executing orders, and each its resident consul, both chosen by the guild.

The contribution of the *Arte di Calimala* to Florentine development and Florentine prestige, directly and indirectly, can hardly be overstated. When these foreign cloths came into Italy they were what the English would call "durable and well made," but they were certainly unpleasing to the eye. Partly because the Florentines still remembered the soft, light, bright-hued fabrics they had seen in the Holy Land, partly because of that innate sense of beauty which dominated every aspect of their lives, this thick, heavy, dark cloth failed to satisfy them as it had satisfied the northerners. Out of such dissatisfaction had developed the craft of the *Calimala*, the turning of the rough but well-woven fabrics into a material unrivalled in all the markets of Europe, and one which taught both East and West to look to the city on the Arno for the finest of *panni*.

Closely allied to the *Arte di Calimala*, since it was born of the need for facilitating trade in foreign wool-cloths, was the *Arte del Cambio*. Wherever the Eagle was established in foreign parts, there came likewise the bankers to transfer and adjust obligations between distant debtors and creditors. By the thirteenth century, Florentine bankers were established in many European cities. In 1199 they were in England, sharing place, however, with the bankers of Siena. Ghibellines though they were, the Sienese had first control of papal finances, and it took some time to establish the superiority of Florentine claims. But because the latter were Guelfs, and so supporters of the Papacy against the Emperor, and also because Florence itself was on the direct highway between Rome and the northern lands, gradually the supremacy of the Florentines was secured. Nor were these shrewd and subtle merchants content with mere exchange transactions. Here, as in the other *Arti*, they left the imprint of their genius by furnishing the world with a coin of such refined superiority as might command respect in all markets. In 1252 they created the florin of gold, a decisive step in the history of finance, for this coin of established value turned the scale definitely in their favor, and by the time the Papacy was removed to Avignon (1309) the bankers of Florence had established themselves as *Campsores Papae*.

Naturally admission to the *Arte del Cambio* was more difficult than to any other. The strictest of rules were laid down and rigidly enforced. Candidates had to enter their names on the matriculation roll; they had to undergo a rigorous examination before the consuls of the guild; and they had to pay a matriculation fee which was not only much higher than that imposed by the other *arti*, but which was also somewhat exorbitant in view of the capital involved in the individual enterprise. Once admitted to membership, they were allowed a table and chair in the Mercato Nuovo or along the Via di Tavolini. This table was covered with a green cloth on which lay a sheet of fresh parchment for recording the day's business, while each "Bank," literally speaking, was a gaily embroidered pouch filled with gold coins, and a bowl of small silver or bronze coins to serve as change. Uncovenanted money dealers were allowed the use of a table in the Mercato under certain conditions, but these were distinguished from those in good and regular standing by the fact that they had neither chair nor green cover to the table. There was an annual conference between the consuls of this guild, the financial officials of the other major *arti*, and the priors of the more important monastic orders to decide on the values of exchange and the rates of loan interest during the coming year, as well as to pass on the qualifications of those allowed to continue their operations in the Mercato.

But the importation of wool cloth to be finished by the *Arte di Calimala* was only the first flowering of the woolen industry in Florence. Extensive production of that cloth within the city itself was to be the second and even greater blooming. Just when the Florentines first engaged in manufacturing wool cloth is as much a matter of conjecture as the origin of the city itself. It seems certain, however, that weaving was but a minor industry until the coming of the Umiliati in 1238.

The Umiliati were a religious brotherhood which originated in Milan early in the eleventh century. According to their own traditions, they had learned from the Flemings the art of weaving the close, heavy cloth of Flanders which hitherto the Florentines had been unable to equal and which they had been buying for years in the fairs and market-places of the north, bringing it back to those same fairs and market-places after the *Arte di Calimala* had dressed and dyed it and selling it at very high prices to the same merchants from whom they had bought it in its cruder state.

Not only did the Umiliati teach the Florentines new methods of weaving and an appreciation of varying qualities in raw material, but the brethren themselves set high standards of business application and of organization. Probably, too, they knew how to market their products more advantageously than the Florentines had hitherto done. For about a century their influence was definitely constructive; then, economically as well as institutionally, they began to suffer change and relaxation. At the same time the Florentines, now qualified to take over their workshops and dyeing vats, were inclined to resent the importance of the Order in the market-place, and began to take over the control of all industry, whether within the city walls or in the outlying *contado*.

By virtue of these technical advances, Florence was able to build up a cloth-making industry which soon surpassed that of her erstwhile rivals. For raw materials she went to England and to Spain, since the fleeces produced on her worn hillsides were of too poor a quality to suit her needs. Agents formerly scattered abroad to buy the heavy northern cloth were now buying bales of wool, while back in the narrow tortuous streets along the banks of the Arno weaver and dyer and dresser worked busily over their *panni*, weaving, dyeing, and dressing it, that it might be taken by the couriers to market-places first of northern Europe and later of the Orient, where they bought both raw material for the artisans and new luxuries for the merchants back in Florence.

Although the *Arte di Calimala* continued to import the rough *panni francesi*,[1] to dress and dye it, and then send it forth, cut and folded, the *panni fiorentini* produced by the *Arte di Lana* increased in quantity and in diversity, and was itself exported to all markets from Scotland to Constantinople. The poorest grade of cloth was that woven from the coarse Tuscan fleece and used by friars and peasants for their tunics. From the foreign wool, that grown in Morocco or in Algarves, was woven the finer fabrics—*garbo* and *panno San Martino*. These

---

1 A term applied to cloth from both Belgium and France.

latter, when dyed with the famous *oricello* dye, were the *panni nobili* reserved for magisterial robes and ecclesiastical hangings, as well as for the special caps worn by those who had the right of entry into the superior courts of the commune. On festive occasions, the magistrate, garbed in his *lucco* of scarlet cloth, like the Lord High Chancellor on his historic wool-sack, felt all that consciousness of moral support born of the patriotic patronage of home industries.

The lowest orders of workmen were the washers and carders, the proletariat of the Republic. Their small wages kept them ever in debt to the *maestro*, who could dismiss them at his pleasure, although, as a matter of fact, he seldom exercised this prerogative. The laborers, on the other hand, were unable, individually at least, to "give notice," regardless of the conditions under which they worked. Because of the wooden clogs which they wore as protection against the wet floors in the washing and carding houses, these laborers were known as *Ciompi*. One of the first labor-strikes on record is their revolt in 1378 under the leadership of Salvestro de' Medici, a revolt against the growing wealth and power of the upper classes. A short but devastating reign of terror marked the uprising. Even the magistracy was threatened for a time. In the end, the demands of the *Ciompi* were granted: taxes were reduced; debtor laws were rescinded; and the municipal franchise was extended to include the members of the lesser guilds. The results of the outbreak were twofold: the lower classes forced themselves into the government, and the Medici first appeared as supporters of the popular party.

The second class of laborers were the spinners, most of whom were women working in their own homes. These were not paid regular wages; they seem to have brought the raw wool and then to have sold the yarn, presumably at an advance. Above these were the weavers, who were paid regular wages, and who were, accordingly, more dependent than the spinners, for they did not own their looms and had little or no control over the conditions of their service. After the Florentines had learned how to weave their own cloth, a new group of artisans appeared: fullers, *gualchieri;* washers, *lavatori;* tenterers, *tiratori;* menders who darned the flaws, *rimendatori; cardatori*, who raised the pile; *affettatori*, who clipped it; and finally the dyers or *tintori*.

Even after the first guilds were established, the laborors had worked with a certain degree of independence. It was not until after the coming of the Umiliati that this freedom was lessened by the precedence given the organization over the laborer. The *Frati*, with that emphasis on centralization of power which characterizes monasticism, developed a policy of concentration, dividing the industry into a hierarchy of laborers, and this policy was continued by the *arti*. The *maestri* seemed to hold the theory that the only way of securing and maintaining prosperity was to keep the price of labor down to an irreducible minimum, a theory neither admirable nor intelligent, but which seems, despite its very obvious flaws, not to have prevented the maintenance of the high quality of their production over an unusually long period of time.

Cloth was rigidly inspected at every stage of its progress for any defects or blemishes, and heavy and inexorable fines were imposed for every infraction of the standards. Each piece of finished cloth had its ticket on which was stated the fixed price, the quantity, the name of the factory where it was made, and the name of the *maestro*. In case of transit all *panni* had to be folded so as not to disturb the nap. As a result of such supervision of production, Florentine cloth set an enviable standard for quality and for accuracy of measurement in all the markets of the civilized world.

The dyeing of woolen fabrics was also controlled by the *Arte di Lana*. If the Florentines excelled in the manufacture of wool, they were unrivalled as dyers. A city of artists, they felt that the creation of a new color or of a new technique, so difficult always, was important enough to be entered into the official records of the commune. Scattered also through the medicinal formulae of the old pharmacy account-books are many items concerned with the chemical composition of colors, instructions for dyeing leather, writing in gold, removing rust, etc. A fifteenth-century book of colors which has been preserved in the *Biblioteca Nazionale* of Florence gives rules for testing *azzuro germanico*, rules for preparing *indaco*, *verde*, *rame*, and *porporo*.

The dyes used were both native and imported. *Guado* or woad, used in dyeing common blue cloth, was cultivated in Tuscany; so, too, was *robbia* or madder, used for the more common type of red. The sale of both of these was restricted to owners of warehouses, while the export of them was absolutely forbidden. Cochineal and Brazil wood were used also for the more common cloths, *lapis lazuli* was the base for the finer blue dyes, and rubies for *rosso*. But the color favored above all others was that rich purplish red known as *scarlatto d'oricello*, made from an admixture of *oricello* and madder. All robes of state and all ecclesiastical hangings were colored with this dye, as well as the special caps worn by those who had right of entry into the superior courts. The bits of this *scarlatto*, wool, silk, and damask, which still exist show as fresh a color and as firm a texture as anything produced since that day. The *oricello* was a lichen discovered in the Orient by a Florentine traveller in 1305, who introduced it into Tuscany, not only making his fortune thereby, but giving his family its name, Rucellai.

As manufacturers of fine silks, velvets, and damasks, the Florentines attained as high excellence as they did in their weaving and finishing of wool, equalling, if indeed they did not outrank, both Persians and Egyptians in the beauty of their designs as in the excellence of their coloring. It was a painstaking craft, and their success was the just reward of patience and skill, not a happy accident. The raw silk was carefully sorted both as to weight and as to quality; the dyeing was supervised with meticulous care; the colored threads were spread to dry on white cloths in sunless lanes and there carefully matched; while the goldsmiths of the city drew fine threads from precious metals to combine with the silk threads in the gold and silver brocades which Crivelli delighted in painting.

The *Arte di Seta* existed as early as 1193; by the fifteenth century it ranked with the *Arte di Lana* in importance and in wealth. In 1473 this *Arte* had no less than eighty-three flourishing *botteghe* in the city, as well as agencies in Rome, Naples, Catalonia, Turkey, Avignon, London, Lyons, and Antwerp. Until the fall of Constantinople most of the raw silk used was imported from the Levant. During the years when the Italian states were at war with the Sultan, mulberry trees were cultivated in the *contado*, and cocoons were brought from Lucca and Pistoia as well as from Spain. Near the turn of the century, however, Bajazet II reopened the markets of Brusa to the western merchants, and raw silk was again imported from that region.

Benedetto Dei, writing to a Venetian in 1472, gave a glowing description of Florence in prosperity.

> Florence is more beautiful and five hundred forty years older than your Venice. We spring from triply noble blood. We are one-third Roman, one-third Frankish, and one-third Fiesolan. . . . We have round about us thirty thousand estates, owned by noblemen and merchants, citizens and craftsmen, yielding us yearly

bread and meat, wine and oil, vegetables and cheese, hay and wood, to the value of nine hundred thousand ducats in cash, as you Venetians, Genoese, Chians, and Rhodians who come to buy them know well enough. We have two trades greater than any four of yours in Venice put together—the trades of wool and silk. Witness the Roman court and that of the King of Naples, the Marches and Sicily, Constantinople and Pera, Broussa and Adrianople, Salonika and Gallipoli, Chios and Rhodes, where, to your envy and disgust, in all of those places there are Florentine consuls and merchants, churches and houses, banks and offices, and whither go more Florentine wares of all kinds, especially silken stuffs and gold and silver brocades, than from Venice, Genoa, and Lucca put together. Ask your merchants who visit Marseilles, Avignon, and the whole of Provence, Bruges, Antwerp, London, and other cities where there are great banks and royal warehouses, fine dwellings, and stately churches; ask those who should know, as they go to fairs every year, whether they have seen the banks of the Medici, the Pazzi, the Capponi, the Buondelmonti, the Corsini, the Falconieri, the Portinari and the Ghini, and a hundred of others which I will not name, because to do so I should need at least a ream of paper. You say we are bankrupt since Cosimo's death. If we have had losses, it is owing to your dishonesty and the wickedness of your Levantine merchants, who have made us lose thousands of florins; it is the fault of those with well-known names who have filled Constantinople and Pera with failures, whereof our great houses could tell many a tale. But though Cosimo is dead and buried, he did not take his gold florins and the rest of his money and bonds with him into the other world, nor his banks and storehouses, nor his woolen and silken cloths, nor his plate and jewelry; but he left them all to his worthy sons and grandsons, who take pains to keep them and to add to them, to the everlasting vexation of the Venetians and other envious foes whose tongues are more malicious and slanderous than if they were Sienese. . . . Our beautiful Florence contains within the city this present year two hundred seventy shops belonging to the wool merchants' guild, from whence their wares are sent to Rome and the Marches, Naples and Sicily, Constantinople and Pera, Adrianople, Broussa and the whole of Turkey. It contains also eighty-three rich and splendid warehouses of the silk merchants' guild, and furnishes gold and silver stuffs, velvet, brocade, damask, taffeta, and satin to Rome, Naples, Catalonia, and the whole of Spain, especially Seville, and to Turkey and Barbary. The principal fairs to which these wares go are those of Genoa, the Marches, Ferrara, Mantua, and the whole of Italy; Lyons, Avignon, Montpellier, Antwerp, and London. The number of banks amounts to thirty-three; the shops of the cabinet-makers, whose business is carving and inlaid work, to eighty-four; and the workshops of the stonecutters and marble workers in the city and its immediate neighborhood, to fifty-four. There are forty-four goldsmiths' and jewelers' shops, thirty gold-beaters, silver wire-drawers, and a wax-figure maker. . . . Go through all the cities of the world, nowhere will you ever be able to find artists in wax equal to those we now have in Florence, and to whom the figures in the Nunziata can bear witness. Another flourishing industry is the making of light and elegant gold and silver wreaths and garlands, which are worn by young maidens of high degree, and which have given their names to the artist family of Ghirlandaio. Sixty-six is the number of the apothecaries' and grocer shops; seventy that of the butchers, beside eight large shops in which are sold fowls of all kinds, as well as game and also the native wine called Trebbiano, from San Giovanni in the upper Arno Valley; it would awaken the dead in its praise.

During the first half of the fifteenth century Florence was still at the peak of her industrial and commercial supremacy, but from 1465 on her remarkable intellectual achievements were accompanied by moral decadence and economic decline. Lorenzo de' Medici has been almost universally blamed for the economic downfall, but rather unjustly so. While he was not so astute a financier as either Giovanni or Cosimo, the situation confronting him was exceedingly complicated, and depended to a very great extent on the machinations of the kings of France, Spain, and England, thus presenting problems which neither of his forbears had been called upon to solve. In Cosimo's day, Florence played an exceedingly important rôle on a fairly small stage. In Lorenzo's day, she was not always the chief actor even on the Italian scene; and by the time of his death, when his weak and incompetent descendants came forward, the stage of operations had been enlarged by the discovery of a New World which soon was to furnish gold to the enemies of the Florentines, as it was also to contribute to the final disappearance of her Levantine trade.

It is true that Lorenzo's interest was in his estates, rather than in the Medici Bank. Because his agents took advantage of this lack of interest and managed his fianances (and incidentally those of the state) so decidedly to their own profit, he has often been accused of wasting the family fortunes and of squandering the funds entrusted to him by the commune. That he spent lavishly is true; that he disliked exceedingly the irksome round of banking and exchange is equally true. But nevertheless it is scarcely fair to burden him with the faithlessness of his agents, nor to hold him responsible for the changing economic and commercial conditions which decreased so appreciably the value of his inheritance.

The elder Cosimo had given away far more than Lorenzo spent, but because he gave to the institution rather than to the individual his generosity has been counted a virtue. In his *Ricordi* Lorenzo says that Giovanni di Averardo left property to the amount of 179,221 *fiorini di suggello*, that Cosimo left 235,137 *fiorini*, Piero, 237,988 *scudi*, and that from 1434 until the end of 1471, that is, from the time of Cosimo's return from exile to shortly after Piero's death, the family gifts, benefices, and donations amounted to 663,755 florins, all of which he (Lorenzo) does not regret, "for though some there are who would consider it better to have part of that money in their own pockets, I consider that the donors, through their gifts, added greatly to the honor of the States, and I am well pleased with the way the money was expended."

Nor has Lorenzo ever been given adequate credit by his biographers for encouraging the great merchants to build their own ships in order to trade freely where they pleased. He believed that a large export trade was an index of the prosperity of a state, and therefore, to encourage commercial expansion, he took shares in many concerns engaged in this enterprise. He also held that the state control of any industry was a bad thing, since it restricted expansion and encouraged inefficiency. He recognized that in the case of wool and silk industries the need for foreign raw materials was dominant; if the manufacture was an essential industry, the re-exportation was certainly a source of a considerable portion of the wealth of the city. He made Pisa a free port, and by his insistence on "free trade" he enormously extended the commercial activities of his people; he made a commercial treaty with Egypt; he maintained close commercial intercourse with the Turks; he sent consuls to far-distant lands to protect Florentine interests, and to report on the best means of extending them.

In common with the long line of merchant-princes who had shaped the foreign policy of the commune, he considered the economic security of the state as the dominant factor in her policy both foreign and domestic. He encouraged friendship with the Marches in order to give the merchants free access to Ancona, ever more important to the Florentines as a port than Venice; Siena must not be antagonized, since she was the bar against intruders from the south; likewise Sarzana and Pietra Santa must be in Florentine hands, since they formed a similar barrier on the north; Pisa must be held at all costs, and must be developed as a port of entry, since the control of that harbor was the very keystone of prosperity and of economic independence. Historians who have accused Florence of having destroyed the liberties of Pisa, and who have hailed Charles VIII of France as the supreme deliverer who threw off the yoke of the oppressor, have forgotten that the period of Florentine control of Pisa is likewise the period of her greatest supremacy; have overlooked the fact that Lorenzo himself refounded her University and spent no little time in residence in the city that he might better understand her needs and her possibilities.

By the end of the fifteenth century, however, the "Golden Age" of Florence was rapidly disappearing. Several conditions combined to make this a period of decadence: all her industry was shrinking through foreign competition, through the introduction of new methods, through the development of new and lesser industries, and through a subtle change in the Florentine character. The discovery of the New World turned men's attention from the Levant; the opening of gold and silver mines in South America revolutionized the existing price relationships. Rising food prices and the demand for higher wages were accompanied by falling prices for cloth both in the domestic and in the foreign markets. In the earlier days the guild had been the manifestation of the spirit of the artist; the famed *panno* was not the work of hirelings, but of artist-craftsmen, who were also statesmen and diplomats. But those days were gone, and with them the passion for perfection, the high standard of the market-place, the honor, loyalty, and patriotism which had once dominated trade. Business was rapidly becoming a mere opportunity for amassing wealth, wealth to be spent on idle pleasures, not devoted to the adornment of the city. The ease and luxury following hard on the heels of enormous profits were bringing a moral decadence which dominated all Florentine life from the early days of the sixteenth century until she fell under the control of the ruthless invader from the north. It was the day foretold by the elder Cosimo, when he grieved that citizenship was becoming a matter of rich raiment, not of honest service.

Florence, being neither a maritime power nor a crusading state, was a late comer to the Levant, and by the time she had arrived, Venice, Amalfi, Genoa, and Pisa had gathered the ripest fruit. Even had she so desired, she could never have prospered commercially, as they had done a century before; nor, on the other hand, could she have reaped rewards as rich as those which came from her trade with the northern lands where she was able to buy much that was necessary to her chief industry—that of finishing, and later of manufacturing, woolen cloth. Save for raw silk and a few dye-stuffs, all that she purchased in the Orient fell in the category of luxuries rather than of commercial necessities. Villani insists that Florentine trade with the East dates from the third crusade, but while there is no way of proving the truth of such a statement, it does seem fairly certain that until the conquest of Pisa in 1409 her trade was indirect rather than direct. She found it rather more to her advantage to sell directly to Genoese and Venetian merchants than to ship her goods through these ports. Moreover, Genoa and Pisa, both Ghibelline states, were political as well as economic rivals of the Guelphic Florence.

By this conquest of Pisa, Florence was not only given a port and a merchant marine, but she automatically took over certain privileges, both commercial and political, in the various cities of the Levant. Through a combination of circumstances, it happened that within a few years she was established there on equal terms with her great rivals, Genoa and Venice. Unfortunately the rise of the Osmanli Turks came within a half century of her arrival in Constantinople, so that she had little time in which to enjoy her newly acquired trade. The fall of Constantinople in 1452 was followed by three decades of petty wars around and about the eastern shores of the Mediterranean. While her merchants did not withdraw from that region, the conditions were anything but favorable to commercial expansion. Still, despite the unfavorable conditions, very few of the merchants withdrew their agencies. As the State no longer maintained a consulship there, however, they were forced to manage their own affairs and assume whatever risks were involved. In 1481 Bajazet II became Sultan. He was far less warlike than his predecessors, and throughout his long reign did whatever he could to further the interests of the western merchants. For some reason, the Florentine *Signoria* quite neglected to send to Constantinople an ambassador bearing the usual compliments; Bajazet waited two years, and then, receiving no sign of recognition, dispatched his own envoy to Florence requesting a renewal of the treaties granted by his father, Mohammed, to the merchants of that city. Not until 1488 did Lorenzo reply; then he dispatched his cousin and favorite, Andrea de' Medici, to arrange the terms of a commercial treaty with Bajazet. The Florentines were to have a resident consul in whose hands should rest the criminal and civil jurisdiction of all cases involving Florentines, whether in their relations with each other, with the Turks, or with citizens of other lands. A fixed impost of two per cent was levied on all goods bought or sold in the market-place; and but one set of duties was to be imposed on merchandise carried through more than one town.

After the death of Lorenzo, in 1492, the general political confusion in Florence resulted in an increased migration to Constantinople, and Geri Risaliti was sent thither in 1499 to renew the concessions granted earlier through the efforts of Andrea de' Medici. By 1507 some sixty or more Florentine firms had their agents over there, and were doing a business estimated at 600,000 ducats a year. But the prosperity was short-lived. Bajazet died in 1512, and his son cared more for war than for commercial alliances. The new routes westward aided in the diversion of commerce away from the Levant, and by 1520, when Raffaello de' Medici was writing his agents in and about Pera regarding the *panni* he was hoping to sell over there, his letters reveal that the end of eastern trade was rapidly approaching. That he himself realized the new conditions is shown by his developing the connections with Spanish merchants established as early as 1470 by his grandfather Giuliano de' Medici.

There has been little change in Florence during the four centuries which have elapsed since Michelangelo threw down his chisel, leaving the unfinished Pietà to plan the fortifications on San Miniato. New bridges have been erected on the piers of the Trinità and of the Carraia, but the Ponte Vecchio and the fringe of weather-worn houses beyond the Arno are unchanged. The grim old prison—the Stinche—has given way to a theater; near the Arno where once stood the *tiratoi* of the merchants the new Biblioteca is being erected, and across the river the *conventi* of the Medici in Via Maggio have entirely disappeared. The gaunt palace of the Pitti, perhaps the most powerful of all those who opposed the power of the Medici, has been completed and houses the kings of Italy when they are in residence in the city, and the Mercato Vecchio, industrial center of the old commune, has been replaced by a hideous modern "square" dominated by an equestrian statue of Victor Emmanuel II.

Nothing that the new Florence has to offer can compare with the lost loveliness of that old market-place, the most picturesque in all Europe. On one side was the low mouldering colonnade of Vasari; opposite was the proud column surmounted by Donatello's *Dovitzia;* about its base clustered the carts of the hucksters. On all sides of the market were the disreputably shabby but still picturesque palaces, topped by low irregular roofs—roofs of brown tiles stained with clumps of grass and weeds and bedecked with wavering lines of laundry (the gay colors bearing witness to the fastness of Florentine dyes). There, too, were the donkey carts of the peasants; and the tumbledown shops full of bric-a-brac of all degrees—beds, crockery, Roman lamps, and assorted broken-nosed saints.

In the unused rooms about the inner cloister of San Marco are collected fragments from the churches, tombs, and portals of this gathering place of the Medici, but the pavings, frescoes, *stemmi,* and fountain heads, tidy and well-ticketed, suggest nothing of the vivid color and the conglomerate life of that happy "*Giardino di Firenze.*"

The palaces of the guilds were hard by the Mercato. So too was the beautiful Or San Michele, which more than any other building of Florence is intimately associated with her commercial prosperity. Erected by the *Arte di Seta* as a storehouse for the grain of the commune, it became, by virtue of the miraculous painting of the Madonna which it held, a shrine for all artisans and consequently the focal point of Florentine commerce at its highest and best. Later, after a fire had more than once wrought havoc, the workers in silk and precious metals erected a new loggia on the pillars of which each guild set up a shrine. Donatello, Ghiberti, Verrocchio, Gianbologna, and other artists of the Renaissance were busy for years on this "tabernacle of commerce," carving the patron saints of the *Arti* for the piety of the citizens and for the everlasting enjoyment of all who came after.

In the dark little *sdruccioli* running off from the Mercato in all directions lived the artisans, grouped according to their various crafts. Today we read in the names of these same streets the industrial geography of the old commune. In the Via Calzaioli lived those who wove the fine, light wool used for stockings; the Corso dei Tintori was the road of the dyers; the Via delle Caldai was the street of the cauldrons where wool was washed; the Vicolo dei Guanti was "glove lane"; the plaiters of straw congregated at Canto alla Paglia; the Via al Fuoco was the grimy street of the furnace-makers; while in the Via degli Speziali perfumers and pharmacists sold their fragrant wares.

City Life in the Dutch Republic

Dr. Richards's book on Florence exposed the socioeconomic and political consequences of Florentine commercial ascendancy. Paul Zumthor's study of *Daily Life in Rembrandt's Holland* takes us into the microcosm of Dutch republican city life. The Dutch dominated world trade and shipping in the late sixteenth and early seventeenth centuries. Antwerp's decline (1550–1585) was Amsterdam's opportunity. The English did not successfully challenge Dutch supremacy until the trade wars of the 1650–1670 period. From about 1570–1670 then, we may well speak of Dutch commercial and city life as the most highly developed in the West. Her naval technology, credit and banking, bourse and bookkeeping supported the prosperous burgher classes who patronized Rembrandt. But Holland's cities also showed the other face of urban-commercial civilization: the grotesque slums, the poverty, overcrowding, and despair of real proletarian life. And in her cities we can also meet the lush undergrowth of whores, pimps, pickpockets, and thieves who made up the "underworld" (lumpen-proletariat). Both the style of city life and its social cost show men wrenched away from the cycles of nature. Zumthor's book thus exhibits the city as less exhuberant than Italian experience teaches and also as more complex. We glimpse in it the social structure of urban capitalism fully developed.

From *Daily Life in Rembrandt's Holland*

Paul Zumthor

The Guilds

The professional existence of artisans, workmen and small businessmen in the Netherlands was governed by guilds. These guilds had emerged from the old medieval "fraternities" and had the function of exercising a complete control over the production of manufactured goods and the distribution of merchandise. Their power was based on a conception of group ethics which was intended to protect the members of a particular guild. Based upon ancient privileges, and complicated by a multitude of more recent regulations, the system laid the ground for a struggle between licensed professionals and outsiders and between tradition and individual initiative. Starting work before the prescribed hour or selling below the agreed price were misdemeanours that the guild's executive committee was authorised to investigate and punish.

The division of jurisdiction between the various guilds and the extent of their powers varied from town to town and were a disruptive factor in the country's developing economy. Utrecht had twenty-one guilds, including five for the garment industry alone: tailors, fur-lining makers, glovers, shoemakers and cobblers. Psychological and social differences were at the root of these arbitrary dis-

SOURCE: Paul Zumthor, *Daily Life in Rembrandt's Holland*, trans. Simon Watson Taylor (New York: The Macmillan Company; London: George Weidenfeld & Nicolson, 1962), pp. 141–148, 248–256.

tinctions; the guild of curriers, working fine leather, considered itself superior to the guild of saddlers working coarse leather. On the other hand, some specialised activities existed side by side in the same guild; the guild of carpenters also included the cabinet-makers and turners, the bakers' guild included the millers, and the shoemakers' included the tanners. All these divisions resulted in harsh regulations by which, for example, a certain workman had the right to sew a new sleeve on to an old doublet but not to make a new doublet. The guild of pewter founders protested against the fact that booksellers were selling ink in pewter inkwells.

A guild member was not permitted to open more than one shop or have more than one market-stall; people were allowed to hawk goods only if their stock was worth less than a certain sum. Certain guilds forbade their members to sell in the markets for fear that they might undercut their fellowmembers; weavers were not permitted to weave or card wool during the summer, or to possess more than three looms; brewers could brew only once a week; pastrycooks were prohibited from baking cakes in any but officially sanctioned shapes. All products were stamped, and anything not bearing the appropriate seal was confiscated and destroyed if found. Thirty regulations governed the preparation of herrings. To some extent this niggling legislation did enhance the quality of products, but at the cost of slowing down the rhythm of production.

A guild's governing body was composed of one or several "doyens," assisted by a few "jurymen" and sometimes inspectors. These officials were appointed by the municipal corporation, and the board personnel was partially changed each year. The board met once a week on the same premises, which would be either a house belonging to the guild, or a designated room in the Clocktower, or a tavern chosen for its comfort and elegance, or perhaps the building of the inspector of weights and measures. These meetings usually developed into boisterous parties. The guilds had their own suites of furniture, and their own table-ware and glassware, all bearing the arms of the guild; each guild possessed its own seal and banner as well.

Workers were subject to guild control from their earliest years, since an apprenticeship could be served only with a guild master. Apprenticeships generally lasted two years, but with surgeons the period was three years, and with the hatters of Amsterdam four; on the other hand, the wood-sawyers only required a six months' apprenticeship. No master was entitled to accept more than two apprentices, and these entered into his service very young, usually at the age of twelve, after having paid an enrolment fee which the master sometimes advanced them against their future wages. In this way they lost all freedom of action: they lodged in their master's house, and if they left him they had to repay him for their board and lodging. They also ran the risk of not being able to find another master to accept them. The choice of one's master was more or less the choice of one's fate; conditions of apprenticeship were not strictly regulated, and some apprentices spent years cleaning the workshop and looking after the tools and equipment before they really had a chance to prepare themselves for their final apprentice's examination.

After passing this examination the apprentice became a journeyman and had to find a new master. Armed with his diploma, he often had to go from town to town in search of employment. This roving existence, which was even more developed in France than in the Netherlands, constituted a period of institutional unemployment in his working life. When a master finally accepted him, the journeyman registered himself with the appropriate guild. Then, after a fairly long period, he could, in most guilds, present a so-called "masterpiece" for examination and judgement, and, if successful, was entitled to call himself a "master" and had the right to keep a shop or open a workshop in his own name. Even so, he had to be in a position to pay his dues and to offer a banquet or at least a toast-drinking reception to his examiners. Many journeymen could not do so and remained salaried workmen for the rest of their lives.

The members of a guild paid a regular subscription to an official whose duties consisted not only in collecting dues but also in performing various secretarial functions, such as sending notifications to attend meetings, arranging the funeral ceremonies of those who had been members, and looking after the upkeep of the guild's place of assembly. This employee's fixed salary was augmented by a percentage of the fines imposed by the doyens. Each year, on the feast-day of its patron saint, the guild gave its official banquet, a celebration that sometimes lasted two days and occasionally led to such excesses that the authorities attempted to ban the tradition outright or at least limit its duration. The wealthier guilds also organised expeditions, to which the ladies were invited, and private parties. A guild's entertainment and recreation fund was always substantial.

But the country's economic development threatened to sweep away the ancient guild structure, despite the support of the dignitaries with life tenure who formed the urban administrations. So long as these gentlemen controlled the guilds they dominated the local economy and were safe from any outside competition; nevertheless, signs of progress became more and more frequent. During the first half of the century the creation of new industries led to the constitution of hitherto nonexistent guilds: linen-weavers in 1614, timber-merchants in 1615, and makers of fustian in 1631. But this development of new guilds was only superficial. Branches of industry in full expansion, such as the textile trade, began to establish factories outside the jurisdiction of the towns, in villages where no guilds operated, with the consequent advantage of cheap manpower. And big contractors took full advantage of the opportunity to exploit the blind rivalry which set at odds identical guilds in different towns. In Amsterdam urban extension played its part in breaking the guilds' influence. Big business, as well as most of the new industries, escaped their domination. The guilds defended themselves by means which inevitably rebounded on them eventually, that is to say, a multiplication of controls and obstacles which finished by making them into exclusive castes barring practically everyone except the sons of deceased members. As a result the volume of "illicit" work increased rapidly despite the most vexatious measures. Within the framework of the guilds the crafts remained predominant, and the regulation of wages placed fairly strict limits on the possibilities of additional payments. The development of large-scale capitalist business enterprises worked doubly against this archaic structure; guilds were either suppressed entirely, as happened to the hatters in 1680, or they became mere insurance agencies in many towns after 1660. Each guild possessed an assistance fund earmarked for the relief of old, sick or needy members. The practice grew, in some places, of selling guild diplomas to people outside the profession; in this way the fund acquired a new subscription and the pseudo-member gained the right to its assistance. Traditionally the guilds imposed certain reciprocal duties on its members, such as looking after the sick and officiating at burials, but this custom of mutual aid had fallen into such neglect that by mid-century fines had to be imposed for failure to perform these duties.

In every town guild members formed a citizens' militia which was originally entrusted with defensive military duties, but during the seventeenth century no longer played any real military role. At the most it gave assistance to the police during civil disturbances or outbreaks of fire. It had really become a friendly society, with its own uniform, parading on major occasions and organising archery competitions. In 1672 the Amsterdam militia numbered at least ten thousand men.

In theory dealers were organised in a "merchants' guild," but in fact only small shopkeepers were dependent on this association. The more a merchant's affairs prospered (especially when he began to deal in international markets, and transit and cash shipments) the less need he had of the guild's controls and assistance. A big merchant's way of life distinguished him immediately from the host of small merchants; he was often a well-educated man, and Sorbière recorded in mid-century that he knew many merchants who spent their evenings in instructive reading. Some of them had been to a university, but they had little opportunity there to learn the elements of their profession; the future merchant spent his apprenticeship as an office-boy in his father's establishment or that of one of his father's colleagues. After spending some months sweeping out the office, replacing the candles and keeping the fire going, he graduated to the position of clerk, sharpening quills, running errands, making book-entries, learning accountancy and familiarising himself with the use of almanacs.

These almanacs, published annually, gave lists of fairs and markets, time-tables of passenger barges and ships and details of high and low tides, and served as the main instrument of commercial "culture." They were produced in great numbers, some being more comprehensive than others and often dealing with a specific town. One of the Dordrecht almanacs even supplied information about the competence of the various town officials. Sometimes, too, the authors supplemented all this information with doctrinal instruction: Gaspar Coolhaas, in his 1606 "Businessmen's Almanac," provided a refutation of the errors of the Catholic Church.

The businessman's office was usually situated in the basement of his house, or occasionally in the loft next to his warehouse, where it would not disturb domestic harmony. He called it his *Kantoor* (a corruption of the French *comptoir*), and the choice of site was dictated less by considerations of comfort than by the wife's refusal to allow the living-rooms to be used for business purposes. So that although a Dutch businessman had his living quarters and his offices under the same roof they were kept rigorously separate, and when the volume of business demanded expanded accommodation, interior walls were put up forming corridors that gave access to the premises, and a special entrance door was let into one of the outside walls—all this to avoid creating any disturbance in the living-rooms!

The merchant's working day started at about ten in the morning. Actual trading activities were only conducted during four hours of a single day. From ten until midday the merchant presided over his office and the apprentices and clerks who had arisen earlier from the attics where they slept and were waiting to start work. Offices were furnished with utmost simplicity: a few sturdy desks, with lead inkwells nailed to them, chairs with leather seats, bookshelves laden with registers, a sand-glass. The boss sat at a raised desk, wearing a night-cap; the clerks sat two by two beneath him, wearing protective oversleeves.

At midday the Stock Exchange opened and was soon crowded with the town's merchants, plus a good number of onlookers and idlers. All important business was conducted here. Brokers hurried to and fro inside the building, carrying a portable writing-desk, and prepared contracts; when two merchants had struck a bargain—perhaps one of them had assigned a cargo of copper still in transit on the high seas at the time, in exchange for a cargo of precious wood plus a cash sum—they signed a contract, went straight off to the bank where their funds were deposited and arranged the transfer of this sum from one account to the other. They had perhaps completed a deal involving thousands of florins, without having had to handle a single coin. All business had to be completed by two o'clock when the Stock Exchange closed. If urgent business demanded a return to the building a little after that time, a fine was levied. This arrangement permitted an advantageous centralisation of major commerce, helped speed up its operations and consequently facilitated credit.

From the sixteenth century onwards each large town involved in trade had its own Exchange; a few had possessed Exchanges since the fifteenth century. They were originally held in the open air, in a square, a street or any other convenient location. The first building designed specifically for the purpose was constructed in Amsterdam in 1611: a huge two-storeyed square edifice in the very centre of town, with a central courtyard measuring nearly five hundred and fifty square yards surrounded by arcades in which the stalls were set up. Free access was provided by a large open porch at each end of the building. Built above a canal, the Exchange served also as a port, and large ships could sail in under its vault after lowering their masts.

Dutch commerce acquired such flexibility under this system, and the credit system became so widespread, especially after 1650, that the Amsterdam Exchange was thenceforward the centre of world trade. When the 1672 crisis occurred, the Austrian ambassador sent his monarch a daily report of exchange quotations. . . .

Paupers and Criminals

A certain proportion of the population was threatened with penury at frequent intervals, as we have already seen. But more unfortunate even than these casual workers were those existing on the fringes of society in a state of permanent economic hardship. The frequency of unemployment and the nomadic tendencies of some workmen made it impossible to distinguish clearly between underprivileged workers and true vagabonds, but together they constituted a collection of humanity sprinkled with antisocial elements whose common denominator was the supreme, ever-present menace of hunger.

This segment of the nation, rejected by the very society whose economic shortcomings had provoked its existence, made its existence known through begging, a curse which afflicted town and country equally and grew worse as the century progressed. The provincial States proscribed mendicancy at regular intervals, but it was a scourge that no edicts could halt. Amsterdam swarmed with beggars and with a horde of imaginary cripples, and after the truce and the disbanding of the mercenaries the evil increased. Bands of dubious characters, leading a wandering existence or living here and there in improvised slums, exercised a variety of minor illegal trades, stole and even killed if necessary, were hunted from one province to another as a result of the banishment pronounced on them at regular intervals, and spoke an incomprehensible thieves' cant that impressed someone sufficiently for him to publish a vocabulary in 1613. It was impossible to distinguish brigands from honest wretches, and the scene was further confused by the presence of gipsies practising palmistry, drawing horoscopes and—reputedly—stealing children. However often the gipsies were imprisoned or expelled they always came back. The forces of law and order were powerless to deal with this situation.

It was dangerous to go about unarmed at night in The Hague's Wood. In 1643 a recently decapitated man's head was found there, but the police were never able to discover the assassins; and in 1661 two young women of good family were abducted in broad daylight. Towns increased their police forces, ordered the imprisonment of all beggars lacking a special permit and offered a reward to any citizen pointing out or apprehending a thief. Teams of "beggar hunters," assisted by dogs, tracked down vagabonds in the rural areas.

The municipalities were more or less aware of the economic origins of this disorderliness and made a practice throughout the century of distributing free rations in years when famine struck or the cost of living suddenly increased. In such cases a shipload of wheat or rye was ordered and a rough bread baked and given away to needy people. In Leyden, in 1634, twenty thousand people received free bread. In Amsterdam, an official allocation of alms to the poor took place once a week. The Reformed Church also dispensed charity at a local level, but only to the needy who had at least a provisional domicile; an official appointed by the deacon or the town provided these deserving poor in winter with a little butter, cheese, bread and peat. But during the summer months these supplies were discontinued and they were left to fend for themselves.

The municipalities and the Church organised regular collections for aid to the poor. Bourgeois citizens were often generous contributors, and a properly launched collection in a large town could well bring in fifteen or even twenty-five thousand florins. Tradition demanded the giving of alms on special occasions such as marriages in wealthy families. In several towns the poor were presented with the pall from the catafalque after a funeral ceremony, and it was rare for a prosperous citizen to die without bequeathing a sum of money for works of charity. The reputation of the Dutch for charity was acknowledged throughout Europe, and Louis XIV himself, on the eve of his invasion of the Netherlands, reassured Charles II in these terms: "Have no fear for the fate of Amsterdam. I live in the certain hope that Providence will save that city, if only in consideration of her charity."

To deal with its large industrial proletariat, Leyden had instituted a body of inspectors of paupers with large assistance funds at its disposal. During the second half of the century cottages were built in several towns with private funds and rented to poor people at very low rates. Poor-houses existed everywhere, some of them taken over from the previous religious orders and now administered by the State, and the largest towns boasted several institutions serving different categories of the needy: Leyden, for example, had an almshouse for homeless paupers, another for old folk, and an orphanage as well. But the main function of

these poor-houses since the beginning of the seventeenth century was the rehabil-
itation of vagabonds. In most districts of any importance tramps could find a
reception centre, sometimes attached to the local hospital, where they received
free board and lodging for three days; on the morning of the fourth day they
were sent on their way again. But in some towns the influx of vagabonds was so
great that the poor-houses were closed from June to October. Amenities were
rudimentary in these establishments: a large communal hall, heated and provided with benches, and two dormitories, one for men, the other for women. At
the end of each meal the knives were counted and locked away for safety, and at
night the dormitory doors were bolted securely.

Besides a large hospital, Amsterdam possessed several orphanages, and an
alms-house providing lodging for as many as four hundred aged female paupers.
In 1613 a "House of Charity" was founded to provide shelter for paupers who
lived by begging, though to be eligible for admission they had to prove several
years' domicile in the town. But this poor-house did provide needy travellers with
a cash subsidy. A provost accompanied by officials attached to the institution
scoured the town every day, rounding up beggars. At a later date the institution
also housed children that the orphanage could not accommodate, and became
responsible for carrying out free burials.

The administration of these various houses of refuge was vested by the town
in committees of distinguished citizens, both men and women, who considered
the appointment an honour. Although the day-to-day running of these charitable
institutions by the junior personnel often left much to be desired, the administration at top level was usually remarkably efficient. They were often financed out of
the revenues from former ecclesiastical properties, and, especially in the second
half of the century, benefited from foundations and legacies. Door-to-door collections were taken once a month, and the churches held a collection every Sunday.
In addition, alms-boxes were to be found at various sites in the town. The muni-
cipality put aside the income it derived from its taxes on imported cereals, auction
sales, banquets and deluxe funerals for the benefit of the various institutions. And
finally, charity lotteries were organised periodically and were extremely popular
with the public; on one occasion, a ticket costing two stuivers won first prize of a
thousand florins.

Towards the end of the century the charitable institutions became too small
and their revenue insufficient. Various remedies had to be devised, including the
creation of special taxes, an increase in the number of lotteries, and sending paupers back to their place of birth.

From the start of the century Amsterdam possessed two institutions for anti-
social elements, combining the functions of prison and reform home, the
Rasphuis for men and the *Spinhuis* for women. A Latin inscription along the
façade of the Rasphuis described its purpose: *Virtutis est domare quae cuncti
pavent.* The inmates were there as a result of sentences imposed by the courts and
were subjected to manual labour, usually rasping logwood (whence the institution's name). Anyone who refused to work was immediately flogged, and if he
repeated the offence he was thrown into a cellar that was gradually filled with
water through a pipe; to escape drowning, the prisoner had to pump without
stopping—an ingenious form of forced labour! Vagabonds, thieves and ne'er-do-
wells of every description shared the premises of the *Rasphuis* with unruly sons
imprisoned on the demand of their fathers. The *Spinhuis* ("spinning house," as

its name implies) harboured prostitutes, daughters who had run away from home and wives whose husbands had them incarcerated for misconduct or drunkenness. The more fortunate women could obtain a private room by paying for it.

A few cases taken at random from Amsterdam's criminal files of the epoch give some idea of the kind of crimes that came up for judgement, and the scale of sentences imposed. Trijntje Pieters, a servant-girl, was condemned to death for killing her illegitimate baby; Laurens Cornelisse, an inmate of the Rasphuis, was hanged for breaking open the bailiff's strong-box on the premises of the law-court which was about to try him for a previous offence; Jean Franchoys was condemned to the pillory for bigamy; the same sentence was imposed on Abraham Frederickszen, employed by a tombstone engraver attached to the Carthusian cemetery, for stripping corpses of their garments; the soldier Ernst Rip was condemned to death for murdering a comrade; Albert Alberta and his wife, notorious receivers of stolen goods, were hanged for making away with the head and right hand of an executed criminal from the pauper's grave (in their case they were doubtless suspected of sorcery as well). The list could be extended indefinitely. Murder, theft, arson and forgery were the commonest offences. One woman, Griet Andries, was sentenced on thirty-two separate occasions. The criminal law was applied with implacable severity and was especially ruthless in dealing with theft, house-breaking and the forging of signatures. Sometimes it showed incredible savagery: sixteen-year-old urchins were branded with red-hot irons; a forty-five-year-old captain, found guilty of murder, was burnt alive. The files record a repeated offender who was flogged eleven times and branded five times during the course of a single year. In the final quarter of the century two hundred and nine death sentences were carried out in Amsterdam.

Grotius justified this cruelty by the necessity for striking fear into the hearts of potential criminals. It can be understood in the context of the general callousness of feeling prevalent throughout Europe at the time, but the element of inhumanity was certainly abetted by the incoherence of the Dutch penal code. Apart from anything else, laws and customs differed from province to province and even from town to town, so that any attempt at unification ran up against the hostility of the local law-courts. In this judicial chaos crime remained wholly undefined, and the criterion, constantly repeated in judges' summings-up, was simply that "such a thing is intolerable in a civilised state." By this standard, blasphemy or passing counterfeit money could rank as acts no less reprehensible than murder or high treason.

Since a confession was considered to be a final and irrevocable proof of guilt, the police had an interest in securing a confession as quickly as possible and torture was used as a matter of course during interrogations, despite the fact that its use was illegal at that juncture. The same "techniques" flourished here as in the rest of Europe: the rack, forced absorption of disgusting substances, whipping and branding. In the province of Holland judicial authorities recognised five degrees of torture; Friesland was more humane and recognised only three. These classifications were used as a basis for the scale of penalties, since torture was often included officially in the sentence after having been used illegally to force a confession.

Refined, cultivated magistrates participated without scruple or disgust in these sessions, sessions which were sometimes so prolonged that they had their meals brought in to them in the torture-chamber where the executioner was tormenting some poor devil. They had been personally responsible for the penalty, choosing it from an established, incongruous, bloodthirsty arsenal: amputation of the right hand, of the nose, or of one or both ears; burning of the tongue; gouging of eyes; slitting of one cheek; branding of the shoulder with a red-hot iron. For the smallest offence the culprit was exposed in the pillory under a wooden bell through which his head projected, with an inscription or symbol indicating his misdemeanour. Or he would be paraded through town in some ignominious garb.

The pedlar found guilty of selling forbidden books was dragged through the streets with a doctor's cap clamped on his head and a package of his merchandise slung around his neck. Petty thieves were taken by the tipstaff and exhibited to the townspeople with the stolen object tied on top of their heads. The pillory, a scaffold set up in front of the town hall, was almost as unpleasant an instrument as the rack, and the public exposure involved a dishonour that only the most miserable could accept with equanimity.

Other penalties were social or economic in character, involving deprivation of civil rights, a ban on exercising a particular profession, or fines and penalties. Banishment allowed the authorities, under some pretext or another, to get rid of individuals whom they considered dangerous; they would be expelled for a month, for life, or even for a hundred years and a day. A culprit guilty of wounding might find himself forbidden to leave his house after eight o'clock in the evening for a year; a drunkard might be prohibited from frequenting taverns for three years. During the periods when big work projects, such as building of fortifications or opening of mines in some distant colony, were under way the courts were quick to condemn culprits to these "public works."

The death penalty was usually carried out by hanging, but this was often preceded by tortures or else replaced by some more cruel method of execution, in which the victim was strapped down in a chair and beheaded with a sword, burnt at the stake, drowned inside a barrel, or even buried alive. Sometimes corpses were executed, as when the bodies of suicides were hanged after being dragged to the foot of the gibbet by a rope attached to a horse. For urban populations these executions provided a free spectacle regarded as a harmless popular diversion.

A gibbet stood at each of Amsterdam's gates, and the outskirts of most towns presented a similar spectacle. They consisted of two strong vertical beams about fifteen feet high, joined by a cross-beam long enough to accommodate easily half a dozen corpses at the end of their ropes. A ladder allowed the executioner to fix the ropes and adjust the nooses.

The executioner was an important official and each province had its officially designated practitioner; in Holland he entitled himself "Master of High Works of Holland, residing at Haarlem." His wages were calculated at piece rates: three florins for a beheading, plus nine florins for burying the corpse. Breaking on the wheel was more profitable; at three florins a blow the total could reach about thirty florins, whereas floggings at the same rate seldom brought in more than twenty-four florins.

Prison sentences were comparatively rare, and imprisonment was considered not so much a punishment as a means of guarding the accused while he awaited trial. Wealthy prisoners could easily alleviate their conditions, but the impecunious suffered terrible privations since prisoners had to pay for their food, and the supply of provisions was a profitable enterprise for the guards. Undernourishment and lack of hygiene made these prisons centres of infection. In Amsterdam the prisons had been established in the cellars of the town hall and in the city's four oldest gate-towers, with walls six feet thick, narrowed barred windows and floors strewn with a thin layer of damp straw. During the course of the century a few audacious prisoners succeeded in escaping; Johannes Palmer achieved this feat in 1652, and the court sentenced him in his absence to perpetual banishment, having failed in its efforts to recapture him.

CHAPTER SEVEN

Modes of Feeling and Thought: The Bases of Medieval Culture

On Time and Nature

No general introduction to Bloch's work is necessary here. It will suffice to point up the central importance of the question he raises: Can the "scientific" historian come to grips with the "modes of feeling and thought" characteristic of men in other epochs than our own? Unless we are willing to exclude all questions of psychology from the historian's craft—and that Bloch would not do—few problems can be more significant. Bloch tried to solve them in his usual way. He applied his extraordinary learning and linguistic equipment to a wide variety of sources. In our own decade "psycho-historians" (Erikson, Lifton, Fromm) are developing the tools to bring us into closer contact with the mental "vicissitudes of the human organism," the "means of measuring the influence" of a cruel environment on men. Bloch would have applauded their effort. His own pioneering you can read here.

From *Feudal Society*

Marc Bloch

MODES OF FEELING AND THOUGHT

Man's Attitude to Nature and Time

The men of the two feudal ages were close to nature—much closer than we are; and nature as they knew it was much less tamed and softened than we see it today. The rural landscape, of which the waste formed so large a part, bore fewer traces of human influence. The wild animals that now only haunt our nursery tales—bears and, above all, wolves—prowled in every wilderness, and even amongst the cultivated fields. So much was this the case that the sport of hunting was indispensable for ordinary security, and almost equally so as a method of supplementing the food supply. People continued to pick wild fruit and to gather honey as in the first ages of mankind. In the construction of implements and tools, wood played a predominant part. The nights, owing to the wretched lighting, were darker; the cold, even in the living quarters of the castles, was more

SOURCE: Marc Bloch, *Feudal Society*, trans. L. A. Manyon (Chicago: University of Chicago Press, 1968), pp. 72–75, 81–87.

intense. In short, behind all social life there was a background of the primitive, of submission to uncontrollable forces, of unrelieved physical contrasts. There is no means of measuring the influence which such an environment was capable of exerting on the minds of men, but it could hardly have failed to contribute to their uncouthness.

A history more worthy of the name than the diffident speculations to which we are reduced by the paucity of our material would give space to the vicissitudes of the human organism. It is very naive to claim to understand men without knowing what sort of health they enjoyed. But in this field the state of the evidence, and still more the inadequacy of our methods of research, are inhibitive. Infant mortality was undoubtedly very high in feudal Europe and tended to make people somewhat callous towards bereavements that were almost a normal occurrence. As to the life of adults, even apart from the hazards of war it was usually short by our standards, at least to judge from the records of princely personages which (inexact though they must often be) constitute our only source of information on this point. Robert the Pious died at about the age of 60; Henry I at 52; Philip I and Louis VI at 56. In Germany the first four emperors of the Saxon dynasty attained respectively the ages of 60 (or thereabouts), 28, 22 and 52. Old age seemed to begin very early, as early as mature adult life with us. This world, which, as we shall see, considered itself very old, was in fact governed by young men.

Among so many premature deaths, a large number were due to the great epidemics which descended frequently upon a humanity ill-equipped to combat them; among the poor another cause was famine. Added to the constant acts of violence these disasters gave life a quality of perpetual insecurity. This was probably one of the principal reasons for the emotional instability so characteristic of the feudal era, especially during its first age. A low standard of hygiene doubtless also contributed to this nervous sensibility. A great deal of effort has been expended, in our own day, in proving that baths were not unknown to seignorial society. It is rather puerile, for the sake of making this point, to overlook so many unhealthy conditions of life: notably under-nourishment among the poor and overeating among the rich. Finally, we must not leave out of account the effects of an astonishing sensibility to what were believed to be supernatural manifestations. It made people's minds constantly and almost morbidly attentive to all manner of signs, dreams, or hallucinations. This characteristic was especially marked in monastic circles where the influence of mortifications of the flesh and the repression of natural instincts was joined to that of a mental attitude vocationally centered on the problems of the unseen. No psychoanalyst has ever examined dreams more earnestly than the monks of the tenth or the eleventh century. Yet the laity also shared the emotionalism of a civilization in which moral or social convention did not yet require well-bred people to repress their tears and their raptures. The despairs, the rages, the impulsive acts, the sudden revulsions of feeling present great difficulties to historians, who are instinctively disposed to reconstruct the past in terms of the rational. But the irrational is an important element in all history and only a sort of false shame could allow its effects on the course of political events in feudal Europe to be passed over in silence.

These men, subjected both externally and internally to so many ungovernable forces, lived in a world in which the passage of time escaped their grasp all the more because they were so ill-equipped to measure it. Water-clocks, which were costly and cumbersome, were very rare. Hour-glasses were little used. The inadequacy of sundials, especially under skies quickly clouded over, was notorious. This resulted in the use of curious devices. In his concern to regulate the course of a notably nomadic life, King Alfred had conceived the idea of carrying with him everywhere a supply of candles of equal length, which he had lit in turn, to mark the passing of the hours, but such concern for uniformity in the division of the day was exceptional in that age. Reckoning ordinarily—after the example of Antiquity—twelve hours of day and twelve of night, whatever the season, people of the highest education became used to seeing each of these fractions, taken one by one, grow and diminish incessantly, according to the annual revolution of the sun. This was to continue till the moment when—towards the beginning of the fourteenth century—counter-poise clocks brought with them at last, not only the mechanization of the instrument, but, so to speak, of time itself.

An anecdote related in a chronicle of Hainault illustrates admirably the sort of perpetual fluctuation of time in those days. At Mons a judicial duel is due to take place. Only one champion puts in an appearance—at dawn; at the ninth hour, which marks the end of the waiting period prescribed by custom, he requests that the failure of his adversary be placed on record. On the point of law, there is no doubt. But has the specified period really elapsed? The county judges deliberate, look at the sun, and question the clerics in whom the practice of the liturgy has induced a more exact knowledge of the rhythm of the hours than their own, and by whose bells it is measured, more or less accurately, to the common benefit of men. Eventually the court pronounces firmly that the hour of "none" is past. To us, accustomed to live with our eyes turning constantly to the clock, how remote from our civilization seems this society in which a court of law could not ascertain the time of day without discussion and inquiry!

Now the imperfection of hourly reckoning was but one of the symptoms, among many others, of a vast indifference to time. Nothing would have been easier or more useful than to keep an accurate record of such important legal dates as those of the births of rulers; yet in 1284 a full investigation was necessary to determine, as far as possible, the age of one of the greatest heiresses of the Capetian realm, the young countess of Champagne. In the tenth and eleventh centuries, innumerable charters and memoranda were undated, although their only purpose was to serve as records. There are exceptional documents which are better in this respect, yet the notary, who employed several systems of reference simultaneously, was often not successful in making his various calculations agree. What is more, it was not the notion of time only, it was the domain of number as a whole which suffered from this haziness. The extravagant figures of the chroniclers are not merely literary exaggeration; they are evidence of the lack of all awareness of statistical realities. Although William the Conqueror certainly did not establish in England more than 5000 knights' fees, the historians of a somewhat later time, and even certain administrators (though it would certainly not have been very difficult for them to obtain the right information), did not hesitate to attribute to him the creation of from thirty-two to sixty thousand of these military tenements. The period had, especially from the end of the eleventh century, its mathematicians who groped their way courageously in the wake of the Greeks and Arabs; the architects and sculptors were capable of using a fairly simple geometry. But among the computations that have come down to us—and this was true till the end of the Middle Ages—there are scarcely any that do not reveal astonishing errors. The inconveniences of the Roman numerical system, ingeniously corrected as they were by the use of the abacus, do not suffice to explain these mistakes. The truth is that the regard for accuracy, with its firmest buttress, the respect for figures, remained profoundly alien to the minds even of the leading men of that age. . . . "Ages of faith," we say glibly, to describe the religious attitude of feudal Europe. If by that phrase we mean that any conception of the world from which the supernatural was excluded was profoundly alien to the minds of that age, that in fact the picture which they formed of the destinies of man and the universe was in almost every case a projection of the pattern traced by a Westernized Christian theology and eschatology, nothing could be more true. That here and there doubts might be expressed with regard to the "fables" of Scripture is of small significance; lacking any rational basis, this crude scepticism, which was not a normal characteristic of educated people, melted in the face of danger like snow in the sun. It is even permissible to say that never was faith more completely worthy of its name. For the attempts of the learned to provide the Christian mysteries with the prop of logical speculation, which had been interrupted on the extinction of ancient Christian philosophy and revived only temporarily and with difficulty during the Carolingian renaissance, were not fully resumed before the end of the eleventh century. On the other hand, it would be wrong to ascribe to these believers a rigidly uniform creed.

Catholicism was still very far from having completely defined its dogmatic system, so that the strictest orthodoxy was then much more flexible than was to be the case later on, after scholastic philosophy and the Counter-Reformation had in turn exercised their influence. Moreover, in the ill-defined border land where Christian heresy degenerated into a religion actively opposed to Christianity, the old Manichaeanism retained a number of votaries in various places. Of these it is not precisely known whether they had inherited their religion from groups who had remained obstinately faithful to this persecuted sect since the first centuries of the Middle Ages, or had received it, after a long interval, from Eastern Europe. But the most notable fact was that Catholicism had incompletely penetrated among the common people. The parish clergy, taken as a whole, were intellectually as well as morally unfit for their task. Recruited with insufficient care, they were also inadequately trained; most commonly instruction consisted in casual lessons given by some priest, himself poorly educated, to a youth who was preparing himself for orders while serving the mass. Preaching, the only effective means of making accessible to the people the mysteries locked up in the Scriptures, was but irregularly practised. In 1031 the Council of Limoges was obliged to denounce the error which claimed that preaching was the prerogative of the bishops, for obviously no bishop would have been capable by himself of preaching the Gospel to the whole of his diocese.

The Catholic mass was recited more or less correctly in all parishes, though sometimes the standard was rather low. The frescoes and bas-reliefs on the walls or the capitals of the principal churches—"the books of the unlettered"— abounded in moving but inaccurate lessons. No doubt the faithful nearly all had a superficial acquaintance with the features most apt to strike the imagination in Christian representations of the past, the present, and the future of the world. But their religious life was also nourished on a multitude of beliefs and practices which, whether the legacy of age-old magic or the more recent products of a civilization still extremely fertile in myths, exerted a constant influence upon official doctrine. In stormy skies people still saw phantom armies passing by: armies of the dead, said the populace; armies of deceitful demons, declared the learned, much less inclined to deny these visions than to find for them a quasi-orthodox interpretation. Innumerable nature-rites, among which poetry has especially familiarized us with the May-day festivals, were celebrated in country districts. In short, never was theology less identified with the popular religion as it was felt and lived.

Despite infinite variations according to environment and regional traditions, some common characteristics of this religious mentality can be discerned. Although it will mean passing over various deep and moving features and some fascinating problems of permanent human interest, we shall be obliged to confine ourselves here to recalling those trends in thought and feeling whose influence on social behaviour seems to have been particularly strong.

In the eyes of all who were capable of reflection the material world was scarcely more than a sort of mask, behind which took place all the really important things; it seemed to them also a language, intended to express by signs a more profound reality. Since a tissue of appearances can offer but little interest in itself, the result of this view was that observation was generally neglected in favour of interpretation. In a little treatise on the universe, which was written in the ninth century and enjoyed a very long popularity, Rabanus Maurus explained how he followed his plan: "I conceived the idea of composing a little work . . . which should treat, not only of the nature of things and the properties of words . . . but still more of their mystic meanings." This attitude explains, in large part, the inadequacy of men's knowledge of nature—of a nature which, after all, was not regarded as greatly deserving of attention. Technical progress—sometimes considerable—was mere empiricism.

Further, this discredited nature could scarcely have seemed fitted to provide its own interpretation, for in the infinite detail of its illusory manifestations it was conceived above all as the work of hidden wills—wills in the plural, in the opinion of simple folk and even of many of the learned. Below the One God and subordinated to his Almighty Power—though the exact significance of this subjection was not, as a rule, very clearly pictured—the generality of mankind imagined the opposing wills of a host of beings good and bad in a state of perpetual strife; saints, angels, and especially devils. "Who does not know," wrote the priest Helmold, "that the wars, the mighty tempests, the pestilences, all the ills, indeed, which afflict the human race, occur through the agency of demons?" Wars, we notice, are mentioned indiscriminately along with tempests; social catastrophes, therefore, are placed in the same class as those which we should nowadays describe as natural. The result was a mental attitude which the history of the invasions has already brought to notice: not exactly renunciation, but rather reliance upon means of action considered more efficacious than human effort. Though the instinctive reactions of a vigorous realism were never lacking, a Robert the Pious or an Otto III could nevertheless attach as much importance to a pilgrimage as to a battle or a law, and historians who are either scandalized by this fact or who persist in discovering subtle political manœuvres in these pious journeys merely prove thereby their own inability to lay aside the spectacles of men of the nineteenth and twentieth centuries. It was not merely the selfish quest of personal salvation that inspired these royal pilgrims. From the patron saints whose aid they went to invoke, they expected for their subjects as well as for themselves, not only the promise of rewards in heaven, but the riches of the earth as well. In the sanctuary, as much as on the field of battle or in the court of law, they were concerned to fulfil their function as leaders of their people.

The world of appearances was also a transitory world. Though in itself inseparable from any Christian representation of the Universe, the image of the final catastrophe had seldom impinged so strongly on the consciousness of men as at this time. They meditated on it; they assessed its premonitory signs. The chronicle of Bishop Otto of Freising, the most universal of all universal histories, began with Creation and ended with the picture of the Last Judgment. But, needless to say, it had an inevitable *lacuna:* from 1146—the date when the author ceased to write—to the day of the great catastrophe. Otto, certainly, expected this gap to be of short duration: "We who have been placed at the end of time . . ." he remarks on several occasions. This was the general conviction among his contemporaries as it had been in earlier times, and it was by no means confined to the clergy; to suppose so would be to forget the profound interpenetration of the two groups, clerical and lay. Even among those who did not, like St. Norbert, go so far as to declare that the event was so close that the present generation would witness it no one doubted of its imminence. In every wicked prince, pious souls believed that they recognized the mark of Antichrist, whose dreadful empire would precede the coming of the Kingdom of God.

But when in fact would it strike—this hour so close at hand? The Apocalypse seemed to supply an answer: "and when the thousand years are expired . . ." Was this to be taken as meaning a thousand years after the death of Christ? Some thought so, thus putting back the great day of reckoning—according to the normal calculation—to the year 1033. Or was it rather to be reckoned from his birth? This latter interpretation appears to have been the most general. It is certain at any rate that on the eve of the year one thousand a preacher in the churches of Paris announced this date for the End of Time. If, in spite of all this, the masses at that time were not visibly affected by the universal terror which historians of the romantic school have mistakenly depicted, the reason is above all that the people of that age, though mindful of the passage of the seasons and the annual cycle of the liturgy, did not think ordinarily in terms of the numbers of the years, still less in figures precisely computed on a uniform basis. How many charters lack any trace of a date! Even among the rest, what diversity there is in the systems of reference, which are mostly unconnected with the life of the Saviour—years of reigns or pontificates, astronomical indications of every kind, or even the fifteen-year cycle of the indiction, a relic of Roman fiscal practices! One entire country, Spain, while using more generally than elsewhere the concept of a definite era, assigned to it—for reasons that are somewhat obscure—an initial date absolutely unrelated to the Gospel, namely the year 38 B.C. It is true that legal documents occasionally and chronicles more frequently adhered to the era of the Incarnation; but it was still necessary to take into account the variations in the beginning of the year. For the Church excluded the first of January as a pagan festival. Thus, according to the province or the chancellery, the year designated the thousandth began at one or other of six or seven different dates, which ranged, according to our calendar, from 25th March 999 to 31st March 1000. What is worse, some of these initial dates, being essentially moveable since they were linked with a particular liturgical moment of the Easter period, could not be anticipated without tables, which only the learned possessed; they were also very apt to lead to permanent confusion in men's minds by making some years longer than others. Thus it was not unusual for the same day of the month, in March or April, or the feast of the same saint to occur twice in the same year. Indeed, for the majority of Western men this expression, "the year 1000," which we have been led to believe was charged with anguish, could not be identified with any precise moment in the sequence of days.

Yet the notion of the shadow cast over men's minds at that time by the supposed imminence of the Day of Wrath is not altogether wrong. All Europe, it is true, did not tremble with fear towards the end of the first millennium, to compose itself suddenly as soon as this supposedly fateful date was past. But, what was even worse perhaps, waves of fear swept almost incessantly over this region or that, subsiding at one point only to rise again elsewhere. Sometimes a vision started the panic, or perhaps a great historic calamity like the destruction of the Holy Sepulchre in 1009, or again perhaps merely a violent tempest. Another time, it was caused by some computation of the liturgists, which spread from educated circles to the common people. "The rumour spread through almost the whole world that the End would come when the Annunciation coincided with Good Friday," wrote the abbot of Fleury a little before the year 1000. Many theologians, however, remembering that St. Paul had said: "the day of the Lord cometh like a thief in the night," condemned these indiscreet attempts to pierce the mystery in which the Divinity chose to veil his dread purpose. But is the period of waiting made less anxious by ignorance of when the blow will fall? In the prevailing disorders, which we should unhesitatingly describe as the ebullience of adolescence, contemporaries were unanimous in seeing only the last convulsions of an "aged" humanity. In spite of everything, an irresistible vitality fermented in men, but as soon as they gave themselves up to meditation, nothing was farther from their thoughts than the prospect of a long future for a young and vigorous human race.

If humanity as a whole seemed to be moving rapidly towards its end, so much the more did this sensation of being "on the way" apply to each individual life. According to the metaphor dear to so many religious writers, the true believer was in his earthly existence like a pilgrim, to whom the end of the road is naturally of more importance than the hazards of the journey. Of course, the thoughts of the majority of men did not dwell constantly on their salvation. But when they did, it was with deep intensity and above all with the aid of vivid and very concrete images, which were apt to come to them by fits and starts; for their fundamentally unstable minds were subject to sudden revulsions. Joined to the penitent mood of a world on the verge of dissolution, the desire for the eternal rewards cut short more than one leader's career by voluntary withdrawal to the cloister. And it ended for good and all the propagation of more than one noble line, as in the case of the six sons of the lord of Fontaines-lès-Dijon who eagerly embraced the monastic life under the leadership of the most illustrious of their number, Bernard of Clairvaux. Thus, in its way, the religious mentality favoured the mixing of the social classes.

Many Christians, nevertheless, could not bring themselves to submit to these austere practices. Moreover, they considered themselves (and perhaps not without reason) to be incapable of reaching heaven through their own merits. They therefore reposed their hopes in the prayers of pious souls, in the merits accumulated for the benefit of all the faithful by a few groups of ascetics, and in the intercession of the saints, materialized by means of their relics and represented by the monks, their servants. In this Christian society, no function exercised in the collective interest appeared more important than that of the spiritual organizations, precisely in so far—let us make no mistake about this—as they were spiritual. The charitable, cultural and economic rôle of the great cathedral chapters and of the monasteries may have been considerable: in the eyes of contemporaries it was merely accessory. The notion of a terrestrial world completely imbued with supernatural significance combined in this with the obsession of the beyond. The happiness of the king and the realm in the present; the salvation of the royal ancestors and of the king himself throughout Eternity: such was the double benefit which Louis the Fat declared that he expected from his foundation when he established a community of Canons Regular at the abbey of St. Victor in Paris. "We believe," said Otto I, "that the protection of our Empire is bound up with the rising fortunes of Christian worship." Thus we find a powerful and wealthy Church, capable of creating novel legal institutions, and a host of problems raised by the delicate task of relating this religious "city" to the temporal "city"; problems ardently debated and destined to influence profoundly the general revolution of the West. These features are an essential part of any accurate picture of the feudal world, and in face of them who can fail to recognize in the fear of hell one of the great social forces of the age?

On Violence as a Characteristic of Life

There were various affinities between Marc Bloch and the great Dutch historian Johan Huizinga. In our context perhaps the most important was their dedication to "total history." Each meant to write about more than the inanimate facts of their "society." Huizinga especially concerned himself with aristocratic and popular *mental* attitudes. He was particularly struck by the violent tenor of medieval life, by the mixture of the sacred and profane, and by the analogies of systems of public social control and the development of emotional restraints in human beings. Huizinga found the "mood of violence" always present but hard to measure. In trying to explain it, he drew attention to two aspects of it. The feeble political organizations of the age did not command adequate police resources. Hence, custom reflected fact; and expressions of violence unacceptable in polite society today were public spectacles invested with the sanction of divine Providence in the 1440s. Both the number of occasions and their publicity mark off acceptable violence then from what is analogous and shameful (at least nominally) today. Huizinga also expressed his controversial view—that human nature then differed from what it has become in industrial society—in seeking to explain medieval modes of thought and expression. In so doing, he made use of phrases such as "child-life." By this he seemed to mean that medieval people gave less restraint to their need for *expression*. They were less self-respective, more apt impulsively to satisfy desires. Huizinga's language often suggests the model of early childhood proposed by some psychoanalysts: ego unrestrained by "internalized" codes of conduct later learned by the child from parents or other socializing agencies (the police function of superego). Huizinga's daring formulations remain controversial. But his evocation of ordinary people's fears and joys is a triumph of historical imagination.

From *The Waning of the Middle Ages*

Johan Huizinga

THE VIOLENT TENOR OF LIFE

To the world when it was half a thousand years younger, the outlines of all things seemed more clearly marked than to us. The contrast between suffering and joy, between adversity and happiness, appeared more striking. All experience had yet to the minds of men the directness and absoluteness of the pleasure and pain of child-life. Every event, every action, was still embodied in expressive and solemn forms, which raised them to the dignity of a ritual. For it was not merely the great facts of birth, marriage and death which, by the sacredness of the sacrament, were raised to the rank of mysteries; incidents of less importance, like a journey, a task, a visit, were equally attended by a thousand formalities: benedictions, ceremonies, formulae.

Source: Johan Huizinga, *The Waning of the Middle Ages* (New York: St. Martin's Press; London: Edward Arnold, 1954), pp. 9–14, 23–29, 30–31.

Calamities and indigence were more afflicting than at present; it was more difficult to guard against them, and to find solace. Illness and health presented a more striking contrast; the cold and darkness of winter were more real evils. Honours and riches were relished with greater avidity and contrasted more vividly with surrounding misery. We, at the present day, can hardly understand the keenness with which a fur coat, a good fire on the hearth, a soft bed, a glass of wine, were formerly enjoyed.

Then, again, all things in life were of a proud or cruel publicity. Lepers sounded their rattles and went about in processions, beggars exhibited their deformity and their misery in churches. Every order and estate, every rank and profession, was distinguished by its costume. The great lords never moved about without a glorious display of arms and liveries, exciting fear and envy. Executions and other public acts of justice, hawking, marriages and funerals, were all announced by cries and processions, songs and music. The lover wore the colours of his lady; companions the emblem of their confraternity; parties and servants the badges or blazon of their lords. Between town and country, too, the contrast was very marked. A medieval town did not lose itself in extensive suburbs of factories and villas; girded by its walls, it stood forth as a compact whole, bristling with innumerable turrets. However tall and threatening the houses of noblemen or merchants might be, in the aspect of the town the lofty mass of the churches always remained dominant.

The contrast between silence and sound, darkness and light, like that between summer and winter, was more strongly marked than it is in our lives. The modern town hardly knows silence or darkness in their purity, nor the effect of a solitary light or a single distant cry.

All things presenting themselves to the mind in violent contrasts and impressive forms, lent a tone of excitement and of passion to everyday life and tended to produce that perpetual oscillation between despair and distracted joy, between cruelty and pious tenderness which characterize life in the Middle Ages.

One sound rose ceaselessly above the noises of busy life and lifted all things unto a sphere of order and serenity: the sound of bells. The bells were in daily life like good spirits, which by their familiar voices, now called upon the citizens to mourn and now to rejoice, now warned them of danger, now exhorted them to piety. They were known by their names: big Jacqueline, or the bell Roland. Every one knew the difference in meaning of the various ways of ringing. However continuous the ringing of the bells, people would seem not to have become blunted to the effect of their sound.

Throughout the famous judicial duel between two citizens of Valenciennes, in 1455, the big bell, "which is hideous to hear," says Chastellain, never stopped ringing. What intoxication the pealing of the bells of all the churches, and of all the monasteries of Paris, must have produced, sounding from morning till evening, and even during the night, when a peace was concluded or a pope elected.

The frequent processions, too, were a continual source of pious agitation. When the times were evil, as they often were, processions were seen winding along, day after day, for weeks on end. In 1412 daily processions were ordered in Paris, to implore victory for the king, who had taken up the oriflamme against the Armagnacs. They lasted from May to July, and were formed by ever-varying orders and corporations, going always by new roads, and always carrying different relics. The Burgher of Paris calls them "the most touching processions in the memory of men." People looked on or followed, "weeping piteously, with many tears, in great devotion." All went barefooted and fasting, councillors of the Parlement as well as the poorer citizens. Those who could afford it, carried a torch or a taper. A great many small children were always among them. Poor country-people of the environs of Paris came barefooted from afar to join the procession. And nearly every day the rain came down in torrents.

Then there were the entries of princes, arranged with all the resources of art and luxury belonging to the age. And, lastly, most frequent of all, one might almost say, uninterrupted, the executions. The cruel excitement and coarse compassion raised by an execution formed an important item in the spiritual food of the common people. They were spectacular plays with a moral. For horrible crimes the law invented atrocious punishments. At Brussels a young incendiary and murderer is placed in the centre of a circle of burning fagots and straw, and made fast to a stake by means of a chain running round an iron ring. He addresses touching words to the spectators, "and he so softened their hearts that every one burst into tears and his death was commended as the finest that was ever seen." During the Burgundian terror in Paris in 1411, one of the victims, Messire Mansart du Bois, being requested by the hangman, according to custom, to forgive him, is not only ready to do so with all his heart, but begs the executioner to embrace him. "There was a great multitude of people, who nearly all wept hot tears."

When the criminals were great lords, the common people had the satisfaction of seeing rigid justice done, and at the same time finding the inconstancy of fortune exemplified more strikingly than in any sermon or picture. The magistrate took care that nothing should be wanting to the effect of the spectacle: the condemned were conducted to the scaffold, dressed in the garb of their high estate. Jean de Montaigu, grand maître d'hôtel to the king, the victim of Jean sans Peur, is placed high on a cart, preceded by two trumpeters. He wears his robe of state, hood, cloak, and hose half red and half white, and his gold spurs, which are left on the feet of the beheaded and suspended corpse. By special order of Louis XI, the head of maître Oudart de Bussy, who had refused a seat in the Parlement, was dug up and exhibited in the market-place of Hesdin, covered with a scarlet hood lined with fur "selon la mode des conseillers de Parlement," with explanatory verses.

Rarer than processions and executions were the sermons of itinerant preachers, coming to shake people by their eloquence. The modern reader of newspapers can no longer conceive the violence of impression caused by the spoken word on an ignorant mind lacking mental food. The Franciscan friar Richard preached in Paris in 1429 during ten consecutive days. He began at five in the morning and spoke without a break till ten or eleven, for the most part in the cemetery of the Innocents. When, at the close of his tenth sermon, he announced that it was to be his last, because he had no permission to preach more, "great and small wept as touchingly and as bitterly as if they were watching their best friends being buried; and so did he." Thinking that he would preach once more at Saint Denis on the Sunday, the people flocked thither on Saturday evening, and passed the night in the open, to secure good seats.

Another Minorite friar, Antoine Fradin, whom the magistrate of Paris had forbidden to preach, because he inveighed against the bad government, is guarded night and day in the Cordeliers monastery, by women posted around the building, armed with ashes and stones. In all the towns where the famous Dominican preacher Vincent Ferrer is expected, the people, the magistrates, the lower clergy, and even prelates and bishops, set out to greet him with joyous songs. He journeys with a numerous and ever-increasing following of adherents, who every night make a circuit of the town in procession, with chants and flagellations. Officials are appointed to take charge of lodging and feeding these multitudes. A large number of priests of various religious orders accompany him everywhere, to assist him in celebrating mass and in confessing the faithful. Also several notaries, to draw up, on the spot, deeds embodying the reconciliations which this holy preacher everywhere brings about. His pulpit has to be protected by a fence against the pressure of the congregation which wants to kiss his hand or habit. Work is at a stand-still all the time he preaches. He rarely fails to move his auditors to tears. When he spoke of the Last Judgment, of Hell, or of the Passion, both he and his hearers wept so copiously that he had to suspend his sermon till the sobbing had ceased. Malefactors threw themselves at his feet, before every one, confessing their great sins. One day, while he was preaching, he saw two persons, who had been condemned to death—a man and a woman—being led to execution. He begged to have the execution delayed, had them both placed under the pulpit, and went on with his sermon, preaching about their sins. After the sermon, only some bones were found in the place they had occupied, and the people were convinced that the word of the saint had consumed and saved them at the same time.

After Olivier Maillard had been preaching Lenten sermons at Orléans, the roofs of the houses surrounding the place whence he had addressed the people had been so damaged by the spectators who had climbed on to them, that the roofer sent in a bill for repairs extending over sixty-four days.

The diatribes of the preachers against dissoluteness and luxury produced violent excitement which was translated into action. Long before Savonarola started bonfires of "vanities" at Florence, to the irreparable loss of art, the custom of these holocausts of articles of luxury and amusement was prevalent both in France and in Italy. At the summons of a famous preacher, men and women would hasten to bring cards, dice, finery, ornaments, and burn them with great pomp. Renunciation of the sin of vanity in this way had taken a fixed and solemn form of public manifestation, in accordance with the tendency of the age to invent a style for everything. . . .

In the blind passion with which people followed their lord or their party, the unshakable sentiment of right, characteristic of the Middle Ages, is trying to find expression. Man at that time is convinced that right is absolutely fixed and certain. Justice should prosecute the unjust everywhere and to the end. Reparation and retribution have to be extreme, and assume the character of revenge. In this exaggerated need of justice, primitive barbarism, pagan at bottom, blends with the Christian conception of society. The Church, on the one hand, had inculcated gentleness and clemency and tried, in that way, to soften judicial morals. On the other hand, in adding to the primitive need of retribution the horror of sin, it had, to a certain extent, stimulated the sentiment of justice. And sin, to violent and impulsive spirits, was only too frequently another name for what their enemies did. The barbarous idea of retaliation was reinforced by fanaticism. The chronic insecurity made the greatest possible severity on the part of the public authorities desirable; crime came to be regarded as a menace to order and society, as well as an insult to divine majesty. Thus it was natural that the late Middle Ages should become the special period of judicial cruelty. That the criminal deserved his punishment was not doubted for a moment. The popular sense of justice always sanctioned the most rigorous penalties. At intervals the magistrate undertook regular campaigns of severe justice, now against brigandage, now against sorcery or sodomy.

What strikes us in this judicial cruelty and in the joy the people felt at it, is rather brutality than perversity. Torture and executions are enjoyed by the spectators like an entertainment at a fair. The citizens of Mons bought a brigand, at far too high a price, for the pleasure of seeing him quartered, "at which the people rejoiced more than if a new holy body had risen from the dead." The people of Bruges, in 1488, during the captivity of Maximilian, king of the Romans, cannot get their fill of seeing the tortures inflicted, on a high platform in the middle of the market-place, on the magistrates suspected of treason. The unfortunates are refused the deathblow which they implore, that the people may feast again upon their torments.

Both in France and in England, the custom existed of refusing confession and the extreme unction to a criminal condemned to death. Sufferings and fear of death were to be aggravated by the certainty of eternal damnation. In vain had the council of Vienne in 1311 ordered to grant them at least the sacrament of penitence. Towards the end of the fourteenth century the same custom still existed. Charles V himself, moderate though he was, had declared that no change would be made in his lifetime. The chancellor Pierre d'Orgemont, whose "forte cervelle," says Philippe de Mézières, was more difficult to turn than a mill-stone, remained deaf to the humane remonstrances of the latter. It was only after Gerson had joined his voice to that of Mézières that a royal decree of the 12th of February, 1397, ordered that confession should be accorded to the condemned. A stone cross erected by the care of Pierre de Craon, who had interested himself in the decree, marked the place where the Minorite friars might assist penitents going to execution. And even then the barbarous custom did not disappear. Etienne Ponchier, bishop of Paris, had to renew the decree of 1311 in 1500.

In 1427 a noble brigand is hanged in Paris. At the moment when he is going to be executed, the great treasurer of the regent appears on the scene and vents his hatred against him; he prevents his confession, in spite of his prayers; he climbs the ladder behind him, shouting insults, beats him with a stick, and gives the hangman a thrashing for exhorting the victim to think of his salvation. The hangman grows nervous and bungles his work; the cord snaps, the wretched criminal falls on the ground, breaks a leg and some ribs, and in this condition has to climb the ladder again.

The Middle Ages knew nothing of all those ideas which have rendered our sentiment of justice timid and hesitating: doubts as to the criminal's responsibility; the conviction that society is, to a certain extent, the accomplice of the individual; the desire to reform instead of inflicting pain; and, we may even add, the fear of judicial errors. Or rather these ideas were implied, unconsciously, in the very strong and direct feeling of pity and of forgiveness which alternated with extreme severity. Instead of lenient penalties, inflicted with hesitation, the Middle Ages knew but the two extremes: the fulness of cruel punishment, and mercy. When the condemned criminal is pardoned, the question whether he deserves it for any special reasons is hardly asked; for mercy has to be gratuitous, like the mercy of God. In practice, it was not always pure pity which determined the question of pardon. The princes of the eleventh century were very liberal of "lettres derémission" for misdeeds of all sorts and contemporaries thought it quite natural that they were obtained by the intercession of quite natural relatives. The majority of these documents, however, concern poor common people.

The contrast of cruelty and of pity recurs at every turn in the manners and customs of the Middle Ages. On the one hand, the sick, the poor, the insane, are objects of that deeply moved pity, born of a feeling of fraternity akin to that which is so strikingly expressed in modern Russian literature; on the other hand, they are treated with incredible hardness or cruelly mocked. The chronicler Pierre de Fenin, having described the death of a gang of brigands, winds up naïvely: "and people laughed a good deal, because they were all poor men." In 1425, an "esbatement" takes place in Paris, of four blind beggars, armed with sticks, with which they hit each other in trying to kill a pig, which is a prize of the combat. On the evening before they are led through the town, "all armed, with a great banner in front, on which was pictured a pig, and preceded by a man beating a drum. . . ."

In the harshness of those times there is something ingenuous which almost forbids us to condemn it. When the massacre of the Armagnacs was in full swing in 1418, the Parisians founded a brotherhood of Saint Andrew in the church of Saint Eustache: every one, priest or layman, wore a wreath of red roses, so that the church was perfumed by them, "as if it had been washed with rose-water." The people of Arras celebrate the annulment of the sentences for witchcraft, which during the whole year 1461 had infested the town like an epidemic, by joyous festivals and a competition in acting "folies moralisées," of which the prizes were a gold fleur-de-lis, a brace of capons, etc.; nobody, it seems, thought any more of the tortured and executed victims.

So violent and motley was life, that it bore the mixed smell of blood and of roses. The men of that time always oscillate between the fear of hell and the most naïve joy, between cruelty and tenderness, between harsh asceticism and insane attachment to the delights of this world, between hatred and goodness, always running to extremes. . . .

Medieval doctrine found the root of all evil either in the sin of pride or in cupidity. Both opinions were based on Scripture texts: *A superbia initium sumpsit omnis perditio.—Radix omnium malorum est cupiditas*. It seems, nevertheless, that from the twelfth century downward people begin to find the principle of evil rather in cupidity than in pride. The voices which condemn blind cupidity, "la cieca cupidigia" of Dante, become louder and louder. Pride might perhaps be called the sin of the feudal and hierarchic age. Very little property is, in the modern sense, liquid, while power is not yet associated, predominantly, with money; it is still rather inherent in the person and depends on a sort of religious awe which he inspires; it makes itself felt by pomp and magnificence, or a numerous train of faithful followers. Feudal or hierarchic thought expresses the idea of grandeur by visible signs, lending to it a symbolic shape, of homage paid kneeling, of ceremonial reverence. Pride, therefore, is a symbolic sin, and from the fact that, in the last resort, it derives from the pride of Lucifer, the author of all evil, it assumes a metaphysical character.

Cupidity, on the other hand, has neither this symbolic character nor these relations with theology. It is a purely worldly sin, the impulse of nature and of the flesh. In the later Middle Ages the conditions of power had been changed by the increased circulation of money, and an illimitable field opened to whosoever was desirous of satisfying his ambitions by heaping up wealth. To this epoch cupidity becomes the predominant sin. Riches have not acquired the spectral impalpability which capitalism, founded on credit, will give them later; what haunts the imagination is still the tangible yellow gold. The enjoyment of riches is direct and primitive; it is not yet weakened by the mechanism of an automatic and invisible accumulation by investment; the satisfaction of being rich is found either in luxury and dissipation, or in gross avarice.

Towards the end of the Middle Ages feudal and hierarchic pride had lost nothing, as yet, of its vigour; the relish for pomp and display is as strong as ever. This primitive pride has now united itself with the growing sin of cupidity, and it is this mixture of the two which gives the expiring Middle Ages a tone of extravagant passion that never appears again.

A furious chorus of invectives against cupidity and avarice rises up everywhere from the literature of that period. Preachers, moralists, satirical writers, chroniclers and poets speak with one voice. Hatred of rich people, especially of the new rich, who were then very numerous, is general. Official records confirm the most incredible cases of unbridled avidity told by the chronicles. In 1436 a quarrel between two beggars, in which a few drops of blood had been shed, had soiled the church of the Innocents at Paris. The bishop, Jacques du Châtelier, "a very ostentatious, grasping man, of a more worldly disposition than his station required," refused to consecrate the church anew, unless he received a certain sum of money from the two poor men, which they did not possess, so that the service was interrupted for twenty-two days. Even worse happened under his successor, Denys de Moulins. During four months of the year 1441, he prohibited both burials and processions in the cemetery of the Innocents, the most favoured of all, because the church could not pay the tax he demanded. This Denys de Moulins was reputed "a man who showed very little pity to people, if he did not receive money or some equivalent; and it was told for truth that he had more than fifty lawsuits before the Parlement, for nothing could be got out of him without going to law."

A general feeling of impending calamity hangs over all. Perpetual danger prevails everywhere. To realize the continuous insecurity in which the lives of great and small alike were passed, it suffices to read the details which Monsieur Pierre Champion has collected regarding the persons mentioned by Villon in his *Testament*, or the notes of Monsieur A. Tuetey to the diary of a Burgher of Paris. They present to us an interminable string of lawsuits, crimes, assaults and persecutions. A chronicle like that of Jacques du Clercq, or a diary such as that of the citizen of Metz, Philippe de Vigneulles, perhaps lay too much stress on the darker side of contemporary life, but every investigation of the careers of individual persons seems to confirm them, by revealing to us strangely troubled lives. . . .

Is it surprising that the people could see their fate and that of the world only as an endless succession of evils? Bad government, exactions, the cupidity and violence of the great, wars and brigandage, scarcity, misery and pestilence—to this is contemporary history nearly reduced in the eyes of the people. The feeling of general insecurity which was caused by the chronic form wars were apt to take, by the constant menace of the dangerous classes, by the mistrust of justice, was further aggravated by the obsession of the coming end of the world, and by the fear of hell, of sorcerers and of devils. The background of all life in the world seems black. Everywhere the flames of hatred arise and injustice reigns. Satan covers a gloomy earth with his sombre wings. In vain the militant Church battles, preachers deliver their sermons; the world remains unconverted. According to a popular belief, current towards the end of the fourteenth century, no one, since the beginning of the great Western schism, had entered Paradise.

Ideal Types in Traditional Society

The tradition of function related to status in three social orders in the early Middle Ages was well established: to pray (the priest), to fight (the knight), to work (the peasant). The permanence of form—"new wine in old bottles"—may explain its survival in later literature. Whatever the explanation, Chaucer received the tradition and passed it on in the "Prologue" to his *Canterbury Tales*.

But the poet (1340–1400) was himself of bourgeois origins. A Londoner, he made his way into courtly society in the entourage of the Duke of Lancaster. The court meant frivolity, but it also meant diplomatic service, political involvement as an M.P., and officeholding in the customs service. Hence idealized writing in Chaucer and Aelfric are quite different things, even setting genius to one side. Chaucer's world is inhabited by courageous knights, pious monks, nuns, and contented plowmen. But it is also well stocked with avaricious merchants, butchers, and bakers; lawless lawyers, corrupt clergymen, cheating estate managers, and dozens of other crafty entrepreneurs familiar enough in London society. These Chaucer delineates with wit, sarcasm, mocking disdain, and acute psychological realism. He is a master of the technique of revealing the tale-teller in his tale. His galaxy of people helps bring alive the commitments men made to God and his heavenly city in the midst of their pursuit of earthly joys. His work also brings vividly to mind the poignant longing for a simpler order of things.

From *The Canterbury Tales*

Geoffrey Chaucer

PROLOGUE

When the sweet showers of April have pierced the dryness of March to its root and soaked every vein in moisture whose quickening force engenders the flower; when Zephyr with his sweet breath has given life to tender shoots in each wood and field; when the young sun has run his half-course in the sign of the Ram; when, nature prompting their instincts, small birds who sleep through the night with one eye open make their music—then people long to go on pilgrimages, and pious wanderers to visit strange lands and far-off shrines in different countries. In England especially they come from every shire's end to Canterbury to seek out the holy blessed martyr St. Thomas à Becket, who helped them when they were sick.

It happened one day at this time of year, while I was lodging at the Tabard in Southwark, ready and eager to go on my pilgrimage to Canterbury, a company of twenty-nine people arrived in the hostelry at nightfall. They were of various sorts, accidentally brought together in companionship, all pilgrims wishing to ride to Canterbury. The rooms and stables were commodious and we were very

SOURCE: Geoffrey Chaucer, *The Canterbury Tales*, trans. David Wright (New York: Random House; London: Barrie & Rockliff, 1964), pp. 3–16.

well looked after. In short, by the time the sun had gone down I had talked with every one of them and soon become one of their company. We agreed to rise early to set out on the journey I am going to tell you about. But nevertheless before I take the story further it seems right to me to describe, while I have the time and opportunity, the sort and condition of each of them as they appeared to me: who they were, of what rank, and how dressed. I shall begin with the Knight.

The KNIGHT was a very distinguished man. From the beginning of his career he had loved chivalry, loyalty, honourable dealing, generosity, and good breeding. He had fought bravely in the king's service, besides which he had travelled further than most men in heathen as well as in Christian lands. Wherever he went he was honoured for his valour. He was at Alexandria when it fell. When he served in Prussia he was generally given the seat of honour above the knights of all other nations; no Christian soldier of his rank had fought oftener in the raids on Russia and Lithuania. And he had been in Granada at the siege of Algeciras, fought in Benmarin and at the conquests of Ayas and Attalia, besides taking part in many armed expeditions in the eastern Mediterranean. He had been in fifteen pitched battles and fought three times for the faith in the lists at Tramassene, and each time killed his foe. This same distinguished Knight had also fought at one time for the king of Palathia against another heathen enemy in Turkey. He was always outstandingly successful; yet though distinguished he was prudent, and his bearing as modest as a maid's. In his whole life he never spoke discourteously to any kind of man. He was a true and perfect noble knight. But, speaking of his equipment, his horses were good, yet he was not gaily dressed. He wore a tunic of thick cotton cloth, rust-marked from his coat of mail; for he had just come back from his travels and was making his pilgrimage to render thanks.

With him came his son, a young SQUIRE; a spirited apprentice-knight, a lover with hair as curly as if it had just been pressed in the tongs—I suppose his age was about twenty. He was of average height, wonderfully active and strong. In Flanders, Artois, and Picardy he had taken part in cavalry forays and in that short space of time had borne himself well, for he hoped to win favour in his lady's eyes. He was decked out like a meadow full of fresh flowers, white and red; he whistled or sang the whole day long, as lively as the month of May. His gown was short, with long wide sleeves. He was a good rider and sat his horse well; he was able to compose songs and set them to music, joust and also dance; and he could draw and write. Being passionately in love, at night he slept no more than a nightingale. Courteous, modest, and willing to serve, he carved for his father at table.

The Knight had a YEOMAN, and no other servants, for that was how he preferred to travel on this occasion. The Yeoman was clad in a green coat and hood and carried a sheaf of sharp bright peacock-feathered arrows slung handily from his belt (he knew how to look after his gear in soldierly fashion; his arrows did not fall short because of poor feathering) while in his hand he carried a mighty bow. He had a brown face, short-cropped hair, and was adept in everything to do with woodcraft. On his arm he wore a gay arm-guard, and a sword and buckler on one side; upon the other a bright well-mounted dagger as sharp as a spearpoint; and on his breast a shining silver image of St. Christopher. He carried a horn slung from a green belt; I'd say he was a forester.

There was also a NUN, a Prioress, who smiled in an unaffected and quiet way; her greatest oath was only, "By St. Loy!" Her name was Madame Eglantine. She sang the divine service prettily, becomingly intoned through the nose. She spoke French elegantly and well but with a Stratford-at-Bow accent, for she did not know the French of Paris. At table she showed her good breeding at every point: she never let a crumb fall from her mouth or wetted her fingers by dipping them too deeply into the sauce; and when she lifted the food to her lips she took care not to spill a single drop upon her breast. Etiquette was her passion. So scrupulously did she wipe her upper lip that no spot of grease was to be seen in her cup after she had drunk from it; and when she ate she reached daintily for her food. Indeed she was most gay, pleasant and friendly. She took pains to imitate courtly behaviour and cultivate a dignified bearing so as to be thought a person deserving of respect. Speaking of her sensibility, she was so tender-hearted and compassionate that she would weep whenever she saw a mouse caught in a trap, especially if it were bleeding or dead. She kept a number of little dogs whom she fed on roast meat, milk, and the best bread. But if one of them died or someone took a stick to it she would cry bitterly, for with her all was sensitivity and tender-heartedness. Her wimple was becomingly pleated; her nose well-shaped, her eyes grey as glass, her mouth small, but soft and red. Certainly she had a fine fore-head, I daresay almost a span in breadth; for indeed there was nothing diminutive about her. I noticed that she wore a most elegant cloak. On her arm she carried a rosary of small coral beads interspersed with large green ones, from which hung a shining golden brooch that had inscribed upon it a crowned A, and underneath. "*Amor vincit omnia.*" [1] With her she had another NUN, her chaplain, and three PRIESTS.

There was a remarkably fine-looking MONK, who acted as estate-steward to his monastery and loved hunting: a manly man, well fitted to be an abbot. He kept plenty of fine horses in his stable, and when he went out riding people could hear the bells on his bridle jingling in the whistling wind as clear and loud as the chapel bell of the small convent of which he was the head. Because the Rule of St. Maur or of St. Benedict was old-fashioned and somewhat strict, this Monk neglected the old precepts and followed the modern custom. He did not give two pins for the text which says hunters cannot be holy men, or that a monk who is heedless of his Rule—that is to say a monk out of his cloister—is like a fish out of water. In his view this saying was not worth a bean; and I told him his opinion was sound. Why should he study and addle his wits with everlasting poring over a book in cloisters, or work with his hands, or toil as St. Augustine commanded? How is the world to be served? Let St. Augustine keep his hard labour for himself! Therefore the Monk, whose whole pleasure lay in riding and the hunting of the hare (over which he spared no expense) remained a hard rider and kept greyhounds swift as birds.

I saw that his sleeves were edged with costly grey fur, the finest in the land. He had an elaborate golden brooch with a love-knot at the larger end to fasten his hood beneath his chin. His bald head shone like glass; and so did his face, as if it had been anointed. He was a plump and personable dignitary, with prominent, restless eyes which sparkled like fire beneath a pot. His boots were supple and his horse in perfect condition. To be sure he was a fine-looking prelate, no pale and wasting ghost! His favourite dish was a fat roast swan. The horse he rode was as brown as a berry.

---

1 "Love conquers all."

A begging FRIAR was there, a gay, pleasant Limiter[2] with an imposing presence; nobody in all the four Orders was so adept with flattery and tittle-tattle. He had had to pay for the marriages of a good many young women;[3] still, he was a noble pillar of his Order. He was well-liked and on easy terms with rich landowners everywhere in the district in which he begged, and also with the wealthy townswomen, for he was a licentiate of his Order and qualified, so he said, to hear confession of graver sins than parish priests were allowed to absolve. He heard confession sweetly, and his absolution was pleasant; when he was sure of a good thank-offering he was an easy man in giving penance. For if a man gives generously to a poor Order it is a sign he has been well shriven. Once a man opened his purse-strings the Friar could vouch for it that he was penitent, for many people —though truly remorseful—are so hardened that they cannot weep. Therefore, instead of prayers and tears, people might as well give money to the poor friars.

The pockets of his hood were always stuffed with knives and pins to give to young women. And he certainly had a pleasant voice; he could sing and play the fiddle well, and was a champion ballad-singer. His neck was white as a lily, but for all that he was as strong as a prize-fighter. In every town he knew all the taverns and innkeepers and barmaids better than the lepers and beggars, for it hardly befitted a man of his ability and distinction to mix with diseased lepers. It is not seemly, and gets a man nowhere, to have dealings with rabble of that type, but only with merchants and the rich. Above all, wherever profit might be had he offered his services with polite submission. Nowhere would you find so capable a man; he was the best beggar in his friary. He paid a fixed fee for the district in which he begged; none of his brethren poached on his preserves. Even if a widow had no shoes, he had only to begin saying the first words of the Gospel of St. John in his pleasant voice and he'd get at least a sixpence out of her before he left. But he made his biggest profits on the side.

He would romp about like a puppy on settling days, when he was of great help as an arbitrator; for on these occasions he did not appear like some poor cloistered student in threadbare vestment, but like a Master of Divinity, or the Pope. His outer cloak was of double-worsted, as rounded as a bell just out of its mould. He lisped a little—this was a mannerism to make his English sound attractive. And when he played the harp and had finished his song his eyes twinkled in his head like stars on a frosty night. This excellent Friar was called Hubert.

Next there was a MERCHANT with a forked beard who rode seated on a high saddle, wearing a many-coloured dress, boots fastened with neat handsome clasps, and upon his head a Flanders beaver hat. He gave out his opinions with great pomposity and never stopped talking about the increase of his profits. In his view the high seas between Harwich and Holland should be cleared of pirates at all costs. He was an expert at the exchange of currency. This worthy citizen used his head to the best advantage, conducting his money-lending and other financial transactions in a dignified manner; none guessed he was in debt. He was really a most estimable man; but to tell the truth his name escapes me.

Also there was a SCHOLAR from Oxford who had long been studying Logic. His horse was as lean as a rake, and I give you my word that he himself was no

---

[2] A Limiter was a begging friar who was assigned a certain district or *limit* in which he was allowed to solicit alms. (Skeat.)
[3] I.e. because he had seduced them himself.

fatter, but looked both melancholy and hollow-cheeked. As he had not yet found himself a benefice and was too unworldly to take secular employment, his overcoat was pretty threadbare. For he preferred his library of Aristotle's philosophical works bound in black calf and red sheepskin at the head of his bed, to fine clothes, the fiddle and psalter; yet for all his philosophy and science he had but little gold in his coffer. He spent everything he could get from his friends on books and learning, and in return prayed assiduously for the souls of those who gave him the money to pursue his studies. Learning was his whole solicitude and care. He never spoke a word more than necessary, and then it was with due formality and respect, brief and to the point, and lofty in theme. His conversation was eloquent with goodness and virtue; he was as glad to learn as to teach.

A sage and cautious SERGEANT-AT-LAW, a well-known figure at the portico of St. Paul's, where the lawyers meet, was also present. He was a man of excellent parts, discreet, and of great distinction—or so he seemed, he spoke with such wisdom. He had often acted as Judge of the Assize by the King's Letters Patent and had authority to hear all types of cases. His skill and great reputation earned him many fees, and robes given in lieu of money. Nowhere was there a better conveyancer; he could untie any entail and get unrestricted possession of the property, while his conveyances were never invalidated. There was no busier man anywhere; and yet he seemed busier than he was. He could quote precisely all the cases and judgments since the Conquest, besides which he could compose and draw up a deed so that none could fault it; and he knew all the statutes word for word. He was dressed simply in a parti-coloured coat girt by a silk belt with narrow stripes. I'll say no more of his appearance.

His companion was a FRANKLIN with a ruddy complexion and daisy-white beard, who was more than partial to a drink of sops-in-wine early in the morning. It was always his custom to live well, for he was a true son of Epicurus who held the opinion that the only real happiness lies in sensual pleasures. The Franklin kept up a magnificent establishment where he was as famous for hospitality as St. Julian, its patron saint. The quality of his bread and wine never varied; his cellar was unsurpassed; his house never lacked food (whether fish or meat) in such plenty that in his house it seemed to snow food and drink and every kind of delicacy one can think of. Different dishes were served according to the season of the year. An abundance of fat partridges filled his coops, while his fishponds were plentifully stocked with bream and pike. His cook found himself in trouble if the sauce was not sharp and piquant or if he were caught unprepared. The hall table was kept laid and ready the livelong day. This Franklin presided at the sessions of justices of the peace and was often Member of Parliament for the shire. A dagger and a silk purse hung from his belt, which was white as milk. He had been Sheriff and Auditor of his county; nowhere would you find a better specimen of the landed gentry.

Among the rest were a HABERDASHER, a CARPENTER, a WEAVER, a DYER, and a TAPESTRY-MAKER, all dressed in uniform livery belonging to a rich and honourable Guild. Their apparel was new and freshly trimmed; their knives were not tipped with brass but finely mounted with wrought silver to match their belts and purses. Each seemed a proper burgess worthy of a place on the dais of a guildhall; and every one of them had the ability and judgement, besides sufficient property and income, to become an alderman. In this they would have the hearty assent of their wives—else the ladies would certainly be much to blame. For it's very pleasant to be called "Madam" and take precedence at church festivals, and have one's mantle carried in state.

They had taken a COOK with them for the occasion, to boil chickens with marrowbones, tart flavouring-powder and spice. Well did he know the taste of London ale! He knew how to roast, fry, seethe, broil, make soup and bake pies. But it was the greatest pity, so I thought, that he'd got an ulcer on his shin. For he made chicken-pudding with the best of them.

A SEA-CAPTAIN, whose home was in the West Country, was also there; as far as I can tell he came from Dartmouth. He rode, after a fashion, upon a farm-horse; and wore a gown of coarse serge reaching to the knee. Under his arm he carried a dagger slung from a lanyard round his neck. The hot summer had tanned him brown; and he was certainly a bit of a lad, for he had lifted any amount of wine from Bordeaux while the merchants were napping. He had no time for the finer feelings; if he fought and got the upper hand, he threw his prisoners overboard and sent them home by water to wherever they came from. From Hull to Cartagena there was none to match his seamanship in calculating tides, currents, and the hazards around him; or his knowledge of harbours, navigation, and the changes of the moon. He was a shrewd and hardy adventurer; his beard had been shaken in many a storm. He knew every harbour there was from Gottland to Cape Finisterre, and every inlet of Brittany and Spain. The name of his ship was the *Magdalen*.

A DOCTOR OF MEDICINE accompanied us. There was none to touch him in matters of medicine and surgery, for he was well grounded in astronomy. His astrological knowledge enabled him to select the most favourable hour to administer remedies to his patients; and he was skilled in calculating the propitious moment to make talismans for his clients. He could diagnose every kind of disease and say in what organ and from which of the four humours—the hot, the cold, the wet, or the dry—the distemper arose. He was a model practitioner. Once he had detected the root of the trouble he gave the sick man his medicine there and then, for he had his apothecaries ready at hand to send him drugs and sirops. In this way each made a profit for the other—their partnership was not new. The Doctor was well versed in the ancient medical authors: Aesculapius, Dioscorides, Rufus, Hali, Galen, Serapion, Rhazes, Avicenna, Averroes, Constantine, Bernard, Gaddesden, and Gilbert. In his own diet he was moderate: it contained nothing superfluous but only what was nourishing and digestible. He seldom read the Bible. The clothes he was dressed in were blood-red and grey-blue, lined with silk and taffeta; yet he was no free spender, but laid by whatever he earned from the plague. In medicine gold is the great restorative; and therefore he was particularly fond of it.

There was among us a worthy WIFE from near Bath, but she was a bit deaf, which was a pity. At cloth-making she beat even the weavers of Ypres and Ghent. There was not a woman in her parish who dared go in front of her when she went to the offertory; if anybody did, you may be sure it put her into such a rage she was out of all patience. Her kerchiefs were of the finest texture; I daresay those she wore upon her head on Sundays weighed ten pounds. Her stockings were of the finest scarlet, tightly drawn up above glossy new shoes; her face was bold, handsome, and florid. She had been a respectable woman all her life, having married five husbands in church (apart from other loves in youth of which there is no need to speak at present). She had visited Jerusalem thrice and crossed many foreign rivers, had been to Rome, Boulogne, the shrine of St. James of

Compostella in Galicia, and Cologne; so she knew a lot about travelling around—the truth is, she was gap-toothed.[4] She rode comfortably upon an ambling horse, her head well covered with a wimple and a hat the size of a shield or buckler. An outer skirt covered her great hips, while on her feet she wore a pair of sharp spurs. In company she laughed and rattled away. No doubt she knew all the cures for love, for at that game she was past mistress.

With us there was a good religious man, a poor PARSON, but rich in holy thoughts and acts. He was also a learned man, a scholar, who truly preached Christ's Gospel and taught his parishioners devoutly. Benign, hardworking, and patient in adversity—as had often been put to the test—he was loath to excommunicate those who failed to pay their tithes. To tell the truth he would rather give to the poor of his parish what had been offered him by the rich, or from his own pocket; for he managed to live on very little. Wide as was his parish, with houses few and far between, neither rain nor thunder nor sickness nor misfortune stopped him from going on foot, staff in hand, to visit his most distant parishioners, high or low. To his flock he set this noble example: first he practised, then he preached. This was a precept he had taken from the Gospel; and to it he added this proverb: "If gold can rust, what will iron do?" For if the priest in whom we trust be rotten, no wonder an ordinary man corrupts. Let priests take note: shame it is to see the shepherd covered in dung while his sheep are clean! It's for the priest to set his flock the example of a spotless life! He did not farm his benefice and leave his sheep to flounder in the mud while he ran off to St. Paul's in London to seek some easy living such as a chantry where he would be paid to sing masses for the souls of the dead, or a chaplaincy in one of the guilds; but dwelt at home and kept watch over his flock so that it was not harmed by the wolf. He was a shepherd, not a priest for hire.

And although he was saintly and virtuous he did not despise sinners. His manner of speaking was neither distant nor severe; on the contrary he was considerate and benign in his guidance. His endeavour was to lead folk to heaven by the example of a good life. Yet if anyone—whatever his rank—proved obstinate, he never hesitated to deliver a stinging rebuke. I'd say that there was nowhere a better priest. He never looked for ceremony and deference, nor was his conscience of the over-scrupulous and specious sort. He taught the Gospel of Christ and His twelve apostles: but first he followed it himself.

With him came his brother, a PLOUGHMAN. Many a load of dung had been carted by this good and faithful labourer, who lived in peace and charity with all. First he loved God with his whole heart, in good times and in bad; next he loved his neighbour as himself. He threshed and dug and ditched, and for Christ's love would do as much for any poor fellow without payment, if he could manage it. He paid his tithes on both his crops and the increase of his livestock fairly and in full. He rode humbly upon a mare and wore a loose labourer's smock.

Finally there was a REEVE, a MILLER, a SUMMONER, a PARDONER, a MANCIPLE, and last of all myself.

The MILLER was a great brawny fellow, big-boned, with powerful muscles which he turned to good account at wrestling matches up and down the land; for he carried off the prize every time. He was thickset, broad, and muscular; there was no door that he couldn't heave off its hinges, or break down by running at it

---

4 This was supposed to be a certain sign of luck and travel; and the physiognomists regarded it as a sign of a lascivious disposition.

with his head. His beard was as red as a fox or a sow, and wide as a spade at that. Upon the tip of his nose, on the right side, was a wart on which stood a tuft of hairs red as bristles in a pig's ear. His nostrils were squat and black. At his side he carried a sword and buckler. He had a great mouth, wide as a furnace-door; and his talk was mostly bawdy and vicious. He was a ribald joker and a chatterer. Well did he know all the tricks of his trade, how to filch corn and charge three times his proper due; yet he was honest enough, as millers go. He wore a white coat and a blue hood and led us out of town lustily blowing and tooting upon the bagpipes.

There was a worthy MANCIPLE of one of the Inns of Court, who might have served as a model to caterers for shrewdness in the purchase of provisions; for whether he paid cash or bought on credit, he watched prices all the time, so that he always got in first and did good business. Now is it not a remarkable example of God's grace that the wit of an uneducated man like this should outmatch the wisdom of a pack of learned men? His superiors numbered more than thirty and were all erudite and expert in the law; there were a dozen men in his college capable of so managing the rents and land of any peer in England as to enable him (unless he were mad) to live honourably and free of debt upon his income, or else as plainly as he pleased; capable too of advising a whole county in any lawsuit that might conceivably arise; yet this Manciple could hoodwink the lot of them!

The REEVE was a slender, choleric man. He shaved his beard as close he could, and cropped his hair short round the ears; the top of his head was shorn in front like a priest's. He had long thin legs like sticks; the calves were invisible. He kept his bins and granaries ably; no auditor could get the better of him. By noting the drought and rainfall he could make a good estimate of the yield of his seed and grain. All his master's livestock, sheep, cattle, dairy, pigs, horses, and poultry was entirely managed by this Reeve, who had undertaken to render accounts ever since his master was twenty years old. None could prove he was in arrears. He knew every dodge and swindle of every one of the bailiffs, herdsmen, and farm-labourers; they feared him like death itself. He lived in a beautiful house standing in a meadow shaded by green trees. At bargaining he was better than his master and had feathered his nest very comfortably. For he was adroit at putting his master under an obligation by a gift or loan of his own property, thus earning his thanks besides the present of a gown or a hood. As a young man he had learned a good trade; he was a skilled workman, a carpenter. This Reeve rode a sturdy dapple-grey cob called Scot. He wore a long blue overcoat with a rusty sword by his side. The Reeve of whom I speak came from near the town of Bawdeswell in Norfolk. He had his coat hitched up by a girdle like a friar, and always rode the hindmost of our party.

Among us at the inn was a SUMMONER[5] with slit eyes and a flaming red visage like a cherub's, all covered with pimples. He was as randy and lecherous as a sparrow. Children were afraid of his face with its scabbed black eyebrows and scraggy beard. No mercury, white lead, sulphur, borax, ceruse, cream of tartar, or other ointments that cleanse and burn could rid him of his white pustules or the

---

[5] A summoner was an officer or constable whose task was to summon delinquents to appear before the ecclesiastical courts, enforce payment of tithes and church dues, etc. He also had power to punish adultery, fornication, and other sins not punishable by common law. "The Friar's Tale" is a satire on the abuses practised by Summoners.

pimply knobs on his cheeks. He had a great love of garlic, onions, and leeks, and of drinking strong wine red as blood, which made him roar and gabble like a madman. When really drunk on wine he'd speak nothing but Latin. He knew two or three tags that he'd learned from some decree or other—and no wonder, for he heard Latin all day long; but as you know a jaybird can call out "Wat" as well as the Pope himself. Yet if you tried him further you found he was out of his depth; all he could do was parrot "*questio quid juris*" [6] over and over. He was a tolerant easy-going dog, as good a fellow as you might hope to find. For a quart of wine he'd allow any rascal of a priest to keep his concubine for a twelvemonth and excuse him altogether; on the other hand he was well able to fleece a greenhorn on the sly. And if he found a fellow with a girl he'd tell him not to worry about the Archdeacon's excommunication in such a case, unless he thought his purse was where his soul was kept, for it was in his purse he'd be punished. "Your purse is the Archdeacon's Hell," he would say. But I am sure he lied in his teeth; the guilty must fear excommunication because it destroys the soul just as absolution saves it—and they should also beware of the writ that sends them to prison. All the young people of his diocese were wholly under his thumb, for he was their confidant and sole adviser. Upon his head the Summoner had set a garland as big as one of those they hang outside alehouses. He had a great round cake which he carried like a shield.

With him rode a worthy PARDONER[7] of Rouncival at Charing Cross, his friend and bosom companion, who had come straight from the Vatican at Rome. He loudly carolled "Come hither, love, to me," while the Summoner sang the bass louder than the loudest trumpet. This Pardoner's hair was waxy yellow and hung down as sleek as a hank of flax; such locks of hair as he possessed fell in meagre clusters spread over his shoulders, where it lay in thinly scattered strands. For comfort he wore no hood; it was packed in his bag. With his hair loose and uncovered except for a cap, he thought he was riding in the latest style. He had great staring eyes like a hare's. Upon his cap he'd sewn a small replica of St. Veronica's handkerchief. His wallet lay on his lap in front of him, chockful of pardons hot from Rome. He'd a thin goatlike voice and no vestige or prospect of a beard; and his skin was smooth as if just shaven. I took him for a gelding or a mare. But as for his profession, from Berwick down to Ware there was not another pardoner to touch him. For in his wallet he kept a pillow-slip which, he said, was Our Lady's veil. He claimed to have a bit of the sail belonging to St. Peter when he tried to walk on the waves and Jesus Christ caught hold of him. He had a brass cross set with pebbles and a glass reliquary full of pig's bones. Yet when he came across some poor country parson he could make more money with these relics in a day than the parson got in two months; and thus by means of barefaced flattery and hocus-pocus he made the parson and the people his dupes. To do him justice, in church at any rate he was a fine ecclesiastic. Well could he read a lesson or a parable; but best of all he sang the offertory hymn, because after it was sung he knew he must preach, as he well knew how, to wheedle money from the congregation with his smooth tongue. Therefore he sang all the louder and merrier.

---

6 "What is the law on this point?"

7 Pardoners were sellers of papal indulgences, i.e., commutation of penances imposed for sins. The profits were supposed to go to religious organizations, or to be used for some pious purpose.

Aristocratic Self-concepts
and Chivalric Order

When Chaucer grasped the conflicting ideals of old nobility, churchmen, and London's teeming City, many aristocrats excluded common people from their view. Huizinga dealt at length with aristocratic self-conceptions, codes of war, and their highly stylized, often ethereal but repressive, views of common people and noblewomen. In the following selections from *The Waning of the Middle Ages*, we can once again see how impressively he joins the solid facts derived from events and those "facts" of the experience of events which constitute our record of mental activity in past cultures. No historian of popular culture can afford to ignore aristocratic culture, simply because the distribution of power forces his attention toward those who hold it. Despite all of the social, political, and economic changes of the Renaissance, neither common people nor the great bourgeoisie replaced aristocracy in power. And the aesthetic trends of the age continued to reflect aristocratic tastes in painting, architecture, music, and literature. Burgesses bought Arthurian romances with the same eagerness they showed in turning merchant capital into landed estates. Thus Huizinga's study is important in our context, even when his subject is decadent chivalric culture.

From *The Waning of the Middle Ages*

Johan Huizinga

THE POLITICAL AND MILITARY VALUE OF CHIVALROUS IDEAS

. . . In warfare, the chivalrous point of honour continues to make itself felt, but when an important question arises for decision, strategic prudence carries the day in the majority of cases. Generals still propose to the enemy to come to an understanding as to the choice of the battlefield, but the invitation is generally declined by the party occupying the better position. In vain did the English in 1333 invite the Scotch to come down from their strong position in order to fight them in the plains; in vain did Guillaume de Hainaut propose an armistice of three days to the king of France, during which a bridge could be built permitting the armies to join battle. Reason, however, is not always victorious. Before the battle of Najera (or of Navarrete), in which Bertrand du Guesclin was taken prisoner, Don Henri de Trastamara desires, at any cost, to measure himself with the enemy in the open field. He voluntarily gives up the advantages offered by the configuration of the ground and loses the battle.

If chivalry had to yield to strategy and tactics, none the less it remained of importance in the exterior apparatus of warfare. An army of the fifteenth century, with its splendid show of rich ornament and solemn pomp, still offered the spectacle of a tournament of glory and honour. The multitude of banners and

SOURCE: Johan Huizinga, *The Waning of the Middle Ages* (New York: St. Martin's Press; London: Edward Arnold, 1954), pp. 100–104, 122–127.

pennons, the variety of heraldic bearings, the sound of clarions, the war-cries resounding all day long, all this, with the military costume itself and the ceremonies of dubbing knights before the battle, tended to give war the appearance of a noble sport.

After the middle of the century, the drum, of Oriental origin, makes its appearance in the armies of the West, introduced by the lansquenets. With its unmusical hypnotic effect it symbolizes, as it were, the transition from the epoch of chivalry to that of the art of modern warfare; together with fire-arms it has contributed towards rendering war mechanical.

The chivalrous point of view still presides over the classification of martial exploits by the chroniclers. They take pains to distinguish, according to technical rules, between a pitched battle and an encounter, for it is imperative that every combat has its appropriate place in the records of glory. "And so, from this day forward"—says Monstrelet—"this business was called the encounter of Mons en Vimeu. And it was declared to be no battle, because the parties met by chance and there were hardly any banners unfurled." Henry V solemnly baptizes his great victory, the battle of Agincourt, "inasmuch as all battles should bear the name of the nearest fortress where they are fought."

In spite of the care taken on all hands to keep up the illusion of chivalry, reality perpetually gives the lie to it, and obliges it to take refuge in the domains of literature and of conversation. The ideal of the fine heroic life could only be cultivated within the limits of a close caste. The sentiments of chivalry were current only among the members of the caste and by no means extended to inferior persons. The Burgundian court, which was saturated with chivalrous prejudice, and would not have tolerated the slightest infringement of rules in a "combat à outrance" between noblemen, relished the unbridled ferocity of a judicial duel between burghers, where there was no code of honour to observe. Nothing could be more remarkable in this respect than the interest excited everywhere by the combat of two burghers of Valenciennes in 1455. The old Duke Philip wanted to see the rare spectacle at any cost. One must read the vivid and realistic description given by Chastellain in order to appreciate how a chivalrous writer who never succeeded in giving more than a vaguely fanciful description of a Passage of Arms, made up for it here by giving full rein to the instincts of natural cruelty. Not one detail of the "very beautiful ceremony" escaped him. The adversaries, accompanied by their fencing masters, enter the lists, first Jacotin Plouvier, the plaintiff, next Mahuot. Their heads are cropped close and they are sewn up from head to foot in cordwain dresses of a single piece. They are very pale. After having saluted the duke, who was seated behind lattice-work, they await the signal, seated upon two chairs upholstered in black. The spectators exchange remarks in a low voice on the chances of the combat: How pale Mahuot is as he kisses the Testament! Two servants come to rub them with grease from the neck to the ankles. Both champions rub their hands with ashes and take sugar in their mouths; next they are given quartersticks and bucklers painted with images of saints, which they hold upside down, having, moreover, in their hands "a scroll of devotion."

Mahuot, a small man, begins the combat by throwing sand into Jacotin's face with the point of his buckler. Soon afterwards he falls to the ground under the formidable blows of Jacotin, who throws himself on him, fills his eyes and mouth with sand, and thrusts his thumb into the socket of his eye, to make him let go of a finger which Mahuot has between his teeth. Jacotin wrings the other's arms, jumps upon his back and tries to break it. In vain does Mahuot cry for mercy, and asks to be confessed. "O my lord of Burgundy," he calls out, "I have served you so well in your war of Ghent! O my lord, for God's sake, I beg for mercy, save my life!" . . . Here some pages of Chastellain's chronicle are missing; we learn elsewhere that the dying man was dragged out of the lists and hanged by the executioner.

Did Chastellain end his lively narrative by a moral? It is probable; anyhow, La Marche tells that the nobility were a little ashamed at having been present at such a spectacle. "Because of which God caused a duel of knights to follow, which was irreproachable and without fatal consequences," adds the incorrigible court poet.

As soon as it is a question of non-nobles, the old and deep-rooted contempt for the villein shows us that the ideas of chivalry had availed but little in mitigating feudal barbarism. Charles VI, after the battle of Rosebeke, wishes to see the corpse of Philip of Artevelde. The king does not show the slightest respect for the illustrious rebel. According to one chronicle, he is said to have kicked the body, "treating it as a villein." "When it had been looked at, for some time"—says Froissart—"it was taken from that place and hanged on a tree."

Hard realities were bound to open the eyes of the nobility and show the false-ness and uselessness of their ideal. The financial side of a knight's career was frankly avowed. Froissart never omits to enumerate the profits which a successful enterprise procured for its heroes. The ransom of a noble prisoner was the back-bone of the business to the warriors of the fifteenth century. Pensions, rents, governor's places, occupy a large place in a knight's life. His aim is "s'avanchier par armes" (to get on in life by arms). Commines rates the courtiers according to their pay, and speaks of "a nobleman of twenty crowns," and Deschamps makes them sigh after the day of payment, in a ballad with the refrain:

Et quant venra le trésorier? [1]

As a military principle, chivalry was no longer sufficient. Tactics had long since given up all thought of conforming to its rules. The custom of making the knights fight on foot was borrowed by the French from the English, though the chivalrous spirit was opposed to this practice. It was also opposed to sea-fights. In the *Debat des Hérauts d'Armes de France et d'Angleterre*, the French herald being asked by his English colleague: Why does the king of France not maintain a great naval force, like that of England? replies very naïvely: In the first place he does not need it, and, then, the French nobility prefer wars on dry land, for several reasons, "for (on the sea) there is danger and loss of life and God knows how awful it is when a storm rages and sea-sickness prevails which many people find hard to bear. Again, look at the hard life which has to be lived, which does not beseem nobility."

Nevertheless, chivalrous ideas did not die out without having borne some fruit. In so far as they formed a system of rules of honour and precepts of virtue, they exercised a certain influence on the evolution of the laws of war. The law of nations originated in antiquity and in canon law, but it was chivalry which

---

[1] And when will the paymaster come?

caused it to flower. The aspiration after universal peace is linked with the idea of crusades and with that of the orders of chivalry. Philippe de Mézières planned his "Order of the Passion" to insure the good of the world. The young king of France—(this was written about 1388, when such great hopes were still entertained of the unhappy Charles VI)—will be easily able to conclude peace with Richard of England, young like himself and also innocent of bloodshed in the past. Let them discuss the peace personally; let them tell each other of the marvellous revelations which have already heralded it. Let them ignore all the futile differences which might prevent peace, if negotiations were left to ecclesiastics, to lawyers, and to soldiers. The king of France may fearlessly cede a few frontier towns and castles. Directly after the conclusion of peace the crusade will be prepared. Quarrels and hostilities will cease everywhere; the tyrannical governments of countries will be reformed; a general council will summon the princes of Christendom to undertake a crusade, in case sermons do not suffice to convert the Tartars, Turks, Jews and Saracens.

THE CONVENTIONS OF LOVE

The courtly code did not serve exclusively for making verses; it claimed to be applicable to life, or at least to conversation. It is very difficult to pierce the clouds of poetry and to penetrate to the real life of the epoch. How far did courting and flirtation during the fourteenth and fifteenth centuries come up to the requirements of the courtly system or to the precepts of Jean de Meun? Autobiographical confessions are very rare at that epoch. Even when an actual love-affair is described with the intention of being accurate, the author cannot free himself from the accepted style and technical conceptions. We find an instance of this in the too lengthy narrative of a love-affair of an old poet and a young girl, which Guillaume de Machaut has given us in *Le Livre du Voir-Dit*. He was approaching his sixtieth year, when Peronnelle d'Armentières, of a noble family in Champagne, sent him, in 1362, her first rondel, in which she offered her heart to the celebrated poet, whom she did not know, and invited him to enter with her into a poetical love correspondence. The poor poet, sickly, blind of one eye, gouty, at once kindles. He replies to her rondel and an exchange of letters and of poems begins. Peronnelle is proud of her literary connection; she does not make a secret of it, and begs the poet to put in writing the true story of their love, inserting their letters and their poetry. Machaut readily complies. "I shall make," he says, "to your glory and praise, something that will be well remembered."

"And, my very sweet heart, are you sorry because we have begun so late? By God, so am I; but here is the remedy: let us enjoy life as much as circumstances permit, so that we may make up for the time we have lost; and that people may speak of our love a hundred years hence, and all well and honourably; for if there were evil, you would conceal it from God, if you could."

The narrative connecting the letters and the poetry teaches us what degree of intimacy was considered compatible with a decent love-affair. The young lady may permit herself extraordinary liberties, provided everything takes place in the presence of third parties, her sister-in-law, her maid or her secretary. At the first interview, which Machaut has been waiting for with misgivings, because of his unattractive appearance, Peronnelle falls asleep, or pretends to sleep, under a cherry tree, with her head on the poet's knees. The secretary covers her mouth with a green leaf and tells Machaut to kiss the leaf. Just when the latter takes courage to do so, the secretary pulls the leaf away.

She grants him other favours. A pilgrimage to Saint Denis, at the time of the fair, provides them with an opportunity of passing some days together. One afternoon, overcome by the heat of mid-June, they fly from the crowd at the fair to take a few hours' rest. A burgher of the town provides them with a double-bedded room. The blinds are closed and the company lies down. The sister-in-law takes one of the two beds. Peronnelle and her maid occupy the other. She orders the bashful poet to lie down between them, which he does, lying very still for fear of disturbing her. On waking, she orders him to kiss her.

At the end of the trip, she permits him to come and wake her, in order to take leave, and the narrative gives us to understand that she refused him nothing. She gives him the golden key of her honour, to guard that treasure, or what was left of it.

The poet's good fortune ended there. He did not see her again, and, for lack of other adventures, he filled the rest of his book with mythological excursions. At last she lets him know that their relations must end, because of a marriage, probably. He resolves to go on loving and revering her till the end of his days. And after their death, he will pray God, to reserve for her, in the glory of Heaven, the name he gave her: *Toute-belle*.

In the *Voir-Dit* of Machaut religion and love are mixed up with a sort of ingenuous shamelessness. We need not be shocked by the fact that the author was a canon of the church of Reims, for, in the Middle Ages, minor orders, which sufficed for a canon (Petrarch was one), did not absolutely impose celibacy. The fact that a pilgrimage was chosen as an occasion for the lovers to meet was not extraordinary either. At this period pilgrimages served all sorts of frivolous purposes. But what astonishes us is that Machaut, a serious and delicate poet, claims to perform his pilgrimage "very devoutly." At mass he is seated behind her:

> . . . Quant on dist: Agnus Dei,
> Foy que je doy à Saint Crepais,
> Doucement me donna la pais,
> Entre deux pilers du moustier.
> Et j'en avoie bien mestier,
> Car mes cuers amoureus estoit
> Troublés, quant si tost se partoit.[1]

He says his hours as he is waiting for her in the garden. He glorifies her portrait as his God on earth. Entering the church to begin a novene, he takes a mental vow to compose a poem about his beloved on each of the nine days— which does not prevent him from speaking about the great devotion with which he said his prayers.

We shall revert elsewhere to the astonishing ingenuousness with which, before the Council of Trent, worldly occupations were mixed up with works of the Faith.

As regards the tone of the love-affair of Machaut and Peronnelle, it is soft, cloying, somewhat morbid. The expression of their feelings remains enveloped in

---

[1] When the priest said: Agnus Dei, Faith I owe to Saint Crepais, Sweetly she gave me the pax Between two pillars of the church. And I needed it indeed, For my armorous heart was Troubled that we had to part so soon.

arguments and allegories. But there is something touching in the tenderness of the old poet, which prevents him from seeing that "Toute-belle," after all, has but played with him and with her own heart.

To grasp what little we can of actual love relations, apart from literature, we should oppose to the *Voir-Dit*, as a pendant, *Le Livre du Chevalier de la Tour Landry pour l'Enseignement de ses Filles*, written at the same epoch. This time we are not concerned with an amorous old poet; we have to do with a father of a rather prosaic turn of mind, an Angevin nobleman, who relates his reminiscences, anecdotes and tales "pour mes filles aprandre à roumancier." This might be rendered, "to teach my daughters the fashionable conventions in love matters." The instruction, however, does not turn out romantic at all. The moral of the examples and admonitions which the cautious father recommends to his daughters tends especially to put them on their guard against the dangers of romantic flirtations. Take heed of eloquent people, always ready with their "false long and pensive looks and little sighs, and wonderful emotional faces, and who have more words at hand than other people." Do not be too encouraging. He himself, when young, was conducted by his father to a castle to make the acquaintance of a young lady to whom they wanted to betroth him. The girl received him very kindly. He conversed with her on all sorts of subjects, so as to probe her character somewhat. They got to talk of prisoners, which gave the knight a chance to pay a neat compliment: " 'Ma demoiselle, it would be better to fall into your hands as a prisoner than into many another's, and I think your prison would not be so hard as that of the English.' She replied that she had recently seen one whom she could wish to be her prisoner. And then I asked her, if she would make a bad prison for him, and she said not at all, and that she would hold him as dear as her own person, and I told her that the man would be very fortunate in having such a sweet and noble prison. What shall I say? She could talk well enough, and it seemed, to judge from her conversation, that she knew a good deal, and her eyes had also a very lively and lightsome expression." When they took leave she begged him two or three times to come back soon, as if she had known him for a long time already. "And when we had departed my lord my father said to me: 'What do you think of her whom you have seen? Tell me your opinion.' 'Monseigneur, she seems to me all well and good, but I shall never be nearer to her than I am now, if you please.' " Her lack of reserve left him without any desire to get better acquainted with her. So they did not get engaged, and of course the author says that he afterwards had reason not to repent it.

It is to be regretted that the chevalier has not given more autobiographical details and fewer moral exhortations, because these personal traits, showing how customs adapted themselves to the ideal, are very rare in the traditions of the Middle Ages.

In spite of his avowed intention to teach his girls "à roumancier," the knight de la Tour Landry thinks, before all things, of a good marriage; and marriage had little to do with love. He reports to them a "debate" between his wife and himself, on the question, whether it is becoming "d'amer par amours." He thinks that a girl may, in certain cases, for example, "in the hope of marrying," love honourably. His wife thinks otherwise. It is better that a girl should not fall in love at all, not even with her betrothed, otherwise piety would suffer in consequence. "For I have heard many women say who were in love in their youth, that when they were in church, their thoughts and fancies made them dwell more on those nimble imaginations and delights of their love-affairs than on the service of God, and the art of love is of such a nature that just at the holiest moments of the service, that is to say, when the priest holds our Lord on the altar, the most of these little thoughts would come to them." Machaut and Peronnelle might have confirmed this.

It is not easy for us to reconcile the general austerity of the Chevalier de la Tour Landry with the fact that this father does not scruple to instruct his daughters by means of stories which would not have been out of place in the *Cent Nouvelles Nouvelles*. Still, even more recent literature, that of the Elizabethan age, for instance, may remind us how completely the world becomes estranged from the erotic forms of a few centuries back. As for betrothals and marriages, neither the graceful forms of the courtly ideal nor the refined frivolity and open cynicism of the *Roman de la Rose* had any real hold upon them. In the very matter-of-fact considerations on which a match between noble families was based there was little room for the chivalrous fictions of prowess and of service. Thus it came about that the courtly notions of love were never corrected by contact with real life. They could unfold freely in aristocratic conversation, they could offer a literary amusement or a charming game, but no more. The ideal of love, such as it was, could not be lived up to, except in a fashion inherently false.

Cruel reality constantly gave the lie to it. At the bottom of the intoxicating cup of the *Roman de la Rose* the moralist exposed the bitter dregs. From the side of religion maledictions were poured upon love in all its aspects, as the sin by which the world is being ruined. Whence, exclaims Gerson, come the bastards, the infanticides, the abortions, whence hatred, whence poisonings?—Woman joins her voice to that from the pulpit: all the conventions of love are the work of men: even when it dons an idealistic guise, erotic culture is altogether saturated by male egotism: and what else is the cause of the endlessly repeated insults to matrimony, to woman and her feebleness, but the need of masking this egotism? One word suffices, says Christine de Pisan, to answer all these infamies: it is not the women who have written the books.

Indeed, medieval literature shows little true pity for woman, little compassion for her weakness and the dangers and pains which love has in store for her. Pity took on a stereotyped and factitious form, in the sentimental fiction of the knight delivering the virgin. The author of the *Quinze Joyes de Mariage*, after having mocked at all the faults of women, undertakes to describe also the wrongs they have to suffer. So far as is known, he never performed this task.

The Counterculture of Medieval Students

We have already discussed on page 315 the significance of universities as both reflections of and aids to the rise of monarchies, church government, and city life. In the following pages from Haskins's *The Rise of the Universities*, there is a marvelous often lyric description of student life in medieval universities. Our own time echoes Haskins's concerns for medieval town-gown riots, the immunities of scholars and schools from police invasions, and the implication of teachers and students in every great issue of political and social controversy. Medieval student life was never life in a cloister away from the cares of the world. Nor was the medieval university content to describe the world as it was; beyond beholding the wonder of its working, doctors, masters, and the "clerks of all the nations" criticized society and sought to change it.

The universities were great melting pots and the chief agencies of upward mobility. But they also made champions of nominalism against realism, the Empire against the papacy, Averroës against Aristotle, and Luther against Roman Christendom. Henry VIII appealed to the universities to justify his divorce of Catherine of Aragon. Students in prison competed for attention with those who made their way up in the councils of kings and the entourages of popes, sinners vied with saints, learned heretics with angelic doctors. Many took to a vagrant life, opposing the culture of youthful buoyancy to that of despair and pessimism which echoed in sermons. Haskins seems to present students as makers of a counterculture, often at war among themselves and with every external authority.

From *The Rise of Universities*

Charles Homer Haskins

Fortunately, out of the scattered remains of mediaeval times, there has come down to us a considerable body of material which deals, more or less directly, with student affairs. There are, for one thing, the records of the courts of law, which, amid the monotonous detail of petty disorders and oft-repeated offences, preserve now and then a vivid bit of mediaeval life—like the case of the Bolognese student who was attacked with a cutlass in a class-room, to the great damage and loss of those assembled to hear the lecture of a noble and egregious doctor of laws; or the student in 1289 who was set upon in the street in front of a lecture-room by a certain scribe, "who wounded him on the head with a stone, so that much blood gushed forth," while two companions gave aid and counsel, saying, "Give it to him, hit him," and when the offence had been committed ran away. So the coroners' rolls of Oxford record many a fatal issue of town and gown riots, while a recently published register of 1265 and 1266 shows the students of Bologna actively engaged in raising money by loans and by the sale of text-books. There are of course the university and college statutes, with their prohibitions and fines, regulating the subjects of conversation, the shape and color of caps and

SOURCE: Charles Homer Haskins, *The Rise of Universities* (Ithaca, N.Y.: Cornell University Press, 1959), pp. 60–92.

gowns, that academic dress which looks to us so mediaeval and is, especially in its American form, so very modern; careful also of the weightier matters of the law, like the enactment of New College against throwing stones in chapel, or the graded penalties at Leipzig for him who picks up a missile to throw at a professor, him who throws and misses, and him who accomplishes his fell purpose to the master's hurt. The chroniclers, too, sometimes interrupt their narrative of the affairs of kings and princes to tell of students and their doings, although their attention, like that of their modern successors, the newspapers, is apt to be caught by outbreaks of student lawlessness rather than by the wholesome routine of academic life.

Then we have the preachers of the time, many of them also professors, whose sermons contain frequent allusions to student customs; indeed if further evidence were needed to dispel the illusion that the mediaeval university was devoted to biblical study and religious nurture, the Paris preachers of the period would offer sufficient proof. "The student's heart is in the mire," says one of them, "fixed on prebends and things temporal and how to satisfy his desires." "They are so litigious and quarrelsome that there is no peace with them; wherever they go, be it Paris or Orleans, they disturb the country, their associates, even the whole university." Many of them go about the streets armed, attacking the citizens, breaking into houses, and abusing women. They quarrel among themselves over dogs, women, or what-not, slashing off one another's fingers with their swords, or, with only knives in their hands and nothing to protect their tonsured pates, rush into conflicts from which armed knights would hold back. Their compatriots come to their aid, and soon whole nations of students may be involved in the fray. These Paris preachers take us into the very atmosphere of the Latin Quarter and show us much of its varied activity. We hear the cries and songs of the streets—

> Li tens s' en veit,
> Et je n' ei riens fait;
> Li tens revient,
> Et je ne fais riens—

the students' tambourines and guitars, their "light and scurrilous words," their hisses and handclappings and loud shouts of applause at sermons and disputations. We watch them as they mock a neighbor for her false hair or stick out their tongues and make faces at the passers-by. We see the student studying by his window, talking over his future with his roommate, receiving visits from his parents, nursed by friends when he is ill, singing psalms at a student's funeral, or visiting a fellow-student and asking him to visit him—"I have been to see you, now come to our hospice."

All types are represented. There is the poor student, with no friend but St. Nicholas, seeking such charity as he can find or earning a pittance by carrying holy water or copying for others, in a fair but none too accurate hand, sometimes too poor to buy books or afford the expense of a course in theology, yet usually surpassing his more prosperous fellows who have an abundance of books at which they never look. There is the well-to-do student, who besides his books and desk will be sure to have a candle in his room and a comfortable bed with a soft mattress and luxurious coverings, and will be tempted to indulge the mediaeval fondness for fine raiment beyond the gown and hood and simple wardrobe prescribed by the statutes. Then there are the idle and aimless, drifting about from master to master and from school to school, and never hearing full courses or regular lectures. Some, who care only for the name of scholar and the income which they receive while attending the university, go to class but once or twice a week, choosing by preference the lectures on canon law, which leave them plenty of time for sleep in the morning. Many eat cakes when they ought to be at study, or go to sleep in the class-rooms, spending the rest of their time drinking in taverns or building castles in Spain (*castella in Hispania*); and when it is time to leave Paris, in order to make some show of learning such students get together huge volumes of calfskin, with wide margins and fine red bindings, and so with wise sack and empty mind they go back to their parents. "What knowledge is this," asks the preacher, "which thieves may steal, mice or moths eat up, fire or water destroy?" and he cites an instance where the student's horse fell into a river, carrying all his books with him. Some never go home, but continue to enjoy in idleness the fruits of their benefices. Even in vacation time, when the rich ride off with their servants and the poor trudge home under the burning sun, many idlers remain in Paris to their own and the city's harm. Mediaeval Paris, we should remember, was not only the incomparable "parent of the sciences," but also a place of good cheer and good fellowship and varied delights, a favorite resort not only of the studious but of country priests on a holiday; and it would not be strange if sometimes scholars prolonged their stay unduly and lamented their departure in phrases which are something more than rhetorical commonplace.

Then the student is not unknown to the poets of the period, among whom Rutebeuf gives a picture of thirteenth-century Paris not unlike that of the sermonizers, while in the preceding century Jean de Hauteville shows the misery of the poor and diligent scholar falling asleep over his books, and Nigel "Wireker" satirizes the English students at Paris in the person of an ass, Brunellus—"Daun Burnell" in Chaucer—who studies there seven years without learning a word, braying at the end as at the beginning of his course, and leaving at last with the resolve to become a monk or a bishop. Best of all is Chaucer's incomparable portrait of the clerk of Oxenford, hollow, threadbare, unworldly—

> For him was lever have at his beddes heed
> Twenty bokes, clad in blak or reed,
> Of Aristotle and his philosophye,
> Than robes riche, or fithele, or gay sautrye.
> Souninge in moral vertu was his speche,
> And gladly wolde he lerne, and gladly teche.

But after all, no one knows so much about student life as the students themselves, and it is particularly from what was written by and for them, the student literature of the Middle Ages, that I wish to draw more at length. Such remains of the academic past fall into three chief classes: student manuals, student letters, and student poetry. Let us consider them in this order.

The manuals of general advice and counsel addressed to the mediaeval scholar do not call for extended consideration. Formal treatises on the whole duty of students are characteristic of the didactic habit of mind of the Middle Ages, but the advice which they contain is apt to be of a very general sort, applicable to one age as well as another and lacking in those concrete illustrations which enliven the sermons of the period into useful sources for university life.

A more interesting type of student manual, the student dictionary, owes its existence to the position of Latin as the universal language of mediaeval education. Text-books were in Latin, lectures were in Latin, and, what is more, the use of Latin was compulsory in all forms of student intercourse. This rule may have been designed as a check on conversation, as well as an incentive to learning, but it was enforced by penalties and informers (called wolves), and the freshman, or yellow-beak as he was termed in mediaeval parlance, might find himself but ill equipped for making himself understood in his new community. For his convenience a master in the University of Paris in the thirteenth century, John of Garlande, prepared a descriptive vocabulary, topically arranged and devoting a large amount of space to the objects to be seen in the course of a walk through the streets of Paris. The reader is conducted from quarter to quarter and from trade to trade, from the book-stalls of the Parvis Notre-Dame and the fowl-market of the adjoining Rue Neuve to the money-changers' tables and goldsmiths' shops on the Grand-Pont and the bow-makers of the Porte S.-Lazare, not omitting the classes of *ouvrières* whose acquaintance the student was most likely to make. Saddlers and glovers, furriers, cobblers, and apothecaries, the clerk might have use for the wares of all of them, as well as the desk and candle and writing-materials which were the special tools of his calling; but his most frequent relations were with the purveyors of food and drink, whose agents plied their trade vigorously through the streets and lanes of the Latin Quarter and worked off their poorer goods on scholars and their servants. There were the hawkers of wine, crying their samples of different qualities from the taverns; the fruit-sellers, deceiving clerks with lettuce and cress, cherries, pears, and green apples; and at night the vendors of light pastry, with their carefully covered baskets of wafers, waffles, and rissoles—a frequent stake at the games of dice among students, who had a custom of hanging from their windows the baskets gained by lucky throws of the six. The *pâtissiers* had also more substantial wares suited to the clerical taste, tarts filled with eggs and cheese and well-peppered pies of pork, chicken, and eels. To the *rôtissiers* scholars' servants resorted, not only for the pigeons, geese, and other fowl roasted on their spits, but also for uncooked beef, pork, and mutton, seasoned with garlic and other strong sauces. Such fare, however, was not for the poorer students, whose slender purses limited them to tripe and various kinds of sausage, over which a quarrel might easily arise and "the butchers be themselves butchered by angry scholars."

A dictionary of this sort easily passes into another type of treatise, the manual of conversation. This method of studying foreign languages is old, as survivals from ancient Egypt testify, and it still spreads its snares for the unwary traveller who prepares to conquer Europe *à la* Ollendorff. To the writers of the later Middle Ages it seemed to offer an exceptional opportunity for combining Instruction in Latin with sound academic discipline, and from both school and university it left its monuments for our perusal. The most interesting of these handbooks is entitled a "Manual of Scholars who propose to attend universities of students and to profit therein," and while in its most common form it is designed for the students of Heidelberg about the year 1480, it could be adapted with slight changes to any of the German universities. "Rollo at Heidelberg," we might call it. Its eighteen chapters conduct the student from his matriculation to his degree, and inform him by the way on many subjects quite unnecessary for either. When the young man arrives he registers from Ulm; his parents are in moderate circumstances; he has come to study. He is then duly hazed after the German fashion, which treats the candidate as an unclean beast with horns and tusks which must be removed by officious fellow-students, who also hear his confession of sin and fix as the penance a good dinner for the crowd. He begins his studies by attending three lectures a day, and learns to champion nominalism against realism and the comedies of Terence against the law, and to discuss the advantages of various universities and the price of food and the quality of the beer in university towns. Then we find him and his room-mate quarrelling over a mislaid book; rushing at the first sound of the bell to dinner, where they debate the relative merits of veal and beans; or walking in the fields beyond the Neckar, perhaps by the famous Philosophers' Road which has charmed so many generations of Heidelberg youth, and exchanging Latin remarks on the birds and fish as they go. Then there are shorter dialogues: the scholar breaks the statutes; he borrows money, and gets it back; he falls in love and recovers; he goes to hear a fat Italian monk preach or to see the jugglers and the jousting in the market-place; he knows the dog-days are coming—he can feel them in his head! Finally our student is told by his parents that it is high time for him to take his degree and come home. At this he is much disturbed; he has gone to few lectures, and he will have to swear that he has attended regularly; he has not worked much and has incurred the enmity of many professors; his master discourages him from trying the examination; he fears the disgrace of failure. But his interlocutor reassures him by a pertinent quotation from Ovid and suggests that a judicious distribution of gifts may do much—a few florins will win him the favor of all. Let him write home for more money and give a great feast for his professors; if he treats them well, he need not fear the outcome. This advice throws a curious light upon the educational standards of the time; it appears to have been followed, for the manual closes with a set of forms inviting the masters to the banquet and the free bath by which it was preceded.

If university students had need of such elementary compends of morals and manners, there was obviously plenty of room for them in the lower schools as well, where they were apt to take the form of Latin couplets which could be readily impressed upon the pupil's memory. Such *statuta vel precepta scolarium* seem to have been especially popular in the later fifteenth century in those city schools of Germany whose importance has been so clearly brought out by recent historians of secondary education. Wandering often from town to town, like the roving scholars of an earlier age, these German boys had good need to observe the moral maxims thus purveyed. The beginning of wisdom was to remember God and obey the master, but the student had also to watch his behavior in church and lift up his voice in the choir—compulsory attendance at church and singing in the choir being a regular feature of these schools—keep his books clean, and pay his school bills promptly. Face and hands should be washed in the morning, but the baths should not be visited without permission, nor should boys run on the ice or throw snowballs. Sunday was the day for play, but this could be only in the churchyard, where boys must be careful not to play with dice or break stones from the wall or throw anything over the church. And whether at play or at home, Latin should always be spoken.

More systematic is a manual of the fifteenth century preserved in a manuscript of the Bibliothèque Nationale at Paris. "Since by reason of imbecility youths cannot advance to a knowledge of the Latin tongue by theory alone," the author has for their assistance prepared a set of forms which contain the expressions most frequently employed by clerks. Beginning with the courtesies of school life, for obedience and due reverence for the master are the beginning of wisdom, the boy learns how to greet his master and to take leave, how to excuse himself for wrong-doing, how to invite the master to dine or sup with his parents—there are half a dozen forms for this! He is also taught how to give proper answers to those who seek to test his knowledge, "that he may not appear an idiot in the sight of his parents." "If the master asks, 'Where have you been so long?' " he must be ready, not only to plead the inevitable headache or failure to wake up, but also to express the causes of delay well known to any village boy. He had to look after the house or feed the cattle or water the horse; he was detained by a wedding, by picking grapes, or making out bills, or—for these were German boys—by helping with the brew, fetching beer, or serving drink to guests.

In school after the "spiritual refection" of the morning singing-lesson comes refection of the body, which is placed after study hours because "the imaginative virtue is generally impeded in those who are freshly sated." In their talk at luncheon or on the playground "clerks are apt to fall from the Latin idiom into the mother tongue," and for him who speaks German the discretion of the master has invented a dunce's symbol called an ass, which the holder tries hard to pass on to another. "Wer wel ein Griffel kouffe[n]?" "Ich wel ein Griffel kouffen." "Tecum sit asinus." "Ach, quam falsus es tu!" Sometimes the victim offers to meet his deceiver after vespers, with the usual schoolboy brag on both sides. As it is forbidden to come to blows in school, the boys are taught to work off their enmities and formulate their complaints in Latin dialogue. "You were outside the town after dark. You played with laymen Sunday. You went swimming Monday. You stayed away from matins. You slept through mass." "Reverend master, he has soiled my book, he shouts after me wherever I go, he calls me names." Besides the formal disputations the scholars discuss such current events as a street fight, a cousin's wedding, the coming war with the duke of Saxony, or the means of getting to Erfurt, whither one of them is going when he is sixteen to study at the university. The great ordeal of the day was the master's quiz on Latin grammar, when every one was questioned in turn (*auditio circuli*). The pupils rehearse their declensions and conjugations and the idle begin to tremble as the hour draws near. There is some hope that the master may not come. "He has guests." "But they will leave in time." He may go to the baths." "But it is not yet a whole week since he was there last." "There he comes. Name the wolf, and he forthwith appears." Finally the shaky scholar falls back on his only hope, a place near one who promises to prompt him.

"When the recitation is over and the lesson given out, rejoicing begins among the youth at the approach of the hour for going home," and they indulge in much idle talk "which is here omitted, lest it furnish the means of offending." Joy is, however, tempered by the contest which precedes dismissal, "a serious and furious disputation for the *palmiterium*," until one secures the prize and another has the *asinus* to keep till next day.

After school the boys go to play in the churchyard, the sports mentioned being hoops, marbles (apparently), ball (during Lent), and a kind of counting game. The author distinguishes hoops for throwing and for rolling, spheres of wood and of stone, but the subject soon becomes too deep for his Latin, and in the midst of this topic the treatise comes to an abrupt conclusion.

In some of its forms the student manual touches on territory already occupied by another type of mediaeval handbook, the manual of manners, which under such titles as "The Book of Urbanity," "The Courtesies of the Table," etc., enjoyed much popularity from the thirteenth century onward. Such manuals have, however, none of the polish of Castiglione's *Courtier* or the elaborateness of the modern book of etiquette. Those who have not mastered the use of knife and fork have little use for the finer points of social intercourse, and the readers of the mediaeval manuals were still at the abc's in the matter of behavior. Wash your hands in the morning and, if you have time, your face; use your napkin and handkerchief; eat with three fingers, and don't gorge; don't be boisterous or quarrelsome at table; don't stare at your neighbor or his plate; don't criticise the food; don't pick your teeth with your knife—such, with others still more elementary, are the maxims which meet us in this period, in Latin and French, in English, German, and Italian, but regularly in verse. Now and then there is a further touch of the age: scrape bones with your knife but don't gnaw them; when you have done with them, put them in a bowl or on the floor!

If the correspondence of mediaeval students were preserved for us in casual and unaffected detail, nothing could give a more vivid picture of university conditions. Unfortunately in some respects for us, the Middle Ages were a period of forms and types in letter-writing as in other things; and for most men the writing of a letter was less an expression of individual feeling and experience than it was the laborious copying of a letter of some one else, altered where necessary to suit the new conditions. And if something fresh or individual was produced, there was small chance of preserving it, since it was on that account all the less likely to be useful to a future letter-writer—"so careful of the type, so careless of the single" letter, history seems. The result is that the hundreds of student letters which have reached us in the manuscripts of the Middle Ages have come down through the medium of collections of forms or complete letter-writers, shorn of most of their individuality but for that very reason reflecting the more faithfully the fundamental and universal phases of university life.

By far the largest element in the correspondence of mediaeval students consists of requests for money; "a student's first song is a demand for money," says a weary father in an Italian letter-writer, "and there will never be a letter which does not ask for cash." How to secure this fundamental necessity of student life was doubtless one of the most important problems that confronted the mediaeval scholar, and many were the models which the rhetoricians placed before him in proof of the practical advantages of their art. The letters are generally addressed to parents, sometimes to brothers, uncles, or ecclesiastical patrons; a much copied exercise contained twenty-two different methods of approaching an archdeacon on this ever-delicate subject. Commonly the student announces that he is at such and such a centre of learning, well and happy but in desperate need of money for books and other necessary expenses. Here is a specimen from Oxford, somewhat more individual than the average and written in uncommonly bad Latin:

> B. to his venerable master A., greeting. This is to inform you that I am studying at Oxford with the greatest diligence, but the matter of money stands greatly in the way of my promotion, as it is now two months since I spent the last of what you sent me. The city is expensive and makes many demands; I have to rent lodgings, buy necessaries, and provide for many other things which I cannot now specify. Wherefore I respectfully beg your paternity that by the promptings of divine pity you may assist me, so that I may be able to complete

what I have well begun. For you must know that without Ceres and Bacchus Apollo grows cold.

If the father was close-fisted, there were special reasons to be urged: the town was dear—as university towns always are!—the price of living was exceptionally high owing to a hard winter, a siege, a failure of crops, or an unusual number of scholars; the last messenger had been robbed or had absconded with the money; the son could borrow no more of his fellows or of the Jews; and so on. The student's woes are depicted in moving language, with many appeals to paternal vanity and affection. At Bologna we hear of the terrible mud through which the youth must beg his way from door to door, crying, "O good masters," and coming home empty-handed. In an Austrian formulary a scholar writes from the lowest depths of prison, where the bread is hard and moldy, the drink water mixed with tears, the darkness so dense that it can actually be felt. Another lies on straw with no covering, goes without shoes or shirt, and eats he will not say what—a tale designed to be addressed to a sister and to bring in response a hundred *sous tournois*, two pairs of sheets, and ten ells of fine cloth, all sent without her husband's knowledge. "We have made little glosses, we owe money," is the terse summary of two students at Chartres.

To such requests the proper answer was, of course, an affectionate letter, commending the young man's industry and studious habits and remitting the desired amount. Sometimes the student is cautioned to moderate his expenses—he might have got on longer with what he had, he should remember the needs of his sisters, he ought to be supporting his parents instead of trying to extort money from them, etc. One father—who quotes Horace!—excuses himself because of the failure of his vineyards. It often happened, too, that the father or uncle has heard bad reports of the student, who must then be prepared to deny indignantly all such aspersions as the unfounded fabrications of his enemies. Here is an example of paternal reproof taken from an interesting collection relating to Franche-Comté:

> To his son G. residing at Orleans P. of Besançon sends greetings with paternal zeal. It is written, "He also that is slothful in his work is brother to him that is a great waster." I have recently discovered that you live dissolutely and slothfully, preferring license to restraint and play to work and strumming a guitar while the others are at their studies, whence it happens that you have read but one volume of law while your more industrious companions have read several. Wherefore I have decided to exhort you herewith to repent utterly of your dissolute and careless ways, that you may no longer be called a waster and your shame may be turned to good repute.

In the models of Ponce de Provence we find a teacher writing to a student's father that while the young man is doing well in his studies, he is just a trifle wild and would be helped by judicious admonition. Naturally the master does not wish it known that the information came through him, so the father writes his son: "I have learned—not from your master, although he ought not to hide such things from me, but from a certain trustworthy source—that you do not study in your room or act in the schools as a good student should, but play and wander about, disobedient to your master and indulging in sport and in certain other dishonorable practices which I do not now care to explain by letter." Then follow the customary exhortations to reform.

Two boys at Orleans thus describe their arrival at this centre of learning:

To their dear and respected parents M. Martre, knight, and M. his wife, M. and S. their sons send greeting and filial obedience. This is to inform you that, by divine mercy, we are living in good health in the city of Orleans and are devoting ourselves wholly to study, mindful of the words of Cato, "To know anything is praiseworthy." We occupy a good dwelling, next door but one to the schools and marketplace, so that we can go to school every day without wetting our feet. We have also good companions in the house with us, well advanced in their studies and of excellent habits—an advantage which we well appreciate, for as the Psalmist says, "With an upright man thou wilst show thyself upright."

Such youths were slow to quit academic life. Again and again they ask permission to have their term of study extended; war might break out, parents or brothers die, an inheritance have to be divided, but the student pleads always for delay. He desires to "serve longer in the camp of Pallas"; in any event he cannot leave before Easter, as his masters have just begun important courses of lectures. A scholar is called home from Siena to marry a lady of many attractions; he answers that he deems it foolish to desert the cause of learning for the sake of a woman, "for one may always get a wife, but science once lost can never be recovered."

The time to leave, however, must come at last, and then the great problem is money for the expenses of commencement, or, as it was then called, inception. Thus a student at Paris asks a friend to explain to his father, "since the simplicity of the lay mind does not understand such things," how at length after much study nothing but lack of money for the inception banquet stands in the way of his graduation. From Orleans D. Boterel writes to his dear relatives at Tours that he is laboring over his last volume of law and on its completion will be able to pass to his licentiate provided they send him a hundred *livres* for the necessary expenses. An account of the inception at Bologna was quoted in the preceding chapter.

Unlike the student letters, which range over the whole of the later Middle Ages, mediaeval student poetry, or rather the best of it, is limited to a comparatively short period comprised roughly within the years 1125 and 1225, and is closely connected with the classical phase of the twelfth-century renaissance. It is largely the work of the wandering clerks of the period—students, ex-students, professors even—moving from town to town in search of learning and still more of adventure, nominally clerks but leading often very unclerical lives. "Far from their homes," says Symonds, "without responsibilities, light of purse and light of heart, careless and pleasure-seeking, they ran a free, disreputable course." "They are wont," writes a monk of the twelfth century, "to roam about the world and visit all its cities, till much learning makes them mad; for in Paris they seek liberal arts, in Orleans classics, at Salerno medicine, at Toledo magic, but nowhere manners and morals." Their chief habitat, however, was northern France, the center of the new literary renaissance.

Possibly from some obscure allusion to Goliath the Philistine, these wandering clerks took the name Goliardi and their verse is generally known as Goliardic poetry. This literature is for the most part anonymous, though recent research has individualized certain writers of the group, notably a Master Hugh, canon of Orleans, ca. 1142, styled the Primate, and the so-called Archpoet. The Primate, mordant, diabolically clever, thoroughly disreputable, became famous for generations as "an admirable improviser, who if he had but turned his heart to the love of God would have had a great place in divine letters and have proved most useful to God's church." The Archpoet is found chiefly in Italy from 1161 to 1165, going "on his own" in spring and summer but when autumn comes on turning to beg shirt and cloak from his patron, the archbishop of Cologne. Ordered to compose an epic for the emperor in a week, he replies he cannot write on an empty stomach—the quality of his verse depends on the quality of his wine:

> Tales versus facio quale vinum bibo.

Good wine he must at times have found, for he composed the masterpiece of the whole school, the Confession of a Goliard, that unforgettable description of the burning temptations of Pavia which contains the famous glorification of the joys of the tavern:

> In the public house to die
>   Is my resolution;
> Let wine to my lips be nigh
>   At life's dissolution;
> That will make the angels cry,
>   With glad elocution,
> "Grant this toper, God on high,
>   Grace and absolution!"

Though written in Latin, the Goliardic verse has abandoned the ancient metrical system for the rhyme and accent of modern poetry, but even the best of modern versions, such as those of John Addington Symonds, from which I am quoting, fail to render the swing, the lilt, the rhythmical flow of the original. Its authors are familiar with classical mythology and especially with the writings of Ovid, whose precepts, copied even in severe Cluny, were freely followed. Most of all is this poetry classical in its frankly pagan view of life. Its gods are Venus and Bacchus, also Decius, the god of dice. Love and wine and spring, life on the open road and under the blue sky, these are the common subjects; the spirit is that of an intense delight in the world that is, a joy in mere living, such as one finds in the Greek and Roman poets or in that sonorous song of a later age which the academic world still cherishes,

> Gaudeamus igitur iuvenes dum sumus.

In general the Goliardic poetry is of an impersonal sort giving us few details from any particular place, but reflecting the gayer, more jovial, less reputable side of the life of mediaeval clerks. The worshipful order of vagrants is described, open to men of every condition and every clime, with its rules which are no rules, late-risers, gamesters, roysterers, proud that none of its members has more than one coat to his back, begging their way from town to town with requests for money which sound like students' letters in verse:

> I, a wandering scholar lad,
>   Born for toil and sadness,
> Oftentimes am driven by
>   Poverty to madness.
>
> Literature and knowledge I
>   Fain would still be earning,
> Were it not that want of pelf
>   Makes me cease from learning.
>
> These torn clothes that cover me
>   Are too thin and rotten;

Oft I have to suffer cold,
    By the warmth forgotten.

Scarce I can attend at church,
    Sing God's praises duly;
Mass and vespers both I miss,
    Though I love them truly.

Oh, thou pride of N——,
    By thy worth I pray thee
Give the suppliant help in need,
    Heaven will sure repay thee.

Take a mind unto thee now
    Like unto St. Martin;
Clothe the pilgrim's nakedness,
    Wish him well at parting.

So may God translate your soul
    Into peace eternal,
And the bliss of saints be yours
    In His realm supernal.

The brethren greet each other at wayside taverns with songs like this:

We in our wandering,
Blithesome and squandering,
    Tara, tantara, teino!
Eat to satiety,
Drink with propriety;
    Tara, tantara, teino!

Laugh till our sides we split,
Rags on our hides we fit;
    Tara, tantara, teino!

Jesting eternally,
Quaffing infernally:
    Tara, tantara, teino!
                        etc.

The assembled topers are described in another poem:

Some are gaming, some are drinking,
Some are living without thinking;
And of those who make the racket,
Some are stripped of coat and jacket;
Some get clothes of finer feather,
Some are cleaned out altogether;
No one there dreads death's invasion,
But all drink in emulation.

Then they sacrilegiously drink once for all prisoners and captives, three times for the living, a fourth time for the whole body of Christians, a fifth for those departed in the faith, and so on to the thirteenth for those who travel by land or water, and a final and unlimited potation for king and Pope. Such poetry is plainly the expression of a "wet" age.

Often bibulous and erotic, the Goliardic verse contains a large amount of parody and satire. Appealing to a public familiar with scripture and liturgy, its authors parody anything—the Bible, hymns to the Virgin, the canon of the mass, as in the "Drinkers' Mass" and the "Office for Gamblers." One of the best-known pieces is a satire on the Papacy under the caption of "The Gospel according to Mark-s of silver." This is only one of many bitter attacks on Rome, while the pride, hardness, and greed of the higher clergy are portrayed in "Golias the Bishop." The point of view in general is that of the lower clergy, especially the looser, wandering, undisciplined element which frequented the schools and the roads, the *jongleurs* of the clerical world, familiar subjects of ecclesiastical legislation since the ninth century.

Poetry of this sort is so contrary to conventional conceptions of the Middle Ages that some writers have denied its mediaeval character. "It is," says one, "mediaeval only in the chronological sense," while others find in it close affinities with the spirit of the Renaissance or of the Reformation. It would be more consonant with the spirit of history to enlarge our ideas of the Middle Ages so as to correspond to the facts of mediaeval life. The Goliardi were neither humanists before the Renaissance nor reformers before the Reformation; they were simply men of the Middle Ages who wrote for their own time. If the writings of these northern and chiefly French clerks seem to anticipate the Italian Renaissance, it may be that the Renaissance began earlier and was less specifically Italian than has been supposed. If the authors are more secular, even more earthy, than we should expect clerks to be, we must learn to expect something different. In lyric poetry, as in the epic and the drama, we are now learning more of the close interpenetration of the lay and ecclesiastical worlds, no longer separated by the air-tight partitions which the imagination of a later day interposed. And whether their spirit was lay or ecclesiastical, the Goliardi were certainly human; they saw and felt life keenly, and they wrote of what they knew.

It is time to redress the balance with a word about a less obtrusive element, the good student. "The life of the virtuous student," says Hastings Rashdall, "has no annals," and in all ages he has been less conspicuous than his more dashing fellows. Thus the ideal scholar of the sermons is a bit colorless but obedient, respectful, eager to learn, assiduous at lectures, and bold in debate, pondering his lessons even during his evening promenades by the river. The ideal student of the manuals is he who practices their precepts. The typical student of the letters has already described himself as devoted wholly to study, though somewhat short of money. The good student of the poems—there is no such person! Student poetry was "not all bacchic or erotic or profane," but much of it was, and we must not look here for the more serious side of academic life. Jean de Hauteville's account of the poor and industrious scholar is representative of a large class of students but not of a large body of poetry. The good student's occupations are best reflected in the course of study, his assiduity best seen in his note-books and disputations. The documents which concern the educational side of the university are also a source for student life! It has been observed that the alumni reunions of our own day are often more prolific in recollections of student escapades than of the daily performance of the allotted task. The studious lad of today never breaks into the headlines as such, and no one has seen fit to produce a play or a film "featuring the good student." Yet everyone familiar with contemporary universities knows that the serious student exists in large numbers, and it has been shown conclusively that the distinction he there achieves reflects itself in his later life. So it was in the Middle Ages. The law students of Bologna insisted on their money's worth of teaching from their professors. The examinations described by Robert de Sorbon required serious preparation. Not only was the vocational motive a strong incentive to study in the mediaeval university, but there was much enthusiasm for knowledge and much discussion of intellectual subjects. The greater universities, at least, were intellectually very much alive, with something of that "religion of learning" which had earlier called Abelard's pupils into the wilderness, there to build themselves huts that they might feed upon his words. The books of the age were in large measure written by its professors, and the students had the advantage of seeing them in the making and thus drinking of learning at its fountainhead. Then as now, the moral quality of a university depended on the intensity and seriousness of its intellectual life.

If we consider the body of student literature as a whole, its most striking, and its most disappointing, characteristic is its lack of individuality. The *Manuale Scholarium* is written for the use of all scholars who propose to attend universities of students. The letters are made as general as possible in order to fit the need of any student who wants money, clothes, or books. Even the poems, where we have some right to expect personal expression of feeling, have the generic character of most mediaeval poetry; they are for the most part the voice of a class, not of individuals.

At the same time it must be remembered that this characteristic of the student productions, if it robs them of something of their interest, increases their historical value. The historian deals with the general rather than the particular, and his knowledge must be built up by a painful collection and comparison of individual facts, which are often too few or to unlike to admit of sound generalization. In the case of these student records, however, that labor has already been performed for him; in the form in which they come down to us they have lost, at the hands of the students themselves, what is local and peculiar and exceptional, and have become, what in view of the nature of our information no historian could hope to make them, the generalized experience of centuries of student life.

Plague and the Culture of Pessimism

We have spoken of student culture as running counter to the culture of pessimism, violence, and despair hallowed by the Church and much secular wisdom. In the studies of Bloch and Huizinga we explored some roots of the dominant culture. Millard Meiss's work on the impact of the Black Death on Florentine and Sienese painting is a classic study of how high culture responded to brutal, disastrous slaughter of people due to "natural causes."

It reminds us that men in Western society through many centuries had no adequate control over their environment. Floods, famine, fires, and epidemics have only recently gained historians' attention. Historical studies of population changes (demography) have led to the rise of demographic history, that is, a history giving prominence to the causal power of changes in population. Economic depressions in the Renaissance and great waves of social unrest are often now attributed to the depopulating terror of the Black Death and subsequent epidemics. Some historians have linked the Death to the renewed persecutions of Jews, the rise of self-flagellation and chiliastic sects, and the vogue of the "Dance of Death" in art and letters. Still others have drawn our attention to the psychological consequences of mass death on the survivors.* In Meiss's study, these various concerns are joined in a powerful examination of two urban cultures in crisis.

From *Painting in Florence and Siena After the Black Death*

Millard Meiss

The Black Death

During the forties, just after the fall of the oligarchy in Florence and several years before it was overthrown in Siena, both cities suffered seriously from lack of grain. The bankruptcies reduced the financial resources of the cities and weakened the influence of the great Florentine houses of the Accaiuoli and the Bardi in Naples, so that normal imports from that region were not forthcoming. On top of this, crops failed throughout Tuscany in 1346, and they were badly damaged in 1347 by exceptionally heavy hailstorms. Weakened and terrified by these deprivations, which they attributed to divine displeasure, the Florentines and Sienese attempted to appease God. They engaged in collective prayer and public displays of repentance. Giovanni Villani tells of great processions, in one of which, continuing for three days, he marched himself. These had scarcely disbanded when an unimaginable catastrophe struck both towns.

Suddenly during the summer months of 1348 more than half the inhabitants of Florence and Siena died of the bubonic plague. By September only some

---

SOURCE: Millard Meiss, *Painting in Florence and Siena After the Black Death* (Princeton: Princeton University Press, 1951), pp. 64–93.

* See Robert J. Lifton's contributions on Jews and citizens of Hiroshima as "survivors" of man-made slaughters.

45,000 of the 90,000 people within the walls of Florence were still living; Siena was reduced from around 42,000 to 15,000. Never before or since has any calamity taken so great a proportion of human life. The plague struck again in 1363 and once more in 1374, though it carried off far fewer people than in the terrible months of 1348.

The survivors were stunned. The Sienese chronicler Agnolo di Tura tells of burying his five children with his own hands. "No one wept for the dead," he says, "because every one expected death himself." All of Florence, according to Boccaccio, was a sepulcher. At the beginning of the Decameron he wrote: ". . . they dug for each graveyard a huge trench, in which they laid the corpses as they arrived by hundreds at a time, piling them up tier upon tier as merchandise is stowed in a ship . . ."—sights not unfamiliar to modern eyes. For years the witnesses were haunted by memories of bodies stacked high in the streets and the unforgettable stench of flesh putrefying in the hot summer sun. Those fortunate few who were able to escape these horrible scenes by fleeing to isolated places were, like Petrarch, overwhelmed by the loss of family and friends.

The momentous nature of the experience is suggested by the words of Matteo Villani as he undertook after a few years to continue the chronicle of his dead brother:

"The author of the chronicle called the Chronicle of Giovanni Villani, citizen of Florence and a man to whom I was closely tied by blood and by affection, having rendered his soul to God in the plague, I decided, after many grave misfortunes and with a greater awareness of the dire state of mankind than the period of its prosperity had revealed to me, to make a beginning with our varied and calamitous material at this juncture, as at a time of renewal of the world. . . . Having to commence our treatise by recounting the extermination of mankind . . . my mind is stupefied as it approaches the task of recording the sentence that divine justice mercifully delivered upon men, who deserve, because they have been corrupted by sin, a last judgment."

Though the rate of mortality was doubtless somewhat greater among the poor, about half of the upper classes succumbed. Agnolo di Tura's comments about Siena are significant: four of the nine members of the oligarchic priorate died, together with two of the four officers of the *Biccherna*, the *podestà*, and the *capitano di guerra*. The painters were not spared. The plague carried off Bernardo Daddi, the most active and influential master in Florence, and perhaps also his associate whom Offner has called the "Assistant of Daddi." It is generally assumed that the disappearance of all references around this time to Ambrogio and Pietro Lorenzetti signifies their death in the epidemic. Undoubtedly a great many minor painters suffered a similar fate. This sudden removal at one stroke of the two great Sienese masters and of Daddi along with numerous lesser painters produced a more than ordinary interruption in pictorial tradition. It gave to the surviving masters, especially the younger ones, a sudden, exceptional independence and a special freedom for the development of new styles.

Economic and Social Consequences of the Plague

In the immediate wake of the Black Death we hear of an unparalleled abundance of food and goods, and of a wild, irresponsible life of pleasure. Agnolo di Tura writes that in Siena "everyone tended to enjoy eating and drinking, hunting, hawking, and gaming," and Matteo Villani laments similar behavior in Florence:

"Those few sensible people who remained alive expected many things, all of which, by reason of the corruption of sin, failed to occur among mankind and actually followed marvelously in the contrary direction. They believed that those whom God's grace had reserved for life, having beheld the extermination of their neighbors, and having heard the same tidings from all the nations of the world, would become better men, humble, virtuous, and Catholic; that they would guard themselves from iniquity and sins; and would be full of love and charity for one another. But no sooner had the plague ceased than we saw the contrary; for since men were few, and since by hereditary succession they abounded in earthly goods, they forgot the past as though it had never been and gave themselves up to a more shameful and disordered life than they had led before. . . . And the common people (*popolo minuto*), both men and women, by reason of the abundance and superfluity that they found, would no longer work at their accustomed trades; they wanted the dearest and most delicate foods for their sustenance; and they married at their will, while children and common women clad themselves in all the fair and costly garments of the illustrious ladies who had died."

This extraordinary condition of plenty did not, of course, last very long. For most people the frenzied search for immediate gratification, characteristic of the survivors of calamities, was likewise short-lived. Throughout the subsequent decades, however, we continue to hear of an exceptional indifference to accepted patterns of behavior and to institutional regulations, especially among the mendicant friars. It seems, as we shall see, that the plague tended to promote an unconventional, irresponsible, or self-indulgent life, on the one hand, and a more intense piety or religious excitement, on the other. Villani tells us, in his very next sentences, of the more lasting consequences of the epidemic:

"Men thought that, through the death of so many people, there would be abundance of all produce of the land; yet, on the contrary, by reason of men's gratitude, everything came to unwonted scarcity and remained long thus . . . most commodities were more costly, by twice or more, than before the plague. And the price of labor, and the products of every trade and craft, rose in disorderly fashion beyond the double. Lawsuits and disputes and quarrels and riots arose everywhere among citizens in every land, by reason of legacies and successions . . . Wars and divers scandals arose throughout the world, contrary to men's expectation."

Conditions were similar in Siena. Prices rose to unprecedented levels. The economy of both Florence and Siena was further disrupted during these years by the defection of almost all the dependent towns within the little empire of each city. These towns seized as an opportunity for revolt the fall of the powerful Florentine oligarchy in 1343, and the Sienese in 1355. The two cities, greatly weakened, and governed by groups that pursued a less aggressive foreign policy, made little attempt to win them back.

The small towns and the countryside around the two cities were not decimated so severely by the epidemic, but the people in these regions felt the consequences of it in another way. Several armies of mercenaries of the sort that all the large states had come to employ in the fourteenth century took advantage of the weakness of the cities. Sometimes in connivance with the great enemy of the north, the Duke of Milan, they invaded, or threatened to invade, the territory of Florence and Siena. These large marauding bands—the companies of the Conte Lando, of John Hawkwood, and of Fra Moriale, the Company of the Hat and the Company of St. George—hovered in the neighborhood throughout the 'fifties and 'sixties, a continual threat to the peace of the *contado* and a serious drain on the already shrunken communal treasuries. Florence, the stronger of the two cities, was less troubled by them. Siena paid out huge ransoms, in twenty years some 275,000 florins, but even then its lands were pillaged and the people on the farms and in the villages lived in constant terror. For them the story of Job, an old paradigm of trial and affliction, acquired a specially poignant meaning. Not only did Job suffer from a disease whose outward symptoms were like those of the plague; his cattle were driven off and his children killed. It is not surprising that his history, rarely if ever represented in earlier Tuscan panels and frescoes, should now have been portrayed several times. In a triptych dated 1365 by a follower of Nardo in S. Croce he appears as a cult figure at the left of the Madonna, and the entire predella is devoted to his trials. Bartolo di Fredi appended five, or probably originally six, scenes of Job's misfortunes to his Old Testament cycle of 1367 in the Collegiata, S. Gimignano. Another of Bartolo's Biblical stories must likewise have evoked an intense response, stirring bitter memories of dislocated families even while it held out a measure of hope: the scene of the Jews fleeing from Egypt and of the pursuing Egyptian army drowned in the Red Sea. In 1362, just a few years before the painting of these frescoes, Bartolo himself had written a report for the Signoria of Siena describing the movement of marauding troops in the countryside.

The ravages of the mercenary companies accelerated a great wave of immigration from the smaller towns and farms into the cities that had been initiated by the Black Death. Most of the newcomers were recruits for the woolen industry, who were attracted by relatively high wages. But the mortality offered exceptional opportunities also for notaries, jurists, physicians, and craftsmen. In both Florence and Siena the laws controlling immigration were relaxed, and special privileges, a rapid grant of citizenship, or exemption from taxes were offered to badly needed artisans or professional men, such as physicians. Throughout the earlier history of Florence and Siena there had been, it is true, a steady influx of people from the *contadi*. Dante complained of a population swollen with rustics from places such as Signa and—ironically enough—Certaldo, Boccaccio's town. But the magnitude of the immigration into Florence after the Black Death was unprecedented. Those historians who estimate the population of the city as 25,000 or 30,000 just after the Black Death postulate 45,000 in 1351, and others state that, despite the successive epidemics, there were 60,000 people by 1375 and 70,000 by 1380. Siena, on the other hand, grew comparatively little.

In addition to bringing into the city great numbers of people from the surrounding towns and country, the Black Death affected the character of Florentine society in still another way. Through irregular inheritance and other exceptional circumstances, a class of *nouveaux riches* arose in the town and also in decimated Siena. Their wealth was accentuated by the impoverishment of many of the older families, such as the Bardi and the Peruzzi, who had lost their fortunes in the financial collapse. In both cities, too, many tradesmen and artisans were enriched to a degree unusual for the *popolo minuto*. Scaramella sees as one of the major conflicts of the time the struggle between the old families and this *gente nuova*. Outcries against both foreigners and the newly rich, never lacking in the two cities, increased in volume and violence. Antagonism to "the aliens and the ignorant" coalesced with antagonism to the new municipal regime; the government, it was said, had been captured by them. Filippo Villani wrote in his chronicle for the year 1363:

"In those days the administration and government of the city of Florence . . . had passed in part—and not a small part—to people newly arrived from the territory of Florence who had little experience in civic affairs, and to people from more distant lands who had settled in the city. Finding themselves after a time enriched with money gotten in the crafts, in trade, and in money-lending, they married into any family that pleased them; and with gifts, feasts, secret and open maneuvers they put themselves forward so much that they were drawn into public offices and were added to the list of those eligible for election to the priorate."

In the *Corbaccio*, written in 1354–1355, Boccaccio compared the rulers of the city to a person lacking liberal studies, and in a letter of about ten years later he said that the reins of government had been given to people who had come, as we would say, from Kalamazoo—"venuti chi da Capalle e quale da Cilicciavole, e quale da Sugame o da Viminiccio"—and who had been "taken from the mason's shop or from the plow and elevated to the highest office in the state."

The Effect upon Culture and Art

The painting of the third quarter of the century, more religious in a traditional sense, more ecclesiastical, and more akin to the art of an earlier time, may reflect these profound social changes in Florence and Siena, or rather the taste and the quality of piety that they brought into prominence. For these *homines novi*, the immigrants, the newly rich and the newly powerful, had not on the whole been closely identified with the growth of the new culture in the first half of the fourteenth century. They adhered to more traditional patterns of thought and feeling, and it may well be that their ideal of a religious art was still the art of the later thirteenth century—an art that was still visible almost everywhere in the cities, and even more in the churches of the *contado*. At other times, less educated, provincial people have clung to an art that had been supplanted for as long as fifty years, and we are given hints that this was so around the middle of the fourteenth century. Writing in 1361 in his will about a Madonna by the "outstanding" painter Giotto, Petrarch says that though its beauty is a source of wonder to the masters of the art, the ignorant do not understand it. Boccaccio wrote that, as a poet and a student of classical antiquity, he was regarded as a sorcerer in the town of Certaldo. Elsewhere he attacked the noisy adversaries of poetry (primarily ancient poetry) who, ignorant of its character and its true meaning, constantly opposed religion to it. There is no reason to suppose that these ready "theologians" or the friars and other groups whom he excoriates for similar reasons were any less antagonistic to certain aspects of the more advanced style of the Trecento in the figure arts. We cannot expect them to have proved anything but hostile to Ambrogio Lorenzetti's fascination with ancient sculpture. One of the known objects of his fascination, a newly discovered statue by Lysippus, was broken into bits by frightened members of the Sienese public. It may be that these same groups were unsympathetic also to the assimilation of ancient forms in the work of Ambrogio or Cavallini or Nicola Pisano. We cannot then suppose them to have been much more friendly to all those dispositions and qualities of the new style that lie behind such assimilations—in other words, to the essence of the new style itself. Because known comments on painting and sculpture in the fourteenth century are few, and limited mostly to scholars, writers, and the artists themselves, such attitudes are difficult to ascertain. It is notable, however, that Fra Simone Fidati seems to disapprove, by implication at least, the warm relationship between man and wife in paintings such as Giotto's. And in the *Meditations on the Life of Christ*, oddly enough in the very treatise that exercised so great an influence on painting, we find the old antagonism to the arts themselves reasserted. Except for their employment in the cult, they are, the Franciscan writer says, stimulants of a "concupiscenza degli occhi."

Whether or not conservative tastes of this sort were felt by the painters at the moment of decisive change and innovation around the middle of the century, their wider diffusion and their prevalence among people who had acquired wealth and influence as patrons at least broadened the acceptance of the new styles and helped to maintain them. A painting such as Giovanni del Biondo's St. John the Evangelist trampling on Avarice, Pride, and Vainglory may thus have struck an immediate response because of its older symbolic pattern and the austere frontality of its chief figure. The panel was doubtless widely approved for its subject, too, particularly the representation of the conquest of avarice. At this time avarice was, as we have mentioned, associated especially with the unlimited desire for profit and accumulation that had already become characteristic of early capitalism. Neither this motive nor the new economic system had been accepted by a considerable part of society. They were often opposed, it seems safe to assume, by the same people who were conservative in other spheres also, particularly religion and the arts. These people clung to the mediaeval belief in a just price and the limitation of wealth to the amounts needed for a livelihood by persons in the several stations of life. Their antagonism to the new system and its values was certainly more assured after the bankruptcies and other events of the 'forties. Segre states that in 1354 twenty-one Florentine bankers were fined for usury, and in tracing the fluctuating policy of Florence on the lending of money at interest, he implies a greater opposition to loans in the second half of the fourteenth century. At this time many people undoubtedly took a special pleasure in the sight of a defeated Avarice sprawled on the ground and ignominiously trodden underfoot.

Though the *gente nuova* was doubtless pleased by this portrayal of the triumph of St. John, it is questionable whether any of them played a leading part in the selection of the painter or the determination of the subject. The work was actually commissioned, as we have seen, by the *Arte della Seta*, one of the seven greater Florentine guilds. Though within it, as in all the other guilds, there were during this period certain shifts of responsibility and power, it was still dominated by established merchants and industrial entrepreneurs. These were, in fact, the chief patrons of painting at the time, just as they had been earlier in the century. It was the Tornaquinci who commissioned the frescoes by Orcagna in the choir of S. M. Novella, the Strozzi who employed Orcagna and Nardo for the altarpiece and murals of their chapel, one of the Guidalotti who provided the funds for the frescoes in the Spanish Chapel, though they were executed after his death. But, as we shall see clearly in the instance of Giovanni Colombini, these wealthy patricians were deeply affected, too, by the events of the 'forties and 'fifties. They had never freed themselves of doubt as to the legitimacy of their occupation. The Church had not explicitly sanctioned either their activities or their profits, and "manifest" usury it strongly condemned. The distinctions between usury of this kind and permissible interest, while precise, were very fine, and in advanced age merchants and bankers quite commonly sought to quiet their conscience and assure their salvation by making regular provision for corporate charity and by a voluntary restitution of wealth that they judged was improperly acquired. As the most eminent living student of Florentine economic history has written: "Reading the last wills of the members of the Company of the Bardi . . . there becomes evident and dramatic the contrast between the practical life of these bold and tenacious men, the builders of immense fortunes, and their terror of eternal punishment for having accumulated wealth by rather unscrupulous methods."

The guilt of the entrepreneurs and bankers was greatly quickened around the middle of the century. The bankruptcies and the depression struck just as hard at their conscience as at their purse. The loss of exclusive control of the state and the terror of the Black Death seemed further punishments for dubious practices. Giovanni Villani attributed the failure of the companies of the Bardi and the Peruzzi, in the second of which he himself had invested money before he joined the Buonaccorsi, to the "avarizia di guadagnare," and his strictures imply a condemnation not so much of poor judgment in this single instance as of all financial practices that involve great risks. Since few large ventures of the time could be undertaken without such risks, Villani comes close to turning his back on capitalism itself, and he had earlier signified that money-lending was a disreputable occupation.

In the wake of the disasters of the 'forties and 'fifties, then, the established wealthy families of Florence very probably welcomed a less worldly and less humanistic art than that of the earlier years of the century. Their own position seriously threatened, they felt sustained by the assertion in art of the authority of the Church and the representation of a stable, enduring hierarchy. Their taste would have tended to converge, then, with that of the *gente nuova*, who emerged from circles that still clung to a pre-Giottesque art in which very similar qualities inhered.

All sections of the middle class were, in any event, clearly united in their desire for a more intensely religious art. It was not only great merchants such as Giovanni Colombini, with their own special reasons for remorse, who vilified and abandoned their former mode of life. Masses of people, interpreting the calamities as punishments of their worldliness and their sin, were stirred by repentance and religious yearning. Some of them joined groups that cultivated mystical experiences or an extreme asceticism; many more sought salvation through the traditional methods offered by the Church. Several painters—Orcagna, Luca di Tommè, Andrea da Firenze—were, as we have seen, moved to a greater piety or a mystical rapture. Others who were less deeply stirred saw fit nevertheless, as the craftsmen of images for the cult, to adopt many of the new values and the new artistic forms, though in a more conventional way. For the painting of the time this religious excitement, and the conflict of values which it entailed, was a crucial cultural event. Hitherto little studied, it will be discussed at some length in the following chapter.

GUILT, PENANCE AND RELIGIOUS RAPTURE

The years following the Black Death were the most gloomy in the history of Florence and Siena, and perhaps of all Europe. The writing of the period, like the painting, was pervaded by a profound pessimism and sometimes a renunciation of life. "Seldom in the course of the Middle Ages has so much been written," Hans Baron has said recently, "concerning the 'miseria' of human beings and human life." Though religious thought throughout the Middle Ages had dwelt on the brevity of life and the certainty of death, no age was more acutely aware of it than this. It was preached from the pulpits, most vividly by Jacopo Passavanti, and set forth in paintings, both altarpieces and murals. In the predella below the Madonna by Giovanni del Biondo in the Vatican there is a representation entirely unprecedented in Tuscan art: a decayed corpse, consumed by snakes and toads. A bearded old hermit points to it with an admonishing gesture while a man and his dog recoil in terror. In the great fresco in the Camposanto at Pisa, painted probably around 1350, Francesco Traini portrayed with intense feeling the suffering of the sick, the horror of rotting flesh, and the sudden, unpredictable coming of death. In part this fresco raises to a monumental scale the theme of the meeting of the quick and the dead which had already been represented in the thirteenth and earlier fourteenth centuries. To this it joins an apparently unprecedented scene of death swooping in like a bird of pray upon its victims. The same scene was painted shortly afterward by Orcagna in S. Croce, Florence, as part of a cycle containing, as at Pisa, the Last Judgment and the tortures of Hell. Only a fragment of Orcagna's scene survives, showing several corpses and a few miserable creatures half alive who vainly implore Death to end their suffering. The cripples fix their eyes upon Death with a marvelously rapt attention, and even the blind man senses its presence.

The style of Traini is essentially unrelated to that of Orcagna and his period, associating itself rather with the art of the earlier fourteenth century. But at this time he, like Orcagna, was engrossed with the actuality of suffering and death in the world, and although he added to his fresco an idyllic scene of the eremitic life, the *vita contemplativa* that banishes the fear of death and secures a triumph over it, it is the triumph of death that is painted with the greater urgency and hence gave its name, borrowed from Petrarch, to the entire composition.

The incidence of the plague, an undeniable triumph of death, was interpreted in diverse ways. Some writers attributed it to astrological influences (the conjunction of the planets), others to climatic conditions (the corruption of the air). But far more common was the belief that, like the Biblical flood, the Black Death was caused by the moral corruption of man and the ensuing wrath of God. This was the opinion, transcending all others, of the chroniclers, of poets and scholars such as Petrarch and Bartolus, of the physicians who wrote treatises on the disease, and of course of the very devout. It is reflected in several novel scenes of the Last Judgment painted in the wake of the epidemic.

In Tuscan representations of the Last Judgment in the late thirteenth century and the first half of the fourteenth, Christ addresses both the blessed and the damned, welcoming one group and rejecting the other. Sometimes he proclaims his will simply by the relative height of his hands—the right higher for the blessed, the left lower for the damned. In other works, however, including Giotto's fresco in Padua, this symbolism of position is disregarded, and he lowers an open hand toward the blessed while extending an averted one toward the damned. His lowered hand expresses acceptance or welcome, and it is often so close to the blessed that it seems to suggest an active assistance. In most of these representations Christ manifests a special concern for the blessed by turning his eyes toward them, or, as in Giotto's fresco and one or two other works, his head or even his body.

In several compositions after the middle of the century, probably beginning once again with the fresco cycle in Pisa, the attitude of Christ is radically different. For the first time in the representation of the Last Judgment he addresses the damned alone, turning on them with an angry mien, his arm upraised in a powerful gesture of denunciation. The Virgin alongside him is also wholly preoccupied, though more compassionately, with the damned, and the apostles do not sit as tense and impartial witnesses, but are moved to pity and to fear by the awesome sentence. One of the archangels at the center of the composition expresses the inevitability and impartiality of the judgment, another cringes, terror-struck with a more human consternation.

Christ appears as a similar angry, denouncing figure in other paintings of the period: a miniature in a Laudario in the Biblioteca Nazionale, Florence, probably painted during the 'fifties, a panel by Niccolò di Tommaso in the Larderel collection, Livorno, and Nardo di Cione's fresco in the Strozzi Chapel, S. M. Novella, though in the latter he permits himself also a rather mild gesture of benediction. In some paintings of a later period, too, Christ is moved to denunciation alone: several panels by Fra Angelico, Giovanni di Paolo's predella in the Siena Gallery, and Michelangelo's fresco in the Sistine Chapel. But these later works do not diminish the significance of the appearance of the action around the middle of the fourteenth century. Like several other novel forms of this period, once created it survives into a later time, retaining in varying degrees its original expressive purpose.

In Italian and northern art of the fifteenth century the conception of an aroused God punishing mankind by pestilence often assumed the form of Christ hurling arrows at the world, like the thunderbolts of Jove. Already at the end of the thirteenth century Jacopo da Voragine had related that when St. Dominic was in Rome he saw Christ in the heavens brandishing three lances against mankind, fully resolved to destroy it because of the prevalence of pride, avarice, and lust. Though this image of the Saviour directing weapons at the world did not, so far as I know, appear in Tuscan art of the Trecento, arrows as symbols of pestilence appear in representations of St. Sebastian. This saint, who had been martyred with arrows, was invoked as an intercessor against plague as early as the seventh century. His cult was unimportant, however, at least in Tuscany, until after 1348. A few years later a relic was brought to Florence from Rome, and his figure began to appear frequently in Florentine painting. His martyrdom was painted also, first perhaps in panels by Giovanni del Biondo and Andrea Vanni. Giovanni's conception is unique. In representations painted later, at the very end of the fourteenth or in the fifteenth century, from three to around fifteen arrows puncture the body of the saint. In Biondo's painting, however, the saint is riddled by dozens of them. He is a relentlessly tortured figure, blood dripping from more than thirty wounds.

In the years following 1348, Florence and Siena were swept by reports of the imminence of a new disaster or the appearance of Anti-Christ. Dire prophecies were uttered. A Franciscan Tertiary, Tommasino da Foligno, proclaimed that new calamities would be wrought by an angry, unappeased God. The pseudo Jacopone da Todi foretold the beginning in 1361 for the "duro male." One Daniele, a Minorite, predicted in 1368 that ten years later there would be "great fears and horrors," that the "little people" would kill all tyrants, that Anti-Christ would appear, that the Turks, Saracens, and other infidels would ravage Italy, and that men would be assailed by tempests and floods, hunger and death. We are not surprised to learn that, in these circumstances, such prophecies were widely believed. Doubtless their primary audience was the uneducated and the poor, to whom the words of Daniele, referring to an uprising of the workers, would even give some hope. But many who in less frightening times would have scoffed listened with heightened anxiety. In any event, the number of sceptics was very shortly reduced, for the plague struck again in 1363 and once more in 1374.

Churches and religious societies were showered with bequests willed to them by people dying, or expecting to die, of the plague. Some of these funds would of course have come to them over the years in any event, but the unusual size of the donations and their concentration all at one time resulted, at least in Florence, in an unprecedented accumulation. Villani estimates that the Company of Or San Michele, a society with religious, social, and philanthropic functions, received the huge sum of 350,000 florins, attaining such spectacular wealth that memory of it was still alive in Vasari's day; he refers to it in his Life of Orcagna. A considerable part of this money was, in fact, given to Orcagna for the splendid marble tabernacle that he made for the church of the fraternity. Ghiberti, who otherwise does not speak of costs in his biographies of the Trecento masters, says that the company paid 86,000 florins for this work. Some of the funds held by the company may have been applied to the frescoes in S. Croce by Giovanni da Milano and the Rinuccini Master. During the height of the epidemic S. M. Novella received from Turino Baldesi the large sum of 1000 lire (or florins) for the painting of the "story of the entire Old Testament from beginning to end." It was perhaps at this time, too, that the Tornaquinci gave to the same church funds for the painting of the choir, undertaken shortly afterward by Orcagna. Like the Company of Or San Michele, the Company of the Bigallo was greatly enriched by the epidemic and in 1352 it undertook to build a new home, the Oratorio that still stands. New hospitals were founded with gifts from wealthy individuals—S. Caterina dei Talani, for instance, in 1349.

Despite the ensuing economic stagnation and the inflation, such gifts did not cease. In the years immediately following the Black Death Buonamico di Lapo Guidalotti, a Florentine merchant, whose wife had died of the plague, gave a considerable portion of his fortune to the Dominicans of S. M. Novella for the construction of a new chapter-house (the "Spanish Chapel"), an unusually costly enterprise for one man. Just before his death in 1355 he added 325 florins, and then 92, for the covering of its walls with frescoes. These were, as we have seen, executed in 1366–1367 by Andrea da Firenze.

In Siena the bequests of the stricken were smaller, and they apparently were not followed by large donations in subsequent years. The city was impoverished to a greater extent than Florence and its economy recovered very slowly, if ever, from the calamity. Barna and Bartolo di Fredi, the chief mural painters of the period, found work outside Siena, in S. Gimignano and Volterra. Construction of the great new cathedral was halted abruptly, never to be resumed. The *operaii* and most of the masters employed on the work had died in the epidemic, and as Agnolo di Tura tells us, there was no wish to begin again "because of the few people that remained in Siena, and also because of their melancholy and grief." Though this fantastically ambitious project had to be abandoned, the fervor of the populace expressed itself in more modest construction. Agnolo di Tura says that the survivors began the Cappella del Campo as an offering to the Virgin and that they constructed the churches of S. M. delle Grazie and S. Onofrio as well as "molti altri oratori e luoghi divoti" throughout the town.

The general feeling of fear, guilt, and sorrow sought expression in other ways also. The pope was besieged with requests from all stricken countries for a "perdono generale" and for the designation of 1350 as a Holy Year. Though in 1300, when the jubilee was instituted, a holy year every hundred years was envisaged, Clement VI decided to proclaim one in 1350, and special indulgences were granted pilgrims to Rome. The number of these was extraordinary. Matteo Villani wrote that on holy days there were in the city as many as a million visitors. This figure is not idly speculative, for Matteo shared with his brother Giovanni the characteristic Florentine interest in quantities and precise calculation. In this instance he goes on to report the number of visitors at various times during the year as well as their places of origin, and he records regretfully the excessive prices of food and lodging, chiding the Romans for cheating the pilgrims at every turn.

The throngs of pilgrims were preceded and accompanied by great congregations of Flagellants. These zealots were addicted to a form of self-chastisement which, from the mid-thirteenth century on, had occasionally swept through the Italian towns. On this occasion they attracted unusually large numbers of adherents not only because self-punishment offered a very tangible kind of penance, an immediate release from an oppressive guilt, but because they claimed a special heavenly intervention in their favor. They possessed a letter which they said was delivered by an angel on Christmas day, 1348, shortly after the pestilence had subsided. It was addressed to the Flagellants by the Virgin Mary, and in it she stated that she had obtained from Christ a pardon of all their sins. Fortified with this document, groups of them set out for Rome, gathering converts along the way as they engaged in their frenzied but methodical practice of whipping themselves, all of them together, twice a day and once during the night for periods of thirty-three days on end.

These violent exhibitions often led to public disorders, accompanied by a display of bitter anticlerical feeling. The town councils tried to control them, and in October 1349, Pope Clement VI issued a bull prohibiting the Flagellants from assembling. By 1351 they were dispersed. Their practices persisted, however, though on a less spectacular scale. In Florence societies of *battuti*, the first of which had been founded around 1330, multiplied during the latter part of the century, and religious leaders, such as St. Catherine, scourged themselves privately.

A second group of zealots, confronting the Church with a far more serious and lasting challenge, likewise increased in numbers and influence from the 'forties on. These were the Fraticelli, dissident Franciscans who descended from the

Spirituals of the thirteenth and early fourteenth centuries. The Spirituals were committed to a strict interpretation of the ideas of St. Francis, especially with relation to property. They maintained that Christ and the apostles had had no possessions either individually or in common, and that therefore a denial of poverty was a denial of Christ. Though in the early fourteenth century this doctrine was advocated, with qualifications, by the general of the Franciscan Order, Michael of Cesena, and by other Franciscan leaders such as William Occam, Pope John XXII declared it heretical in a bold and decisive series of bulls issued from 1318 to 1324. Thereupon most of the Spirituals, including those of Tuscany, fled to regions where they would be less accessible to papal authority. But their ideas persisted—even Fra Simone Fidati, preaching in S. Croce in the 'thirties, showed sympathy with them. The extremist friars, now called Fraticelli, increased greatly in number from the 'forties on, especially in Florence. This city, in fact, became a center of the sect in the third quarter of the century.

The rise of the Fraticelli was facilitated by the "Babylonian Captivity." The residence of the popes in Avignon, together with the corruption of the Curia, weakened ecclesiastical control in Italy. The authority of the Papacy and the Church, along with that of all institutions, was shaken also by the Black Death and the developing social conflicts. The Fraticelli, preaching a doctrine that had been declared heretical, now boldly claimed that Pope John XXII and his successors were heretics themselves. They rejected the sacraments offered by priests other than their own. Under the influence of the urban social and economic crisis, their original conception of the sinfulness of property tended to shift to a belief in the sinfulness of the rich and the justice of taking from them to give to the poor. These ideas were especially attractive to the oppressed woolworkers and the poorer artisans and shopkeepers; they provided a religious stimulus and sanction for their struggle for political rights and economic betterment. It is not surprising therefore that the great merchants, bankers, and industrialists, once again in complete control of the government of Florence in 1382 and eager to please the pope, who protected their commerce abroad, immediately had the old laws against heresy reenacted. And in 1389 one of the Fraticelli, Michele da Calci, was burnt at the stake.

The ideas of an extremist group such as the Fraticelli were not likely to have affected directly an art so closely connected as that of the fourteenth century with the Church. Their own houses were poor and impermanent, and they did not, so far as we know, commission altarpieces and frescoes. On the other hand, their zeal undoubtedly quickened the religious life of the age, and I believe that we can discern their influence upon painting, no less important because it is indirect and in a sense even negative. The challenge which they, and to a lesser extent the Flagellants, presented to the ecclesiastical institutions provided one more stimulus for that affirmation of the authority of the Church which we have discovered in the painting of the time, and of which we shall speak again later.

These heretical or heterodoxical movements represent only one—and a secondary—aspect of the religious life of the time. The majority of the people expressed their devotion within the traditional patterns established by the Church. In Florence they were inspired chiefly by the great Dominican preacher, Jacopo Passavanti. Passavanti, who was born in Florence at the beginning of the century, became a preacher at S. M. Novella in 1340 after a period of study and teaching in Paris and Rome. From the late 'forties on he was in charge of maintenance and construction, and from 1354 to his death in 1357 he served as prior. During this period the new church was completed, the Spanish Chapel was built, and funds were received for its frescoes. Orcagna, furthermore, painted the altarpiece of the Strozzi Chapel and Nardo its walls, and the former may also have undertaken the mural cycle in the choir. Much of this activity was due to Passavanti's influence, and all of it was under his supervision. The existence of iconographic similarities between the paintings in the Strozzi and Spanish Chapels is therefore not surprising, even though the latter were not executed until some years after his death.

Passavanti's thought has been preserved in a treatise called "Lo Specchio di Vera Penitenza." In it he brought together in a systematic way ideas expressed in his sermons, especially those delivered in 1354. They center on the necessity, or rather the urgency, of penitence, for which he accumulates evidence among the writers of the past, and which he makes vivid for his contemporaries by numerous "esempi," brief but highly colored images or stories interspersed through his more formal discourse. His thought and his mood are closely related to contemporary painting. Man appears a perpetual sinner, hovering in the shadow of death and judgment. His only hope lies in penitence, a mordant contrition that continually torments the spirit. Recalling the common experiences of 1348, Passavanti describes for his audience the frightful decay of human flesh, more odorous, he insists, than that of putrefying dogs or asses. He tells of sceptics who, uttering words of doubt, were struck by lightning and immediately reduced to ashes. He follows them to hell, lingering with their tortures and listening for their anguished shrieks, as though he were under the spell of Orcagna's fresco in S. Croce, where the faces of the damned are torn by the claws and teeth of wildly sadistic devils. Passavanti tries to convey by calculation the incalculable endlessness of their doom. "One day," he says "a French nobleman wondered if the damned in hell would be freed after a thousand years, and his reason told him no; he wondered again if after a hundred thousand years, if after a million years, if after as many thousands of years as there are drops of water in the sea, and once again his reason told him no."

To Passavanti sin is a daemonic power and life a relentless struggle against it, grimly maintained because of fear of something worse. And in many paintings of the time, devils, the embodiments of guilt, become more prominent, more aggressive and more vengeful, attacking the lost souls in Orcagna's Hell, menacing the patriarchs in Andrea's Limbo and the unrepentant thief in Barna's Crucifixion, the most intense image of guilt and fear in the art of the fourteenth century. In the fresco of Limbo one devil munches ecstatically on a human head, and in Orcagna's Hell a group of them lacerate the faces and bodies of their victims. For Giotto, on the other hand, hell is a place more of disorder than of excruciating pain. The damned and all the devils except Satan are reduced in scale. Hell is more remote and less substantial than heaven. Indeed Giotto's outlook, serene and assured, differs from that of these later masters in much the same way as Giotto's Dominican contemporary Domenico Cavalca (ca. 1270–1342) differs from Jacopo Passavanti. Of these two preachers De Sanctis wrote: "Cavalca's muse is love, and his material is Paradise, of which we get a foretaste in that spirit of charity and gentleness which gives his prose so much softness and color. But Passavanti's muse is terror, and his material is vice and hell, pictured less from their grotesque and mythological sides than in their human aspects, as in remorse and the cry of conscience. . . ." Can we avoid thinking of Giotto and Orcagna?

Many of Passavanti's "esempi" are drawn from the eremitic life, a life that was actually exemplified in Florence at this time by the Beato Giovanni dalle Celle. Born of a noble Tuscan family around 1310, Giovanni entered the monastery of Vallombrosa before 1347, became abbot of S. Trinita in Florence in 1351, and shortly afterward retired to the "eremo delle celle" above Vallombrosa. Here numerous disciples gathered around him, among them Agnolo Torini, author of a treatise on human misery.

Giovanni dalle Celle extended his influence beyond the circle of Vallombrosa to all of Tuscany by an indefatigable correspondence. His letters were preserved and copied, and his reputation became so great that they continued to be read in the Quattrocento as models of sanctity and pious asceticism. He addressed these letters, moral and hortatory in character, to friends and acquaintances, some of whom were members of the Florentine government, and he communicated with all the Tuscan religious leaders of the time—Colombini, St. Catherine, William Fleet, St. Bridget (in Rome), and even the Fraticelli. He tried to induce the latter to confess to their presumptuousness and to renounce their heretical convictions.

As far as his own life was concerned Giovanni was uncompromising—"we enter the world," he said, "by dying." Although he was also a firm supporter of the Church, he was tolerant and discreet in judging others. His ardor was tempered by rational analysis and by consideration of the practical consequences of religious beliefs and acts. He was wary of ecstasy and visions, attributing them to pride. On the other hand, he showed sympathy if not enthusiasm for Colombini and his followers, counseling them to ignore the scorn of their opponents and assuring them in the words of the Gospels, Job, and Seneca, as well as in eloquent phrases of his own that poverty is the only way to salvation. He admired St. Catherine and attacked her opponents; occasionally, however, his views differed sharply from hers. At the time of the War of the Eight Saints (1374–1378) Catherine declared that the Florentines were sinful to take arms against the pope, but Giovanni felt they were justified in defending their "patria" against oppression, as he put it, and he did not insist on strict observance of the papal interdict, holding that such decrees were applicable only to mortal sin. On one occasion he even countered Catherine's advice. She had urged a young Florentine follower, Suora Domitilla, to join a crusade. Hearing of this, Giovanni wrote a letter and cautioned her from undertaking the journey, pointing to the constant temptation to which travel with young men would expose her. He added: "If you have Christ in the sacrament of the altar . . . why would you abandon him to go see a stone?"

These differences between Giovanni and Catherine arise from dissimilar personalities and religious conceptions. In the relationship of these two figures as well as of the Florentine Passavanti with the Sienese Bianco and Colombini we sense differences in the religious culture of the two cities that correspond to those in the realm of art, at this time as well as others. It is true that Giovanni dalle Celle had a sort of lesser counterpart in the region of Siena, but he was an Englishman, William Fleet, a hermit who settled at Lecceto apparently in 1362 and who acquired a great reputation for piety and learning. He held a degree from Cambridge, and was respectfully known as "il Baccelliere." Like Giovanni, with whom he corresponded, Fleet took an active part in the religious movements of the time, and became a disciple of St. Catherine. The two never met, but they felt they knew each other "per istinto di Spirito Santo." It is noteworthy that when Catherine's opinions were contradicted by Giovanni dalle Celle, Fleet immediately sided with her, and made angry protests to Giovanni.

At Lecceto, too, another voice was raised a few years later which was reminiscent of Passavanti—that of the Augustinian Fra Filippo. It is true also that the devotion of all the leaders of the time, in Siena as well as Florence, was distinguished by strenuousness, excitement, and a sense of urgency. But there remains nevertheless a significant difference between the stern, terrifying but logical preacher of S. M. Novella or the discreet speculative hermit of Vallombrosa and the impulsive, ecstatic and more essentially mystical figures of Siena. The Florentines exhibit self-denial, the Sienese, self-abandon. In religion as in art the Florentines inclined toward rationality and speculation, the Sienese toward an immediacy of sentiment and emotion.

Historians, even of religion, have paid little attention to the Beato Giovanni Colombini. He has remained in the shadow of his far greater and rather like-minded Sienese contemporary, Catherine, but his life gives us an uncommonly clear insight into many aspects of the religious sentiment of this troubled time. Colombini was a wealthy Sienese merchant. He held high political office, probably that of prior of the commune. In 1355, apparently in July and thus just two months after the fall of the government of the great merchants and bankers, Colombini suddenly dedicated himself to a different life. Haunted by guilt, he gave all his possessions to the poor, to the convent of Santa Bonda, and to the Hospital of the Scala. His wife, eminently pious but not, in the salty words of his Quattrocento Florentine biographer, Feo Belcari, "equally in love with poverty," reproved him for his abandon. When Colombini remonstrated that she had always prayed for a greater charitableness on his part, she replied with a burgher's pungent sense of the proper and practical mean: "io pregavo che piovesse, ma non che venisse il diluvio" (I prayed for rain, but not for the flood).

Colombini, wishing to demean his former mode of life and to humiliate himself publicly, undertook manual labor in the Town Hall where formerly he had sat with the rulers. Together with the first of his followers, fellow members of the oligarchic party of the *Nove*, he carried wood and water, cleaned the stairs, and helped in the priors' kitchen. He had himself whipped in front of the Town Hall, and he sought to expiate his sense of guilt about his former business practices by asking his companions to shout accusations at him in the great piazza that he had sold poor grain high and bought good grain low. To growing audiences, who watched him riding about the city on an ass, he advocated democracy and, like the Fraticelli, the elimination of the privileges of the wealthy. Neither the Church nor the ruling council of Siena, though now dominated by the smaller merchants and bankers together with shopkeepers and artisans, were sympathetic to these views. The government soon condemned him to perpetual exile as "a dangerous innovator, who might ruin the peace of families and cause a rebellion among the populace." Colombini went to Arezzo, where one of his followers was hung as a heretic. When the plague appeared for the second time, however, in 1363, so many Sienese interpreted it as a sign of God's displeasure with the ban that it was hastily lifted and Colombini was recalled to the city.

Colombini continued to be opposed by both civil and ecclesiastical authorities, but the number of his followers increased in Siena and in the other Tuscan towns, Florence among them, where he preached. They all lived in a state of exaltation, identifying themselves with Christ and St. Francis, and praying God to grant death and martyrdom. They urged the populace to shoulder the Cross of Christ—"prendiamo la croce di Cristo," Colombini wrote in his letters. This is the exhortation of Christ himself in a unique group of contemporary paintings. In them the Child holds a cross and sometimes a scroll on which is inscribed the passage from the Gospels: "If any man will come after me, let him deny himself, and take up his cross, and follow me."

As Colombini and his followers moved about they sang mystical *laudi*. Some of the songs were composed by members of the band, and those of Bianco da Siena, who was the most ardent and the most eloquent of the disciples, have taken a place in the religious literature of the century. Though the group prayed for long hours, Colombini rejected the mass, citing as predecessors in this respect Mark, Anthony, and Paul. He was intolerant of intermediation and of formal worship. Like the extreme Spirituals and the Fraticelli, he was opposed to books and learning. A certain Domenico, to whom he had written of mystical love, replied: "I understand clearly by your letter that all sciences, natural, ethical, political, metaphysical, economic, liberal, mechanical . . . are merely a dark cloud over the soul. . . ." Colombini said that he was "burning with the love of the Holy Spirit"; the core of his faith was the conviction that the "fire of love" would renew the world. The wide acceptance of these views finally reversed the position of the Church. In 1367, a few days before Colombini died, his activity was approved by Pope Urban V, despite the hostility of members of the Curia. He was welcomed as the founder of a new lay congregation, called the Gesuati, and his followers became its first members.

Colombini's life was to some extent a model for the youthful Catherine Benincasa, the last and the greatest religious figure of the period, and the only one, too, who is generally known, so that our account may be proportionately brief. Like Colombini and the other leaders whom we have discussed, her religious impulses were stimulated by an acute sense of personal and human misery, deepened by the plague, inflation, and poverty, the devastation of the countryside by marauding condottieri, and social and political conflicts in the commune. Born the youngest of the twenty-five children of a Sienese dyer, she was only an infant at the time of the Black Death, but we know that at least one of her brothers died of the "mortalità" in 1363, two others in 1374, together with a sister and with eight of the eleven grandchildren in her mother's house. She buried all the latter herself. During the epidemics she was one of the few who moved about the city, helping to bury the dead, comfort the stricken, and on one occasion she was said to have cured a dying victim—the rector of the Hospital of the Scala. Her family acquired political influence and a measure of prosperity in 1355 by the victory of the *Dodicini*, the political party to which her brothers belonged; but this party was driven from power thirteen years later, her brothers became the objects of attack, and Catherine, who by then had become an inviolable person, had to conduct them through those parts of the city where their enemies were concentrated.

Catherine devoted herself at an early age to strenuous ascetic practices. At six or seven she began to have visions, and she showed so much independence and so firm a conviction of her religious insight and her mission that, like Colombini, she incurred the hostility of several ecclesiastical authorities in Siena. Though she became a Dominican tertiary in 1363–1364, she wanted to live outside any fixed rule. Like Colombini she formed a *cenacolo* or lay society, one member of which, Cristofano Guidini, wrote the first biography of Colombini. The rule of her group, she said, came straight from God.

Catherine's devotion, especially during these early years, assumed an intensity greater even than Colombini's. To Andrea of Lucca she wrote: "Drown you in the Blood of Christ, and may our own will die in all things." The Blood was an obsessive symbol for her, and almost every letter begins: "I write thee in the precious blood of Jesus Christ." She enjoined her friends and followers to sacrifice themselves for the sins of humanity, and indeed one of the accusations of her opponents was that she had undertaken to do penance for others. Though she believed, with Passavanti, that sin had to be overcome by strenuous striving, it was not, as for him, a constant, oppressive, tormenting threat. She thought that knowledge of self, which everyone could gain by assiduous introspection, would reveal love. And love rather than penitence was essential to perfection.

Some of the ideas which Catherine advanced in her youth made her, as we have mentioned, a controversial figure. Her visions were mistrusted, and she was openly attacked by several friars and clerics in positions of authority. In 1374 she was called to S. M. Novella in Florence for interrogation by the provincial chapter of the Dominican Order. After an examination, the record of which has been lost, she was assigned a spiritual director, Raymond of Capua, a learned Dominican, who remained with her till her death. From about this time on, Catherine's career became more public, and her activities reached beyond Siena to Rome, Avignon, and the whole of Italy. Opposed now to mere asceticism and contemplation, and preaching a militant Christianity, she advocated a crusade in the east, and attempted to persuade the condottiere John Hawkwood to abandon his freebooting for generalship of the Christian forces. She tried to end the war between Florence and the pope, and her persistent attempt to induce the latter to quit Avignon for Rome is a familiar chapter in European history.

St. Catherine, Giovanni dalle Celle, Colombini, Passavanti and several of their more inspired adherents compose an exceptional group of religious leaders. There was no comparable group in Tuscany at any other period within the fourteenth century; indeed, they made the two decades immediately after the plague one of the most important moments in the religious history of central Italy. They communicated their ardor to a people that had already been made fervent by the trials and disasters of the time. We feel their influence in the spiritual intensity of contemporary painting, in its reversion, not without ambivalence and conflict, from the natural and the human to the unnatural and the divine. The mystical rapture of Colombini and Catherine is more evident in the painting of Siena; in Florentine art there is more of Passavanti's grim penance, of his insistence upon the curative power of good works and participation in the rites of the Church.

The zeal of these religious figures affected the Church and the orders as well as the laity, and it embodied itself in a series of lasting institutional innovations and reforms. We have already seen that the movement led by Giovanni Colombini, similar to the Spiritual Franciscans in character, was eventually recognized by the Church and in 1367 established as the Order of the Gesuati. St. Catherine provided the initial and the major stimulus for the reform of the Preaching Friars. In the later fourteenth century many Dominican as well as Franciscan houses had become lax, partly because of the plague, which promoted an indifference to religion and to ecclesiastical authority among some people at the same time that it stimulated a more intense devotion in others. Catherine induced her order to commit several monasteries to strict observance. Their number was greatly increased by a formal program of reform launched shortly afterward by her spiritual director, Raymond of Capua, whom she had deeply influenced and who became general of the order in 1390. He was aided in this work by Giovanni Dominici, who began his novitiate at S. M. Novella the year Catherine was in Florence.

Among the Minorites a similar movement of reform gained momentum and led to an equally important institutional innovation. Since their suppression by Bonaventura in the late thirteenth century, the Spirituals had been striving for recognition and for the grant of houses where they might live according to their own beliefs. In 1334 one group was permitted to live apart, and in 1352 four additional hermitages were granted, into which even a few of the Fraticelli came. In 1368 some of the more moderate under Paolo da Trinci were formally recognized by the Franciscan general as a separate section of the order and they founded the first monastery of the Observants or Osservanti at Brugliano in Umbria. There they adhered more closely to the original rule of St. Francis, without reference to subsequent papal interpretations and dispensations. In 1374 the general permitted them to form communities outside Umbria, and their houses multiplied rapidly during the later fourteenth and fifteenth centuries.

During the period when they lived as marginal members of the Franciscan order the Spirituals maintained close relationship with the very numerous devout laity existing within the Third Order, and with the anticlerical and antipapal Fraticelli. After the recognition of the Osservanti in 1368, the Fraticelli began to lose their influence. Some of them compromised their views and joined the new sect; the rest were persecuted as heretics by the Osservanti themselves.

It is evident that this broad movement of reform was not caused simply by laxity in the orders and the desire of their leaders for a stricter observance. In the 'sixties and 'seventies the Church and the mendicant orders felt the necessity of redirecting and absorbing the intense religious impulses of the time, many of which had assumed forms independent of, or hostile to, ecclesiastical institutions. Thus in 1367 Pope Urban V, against the advice of several cardinals, transformed Colombini's group into the Gesuati. The moderate Spirituals were accepted as the Osservanti in 1368. In 1370 the pope confirmed another new order, likewise with a strict rule—that of St. Bridget, the Swedish mystic who had been living in Rome since the end of the 'forties. And a few years later Catherine began the reform of the Preaching Friars. These institutional innovations and changes are in themselves ample proof of the scope and intensity of the religious movements of the preceding years.

The role of the mendicant orders in the religious movements of the time was greater than that of any other branch of the Church. They were much more active than the secular clergy, especially during this period of the Babylonian Captivity of the popes, and they were in closer touch with the people, as indeed they had been for over a hundred years. The Dominicans, moreover, were more prominent than the Franciscans. They provided two of the leaders, Catherine and Passavanti. But it would be erroneous to understand the quickening of religious life as exclusively, or even primarily, a Dominican phenomenon. It was more broadly rooted, stimulated by the economic collapse, the plague, and the general feeling of guilt and despair. It is clear, too, that Colombini and the Gesuati were inspired by Franciscan patterns of devotion, and that the dissident Franciscans, the Spirituals and the Fraticelli, attracted widespread sympathy. Furthermore, Giovanni dalle Celle was Vallombrosan, William Fleet Augustinian, and within the *cenacolo* of Catherine, at first only loosely connected with the Dominican Order, there were two Franciscans and an Augustinian.

Similarly the more important paintings of the period were neither commissioned nor inspired by any one order or institution. Again the Dominicans are most conspicuous—three of the best surviving works are in the Strozzi and Spanish Chapels of S. M. Novella. But Franciscan S. Croce can show a chapel frescoed by Giovanni da Milano and a collaborator and fragments of a large cycle by Orcagna—the Triumph of Death, Last Judgment, and Hell. Orcagna, it is true, painted the choir of S. M. Novella, too, but he also painted, according to Ghiberti, a chapel in S. Croce, two chapels in the Church of the Servites, and the refectory of the Augustinian church of S. Spirito, all but the last now lost. His great sculptured tabernacle was made for the company of Or San Michele. In Siena itself nothing comparable to these Florentine works has come down to us, and indeed little monumental painting was originally undertaken. The large cycles by Barna and Bartolo di Fredi are in S. Gimignano, in the Collegiata and the church of S. Agostino. The lesser Florentine and Sienese paintings of the period show an equally wide distribution.

In one form or another the religious revival of the third quarter of the century pervaded the various levels of society and the numerous orders or lay bodies existing within, or on the margins of, the organized Church. What we hear of the size of these religious groups proves that the masses of the people were included. The participation of the well-to-do is indicated by the social status of the leaders and their closest followers. St. Catherine, the daughter of a dyer, belonged to a lower middle class family. In her immediate circle of some fifteen or twenty disciples there were nine members of the most prominent Sienese noble families and two of the greater guild of the judges and notaries. Many of them joined her *cenacolo* during her early independent, pre-ecclesiastical years. Another of her adherents was the painter Andrea Vanni; three letters that she wrote to him while he held communal offices have been preserved. In Florence she won followers in all classes; the center of her group was Francesco di Pippino, a tailor and a recent immigrant from S. Miniato al Tedesco. Jacopo Passavanti came from an old and eminent Florentine family, Giovanni dalle Celle from the nobility. Colombini was a merchant and banker, one of the Sienese oligarchy. His first companion, Francesco di Mino Vincenti, and a later follower, Tommaso di Guelfaccio, had likewise been leading Noveschi, and three "grandi" of the noble Piccolomini family joined his group. Of the identity of the leading Flagellanti and Fraticelli we know almost nothing, but it seems probable that more of them originated in the lower middle class or among the workers in the woolen industry.

Not all the populace responded to the events of the 'forties and 'fifties with increased devotion. That "unexpected" and "marvelously contrary direction" described by Matteo Villani—religious indifference, and what to him seemed immorality and irresponsibility—persisted for many years. Within the religious orders themselves we hear of considerable laxity. The plague, joined by the other disturbances, tended in fact to polarize society toward strenuous religiosity on the one hand and moral and religious dissidence on the other. The culture of the time is characterized by a heightened tension between the two. They constitute a kind of dialectic that is reflected in the conflicts of contemporary painting, and they gave a more highly charged polemical and evangelical meaning to its celebration of the Church, the ritual, and the priest.

A B C D E F G H I J   9 8 7 6 5 4 3 2 1